FLORA OF THE USSR

Alphabetical Indexes to Volumes I-XXX

FLORA OF THE USSR

Initiated under the supervision and chief editorship
of Academician V.L. Komarov

ALPHABETICAL INDEXES TO VOLUMES I-XXX

Editors

E.G. BOBROV and N.N. TZVELEV

General Scientific Editor

STANWYN G. SHETLER

Translated from Russian

CRC Press
Taylor & Francis Group
Boca Raton London New York

CRC Press is an imprint of the
Taylor & Francis Group, an **informa** business
A SCIENCE PUBLISHERS BOOK

Translation of:
Flora SSSR, Alfavitnye Ukazateli k tt. I-XXX
Nauka Publishers, Moscow, Leningrad, 1964

Support for this volume was provided by The Robert
Bateman Arctic Fund of the Smithsonian Institution

Translator & General Editor: Dr. V.S. Kothekar

CRC Press
Taylor & Francis Group
6000 Broken Sound Parkway NW, Suite 300
Boca Raton, FL 33487-2742

First issued in hardback 2019

ISBN-13: 978-1-57808-334-3 (hbk)

Visit the Taylor & Francis Web site at
http://www.taylorandfrancis.com

and the CRC Press Web site at
http://www.crcpress.com

INSTITUTUM BOTANICUM nomine V. KOMAROVII ACADEMIAE SCIENTIARUM URSS

FLORA URSS

(FLORA UNIONIS RERUMPUBLICARUM SOCIALISTICARUM SOVIETICARUM)

INDICES ALPHABETICI I—XXX

„NAUKA"
MOSQUA · 1964 · LENINGRAD

SCIENTIFIC EDITOR'S PREFACE

The Alphabetical Indexes are cumulative indexes to all 30 volumes of the "Flora of the USSR" and together constitute the thirty-first and final volume of this monumental work. This volume consists mainly of separate indexes to the scientific (Latin) names of (1) the families and (2) the genera and species in the Flora, but it also includes a list or index of the authors of the Latin names, alphabetized by the standard abbreviations used for them in the text, and summaries of other commonly used abbreviations. The Latin-name indexes have been reproduced page for page from the original Russian edition. The original Russian volume also included cumulative indexes to the Russian vernacular names, but these indexes have been omitted from the English translation, because the Russian vernacular names were never included in the English translations of volumes I-XXX.

Grateful thanks are extended once again to all those who have had a hand in the translation, review, and publication of this work during the many years that it has been in progress.

30 June 2003

STANWYN G. SHETLER
Curator of Botany Emeritus
National Museum of Natural History
Smithsonian Institution, Washington DC 20013

PREFACE

Given that detailed indexes were published for the first four volumes of the *Flora of the USSR* and for each of the subsequent twenty-six volumes, it is sufficient to give here brief indexes for the entire series to facilitate its use as a whole.

This volume comprises the following parts: 1 – Index of Russian Names of Families [omitted in translation]; 2 – Index of Russian Names of Genera [also omitted in translation]; 3 – Index of Latin Names of Genera and Species of Plants, with the most important generic synonyms.

The general trend of development in the nomenclature of plants toward greater stability is reflected to a certain extent in the present index. This is expressed by the use of a different name for many of the families in conformity with what was adopted at the Ninth International Botanical Congress, by the use of some more recent orthographic variants of specific epithets, and by the more current spelling of the authors' names. The abbreviation of authors' names follows the list compiled by M.E. Kirpicznikov, which is included here as an appendix.

In these indexes the first number following the names of families and genera indicates the volume and the second number – the page; the third number, if present, indicates the page number of the diagnosis in Latin (Addenda). A further number indicates a plate [tab.] on which the plant or its parts have been illustrated.

The present index volume has, first of all, an entirely practical purpose—to facilitate the use of the whole series. At the same time, it also has some value in itself—the main index is an alphabetical list of the principal names of families, genera, and species of higher plants of the Soviet Union described in the *Flora of the USSR*.

The indexes were compiled by: A.E. Matsenko (Vols. I-X), L.I. Ivanova (Vols. XI-XX), T.G. Leonova (Vols. XXI-XXX).

Editorial Board

CONTENTS

SCIENTIFIC EDITOR'S PREFACE	...	VII
PREFACE	...	ix
INDEX OF RUSSIAN NAMES OF FAMILIES*	...	
INDEX OF RUSSIAN NAMES OF GENERA*	...	
INDEX OF LATIN NAMES OF FAMILIES	...	1
INDEX OF LATIN NAMES OF GENERA AND SPECIES	...	3
ALPHABETICAL LIST OF AUTHORS OF THE TAXA	...	205
ABBREVIATIONS		
Abbreviations of Floristic Regions of USSR	...	237
Abbreviations of Regions Used to Indicate General Distribution of Species in *"Flora of the USSR"*	...	238
Other Geographical Abbreviations	...	239
General Abbreviations	...	240

*Omitted in translation—General Editor.

INDEX OF LATIN NAMES OF FAMILIES*

INDEX NOMINUM FAMILIARUM

*The Latin indexes are photo-reproduced page for page from the Russian edition, and the page number on the line with the running head is the number in the Russian edition.

Aceraceae Juss. 14, 580.
Actinidiaceae Hutch. 15, 185.
Adoxaceae J. Agardh 23, 584.
Aizoaceae Rudolphi 6, 374.
Alismataceae Vent. 1, 278.
Amaranthaceae Juss. 6, 354.
Amaryllidaceae Jaume 4, 475.
Anacardiaceae Lindl. 14, 518.
Annonaceae Juss. 7, 567.
Apocynaceae Juss. 18, 645.
Aquifoliaceae Bartl. 14, 539.
Araceae Juss. 3, 478.
Araliaceae Juss. 16, 1.
Aristolochiaceae Juss. 5, 431.
Asclepiadaceae R. Br. 18, 663.

Balsaminaceae A. Rich. 14, 624.
Berberidaceae Juss. 7, 539.
Betulaceae S. Gray 5, 252.
Bignoniaceae Juss. 23, 1.
Boraginaceae Juss. 19, 97.
Butomaceae Rich. 1, 291.
Buxaceae Dum. 14, 503.

Callitrichaceae Link 14, 495.
Campanulaceae Juss. 24, 126.
Capparaceae Juss. 8, 1.
Caprifoliaceae Juss. 23, 419.
Caryophyllaceae Juss. 6, 386.
Celastraceae R. Br. 14, 546.
Ceratophyllaceae S. Gray 7, 14.
Chenopodiaceae Vent. 6, 2.
Chloranthaceae R. Br. ex Lindl. 5, 20.
Cistaceae Juss. 15, 327.
Combretaceae E. Br. 15, 554.
Commelinaceae R. Br. 3, 498.
Compositae Giseke 25, 1.
Convolvulaceae Juss. 19, 1.
Cornaceae Dum. 17, 315.
Crassulaceae DC. 9, 8.
Cruciferae Juss. 8, 14.
Cucurbitaceae Juss. 24, 91.
Cupressaceae Neger 1, 173.
Cuscutaceae Dum. 19, 37.
Cycadaceae Lindl. 1, 129.
Cynomoriaceae Lindl. 15, 670.
Cyperaceae Juss. 3, 1.

Datiscaceae Lindl. 15, 479.
Diapensiaceae Lindl. 18, 105.
Dioscoreaceae R. Br. 4, 494.
Dipsacaceae Juss. 24, 10.
Droseraceae Salisb. 9, 1.

Ebenaceae Guerke 18, 475.
Elaeagnaceae Juss. 15, 515.
Elatinaceae Dum. 15, 259.
Empetraceae S. Gray 14, 511.
Equisetaceae Rich. 1, 101.
Ephedraceae Wettst. 1, 195.
Ericaceae Juss. 18, 22.
Eriocaulaceae Desv. 3, 494.
Eucommiaceae Engl. 9, 273.
Euphorbiaceae Juss. 14, 266.

Fagaceae Dum. 5, 319.
Frankeniaceae S. Gray 15, 271.

Gentianaceae Juss. 18, 525.
Geraniaceae Juss. 14, 1.
Ginkgoaceae Engl. et Prantl 1, 130.
Globulariaceae DC. 23, 128.
Gramineae Juss. 2, 1.
Guttiferae Juss. 15, 201.

Haloragaceae R. Br. 15, 662.
Hamamelidaceae R. Br. 9, 271.

Hippocastanaceae DC. 14, 623.
Hippuridaceae Link 15, 668.
Hydrocaryaceae Raimann 15, 637.
Hydrocharitaceae Juss. 1, 293.
Hydrophyllaceae R. Br. ex Edwards 19, 95.
Hymenophyllaceae Gaudich. 1, 14.

Iridaceae Juss. 4, 498.
Isoetaceae Rich. 1, 127.

Juglandaceae A. Rich. ex Kunth 5, 247.
Juncaceae Juss. 3, 504.
Juncaginaceae Rich. 1, 275.

Labiatae Juss. 20, 1.
Lauraceae Juss. 7, 571.
Leguminosae Juss. 11, 1.
Lemnaceae S. Gray 3, 491.
Lentibulariaceae Rich. 23, 117.
Liliaceae Juss. 4, 1.
Linaceae S. Gray 14, 84.
Lobeliaceae R. Br. 24, 450.
Loranthaceae Juss. 5, 406.
Lycopodiaceae Rich. 1, 112.
Lythraceae Jaume 15, 525.

Magnoliaceae Juss. 7, 564.
Malvaceae Juss. 15, 23.
Marsileaceae R. Br. 1, 90.
Martyniaceae Stapf 23, 18.
Meliaceae Juss. 14, 244.
Menispermaceae Juss. 7, 560.
Menyanthaceae Dum. 18, 640.
Monotropaceae Nutt. 18, 19.
Moraceae Link 5, 376.
Morinaceae (Dum.) Van-Tiegh. 24, 1.
Myricaceae Blume 5, 242.
Myrtaceae Juss. 15, 554.

Najadaceae Juss. 1, 269.
Nyctaginaceae Juss. 6, 370.
Nymphaeaceae Salisb. 7, 2.

Oleaceae Hoffmgg. et Link 18, 483.
Onagraceae Juss. 15, 565.
Ophioglossaceae C. Presl 1, 92.
Orchidaceae Juss. 4, 589.
Orobanchaceae Vent. 23, 19.
Osmundaceae R. Br. 1, 87.
Oxalidaceae R. Br. 14, 76.

Palmae Juss. 3, 464.
Papaveraceae Juss. 7, 573.
Pedaliaceae R. Br. 23, 11.
Phrymaceae Schauer 23, 131.
Phytolaccaceae R. Br. 6, 372.
Pinaceae Lindl. 1, 133.
Pittosporaceae R. Br. 9, 270.
Plantaginaceae Juss. 23, 133.
Platanaceae Dum. 9, 274.
Plumbaginaceae Juss. 18, 292.
Polemoniaceae Juss. 19, 77.
Polygalaceae R. Br. 14, 246.
Polygonaceae Juss. 5, 442.
Polypodiaceae R. Br. 1, 18.
Pontederiaceae Kunth 3, 502.
Portulacaceae Juss. 6, 376.
Potamogetonaceae Dum. 1, 229.
Primulaceae Vent. 18, 108.
Punicaceae Horan. 15, 553.
Pyrolaceae Dum. 18, 7.

Rafflesiaceae Dum. 5, 441.
Ranunculaceae Juss. 7, 20.
Resedaceae S. Gray 8, 606.
Rhamnaceae Juss. 14, 634.
Rosaceae Juss. 9, 279.
Rubiaceae Juss. 23, 164.
Rutaceae Juss. 14, 198.

Salicaceae Mirbel 5, 21.
Salviniaceae Dum. 1, 89.
Santalaceae R. Br. 5, 412.
Saururaceae E. Mey. 5, 20.
Saxifragaceae Juss. 9, 134.
Scrophulariaceae Juss. 22, 117.
Selaginellaceae Mett. 1, 122.
Simaroubaceae DC. 14, 241.
Solanaceae Juss. 22, 1.
Sparganiaceae Rudolphi 1, 216.
Staphyleaceae Lindl. 14, 577.

Tamaricaceae Link 15, 276.
Taxaceae Lindl. 1, 131.
Theaceae D. Don 15, 197.
Theligonaceae Dum. 6, 371.
Thymelaeaceae Juss. 15, 481.
Tiliaceae Juss. 15, 1.
Tropaeolaceae DC. 14, 83.
Typhaceae Juss. 1, 209.

Ulmaceae Mirbel 5, 360.
Umbelliferae Juss. 16, 36.
Urticaceae Juss. 5, 384.

Vacciniaceae S. Gray 18, 93.
Valerianaceae Batsch 23, 585.
Verbenaceae Jaume 19, 692.
Violaceae Batsch 15, 350.
Vitaceae Juss. 14, 674.

Zosteraceae Dum. 1, 265.
Zygophyllaceae R. Br. 14, 146.

INDEX OF LATIN NAMES OF GENERA AND SPECIES

INDEX GENERUM SPECIERUMQUE GENERALIS

Abelia R. Br. 23, 460.
— coreana Nakai 463.
— corymbosa Rgl. et Schmalh. 461.
Abies Mill. 1, 134.
— alba Mill. 136.
— balsamea Mill. 138.
— Fraseri (Pursch) Poir. 138.
— gracilis Kom. 142.
— holophylla Maxim. 137.
— nephrolepis Maxim. 140.
— Nordmanniana (Stev.) Spach 136.
— sachalinensis Mast. 141.
— Semenovii B. Fedtsch. 140.
— sibirica Ledeb. 139; tab. 7.
Abutilon Adans. 15, 146.
— Theophrasti Medic. 146.
Acacia Willd. 11, 11.
— dealbata Link 12.
— melanoxylon R. Br. 13.
— retinoides Schlecht. 12.
Acalypha L. 14, 298.
— australis L. 299; tab. 17.
— indica L. 299.
Acanos Adans. 28, 228.
Acanthocephalus Kar. et Kir. 29, 593.
— amplexifolius Kar. et Kir. 594.
— Benthamianus Rgl. 593.
Acantholepis Less. 27, 54.
— orientalis Less. 54; tab. 4.
Acantholimon Boiss. 18, 301.
— alaicum Czerniak. 316, 732.
— alatavicum Bge. 356.
— Alberti Rgl. 361; tab. 17.
— Alexandri Fed. 320.
— araxanum Bge. 345.
— armenum Boiss. et Huet 347.
— aulieatense Czerniak. 332; tab. 17.
— avenaceum Bge. 327.
— balchanicum Korov. 369.
— blandum Czerniak. 352.
— Borodinii Krassn. 317.
— bracteatum (Girard) Boiss. 312; tab. 16.
— caryophyllaceum Boiss. 346.
— compactum Korov. 322.
— diapensioides Boiss. 318; tab. 17.
— Ekatherinae (B. Fedtsch.) Czerniak. 326.
— erinaceum (Jaub. et Spach) Lincz. 370; tab. 18.
— erythreum Bge. 339.
— Fetissovii Rgl. 321.
— Fominii Kusn. 344.
— gaudanense Czerniak. 339.
— glumaceum (Jaub. et Spach) Boiss. 350; tab. 18.
— Gontscharovii Czerniak. 335.
— Hedinii Ostenf. 317.
— hissaricum Lincz. 335, 736.
— Hohenackeri (Jaub. et Spach) Boiss. 348.
— karatavicum Pavl. 361.
— Karelinii (Stschegl.) Bge. 366.
— khorassanicum Czerniak. 328.
— kokandense Bge. 359.
— Komarovii Czerniak. 365, 741.
— Korolkovii (Rgl.) Korov. 363.
— Korovinii Czerniak. 324.
— Knorringianum Lincz. 364, 740.
— langaricum O. et B. Fedtsch. 340.
— laxum Czerniak. 342.
— lepturoides (Jaub. et Spach) Boiss. 343; tab. 18.
— lycopodioides (Girard) Boiss. 314.
— Majevianum Rgl. 338.
— Margaritae Korov. 333, 735.
— Mikeschinii Lincz. 327, 733.

— minshelkense Pavl. 333.
— mirandum Lincz. 371, 742; tab. 19.
— mirum Lincz. 372, 743.
— Nikitinii Lincz. 368, 742; tab. 18.
— nuratavicum Zak. 334, 736.
— pamiricum Czerniak. 358.
— parviflorum Rgl. 364.
— procumbens Czerniak. 331.
— pskemense Lincz. 325, 734.
— pterostegium Bge. 311; tab. 16.
— pulchellum Korov. 351.
— purpureum Korov. 332.
— quinquelobum Bge. 367; tab. 18.
— Raddeanum Czerniak. 355.
— Ruprechtii Bge. 321.
— Sackenii Bge. 357.
— sahendicum Boiss. et Buhse 350.
— saravschanicum Rgl. 336.
— schemachense Grossh. 349.
— squarrosum Pavl. 314.
— strictum Czerniak. 323, 734.
— tarbagataicum Gamajun. 357.
— tataricum Boiss. 313; tab. 17.
— tenuiflorum Boiss. 342.
— tianschanicum Czerniak. 319.
— Titovii Lincz. 362, 740.
— Varivtzevae Czerniak. 341, 738.
— velutinum Czerniak. 360, 739.
— virens Czerniak. 337, 737.
— Zaprjagaevii Lincz. 315, 732; tab. 17.
Acanthopanax Seem. 16, 19.
— sessiliflorum (Rupr. et Maxim.) Seem. 19; tab. 2.
Acanthophyllum C. A. Mey. 6, 780.
— acerosum Sosn. 791.
— aculeatum Schischk. 791, 895.
— adenophorum Freyn 789; tab. 49.
— albidum Schischk. 787, 893.
— Borszowii Litv. 792; tab. 49.
— brevibracteatum Lipsky 790.
— Bungei (Boiss.) Trautv. 800; tab. 50.
— coloratum (Preobr.) Schischk. 799.
— elatius Bge. ex Boiss. 792; tab. 49.
— elongatum Preobr. 796, 895; tab. 49.
— glandulosum Bge. ex Boiss. 794; tab. 49.
— gypsophiloides Rgl. 801; tab. 50.
— Knorringianum Schischk. 794.
— Korolkowii Rgl. et Fr. Schmidt 795; tab. 50.
— Korshinskyi Schischk. 789; tab. 49.
— Krascheninnikovii Schischk. 790, 894.
— lilacinum Schischk. 787, 892.
— microcephalum Boiss. 793; tab. 49.
— mucronatum C. A. Mey. 793; tab. 49.
— paniculatum Rgl. 800; tab. 50.
— pulchrum Schischk. 791.
— pungens (Bge.) Boiss. 784; tab. 49.
— schugnanicum (Preobr.) Schischk. 795, 895.
— serawschanicum Golenk. 796; tab. 49.
— sordidum Bge. ex Boiss. 793; tab. 49.
— squarrosum Boiss. 793.
— stenostegium Freyn 768; tab. 49.
— subglabrum Schischk. 787, 893.
— tadshikistanicum Schischk. 801; tab. 50.
— tenuifolium Schischk. 788, 894.
— transhyrcanum Preobr. 800; tab. 50.
— turcomanicum Schischk. 789, 894.
— versicolor Fisch. et Mey. 799; tab. 50.
Acanthopleura C. Koch 17, 278.
Accorombona Endl. 11, 303.
Acelidanthus Trautv. et Mey. 4, 9.
— anticleoides Trautv. et Mey. 10; tab. 1.
Acer L. 14, 580.
— campestre L. 596.
— barbinerve Maxim. 618; tab. 32.
— divergens C. Koch et Pax 593.
— ginnala Maxim. 601.
— hyrcanum Fisch. et Mey. 613.
— ibericum M. B. 615; tab. 32.
— Komarovii Pojark. 611, 746; tab. 32.
— laetum C. A. Mey. 591.
— mandschuricum Maxim. 617.
— mono Maxim. 590.
— negundo L. 621.
— pennsylvanicum L. 610.
— pictum Thunb. 589.
— platanoides L. 592.
— pseudoplatanus L. 598.
— pseudosieboldianum (Pax) Kom. 608.
— pubescens Franch. 594; tab. 31.
— Regelii Pax 595; tab. 31.
— rubrum L. 619.
— saccharinum L. 620.
— saccharum Marsh. 612.
— Semenovii Rgl. et Herd. 602.
— Stevenii Pojark. 614.
— tataricum L. 603.
— tegmentosum Maxim. 609.
— Trautvetteri Medw. 600.
— turcomanicum Pojark. 616.
— turkestanicum Pax 590.
— ukurunduense Trautv. et Mey. 607.
— velutinum Boiss. 597.
Achillea L. 26, 70.
— acuminata (Ledeb.) Sch. Bip. 108.
— alpina L. 119.
— asiatica Serg. 85.

Achillea Biebersteinii Afan. 97.
— biserrata M. B. 106.
— borealis Bong. 81.
— bucharica C. Winkl. 86.
— camtschatica Rupr. ex Heimerl 122.
— cartilaginea Ledeb. 114.
— coarctata Poir. 91.
— cuneatiloba Boiss. et Buhse 99.
— distans Waldst. et Kit. ex Willd. 87.
— filipendulina Lam. 90.
— glaberrima Klok. 95.
— griseo-virens Albov 107.
— impatiens L. 116.
— japonica Heimerl 120.
— kermanica Gand. 101.
— latiloba Ledeb. ex Nordm. 89.
— Ledebouri Heimerl 117.
— leptophylla M. B. 94.
— lingulata Waldst. et Kit. 105.
— macrocephala Rupr. 111.
— micrantha Willd. 96.
— micranthoides Klok. 100.
— millefolium L. 78.
— Neilreichii A. Kerner 77.
— nobilis L. 76.
— ochroleuca Ehrh. 93.
— pannonica Scheele 82.
— ptarmica L. 108.
— ptarmicifolia (Willd.) Rupr. ex Heimerl 109.
— ptarmicoides Maxim. 121.
— Sachokiana Sosn. 110.
— salicifolia Bess. 112.
— Schischkinii Sosn. 103.
— Schurii Sch. Bip. 123.
— Sedelmeyeriana Sosn. 113.
— septentrionalis (Serg.) Botsch. 115.
— setacea Waldst. et Kit. 83.
— stricta Schleich. 88.
— sudetica Opiz 80.
— tenuifolia Lam. 103.
— vermicularis Trin. 102.
— Wilhelmsii C. Koch 100.
Achnodon Nees 2, 128.
Achnodonton Beauv. 2, 128.
Achroanthes Raf. 4, 600.
Achudemia Blume 5, 397.
— japonica Maxim. 397.
Achyranthes L. 6, 367.
— bidentata Blume 367; tab. 18.
Achyrophorus Scop. 29, 201.
— ciliatus (Thunb.) Sch. Bip. 202.
— maculatus (L.) Scop. 202; tab. 12.
— uniflorus (Vill.) Bluff et Fingerh. 203.
Achyroseris Sch. Bip. 29, 54, 61.
Acinos Moench 21, 441.
— alpinus (L.) Moench 447.
— Fominii Schost. 445.
— graveolens (M. B.) Link 446; tab. 23.
— thymoides (L.) Moench 442; tab. 23.
Acomastylis Greene 10, 249.
— Rossii (R. Br.) Greene 249; tab. 18.
Aconitum L. 7, 183.
— ajanense Steinb. 197, 727.
— albo-violaceum Kom. 192.
— altaicum Steinb. 222, 731.
— ambiguum Reichb. 220.
— angusticassidatum Steinb. 231, 734.
— anthora L. 190; tab. 12.
— apetalum (Huth) B. Fedtsch. 200; tab. 13.
— arcuatum Maxim. 215.
— baicalense Turcz. 223, 732.
— barbatum Pers. 204; tab. 13.
— biflorum Fisch. 226.
— crassifolium Steinb. 199, 728.
— cymbulatum (Schmalh.) Lipsky 212.
— Czekanovskyi Steinb. 224, 733.
— delphinifolium DC. 225.
— Desoulavyi Kom. 193.
— excelsum Reichb. 201; tab. 13.
— Fischeri Reichb. 214.
— flagellare (Fr. Schmidt) Steinb. 224.
— Flerovii Steinb. 221, 730.
— jaluense Kom. 219.
— karakolicum Rapaics 233.
— kirinense Nakai 205.
— Komarovii Steinb. 191; tab. 12.
— Krylovii Steinb. 206, 729.
— Kusnezofii Reichb. 217.
— lasiostomum Reichb. 205.
— macrorhynchum Turcz. 218.
— maximum Pall. ex DC. 216.
— monticola Steinb. 209, 730.
— nasutum Fisch. ex Reichb. 212.
— nemorum M. Pop. 236.
— orientale Mill. 203.
— pubiceps (Rupr.) Trautv. 211.
— Raddeanum Rgl. 215.
— ranunculoides Turcz. 194; tab. 13.
— rotundifolium Kar. et Kir. 229.
— sachalinense Fr. Schmidt 217.
— Saposhnikovii B. Fedtsch. 234, 734.
— seravschanicum Steinb. 230, 733; tab. 14.
— sichotense Kom. 219.
— Smirnovii (Sukacz.) Steinb. 222, 731; tab. 14.
— soongoricum Stapf 232.

— Stoerkianum Reichb. 210.
— Sukaczevii Steinb. 198, 728; tab. 13.
— talassicum M. Pop. 231.
— Tranzschelii Steinb. 235, 735.
— umbrosum (Korsh.) Kom. 199; tab. 13.
— variegatum L. 210.
— villosum Reichb. 213.
— volubile Pall. ex Koelle 213.
Aconogon Tourn. 5, 661.
Aconopogon Mori 5, 661.
Acontogonum Aschers. et Graebn. 5, 661.
Acorellus Palla 3, 22.
— distachyus (All.) Palla 23; tab. 1.
— laevigatus (L.) Palla 23.
— pannonicus (Jacq.) Palla 22.
Acorus L. 3, 479.
— calamus L. 479.
Acrocentron Cass. 28, 370, 493.
Acrolophus Cass. 28, 370, 512.
Acroptilon Cass. 28, 343.
— australe Iljin 344.
— repens (L.) DC. 345.
Actaea L. 7, 79.
— acuminata Wall. 80.
— erythrocarpa Fisch. 81.
— spicata L. 80.
Actinanthus Ehrenb. 16, 526.
Actinidia Lindl. 15, 186.
— arguta (Sieb. et Zucc.) Planch. 190.
— Giraldii Diels 192.
— kolomikta Maxim. 188.
— polygama (Sieb. et Zucc.) Maxim. 194.
— rubricaulis Dunn 196.
— Sugawarana Koidz. 195.
Actinolema Fenzl 16, 71.
— eryngioides Fenzl 72; tab. 3.
— macrolema Boiss. 72; tab. 3.
Actinostemma Griff. 24, 93.
— lobatum Maxim. 94.
Adamsia Willd. 4, 394.
Adelocaryum Brand 19, 627.
Adenarium Raf. 6, 506.
Adenocaulon Hook. 25, 506.
— adhaerescens Maxim. 507.
Adenolinum Reichb. 14, 103.
Adenonema Bge. 6, 419.
Adenophora Fisch. 24, 343.
— coronopifolia Fisch. 368.
— crispata (Korsh.) Kitag. 366.
— Gmelinii (Spreng.) Fisch. 367.
— Golubinzevaeana Reverd. 353.
— himalayana Feer 370.
— jacutica Fed. 356, 470.
— Lamarckii Fisch. 354.
— liliifolia (L.) Bess. 348.
— pereskiifolia (Fisch. ex Roem. et Schult.) G. Don 362; tab. 21.
— rupestris Reverd. 369.
— stenanthina (Ledeb.) Kitag. 364.
— sublata Kom. 355.
— taurica (Sukacz.) Juz. 351, 469; tab. 21.
— tetraphylla (Thunb.) Fisch. 361.
— trachelioides Maxim. 359.
— tricuspidata (Fisch. ex Roem. et Schult.) A. DC. 352.
Adenostyles Cass. 25, 22.
— alliariae (Gouan) A. Kerner 22.
Adiantum L. 1, 79.
— capillus-veneris L. 80.
— pedatum L. 79.
Adlumia Raf. 7, 647.
— asiatica Ohwi 647; tab. 45.
Adonis L. 7, 528.
— aestivalis L. 537.
— amurensis Rgl. 532.
— autumnalis L. 538.
— chrysocyathus Hook. f. et Thoms. 535; tab. 36.
— flammeus Jacq. 538.
— sibiricus Patr. ex Ledeb. 530.
— tianschanicus (Adolf) Lipsch. 531.
— turkestanicus (Korsh.) Adolf 532; tab. 36.
— vernalis L. 535; tab. 36.
— villosus Ledeb. 530.
— wolgensis Stev. 536; tab. 36.
Adoxa L. 23, 584.
— moschatellina L. 584.
Aegilops L. 2, 669.
— biuncialis Vis. 673; tab. 48.
— crassa Boiss. 671; tab. 48.
— cylindrica Host 671; tab. 48.
— juvenalis (Thell.) Eig. 672.
— Kotschyi Boiss. 673.
— ovata L. 674; tab. 48.
— squarrosa L. 671; tab. 48.
— triaristata Willd. 674; tab. 48.
— triuncialis L. 672; tab. 48.
Aegirus Aschers. 5, 228.
Aegopodium L. 16, 452.
— alpestre Ledeb. 457.
— brachycarpum (Kom.) Schischk. 457.
— latifolium Turcz. 456.
— podagraria L. 452.
— tadzhikorum Schischk. 456, 600.
Aegoseris (Reichb.) Steud. 29, 595, 697.
Aellenia Ulbr. 6, 270.
— auricula (Moq.) Ulbr. 270; tab. 12.

Aeluropus Trin. 2, 357.
— litoralis (Gouan) Parl. 357; tab. 27.
— repens (Desf.) Parl. 358; tab. 27.
Aesculus L. 14, 622.
— hippocastanum L. 622.
Aetheopappus Cass. 28, 353.
— caucasicus Sosn. 355.
— pulcherrimus (Willd.) Cass. 355.
— Vvedenskii (Sosn.) Sosn. 354.
Aethionema R. Br. 8, 552.
— arabicum (L.) Andrz. ex DC. 563; tab. 31.
— armenum Boiss. 559.
— caespitosum Boiss. 568; tab. 30.
— cardiophyllum Boiss. et Heldr. 564; tab. 30.
— carneum (Soland.) B. Fedtsch. 563; tab. 31.
— diastrophis Bge. 559; tab. 31.
— edentulum N. Busch 560; tab. 14.
— elongatum Boiss. 556; tab. 30.
— heterophyllum Boiss. 567.
— grandiflorum Boiss. et Hohen. 558.
— Levandovskyi N. Busch 558.
— Lipskyi N. Busch 555.
— membranaceum DC. 560; tab. 31.
— pulchellum Boiss. et Huet 558; tab. 31.
— sagittatum Boiss. 567.
— salmasium Boiss. 564; tab. 31.
— schistosum Boiss. et Kotschy 559.
— spinosum (Boiss.) N. Busch 557; tab. 30.
— Szovitsii Boiss. 555, tab. 30.
— transhyrcanum (Czerniak.) N. Busch 557; tab. 31.
— trinervium (DC.) Boiss. 567; tab. 30.
— Voronovii Schischk. 556.
Aethiopis Benth. 21, 298.
Aethusa L. 16, 538.
— cynapium L. 539.
Agastache Clayt. ex Gronov. 20, 273.
— rugosa (Fisch. et Mey.) Kuntze 274; tab. 14.
Agasyllis Spreng. 17, 33.
— latifolia (M. B.) Boiss. 34; tab. 3.
Agathyrsus D. Don 29, 279, 351.
Agatophyton Moq. 6, 51.
Ageratum L. 25, 17.
— Houstonianum Mill. 17.
Agostana S. F. Gray 16, 333.
Agrimonia L. 10, 411.
— asiatica Juz. 412; tab. 25.
— eupatoria L. 413; tab. 25.
— granulosa Juz. 415, 635.
— japonica (Miq.) Koidz. 419.
— odorata Mill. 411; tab. 25.
— pilosa Ledeb. 416; tab. 25.
— velutina Juz. 420, 636; tab. 25.
Agriophyllum M. B. 6, 158.
— arenarium M. B. 159; tab. 7.
— lateriflorum (Lam.) Moq. 160; tab. 7.
— latifolium Fisch. et Mey. 161; tab. 7.
— minus Fisch. et Mey. 161; tab. 7.
— Paletzkianum Litw. 161; tab. 7.
Agropyrum Gaertn. 2, 627.
— alaicum Drob. 633.
— alatavicum Drob. 633.
— amgunense Nevski 636.
— angulare Nevski 639.
— armenum Nevski 640.
— attenuatiglume Nevski 636.
— badamense Drob. 658.
— Batalinii (Krassn.) Roshev. 632.
— caespitosum C. Koch 646.
— ciliolatum Nevski 650.
— cimmericum Nevski 655; tab. 47a.
— cristatum (L.) Gaertn. 661; tab. 47a.
— dasyanthum Ledeb. 655; tab. 47a.
— desertorum (Fisch.) Schult. 657; tab. 47a.
— dshungaricum Nevski 641.
— elongatiforme Drob. 651.
— elongatum (Host) Beauv. 647.
— ferganense Drob. 641.
— firmiculmis Nevski 646.
— fragile (Roth) Nevski 656.
— geniculatum (Trin.) Korsh. 646; tab. 47.
— Gmelinii Schrad. 635.
— gracillimum Nevski 638.
— imbricatum (M. B.) Roem. et Schult. 660.
— intermedium (Host) Beauv. 648.
— jacutorum Nevski 636.
— junceum (L.) Beauv. 647; tab. 47.
— Krylovianum Schischk. 650.
— lolioides (Kar. et Kir.) Roshev. 651; tab. 47.
— Michnoi Roshev. 656.
— nodosum (Stev.) Nevski 646.
— pectiniforme Roem. et Schult. 659.
— pertenue (C. A. Mey.) Nevski 640.
— pinifolium Nevski 659.
— ponticum Nevski 658.
— pulcherrimum Grossh. 649.
— pumilum (Steud.) Nevski 650.
— propinquum Nevski 634.
— pruiniferum Nevski 640.
— reflexiaristatum Nevski 634.
— repens (L.) Beauv. 652; tab. 47.
— Roshevitzii Nevski 635.
— scythicum Nevski 638.

— setuliferum Nevski 642.
— sibiricum (Willd.) Beauv. 657; tab. 47a.
— sinuatum Nevski 639; tab. 47.
— Sosnovskyi Hack. 645.
— stipaefolium Czern. 637.
— strigosum (M. B.) Boiss. 633.
— tanaiticum Nevski 655.
— trichophorum (Link.) Richt. 648; tab. 47
Agrostemma L. 6, 574.
— githago L. 575.
— linicola Terech. 575.
Agrostis L. 2, 170.
— alba L. 183.
— albida Trin. 180.
— anadyrensis Socz. 176, 746; tab. 13.
— Biebersteiniana Claus 173; tab. 13.
— borealis Hartm. 174; tab. 13.
— canina L. 174.
— capillaris L. 185; tab. 13.
— clavata Trin. 178; tab. 13.
— exarata Trin. 186; tab. 13.
— hiemalis (Walt.) Britt. 186; tab. 13.
— hissarica Roshev. 178; tab. 13.
— jacutica Schischk. 179, 747.
— karsensis Litv. 187.
— lazica Bal. 177.
— macrantha Schischk. 177, 746.
— macrothyrsa Hack. 187.
— maritima Lam. 185.
— mongholica Roshev. 184.
— Paulsenii Hack. ex Pauls. 178; tab. 13.
— pisidica Boiss. 173.
— planifolia C. Koch 175; tab. 13.
— salsa Korsh. 184; tab. 13.
— sibirica V. Petrov 180.
— stolonizans Bess. 184.
— teberdensis Litv. 179.
— tenuifolia M. B. 176; tab. 13.
— transcaspica Litv. 187; tab. 13.
— Trinii Turcz. 175; tab. 13.
— verticillata Vill. 188; tab. 13.
Ailanthus Desf. 14, 241.
— altissima (Mill.) Swingle 242.
Aipyanthus Stev. 19, 167, 171.
Aira L. 2, 239.
— capillaris Host 240; tab. 18.
— caryophyllea L. 240; tab. 18.
— praecox L. 239; tab. 18.
Aizoon L. 6, 376.
— hispanicum L. 376; tab. 19.
Ajania Poljak. 26, 398.
— fastigiata (C. Winkl.) Poljak. 405.
— fruticulosa (Ledeb.) Poljak. 406.
— gracilis (Hook. f. et Thoms.) Poljak. 407.
— kokanica (Krasch.) Tzvel. 408; tab. 17.
— manshurica Poljak. 401.
— Pallasiana (Fisch. ex Bess.) Poljak. 402; tab. 31.
— Scharnhorstii (Rgl. et Schmalh.) Tzvel. 409; tab. 18.
— tibetica (Hook. f. et Thoms.) Tzvel. 410.
— trilobata Poljak. 409, 880.
Ajuga L. 20, 17.
— chamaecistus Ging. ex Benth. 27; tab. 2.
— chamaepitys (L.) Schreb. 25; tab. 1.
— chia Schreb. 20; tab. 1.
— genevensis L. 32.
— Laxmannii (L.) Benth. 29; tab. 2.
— multiflora Bge. 36.
— oblongata M. B. 26; tab. 2.
— orientalis L. 38.
— pseudochia Schost. 21; tab. 1.
— pyramidalis L. 37.
— reptans L. 30.
— salicifolia (L.) Schreb. 26; tab. 2.
— shikotanensis Miyabe et Tatewaki 32.
— turkestanica (Rgl.) Briq. 28; tab. 2.
Alacospermum Neck. 16, 380.
Albertia Rgl. et Schmalh. 16, 151.
— paleacea Rgl. et Schmalh. 152; tab. 15.
Albizzia Durazz. 11, 10.
— Julibrissin Durazz. 10; tab. 1.
Albovia Schischk. 16, 450.
— tripartia (Kalenicz.) Schischk. 450; **tab. 17.**
Alboviodoxa Woronow ex Grossh. 27, 75.
Alcanna Orph. 19, 231.
Alcea L. 15, 84.
— abchazica Iljin 94.
— angulata Freyn et Sint. 116; tab. 6.
— Antoninae Iljin 117, 678.
— baldshuanica (Bornm.) Iljin 124; tab. 6.
— excubita Iljin 108.
— flavovirens Boiss. et Buhse 100.
— Freyniana Iljin 125, 678.
— Froloviana (Fisch.) Iljin 111.
— Grossheimii Iljin 99.
— Heldreichii Boiss. 121.
— hyrcana Grossh. 123.
— karakalense Freyn 114.
— karsiana Litv. 107.
— kopetdaghensis Iljin 113, 677.
— kusariensis Iljin 92.
— lenkoranica Iljin 96.
— Litwinowii Iljin 111.
— Novopokrovskii Iljin 93, 676.
— nudiflora (Lindl.) Boiss. 108.
— pallida (Waldst. et Kit.) Bess. 118.

Alcea rhyticarpa (Trautv.) Iljin 115.
— rosea L. 126.
— rugosa Alef. 90; tab. 6.
— sachsachanica Iljin 99, 676.
— Sophiae Iljin 106.
— Sosnovskyi Iljin 102.
— sycophylla Iljin et Nikit. 95, 676;
— tabrisiana Boiss. et Buhse 98.
— talassica Iljin 115, 677.
— taurica Iljin 92, 675.
— transcaucasica Iljin 105.
— turcomanica Iljin 96, 677.
— Turkeviczii Iljin 120.
— Woronowii Iljin 122.
Alchimilla L. 10, 289.
— abchasica Bus. 405.
— acutangula Bus. 353.
— aemula Juz. 329.
— Alexandri Juz. 372.
— alpestris Fr. Schmidt 391.
— alpina L. 308; tab. 22.
— altaica Juz. 371.
— anisopoda Juz. 373.
— appressipila Juz. 380.
— arcuatiloba Juz. 347, 621.
— argutiserrata Lindb. f. 332.
— barbatiflora Juz. 405.
— barbulata Juz. 333.
— biquadrata Juz. 368, 627.
— brevidens Juz. 355.
— breviloba Lindb. f. 341.
— Bungei Juz. 327.
— Buschii Juz. 379.
— camptopoda Juz. 378.
— capillacea Juz. 633.
— cartalinica Juz. 389.
— caucasica Bus. 318.
— chlorosericea Bus. 310.
— circassica Juz. 398, 634.
— compactilis Juz. 343.
— conglobata Lindb. f. 338.
— crebridens Juz. 379.
— cymatophylla Juz. 345.
— cyrtopleura Juz. 342, 620.
— daghestanica Juz. 384.
— debilis Juz. 407.
— decalvans Juz. 360.
— denticulata Juz. 349, 622.
— depexa Juz. 383.
— diglossa Juz. 368, 626.
— divaricans Bus. 403.
— diversipes Juz. 372.
— dura Bus. 408.
— egens Juz. 325.
— elata Bus. 409.
— Elisabethae Juz. 314.
— epipsila Juz. 398.
— erythropoda Juz. 318.
— exsanguis Juz. 320.
— exuens Juz. 358.
— filicaulis Bus. 361.
— flavescens Bus. 386.
— fontinalis Juz. 369, 628.
— frondosa Juz. 384, 631.
— georgica Juz. 387.
— gibberulosa Lindb. f. 329.
— glabricaulis Lindb. f. 376.
— glomerulans Bus. 381.
— grandidens Juz. 402, 634.
— Grossheimii Juz. 328.
— Haraldi Juz. 404.
— hebescens Juz. 330.
— heptagona Juz. 359; tab. 23.
— heteroscnista Juz. 311, 617.
— hians Juz. 348, 621.
— hirsuticaulis Lindb. f. 323; tab. 22.
— hirsutissima Juz. 333.
— hirtipedicellata Juz. 394.
— holotricha Juz. 394, 632; tab. 24.
— humilicaulis Juz. 362, 624.
— hypochlora Juz. 310, 617.
— hyrcana Bus. 361.
— imberbis Juz. 347.
— insignis Juz. 370.
— jailae Juz. 324; tab. 22.
— Juzepczukii Alech. 339.
— Kozlowski Juz. 354.
— Krylovii Juz. 375.
— laeta Juz. 397, 634.
— laeticolor Juz. 358, 623.
— languescens Juz. 337.
— languida Bus. 319.
— leiophylla Juz. 354.
— Lindbergiana Juz. 350.
— Lipschitzii Juz. 367, 625.
— lithophila Juz. 324.
— Litwinowii Juz. 344.
— Lydiae Zam. 408.
— micans Bus. 348.
— microdonta Juz. 351.
— minor Huds. 316; tab. 22.
— minusculiflora Bus. 387.
— Murbeckiana Bus. 388.
— nebulosa Sam. 388.
— nemoralis Alech. 352.
— neo-Stevenii Juz. 335.
— obtegens Juz. 378.
— obtusa Bus. 390.

— obtusiformis Alech. 381.
— oligotricha Juz. 376, 629.
— omalophylla Juz. 336.
— orbicans Juz. 363.
— oxysepala Juz. 395; tab. 24.
— pachyphylla Juz. 344, 620.
— pastoralis Bus. 334.
— pilosiplica Juz. 386, 631.
— plicata Bus. 319.
— pogonophora Juz. 377, 630.
— porrectidens Juz. 396, 633.
— propinqua Lindb. f. ex Juz. 336; tab. 23.
— pseudo-cartalinica Juz. 391.
— pseudomollis Juz. 397, 633.
— psilocaula Juz. 385, 631.
— psiloneura Juz. 390.
— purpurascens Juz. 367, 626.
— pycnantha Juz. 328.
— pycnotricha Juz. 331.
— Raddeana Bus. 311.
— retinervis Bus. 407; tab. 24.
— retropilosa Juz. 342.
— rigescens Juz. 352.
— rigida Bus. 314.
— rubens Juz. 326, 618.
— sanguinolenta Juz. 362, 624.
— sarmatica Juz. 340.
— Sauri Juz. 374, 629.
— Schischkinii Juz. 364.
— schistophylla Juz. 335.
— Sedelmeyeriana Juz. 351.
— semilunaris Alech. 346.
— sericata Reichb. 312.
— sericea Willd. 309; tab. 22.
— sevangensis Juz. 401.
— sibirica Zam. 338.
— Smirnovii Juz. 394, 633.
— speciosa Bus. 393.
— stellaris Juz. 359.
— stellulata Juz. 401.
— Stevenii Bus. 339; tab. 23.
— subcrenata Bus. 355.
— subcrenatiformis Juz. 356.
— suberectipila Juz. 382.
— subsplendens Bus. 403.
— substrigosa Juz. 341.
— supina Juz. 316.
— Tamarae Juz. 380.
— taurica Juz. 317.
— tephroserica Bus. 313; tab. 22.
— tianschanica Juz. 357, 622.
— transiliensis Juz. 369, 627.
— tredecimloba Bus. 506; tab. 24.
— turuchanica Juz. 374.
— tytthantha Juz. 347; tab. 23.
— undecimloba Juz. 406, 635.
— urceolata Juz. 383, 630; tab. 23.
— valdehirsuta Bus. 331.
— venosa Juz. 404.
— vulgaris L. 356.
— Woronowii Juz. 315, 618.
Aldrovanda Monti 9, 1.
— vesiculosa L. 1; tab. 1.
Alectorolophus H. M. Hall 22, 659.
Aleurites Forst. 14, 303.
— Fordii Hemsl. 303.
Alexandra Bge. 6, 198.
— Lehmanni Bge. 198, tab. 10.
Alfredia Cass. 28, 39.
— acantholepis Kar. et Kir. 41.
— cernua (L.) Cass. 40.
— nivea Kar. et Kir. 42.
Alhagi Adans. 13, 367.
— canescens (Rgl.) Shap. ex Keller et Shap. 369; tab. 19.
— kirghisorum Schrenk 369; tab. 19.
— persarum Boiss. et Buhse 370; tab. 19.
— pseudalhagi (M. B.) Desv. 363; tab. 19.
— sparsifolia Shap. 370; tab. 19.
Alisma L. 1, 279.
— lanceolatum With. 281; tab. 14.
— Loeselii Gorski 282; tab. 14.
— orientale (Sam.) Juz. 281.
— plantago-aquatica L. 280; tab. 14.
— Wahlenbergii Holmb. 283.
Alkanna Tausch 19, 231.
— cordifolia C. Koch 233.
— orientalis (L.) Boiss. 232.
Allardia Decne. 26, 263, 267.
Alliaria Scop. 8, 31.
— brachycarpa M. B. 32.
— officinalis Andrz. ex M. B. 31.
Allinum Neck. 16, 549, 560.
Allionia Loefl. 6, 370.
Allium L. 4, 112.
— affine Ledeb. 239.
— aflatunense B. Fedtsch. 268.
— akaka Gmel. ex Roem. et Schult. 260.
— alaicum Vved. 267; tab. 15.
— albanum Grossh. 213.
— albidum Fisch. ex Bess. 164.
— Albovianum Vved. 187.
— Alexandrae Vved. 186.
— Alexejanum Rgl. 259.
— altaicum Pall. 196.
— altissimum Rgl. 269.
— ampeloprasum L. 253.
— amphibolum Ledeb. 152.

Allium angulosum L. 164.
— anisopodium Ledeb. 174.
— anisotepalum Vved. 215; tab. 13.
— aroides M. Pop. et Vved. 258.
— atroviolaceum Boiss. 252.
— Aucheri Boiss. 238; tab. 14.
— Barszczewskii Lipsky 158.
— bellulum Prokh. 173.
— bidentatum Fisch. ex Prokh. 172.
— Bodeanum Rgl. 261.
— bogdoicolum Rgl. 151.
— Borszczowii Rgl. 233.
— brachyodon Boiss. 153.
— brachyscapum Vved. 261.
— brevidens Vved. 233; tab. 14.
— caesium Schrenk 222.
— caespitosum Sievers ex Bong. et Mey. 174.
— callidictyon C. A. Mey. ex Kunth 201.
— Candolleanum Albov 255; tab. 15.
— cardiostemon Fisch. et Mey. 263.
— caricoides Rgl. 181.
— carinatum L. 204.
— caspium (Pall.) M. B. 275.
— cepa L. 199.
— Christophii Trautv. 261.
— clathratum Ledeb. 145.
— coeruleum Pall. 221.
— condensatum Turcz. 182.
— confragosum Vved. 227.
— convallarioides Grossh. 208.
— crystallinum Vved. 234.
— cupuliferum Rgl. 277.
— daghestanicum Grossh. 188.
— darvasicum Rgl. 276.
— dasyphyllum Vved. 264.
— decipiens Fisch. ex Roem. 265.
— delicatulum Siev. ex Roem. et Schult. 219.
— Derderianum Rgl. 259.
— dictyoprasum C. A. Mey. ex Kunth 236; tab. 14.
— dictyoscordum Vved. 236.
— Dioscoridis Sibth. et Smith 280.
— dolichomischum Vved. 160; tab. 11.
— dolichostylum Vved. 160.
— drepanophyllum Vved. 157.
— Drobovii Vved. 144.
— elatum Rgl. 274.
— elegans Drob. 223, tab. 13.
— eremoprasum Vved. 226.
— erubescens C. Koch 246.
— Eugenii Vved. 257.
— ferganicum Vved. 229.
— Fetissovii Rgl. 263.
— fibrosum Rgl. 209.
— filidens Rgl. 235.
— filifolium Rgl. 180.
— firmotunicatum Fomin 240.
— Fischeri Rgl. 145.
— fistulosum L. 196.
— flavescens Bess. 165; tab. 10.
— flavidum Ledeb. 146.
— Fominianum Miscz. ex Grossh. et Schischk. 244.
— fuscoviolaceum Fomin 240.
— galanthum Kar. et Kir. 197.
— giganteum Rgl. 274.
— glaciale Vved. 177.
— globosum M. B. ex Redoute 185.
— glomeratum Prokh. 220.
— gracilescens Somm. et Lev. 245.
— grande Lipsky 267.
— Griffithianum Boiss. 217.
— gultschense B. Fedtsch. 268.
— gunibicum Miscz. ex Grossh. 188.
— guttatum Stev. 238.
— guzaricum Rgl. 156.
— gypsaceum M. Pop. et Vved. 156 tab. 16.
— helicophyllum Vved. 256.
— hymenorrhizum Ledeb. 176.
— iliense Rgl. 278.
— inaequale Janka 200.
— inconspicuum Vved. 156.
— inderiense Fisch. ex Bge. 161
— inops Vved. 218.
— insufficiens Vved. 264.
— Jacquemontii Kunth 215.
— jajlae Vved. 247.
— jucundum Vved. 179.
— karataviense Rgl. 262.
— karsianum Fomin 207.
— kaschianum Rgl. 177.
— kirindicum Bornm. 209.
— kokanicum Rgl. 180.
— Komarovianum Vved. 143.
— Komarovii Lipsky 270.
— kopetdagense Vved. 223.
— Korolkovii Rgl. 187.
— kossoricum Fomin 202.
— kujukense Vved. 225; tab. 13.
— Kunthianum Vved. 207; tab. 12
— lacerum Freyn 201.
— lasiophyllum Vved. 219.
— Ledebourianum Roem. et Schult. 194.
— Lehmannianum Merckl. 230.
— lencoranicum Miscz. ex Grossh. 206.
— leucanthum C. Koch 252.
— leucocephalum Turcz. 146.

— lineare L. 150.
— longicuspis Rgl. 243.
— longiradiatum Vved. 159.
— lutescens Vved. 159.
— Maackii Prokh. ex Kom. 149; tab. 10.
— macrostemum Bge. 221.
— Margaritae B. Fedtsch. 199.
— Mariae Bordz. 262.
— Marschallianum Vved. 184.
— materculae Bordz. 260.
— Maximowiczii Rgl. 194, tab. 11.
— microbulbum Prokh. 195.
— minutum Vved. 216.
— monadelphum Less. ex Kunth 189.
— monanthum Maxim. 254.
— monophyllum Vved. 259; tab. 15.
— moschatum L. 200.
— nerinifolium Baker 279.
— nutans L. 171.
— obliquum L. 175.
— odorum L. 163.
— oleraceum L. 205.
— oliganthum Kar. et Kir. 195.
— ophiophyllum Vved. 218.
— oreodictyum Vved. 154.
— oreophiloides Rgl. 224.
— oreophilum C. A. Mey. 255.
— oreoprasoides Vved. 154.
— oreoprasum Schrenk 162.
— oreoscordum Vved. 144.
— Oschanini O. Fedtsch. 198.
— Pallasii Murr. 220.
— paniculatum L. 205.
— paradoxum (M. B.) G. Don 254.
— parvulum Vved. 216.
— petraeum Kar. et Kir. 183.
— platyspathum Schrenk 175.
— polyphyllum Kar. et Kir. 176.
— polyrrhizum Turcz. ex Rgl. 172; tab 10.
— ponticum Miscz. ex Grossh. 244.
— Popovii Vved. 226.
— porrum L. 254.
— prostratum Trev. 166.
— pseudoampeloprasum Miscz. ex Grossh. 251.
— pseudoflavum Vved. 203; tab. 12.
— pseudoseravschanicum M. Pop. et Vved. 273.
— pskemense B. Fedtsch. 197.
— pulchellum G. Don 204.
— pumilum Vved. 171.
— Regelianum A. Beck. ex Iljin 241.
— Regelii Trautv. 278.
— robustum Kar. et Kir. 265.
— Rosenbachianum Rgl. 270.
— rotundum L. 246.
— rubellum M. B. 213.
— rubens Schrad. ex Willd. 166.
— rupestre Stev. 208.
— sabulosum Stev. ex Bge. 228.
— sacculiferum Maxim. 143.
— sativum L. 243.
— saxatile M. B. 184.
— scabrellum Boiss. et Buhse 214; tab. 13.
— scabriscapum Boiss. et Kotschy, 155.
— schoenoprasoides Rgl. 224.
— schoenoprasum L. 190.
— Schubertii Zucc. 275; tab. 16.
— scorodoprasum L. 242.
— scrobiculatum Vved. 227.
— Semenovii Rgl. 189.
— senescens L. 169.
— seravschanicum Rgl. 273.
— Sergii Vved. 257.
— setifolium Schrenk 178.
— Severtzovii Rgl. 265.
— sphaerocephalum L. 241.
— splendens Willd. ex Roem. et Schult. 150.
— staminum Boiss. 202.
— Stellerianum Willd. 165.
— stephanophorum Vved. 161.
— stipitatum Rgl. 267.
— strictum Schrad. 151.
— subtilissimum Ledeb. 178.
— Suvorovii Rgl. 266.
— syntamanthum C. Koch 210.
— Szovitsii Rgl. 153.
— taeniopetalum M. Pop. et Vved. 269.
— talassicum Rgl. 183; tab. 11.
— talyschense Miscz. ex Grossh. 251.
— tenuicaule Rgl. 162.
— tenuissimum L. 173.
— teretifolium Rgl. 186.
— tianschanicum Rupr. 182.
— trachyscordum Vved. 155.
— transvestiens Vved. 228; tab. 12.
— Trautvetterianum Rgl. 274.
— tripedale Trautv. 279.
— turcomanicum Rgl. 234.
— turkestanicum Rgl. 229.
— tytthanthum Vved. 179.
— tytthocephalum Roem. et Schult. 169; tab. 10.
— ursinum L. 142.
— Vavilovii M. Pop. et Vved. 198.
— verticillatum Rgl. 258; tab. 15.
— Veschnjakovii Rgl. 188.
— victorialis L. 141.

Allium vineale L. 237.
— viride Grossh. 236.
— Waldsteinii G. Don 248.
— Winklerianum Rgl. 276; tab. 16.
— xiphopetalum Aitch. et Baker 157.
Allocarya Greene 19, 526.
— orientalis (L.) Brand 526.
Allochrusa Bge. 6, 799.
Alnaster Spach 5, 308.
Alnus Gaertn. 5, 306.
— barbata C. A. Mey. 315.
— fruticosa Rupr. 308; tab. 16.
— glutinosa (L.) Gaertn. 312.
— hirsuta Turcz. 318; tab. 16.
— incana (L.) Moench 316.
— japonica Sieb. et Zucc. 311; tab. 16.
— kamtschatica (Call.) Kom. 310.
— manshurica (Call.) Hand.-Mazz. 309.
— Maximowiczii Call. 310.
— sibirica Fisch. 317.
— subcordata C. A. Mey. 311; tab. 16.
— tinctoria Sarg. 319.
Alopecias Stev. 12, 383.
Alopecurus L. 2, 135.
— aequalis Sobol. 158; tab. 10.
— alpinus Smith 155; tab. 10.
— amurensis Kom. 158; tab. 10.
— apiatus Ovcz. 146, 744; tab. 10.
— Aucheri Boiss. 144.
— borealis Trin. 155; tab. 10.
— brachystachyus M. B. 144; tab. 10.
— dasyanthus Trautv. 143; tab. 11.
— geniculatus L. 157; tab. 10.
— Gerardii Vill. 150.
— glacialis C. Koch 145; tab. 11.
— glaucus Less. 153; tab. 10.
— himalaicus Hook. f. 144.
— laxiflorus Ovcz. 151, 744; tab. 11.
— longiaristatus Maxim. 157; tab. 10.
— mucronatus Hack. ex Pauls. 146; tab. 10.
— myosuroides Huds. 156; tab. 10.
— pratensis L. 150; tab. 11.
— pseudobrachystachyus Ovcz. 153, 745; tab. 10.
— Roshevitzianus Ovcz. 154, 746; tab. 10.
— seravschanicus Ovcz. 151, 745; tab. 11.
— sericeus Albov 149; tab. 11.
— soongoricus (Roshev.) V. Petrov 152; tab. 10.
— Stejnegeri Vasey 156; tab. 10.
— tenuis Kom. 154.
— textilis Boiss. 143; tab. 11.
— tiflisiensis (Westb.) Grossh. 140; tab. 11.
— vaginatus Pall. 140; tab. 11.
— ventricosus Pers. 149.
Alschingera Vis. 16, 223.
Alsibene Griseb. 22, 412.
Alsinanthus Reichb. 6, 507.
Alsine Gaertn. 6, 482.
Alsine L. 6, 562.
— segetalis L. 562; tab. 32.
Alsinopsis Small 6, 482.
Alternanthera Forsk. 6, 368.
— sessilis (L.) R. Br. 368; tab. 18
Althaea L. 15, 127.
— armeniaca Ten. 134; tab. 7.
— broussonetiifolia Iljin 139, 678; tab. 7.
— cannabina L. 141; tab. 7.
— hirsuta L. 129.
— Ludwigii L. 130; tab. 1.
— narbonensis Pourr. 142; tab. 7.
— officinalis L. 131.
— taurinensis DC. 137; tab. 7.
Alyssopsis Boiss. 8, 56.
— Kotshyi Boiss. 57.
Alyssum L. 8, 340.
— andinum Rupr. 350.
— armenum Boiss. 352.
— artvinense N. Busch 345.
— biovulatum N. Busch 346.
— bracteatum Boiss. et Buhse 344.
— Buschianum Grossh. 358.
— calycinum L. 358.
— calycocarpum Rupr. 350.
— campestre L. 357.
— daghestanicum Rupr. 352.
— dasycarpum Steph. 358.
— desertorum Stapf 354.
— Fedtschenkoanum N. Busch 347.
— Gmelinii Jord. et Fourr. 348.
— hirsutum M. B. 356.
— lenense Adam 351.
— macrostylum Boiss. 353.
— marginatum Steud. 355.
— minutum Schlecht. 353.
— Muelleri Boiss. 352.
— murale Waldst. et Kit. 346.
— obtusifolium Stev. 345.
— parviflorum M. B. 357.
— peltarioides Boiss. 347.
— persicum Boiss. 353.
— rostratum Stev. 349.
— saxatile L. 348.
— schemachense N. Busch 350.
— strictum Willd. 356.
— Szovitsianum Fisch. et Mey. 355.
— tortuosum Waldst. et Kit. 344.

— trichostachyum Rupr. 349.
— umbellatum Desv. 355.
Amaracus Gled. 21, 447.
— rotundifolius (Boiss.) Briq. 448; tab. 21.
Amaranthus L. 6, 356.
— albus L. 364; tab. 18.
— blitoides S. Wats. 363; tab. 18.
— blitum L. 364; tab. 18.
— caudatus L. 360; tab. 18.
— deflexus L. 366; tab. 18.
— graecizans L. 365; tab. 18.
— leucospermus S. Wats. 360.
— lividus L. 366; tab. 18.
— paniculatus L. 361; tab. 18.
— retroflexus L. 362; tab. 18.
— Thellungianus Nevski 365; tab. 18.
— tricolor L. 363; tab. 18.
Amarenus C. Presl 11, 218.
Amaryllis Pall. 4, 489.
Amatula Medic. 22, 42.
Amberboa (Pers.) Less. 28, 323.
— amberboi (L.) Tzvel. 331.
— bucharica Iljin 332; tab. 18, 20.
— glauca (Willd.) Sosn. 326; tab. 18.
— moschata (L.) DC. — 325.
— nana (Boiss.) Iljin 328.
— Sosnovskyi Iljin 327.
— turanica Iljin 330.
Amblirion Raf. 4, 302.
Amblynotus Johnst. 19, 527.
— obovatus (Ledeb.) Johnst. 528.
Amblyocarpum Fisch. et Mey. 25, 497.
— inuloides Fisch. et Mey. 497.
Amblyopogon (DC.) Jaub. et Spach 28, 370, 472.
Ambrosia L. 25, 517.
— aptera DC. 520.
— artemisifolia L. 519.
— psilostachya DC. 519.
— trifida L. 520.
Amelanchier Medic. 9, 408.
— canadensis (L.) Medic. 410.
— integrifolia Boiss. et Hohen. 409.
— rotundifolia (Lam.) Dum.-Cours. 408; tab. 27.
— spicata (Lam.) C. Koch 413.
— turkestanica Litw. 410.
Amethystea L. 20, 69.
— coerulea L. 70.
Ammannia L. 15, 527.
— aegyptiaca Willd. 531.
— arenaria H. B. K. 529.
— multiflora Roxb. 529.
— pubiflora (Koehne) Sosn. 530.
— verticillata (Ard.) Lam. 530.
— viridis Hornem. 532.
Ammi L. 16, 381.
— visnaga (L.) Lam. 381.
Ammios Moench 16, 378.
Ammochloa Boiss. 2, 298.
— palaestina Boiss. 298; tab. 23.
Ammodendron Fisch. 11, 30.
— argenteum (Pall.) Kuntze 32.
— Conollyi Bge. ex Boiss. 31; tab. 2.
— Eichwaldii Ledeb. 30.
— Karelinii Fisch. et Mey. 31.
— Lehmannii Bge. 32.
— longiracemosum Raik. 31; tab. 2.
Ammodenia Patr. 6, 516.
— peploides (L.) Rupr. 517; tab. 26.
Ammodytes Stev. 12, 528.
Ammolirion Kar. et Kir. 4, 42.
Ammophila Host 2, 231.
— arenaria (L.) Link 231, tab. 17.
Ammothamnus Bge. 11, 28.
— Lehmannii Bge. 29.
— songoricus (Schrenk) Lipsky 29.
Amoria C. Presl 11, 206.
Amorpha L. 1, 302.
— fruticosa L. 302.
Ampelopsis Michx. 14, 704.
— brevipedunculata (Maxim.) Trautv. 705.
— heterophylla (Thunb.) Sieb. et Zucc. 706.
— japonica (Thunb.) Makino 706; tab. 39.
— vitifolia (Boiss.) Planch. 705.
Amphicarpaea Ell. 13, 526.
Amphoricarpus Vis. 27, 74.
— elegans Albov 75; tab. 6.
Amphorocarpus Post et Kuntze 27, 74.
Amygdalopsis Carr. 10, 545.
Amygdalus L. 10, 522.
— bucharica Korsh. 529.
— communis L. 524; tab. 33.
— Fenzliana (Fritsch) Lipsky 531; tab. 33.
— georgica Desf. 538.
— Ledebouriana Schlecht. 537.
— nairica Fed. et Takht. 542; tab. 34.
— nana L. 535; tab. 33.
— pedunculata Pall. 546.
— Petunnikowii Litv. 539; tab. 33.
— pseudopersica Tamamsch. 532.
— scoparia Spach 544; tab. 34.
— spinosissima Bge. 540; tab. 34.
— turcomanica Lincz. 542, 639.
— ulmifolia (Franch.) M. Pop. 545.
— Vavilovii M. Pop. 543; tab. 34.
— zangezura Fed. et Takht. 532; tab. 33.
Anabasis L. 6, 281.

Anabasis Abolinii Iljin 287, 878.
— affinis Fisch. et Mey. 286.
— aphylla L. 297.
— balchaschensis Iljin 299.
— brachiata Fisch. et Mey. ex Kar. et Kir. 292; tab. 15.
— brevifolia C. A. Mey. 287; tab. 15.
— cretacea Pall. 295.
— elatior (C. A. Mey.) Schischk. 300.
— eriopoda (Schrenk) Benth. 290; tab. 15.
— ferganica Drob. 291.
— gypsicola Iljin 299; tab. 15.
— hispidula (Bge.) Benth. 301; tab. 15.
— jaxartica (Bge.) Benth. 301.
— Korovinii Iljin 300; tab. 15.
— macroptera Moq. 196.
— micradena Iljin 284, 878.
— pauciflora M. Pop. 299, 879.
— Pelliotii Danguy ex Lecomte 285; tab. 15.
— ramosissima Minkw. 289.
— salsa (C. A. Mey.) Benth. 288.
— truncata (Schrenk) Bge. 296; tab. 15.
— turgaica Iljin et Krasch. 286.
— turkestanica Korov. 292. 878; tab. 15.
Anacamptis Rich. 4, 718.
— pyramidalis (L.) Rich. 718; tab. 43.
Anacyclus L. 26, 68.
— ciliatus Trautv. 68.
— officinarum Hayne 69.
Anagallidium Griseb. 18, 622.
— dichotomum (L.) Griseb. 622.
Anagallis L. 18, 275.
— arvensis L. 276.
— coerulea Schreb. 277.
Anandria Less. 28, 592.
Anaphalis DC. 25, 361.
— darvasica Boriss. 375, 591.
— depauperata Boriss. 371, 590.
— garanica Boriss. 373, 591.
— margaritacea (R. Br.) Benth. et Hook. f. 365.
— possietica Kom. 366.
— racemifera Franch. 372; tab. 23.
— roseo-alba Krasch. 374.
— scopulosa Boriss. 376, 592.
— serawschanica (C. Winkl.) B. Fedtsch. 364; tab. 23.
— tenuicaulis Boriss. 367, 589.
— velutina Krasch. 368.
— virgata Thoms. 374.
Anaspis Rech. f. 20, 207.
Anatropa Ehrenb. 14, 151.
Ancathia DC. 28, 222.
— igniaria (Spreng.) DC. 222; tab. 10.
Anchonium DC. 8, 320.
— brachycarpum (Trautv.) Vass. 321, 635; tab. 21.
— elichrysifolium (DC.) Boiss. 321.
— sterigmoides Lipsky 322, 636.
Anchusa L. 19, 298.
— Barrelieri (All.) Vitm. 310.
— Gmelinii Ledeb. 302.
— incana Ledeb. 301.
— italica Retz. 306.
— leptophylla Roem. et Schult. 300.
— ochroleuca M. B. 303.
— officinalis L. 304.
— procera Bess. 305.
— pseudoochroleuca Schost. 304.
— pusilla Gusul. 309.
— stylosa M. B. 309.
— thessala Boiss. et Sprun. 307.
Andrachne L. 14, 269.
— Buschiana Pojark. 277.
— Fedtschenkoi C. Koss. 281; tab. 16.
— filiformis Pojark. 280.
— pusilla Pojark. 278, 733; tab. 16.
— pygmaea C. Koss. 279.
— rotundifolia C. A. Mey. 272.
— stenophylla C. Koss. 279; tab. 16.
— telephioides L. 272; tab. 15.
— virga-tenuis Nevski 274.
Androcera Nutt. 22, 40.
Andromeda L. 18, 74.
— polifolia L. 74.
Andropogon L. 2, 14.
— caucasicus Trin. 15.
— ischaemum L. 14; tab. 1.
Androsace L. 18, 217.
— acrolasia Ovcz. 228.
— akbaitalensis Derg. 230.
— albana Stev. 236; tab. 13.
— angrenica Ovcz. 228, 731.
— arctica Cham. et Schlecht. 234.
— armeniaca Duby 237.
— barbulata Ovcz. 225, 729.
— bidendata C. Koch 237.
— bryomorpha Lipsky 234; tab. 13.
— Bungeana Schischk. et Bobr. 231; tab. 13.
— caduca Ovcz. 222.
— capitata Willd. ex Roem. et Schult. 229.
— dasyphylla Bge. 227.
— elongata L. 239.
— Fedtschenkoi Ovcz. 240.
— filiformis Retz. 241.
— Gmelinii (Gaertn.) Roem. et Schult. 235.
— incana Lam. 225; tab. 13.
— intermedia Ledeb. 238.

— Koso-Poljanskii Ovcz. 227, 730.
— lactiflora Pall. 239.
— Lehmanniana Spreng. 232.
— macrantha Boiss. et Huet 238.
— ochotensis Willd. ex Roem. et Schult. 233; tab. 13.
— Olgae Ovcz. 230.
— Ovczinnikovii Schischk. et Bobr. 221, 729.
— Raddeana Somm. et Lévier 238.
— septentrionalis L. 240.
— sericea Ovcz. 221, 729.
— taurica Ovcz. 226, 730.
— trifoliata Adam 232.
— Turczaninovii Freyn 242.
— umbellata (Lour.) Merr. 235.
Androsaemum Adans. 15, 203, 214.
— Necker 15, 214.
Andrzeiowskia Reichb. 8, 500.
— cardaminifolia (DC.) Prantl 500.
Aneilema R. Br. 3, 501.
— Keisak Hassk. 502; tab. 26.
Anemone L. 7, 236.
— almaatensis Juz. 263, 736.
— altaica Fisch. ex C. A. Mey. 248; tab. 15.
— amurensis (Korsh.) Kom. 247.
— baicalensis Turcz. 253; tab. 15.
— biarmiensis Juz. 272, 738; tab. 17.
— blanda Schott et Kotschy 251.
— brevipedunculata Juz. 280, 740.
— bucharica Rgl. ex Fin. et Gagnep. 258; tab. 16.
— calva Juz. 279.
— caucasica Willd. ex Rupr. 251; tab. 15.
— coerulea DC. 243.
— crinita Juz. 274, 739.
— debilis Fisch. 244.
— dichotoma L. 268.
— eranthioides Rgl. 264.
— fasciculata L. 270; tab. 17.
— flaccida Fr. Schmidt 252.
— glabrata (Maxim.) Juz. 254; tab. 15.
— Gortschakovii Kar. et Kir. 260; tab. 16.
— impexa Juz. 271, 737.
— jenisseensis (Korsh.) Kryl. 242.
— Kuznetzowii Woronow ex Grossh. 257.
— laxa (Ulbr.) Juz. 269; tab. 17.
— litoralis (Litw.) Juz. 254.
— nemorosa L. 247; tab. 15.
— obtusiloba D. Don 256; tab. 16.
— ochotensis Fisch. 267.
— oligotoma Juz. 263, 737.
— parviflora Michx. 267.
— petiolulosa Juz. 259, 736; tab. 16.
— protracta (Ulbr.) Juz. 273; tab. 17.
— Raddeana Rgl. 249; tab. 15.
— ranunculoides L. 242; tab. 15.
— reflexa Steph. 250; tab. 15.
— Richardsonii Hook. 255; tab. 16.
— sachalinensis (Miyabe et Miyake) Juz. 280, 740.
— Schrenkiana Juz. 273, 738.
— serawschanica Kom. 265.
— sibirica L. 276; tab. 17.
— silvestris L. 266; tab. 16.
— speciosa Adam 281.
— Tschernjaewii Rgl. 264; tab. 16.
— udensis Trautv. 250; tab. 15.
— umbrosa C. A. Mey. 248.
— uralensis Fisch. ex DC. 243.
— villosissima (DC.) Juz. 275.
Anethum L. 17, 209.
— graveolens L. 209
— involucratum Korov. 210.
Aneurolepidium Nevski 2, 697.
— aemulans Nevski 711.
— alaicum (Korsh.) Nevski 704.
— akmolinense (Drob.) Nevski 708.
— angustum (Trin.) Nevski 700; tab. 49.
— badshuanicum (Roshev.) Nevski 703 tab. 49.
— dasystachys (Trin.) Nevski 706.
— divaricatum (Drob.) Nevski 709.
— fasciculatum (Roshev.) Nevski 709.
— flexile Nevski 705.
— karataviense (Roshev.) Nevski 705.
— kopetdaghense (Roshev.) Nevski 703; tab. 49.
— multicaule (Kar. et Kir.) Nevski 708; tab. 50.
— ovatum (Trin.) Nevski 707.
— Paboanum (Claus) Nevski 707.
— petraeum Nevski 705.
— pseudoagropyrum (Trin.) Nevski 710.
— ramosum (Trin.) Nevski 710.
— Regelii (Roshev.) Nevski 709.
— tianschanicum (Drob.) Nevski 703.
— ugamicum (Drob.) Nevski 704.
Angelica L. 17, 11.
— amurensis Schischk. 19, 352; tab. 8.
— anomala Avé-Lall. 19.
— brevicaulis (Rupr.) B. Fedtsch. 25; tab. 2, 8.
— Czernaevia (Fisch. et Mey.) Kitag. 26.
— dahurica (Fisch.) Benth. et Hook. ex Franch. et Sav. 23.
— decursiva (Miq.) Franch. et Sav. 27; tab. 8.

Angelica jaluana Nakai 20.
— koreana Maxim. 22.
— Maximowiczii (Fr. Schmidt.) Benth. ex Maxim. 22.
— pachyptera Avé-Lall. 14.
— refracta Fr. Schmidt 17.
— sachalinensis Maxim. 18; tab. 8.
— silvestris L. 13.
— ternata Rgl. et Schmalh. 25; tab. 2.
— ursina (Rupr.) Rgl. et Schmalh. 24.
— viridiflora (Turcz.) Benth. ex Maxim. 21.
Angelocarpa Rupr. 17, 11, 25.
Angelophyllum Rupr. 17, 11, 24.
Anidrum Neck. 16, 198.
Anisactis Dulac 16, 373, 385.
Anisanthera Raf. 19, 685.
Anisoderis Cass. 29, 595, 687.
Anisodus Link et Otto 22, 99.
Anisophyllum Haw. 14, 482.
Anisoramphus DC. 29. 595.
Anisum Gaertn. 16, 444.
— vulgare Gaertn. 445.
Anithista Raf. 3, 403.
Ankylotus Stev. 12, 302.
Anogramma Link 1, 73.
— leptophylla (L.) Link 73; tab. 4.
Anoplocaryum Ledeb. 19, 524.
— compessum (Turcz.) Ledeb. 525.
Anosmia Bernh. 16, 218.
Anserina Dum. 6, 51.
Antennaria Gaertn. 25, 326.
— alpina (L.) Gaertn. 336.
— carpatica (Wahl.) R. Br. 338.
— caucasica Boriss. 330.
— dioica (L.) Gaertn. 329.
— dioiciformis Kom. 334; tab. 21.
— Friesiana (Trautv.) Ekman 337.
— Komarovii Juz. 335.
— monocephala DC. 334.
— villifera Boriss. 340; tab. 21.
Anthemis L. 26, 9.
— abagensis Fed. 56, 869.
— altissima L. 58.
— Anahytae Woronow ex Sosn. 19, 865.
— anatolica Boiss. 24.
— arvensis L. 35.
— austriaca Jacq. 59.
— calcarea Sosn. 30.
— candidissima Willd. ex Spreng. 37; tab. 4.
— carpatica Waldst. et Kit. ex Willd. 20.
— coelopoda Boiss. 59.
— cotula L. 63; tab. 4.
— cretacea Zefir. 43.
— debilis Fed. 44, 867.
— deserticola Krasch. et M. Pop. 65.
— dubia Stev. 51.
— dumetorum Sosn. 47; tab. 3.
— Emiliae Sosn. 25.
— euxina Boiss. 53.
— fruticulosa M. B. 28; tab. 2.
— Grossheimii Sosn. 38.
— hirtella C. Winkl. 38.
— iberica M. B. 18; tab. 1.
— jailensis Zefir. 47.
— Linczevskyi Fed. 48, 868.
— lithuanica (DC.) Trautv. 64.
— macroglossa Somm. et Levier 54.
— maris-nigri Fed. 42, 866.
— markhotensis Fed. 42, 866.
— Marschalliana Willd. 30; tab. 2.
— melanoloma Trautv. 57.
— microcephala (Schrenk) B. Fedtsch. 65.
— monantha Willd. 43.
— odontostephana Boiss. 64.
— ptarmiciformis C. Koch 23.
— rigescens Willd. 46.
— ruthenica M. B. 36.
— saguramica Sosn. 42.
— Saportana Albov 17; tab. 1.
— Schischkiniana Fed. 55, 870.
— Sosnovskyana Fed. 31.
— sterilis Stev. 25.
— subtinctoria Dobrocz. 40.
— talyschensis Fed. 53, 869.
— Tempskyana Freyn et Sint. 27.
— tinctoria L. 39.
— Tranzscheliana Fed. 27, 866.
— Trotzkiana Claus ex Bge. 29.
— Wiedemanniana Fisch. et Mey. 60.
— Woronowii Sosn. 52.
— Zephyrovii Dobrocz. 45.
— zyghia Woronow 56; tab. 3.
Anthericum L. 4, 53.
— ramosum L. 53; tab. 3.
Anthochlamys Fenzl 6, 162.
— polygaloides (Fisch. et Mey.) Fenzl. 163; tab. 7.
— turcomanica Iljin 165, 875; tab. 7.
Anthochytrum Reichb. f. 29, 595, 687.
Anthopetitia A. Richt. 13, 245.
Anthosachne Steud. 2, 597.
— Jacquemontii (Hook. f.) Nevski 598; tab. 44.
— longiaristata (Boiss.) Nevski 598; tab. 44.
Anthoxanthum L. 2, 55.
— aristatum Boiss. 59; tab. 4.
— odoratum L. 56; tab. 4.

Anthriscus Bernh. 16, 153.
Anthriscus (Pers.) Hoffm. 16, 127.
— aemula (Woronow) Schischk. 129.
— cerefolium (L.) Hoffm. 136.
— glacialis Lipsky 131.
— longirostris Bertol. 137; tab. 9.
— nemorosa (M. B.) Spreng. 127.
— nitida (Wahl.) Garcke 131.
— Ruprechtii Boiss. 132; tab. 9.
— scandicina (Web.) Mansf. 138; tab. 9.
— Schmalhausenii (Albov) K.-Pol. 135; tab. 9.
— silvestris (L.) Hoffm. 128.
— Sosnovskyi Schischk. 135; tab. 9.
— velutina Somm. et Lev. 130.
Anthyllis L. 11, 262.
— affinis Britt. 280.
— alpestris Reichb. 279.
— arenaria (Rupr.) Juz. 270; tab. 17.
— Biebersteiniana (Taliev) Popl. 267; tab. 17.
— caucasica (Grossh.) Juz. 277.
— coccinea (L.) Beck 266.
— colorata Juz. 277, 393.
Anthyphylla Haw. 9, 197.
Antirrhinum L. 22, 225.
— majus L. 226.
— orontium L. 226.
Antitoxicum Pobed. 18, 674.
— acuminatum (Decne.) Pobed. 706.
— Albovianum (Kusn.) Pobed. 694.
— amplexicaule (Sieb. et Zucc.) Pobed. 705.
— atratum (Bge.) Pobed. 704; tab. 38.
— cretaceum Pobed. 702, 753; tab. 38.
— darvasicum (B. Fedtsch.) Pobed. 693; tab. 37.
— funebre (Boiss. et Kotschy) Pobed. 691.
— inamoenum (Maxim.) Pobed. 707.
— intermedium (Taliev) Pobed. 689.
— jailicola (Juz.) Pobed. 696.
— Juzepczukii Pobed. 683, 752.
— laxum (Bartl.) Pobed. 701.
— maeoticum (Kleop.) Pobed. 687.
— minus (C. Koch) Pobed. 687.
— mugodsharicum Pobed. 690.
— officinale (Moench) Pobed. 697; tab. 38.
— pumilum (Decne.) Pobed. 692.
— Raddeanum (Albov) Pobed. 695.
— Rehmannii (Boiss.) Pobed. 682; tab. 37.
— rossicum (Kleop.) Pobed. 690.
— scandens (Somm. et Lév.) Pobed. 680.
— Schmalhausenii (Kusn.) Pobed. 684; tab. 37.
— sibiricum (L.) Pobed. 707; tab. 38.
— stauropolitanum Pobed. 695.
— stepposum Pobed. 702, 752; tab. 37.
— tauricum Pobed. 696.
— volubile (Maxim.) Pobed. 703.
Anura (Juz.) Tschern. 27, 107.
— pallidivirens (Kult.) Tschern. 107; tab. 29.
Apargia Scop. 29, 207.
Apeltanthus Nevski ex Pavl. 20, 217.
Apenula Neck. 24, 427.
Apera Adans. 2, 232.
— intermedia Hack. 233; tab. 17.
— interrupta (L.) Beauv. 233; tab. 17.
— spica-venti (L.) Beauv. 233; tab. 17.
Aphananthemum Spach 15, 336.
Aphanes L. 10, 410.
— arvensis L. 410.
Aphanopleura Boiss. 16, 365.
— capillifolia (Rgl. et Schmalh.) Lipsky 367.
— leptoclada (Aitch. et Hemsl.) Lipsky 366; tab. 12.
— trachysperma Boiss. 366.
Aphragamus Andrz. 8, 57.
— involucratus (Bge.) O. E. Schulz 58.
Apinella Caruel 16, 350.
Apinella Neck. 16, 349.
Apium Caruel 16, 385.
Apium L. 16, 370.
— graveolens L. 371, 373.
Aplotaxis DC. 27, 361.
Apluda L. 2, 12.
— inermis Rgl. 12; tab. 1.
Apocynum L. 18, 652.
Aposeris Neck. 29, 404.
— foetida (L.) Less. 404; tab. 1.
Aquilegia L. 7, 86.
— amurensis Kom. 97; tab. 7.
— Borodinii Schischk. 91; tab. 7.
— colchica Kom.-Nath. 98.
— flabellata Sieb. et Zucc. 97; tab. 7.
— formosa Fisch. 98.
— glandulosa Fisch. ex Link 95; tab. 7.
— hybrida Sims 98.
— Karelinii (Baker) O. et B. Fedtsch. 93; tab. 7.
— lactiflora Kar. et Kir. 92; tab. 7.
— leptoceras Fisch. et Mey. 88; tab. 7.
— olympica Boiss. 94; tab. 7.
— oxysepala Trautv. 93; tab. 7.
— parviflora Ledeb. 88; tab. 7.
— sibirica Lam. 96; tab. 7.
— viridiflora Pall. 91; tab. 7.
— vulgaris L. 94; tab. 7.

Arabidopsis Heynh. 8, 76.
-- mollissima (C. A. Mey.) N. Busch 79.
— parvula (Schrenk) O. E. Schulz 78.
— pumila (Steph.) N. Busch 78.
— Thaliana (L.) Heynh. 77.
— toxophylla (M. B.) N. Busch 79.
- verna (C. Koch) N. Busch 80.
Arabis L. 8, 172.
— alpina L. 194.
— amurensis N. Busch 192.
— arenosa (L.) Scop. 191.
— auriculata Lam. 185; tab. 10.
— Billardieri DC. 197.
— borealis Andrz. 184.
— brachycarpa Rupr. 177; tab. 9.
— bucharica (Lipsky) N. Busch 188, 640; tab. 10.
— caucasica Willd. 194.
— Christianii N. Busch 176.
— dolichothrix N. Busch 195, 641.
— erysimoides Kar. et Kir. 186.
— farinacea Rupr. 196.
— flaviflora Bge. 196.
— fruticulosa C. A. Mey. 185.
— Gerardii Bess. 183.
— hirsuta Scop. 183.
— ionocalyx Boiss. 195.
— kamtschatica Fisch. 192.
— karategina Lipsky 187; tab. 10.
— kokanica Rgl. et Schmalh. 187; tab. 10.
— laxa Sibth. et Smith 181; tab. 9.
— Maximoviczii N. Busch 191.
— media N. Busch 193.
— Montbretiana Boiss. 184.
— nepetaefolia Boiss. 177.
— Nordmanniana Rupr. 176; tab. 9.
— pachyrhiza Kar. et Kir. 186; tab. 10.
— pendula L. 181.
— secunda N. Busch 178.
— septentrionalis N. Busch 193.
— sinuata Turcz. 188.
— Stelleri DC. 182.
— trichopoda Turcz. 187.
— Turczaninovii Ledeb. 182.
— turrita L. 182; tab. 9.
Arachis L. 13, 373.
— hypogaea L. 373.
Arachne Neck. 14, 269.
— colchica (Fisch. et Mey.) Pojark. 269; tab. 15.
Aracium Neck. 29, 594, 606.
Aralia L. 16, 24.
— continentalis Kitag. 32.
— cordata Thunb. 31.
— elata (Miq.) Seem. 25.
— mandshurica Rupr. et Maxim. 27.
— Schmidtii Pojark. 33, 588; tab. 1.
Araujia Brot. 18, 667.
-- sericifera Brot. 667.
Arbutus L. 18, 79.
— andrachne L. 80.
— unedo L. 80.
Arceutobium M. B. 5, 407.
Archangelica Hoffm. 17, 28.
— decurrens Ledeb. 30; tab. 8.
— Komarovii Schischk. 31, 353.
— litoralis (Fries) Agardh 30.
— officinalis (Moench) Hoffm. 28.
— tschimganica (Korov.) Schischk. 32.
Archieracium Fries 30, 10.
Arctagrostis Griseb. 2, 166.
— arundinacea (Trin.) Beal 168; tab. 12.
— festucacea V. Petrov 167.
— latifolia (R. Br.) Griseb. 167; tab. 12.
— ursorum Kom. 168.
Arcterica Cov. 18, 71.
— nana (Maxim.) Makino 72.
Arctium L. 27, 93.
— lappa L. 97; tab. 7.
— leiospermum Juz. et E. Serg. 98.
— minus Bernh. 100; tab. 7.
— nemorosus Lej. 99; tab. 7.
— Palladinii (Marc.) Grossh. 103; tab. 7.
— platylepis (Boiss. et Bal.) Sosn. ex Grossh. 95; tab. 7.
— radula Juz. et E. Serg. 105.
— tomentosum Mill. 104; tab. 7.
Arctocrania Nakai 17, 324.
Arctogeron DC. 25, 135.
— gramineum (L.) DC. 136; tab. 8.
Arctophila Rupr. 2, 433.
— fulva (Trin.) Anders. 433; tab. 33.
Arctostaphylos Adans. 18, 83.
— uva-ursi (L.) Spreng. 83.
Arctous (A. Gray) Nied. 18, 84.
— alpina (L.) Nied. 85.
— erythrocarpa Small 87; tab. 2.
— japonica Nakai 86.
Arenaria L. 6, 517.
— asiatica Schischk. 527.
— capillaris Poir. 530; tab. 30.
— cephalotes M. B. 521; tab. 29.
— cucubaloides Smith 528.
— dianthoides Smith 520.
— ferganica Schischk. 532.
— formosa Fisch. ex DC. 529.
— graminea C. A. Mey. 526.
— graminifolia Schrad. 525.

— Criffithii Boiss. 536; tab. 30.
— gypsophiloides L. 528; tab. 29.
— holostea M. B. 522; tab. 29.
— humifusa (Sr.) Wahl. 537.
— insignis Litw. 535; tab. 30.
— juncea M. B. 529; tab. 29.
— Koriniana Fisch. ex Fenzl 525.
— Ledebouriana Fenzl 535.
— leptoclados Guss. 539.
— longifolia M. B. 527; tab. 29.
— lychnidea M. B. 530; tab. 29.
— macrantha Schischk. 522, 886.
— Meyeri Fenzl. 531.
— mongolica Schischk. 531.
— Paulsenii H. Winkl. 535.
— polaris Schischk. 526, 887.
— Potaninii Schischk. 536; tab. 30.
— pseudofrigida (Ostenf. et Dahl) Juz. 537; tab. 30.
— Redowskii Cham. et Schlecht. 538.
— rigida M. B. 521.
— rotundifolia M. B. 538; tab. 30.
— serpyllifolia L. 539; tab. 30.
— Steveniana Boiss. 527.
— Szowitsii Boiss. 522.
— tschuktschorum Rgl. 532.
— turkestanica Schischk. 538.
Arctia L. 18, 233.
Argentina Lam. 10, 220.
Arguzia Stev. 19, 116.
Argyranthemum Webb ex Sch. Bip. 26, 129.
— frutescens (L.) Sch. Bip. 130.
Argyrolobium Eckl. et Zeyh. 11, 51.
— calycinum (M. B.) Jaub. et Spach 51.
— lotoides Bge. 53.
— trigonelloides Jaub. et Spach 52.
Arisaema Mart. 3, 490.
— amurense Maxim. 490; tab. 25.
— japonicum Blume 490.
Aristida L. 2, 65.
— adscensionis L. 66; tab. 5.
— arachnoidea Litv. 71; tab. 5.
— Karelinii (Trin. et Rupr.) Roshev. 67. 740; tab. 5.
— pennata Trin. 67; tab. 5.
— plumosa L. 68; tab. 5.
Aristolochia L. 5, 435.
— Bottae Jaub. et Spach 438.
— clematitis L. 438; tab. 25.
— contorta Bge. 438; tab. 25.
— iberica Fisch. et Mey. 437.
— macrophylla Lam. 441.
— manshuriensis Kom. 436; tab. 25.
— pontica Lam. 436.
— Steupii Woronow 437; tab. 25.
Armeniaca Mill. 10, 584.
— dasycarpa (Ehrh.) Pers. 598.
— Davidiana Carr. tab. 38.
— manshurica (Koehne) Skvortz. 596; tab. 38.
— sibirica (L.) Lam. 590; tab. 38.
— vulgaris Lam. 586; tab. 38.
Armeria Willd. 18, 408.
— arctica (Cham.) Wallr. 411; tab. 21.
— elongata (Hoffm.) C. Koch 410; tab. 21.
— sibirica Turcz. ex Boiss. 409.
Armeriastrum Kuntze 18, 301.
Armoracia Gaertn., Mey. et Scherb. 8, 141.
— rusticana (Lam.) Gaertn., Mey. et Scherb. 142.
— sisymbrioides (DC.) Cajand. 141.
Arnebia Forsk. 19, 173.
— baldschuanica (Lipsky) Schischk. 179, 705.
— coerulea Schipcz. 178.
— decumbens (Vent.) Coss. et Kral. 180.
— grandiflora (Trautv.) M. Pop. 179, 705.
— guttata Bge. 176.
— linearifolia DC. 182.
— obovata Bge. 175.
— Thomsonii Clarke 177.
— tibetana Kurz 177.
— transcaspica M. Pop. 180, 705.
Arnica L. 26, 655.
— alpina (L.) Olin et Ladan 657.
— frigida C. A. Mey. et Iljin 660.
— Iljinii (Maquire) Iljin 658.
— intermedia Turcz. 659.
— Lessingii Green 665, tab. 32.
— montana L. 663.
— sachalinensis (Rgl.) A. Gray 668.
— unalaschcensis Less. 666; tab. 32.
Arnoseris Gaertn. 29, 20.
— minima (L.) Schweigg. et Korte 20; tab. 1.
Aronicum Neck. 26, 669.
Arpitium Neck. 16, 579.
Arrhenatherum Beauv. 2, 281.
— elatius (L.) Mert. et Koch 281.
— Kotschyi Boiss. 282.
Artemisia L. 26, 425.
— abrotanum L. 456; tab. 23.
— absinthium L. 515.
— Adamsii Bess. 463; tab. 24.
— albicerata Krasch. 542.
— altaiensis Krasch. 514.
— amoena Poljak. 600.
— anethifolia Web. 522; tab. 27.

Artemisia annua L. 489.
— aralensis Krasch. 594.
— arctica Less. 479.
— arenaria DC. 540.
— Argyi Lévl. et Vaniot 451.
— armeniaca Lam. 471.
— Aschurbajevii C. Winkl. 503.
— aurata Kom. 492.
— austriaca Jacq. 498.
— badhysi Krasch. et Lincz. ex Poljak. 619.
— balchanorum Krasch. 617.
— baldschuanica Krasch. et Zapr. 624.
— bargusinensis Spreng. 555.
— borealis Pall. 557; tab. 29.
— caespitosa Ledeb. 499.
— camelorum Krasch. 595.
— campestris L. 553.
— capillaris Thunb. 550; tab. 30.
— caucasica Willd. 492.
— chamaemelifolia Vill. 458.
— cina Berg. ex Poljak. 601.
— ciniformis Krasch. et M. Pop. ex Poljak. 603.
— commutata Bess. 551; tab. 28.
— compacta Fisch. ex DC. 585.
— cuspidata Krasch. 500.
— daghestanica Krasch. et Poretzky 535.
— demissa Krasch. 562; tab. 29.
— deserti Krasch. 626.
— desertorum Spreng. 549; tab. 30.
— Dimoana M. Pop. 536.
— dracunculiformis Krasch. 534.
— dracunculus L. 529.
— Dubjanskyana Krasch. ex Poljak. 587.
— Fedtschenkoana Krasch. 600.
— ferganensis Krasch. ex Poljak. 607.
— fragrans Willd. 576.
— Freyniana (Pampan.) Krasch. 467.
— frigida Willd. 494.
— furcata M. B. 483.
— glanduligera Krasch. ex Poljak. 604.
— glauca Pall. 535; tab. 28.
— globularia Bess. 481.
— glomerata Ledeb. 482.
— Gmelinii Web. ex Stechm. 464; tab. 23.
— gracilescens Krasch. et Iljin 591.
— gypsacea Krasch., M. Pop. et Lincz. ex Poljak. 610.
— halodendron Turcz. ex Bess. 539; tab. 28.
— halophila Krasch. 581.
— Henriettae Krasch. 558.
— hololeuca M. B. ex Bess. 507.
— incana (L.) Druce 524.
— insulana Krasch. 484.
— integrifolia L. 449.
— jacutica Drob. 522.
— japonica Thunb. 545.
— juncea Kar. et Kir. 629.
— kaschgarica Krasch. 599.
— Keiskeana Miq. 487.
— Kelleri Krasch. 537.
— kemrudica Krasch. 614.
— Knorringiana Krasch. 588.
— kochiiformis Krasch. et Lincz. ex Poljak. 615.
— Koidzumii Nakai 447.
— Komarovii Poljak. 449.
— kopetdaghensis Krasch. ex Poljak. 617.
— Korovinii Poljak. 589.
— kulbadica Boiss. et Buhse 519.
— Kuschakewiczi C. Winkl. 544.
— laciniata Willd. 473.
— laciniatiformis Kom. 477.
— lagocephala (Bess.) DC. 504; tab. 26.
— lagopus Fisch. ex Bess. 486.
— latifolia Ledeb. 472.
— lavandulifolia DC. 453.
— Ledebouriana Bess. 555; tab. 29.
— Lehmanniana Bge. 622.
— leontopodioides Fisch. ex Bess. 485.
— Lercheana Web. ex Stechm. 579.
— Lessingiana Bess. 594.
— leucodes Schrenk 630.
— leucophylla Turcz. ex Clarke 443.
— limosa Koidz. 559.
— Lipskyi Poljak. 544.
— littoricola Kitam. 546.
— lobulifolia Boiss. 625.
— macilenta (Maxim.) Krasch. 550.
— macrantha Ledeb. 462.
— macrocephala Jacq. 518.
— macrorhiza Turcz. 478.
— maritima L. 573.
— Martjanovii Krasch. ex Poljak. 513.
— Maximovicziana Krasch. ex Poljak. 409; tab. 25.
— medioxima Krasch. ex Poljak. 468; tab. 25.
— mongolica Fisch. ex Nakai 444.
— monogyna Waldst. et Kit. 574.
— montana Pampan. 452.
— namanganica Poljak. 611.
— nitrosa Web. ex Stechm. 580.
— norvegica Fries 480.
— obtusiloba Ledeb. 509.
— opulenta Pampan. 439.
— palustris L. 491.
— pamirica C. Winkl. 534.

— pannosa Krasch. 556; tab. 28.
— pauciflora Web. 589.
— persica Boiss. 506; tab 21.
— phaeolepis Krasch. 474.
— pontica L. 461.
— porrecta Krasch. ex Poljak. 607.
— prasina Krasch. ex Poljak. 620.
— prolixa Krasch. ex Poljak. 587.
— punctigera Krasch. ex Poljak. 480.
— pycnorhiza Ledeb. 559.
— quinqueloba Trautv. 540.
— remotiloba Krasch. ex Poljak. 470.
— rhodantha Rupr. 586.
— rubripes Nakai 450.
— rupestris L. 508.
— rutifolia Steph. ex Spreng. 505; tab. 22.
— Saitoana Kitam. 448.
— salsoloides Willd. 538.
— samoiedorum Pamp. 520.
— santolina Schrenk 625.
— santolinifolia Turcz. ex Bess. 465.
— Saposhnikovii Krasch. ex Poljak. 542; tab. 30.
— Schmidtiana Maxim. 501.
— Schrenkiana Ledeb. 582.
— scopaeformis Ledeb. 581.
— scoparia Waldst. et Kit. 560.
— scotina Nevski 614.
— selengensis Turcz. ex Bess. 454.
— Senjaviensis Bess. 485.
— sericea Web. ex Stechm. 501.
— serotina Bge. 608.
— Sieberi Bess. 621.
— Sieversiana Willd. 517.
— silvatica Maxim. 445; tab. 21.
— sogdiana Bge. 616.
— songarica Schrenk 543.
— specigera C. Koch 620.
— splendens Willd. 493.
— Stelleriana Bess. 455; tab. 20.
— stenocephala Krasch. ex Poljak. 616.
— stolonifera (Maxim.) Kom. 446; tab. 22
— sublessingiana (B. Keller) Krasch. ex Poljak. 596.
— subulata Nakai 450.
— subviscosa Turcz. 510.
— succulenta Ledeb. 521.
— Szowitziana (Bess.) Grossh. 575.
— tanacetifolia L. 468.
— taurica Willd. 578.
— tenuisecta Nevski 598.
— terrae-alba Krasch. 592.
— Tilesii Ledeb. 443.
— tomentella Trautv. 554.
— Tournefortiana Reichb. 488.
— transiliensis Poljak. 597.
— Trautvetteriana Bess. 537.
— Triniana Bess. 487.
— turanica Krasch. 612.
— turcomanica Gand. 618.
— umbrosa Turcz. ex DC. 453.
— unalaskensis Rydb. 440.
— ussuriensis Poljak. 446.
— vachanica Krasch. ex Poljak. 623.
— valida Krasch. ex Poljak. 610.
— vulgaris L. 438.
Arthraxon Beauv. 2, 13.
— Langsdorfii (Trin.) Hochst. 13; tab. 1.
Arthrophytum Schrenk 6, 302.
— betpakdalense Korov. 307.
— iliense Iljin 303; tab. 10.
— Lehmannianum Bge. 304.
— leptocladum M. Pop. ex Iljin 308, 879
— Litwinowii Korov. 306.
— subulifolium Schrenk 306; tab. 10.
— wakhanicum (Pauls.) Korov. 308.
Arum L. 3, 485.
— albispathum Stev. 487.
— elongatum Stev. 486.
— Korolkowii Rgl. 486.
— maculatum L. 487.
— orientale M. B. 485; tab. 26.
Aruncus Adans. 9, 309.
— asiaticus Pojark. 311, 491.
— kamtschaticus Rydb. 311; tab. 18.
— parvulus Kom. 312.
— vulgaris Raf. 310; tab. 18.
Arundinaria Michx. 2, 732.
— Simonii A. et C. Rivière 732.
Arundinella Raddi 2, 24.
— anomala Steud. 24; tab. 1.
Arundo L. 2, 302.
— donax L. 303; tab. 23.
Asaphes Spreng. 24, 2.
Asarum L. 5, 432.
— europaeum L. 433.
— ibericum Stev. ex Ledeb. 434; tab. 25.
— Sieboldii Miq. 432; tab. 25.
Ascalea Hill 28, 52.
Asclepias L. 18, 669.
— syriaca L. 670; tab. 36.
Asimina Adans. 7, 568.
— triloba (L.) Dun. 568; tab. 39.
Asparagus L. 4, 424.
— angulofractus Iljin 432, 746; tab. 25
— brachyphyllus Turcz. 437; tab. 25.
— Breslerianus Schult. 431; tab. 25.
— bucharicus Iljin 430, 746; tab. 25.

Asparagus caspius Hohen. 440.
— dahuricus Fisch. 433.
— kasakstanicus Iljin 434, 747.
— Leviniae Klok. 439.
— litoralis Stev. 442.
— Misczenkoi Iljin 431, 746.
— neglectus Kar. et Kir. 430; tab. 25.
— officinalis L. 439.
— oligochlonos Maxim. 441.
— persicus Baker 438.
— polyphyllus Stev. 441; tab. 25.
— Popovii Iljin 430, 745; tab. 25.
— pseudoscaber Grec. 439; tab. 25.
— schoberioides Kunth 428; tab. 25.
— setiformis Kryl. 440.
— soongoricus Iljin 432, 747; tab. 25.
— tenuifolius Lam. 429; tab. 25.
— tuberculatus Bge. 433, 747.
— turkestanicus M. Pop. 429.
— verticillatus L. 428.
Asperella Humb. 2, 692.
— coreana (Honda) Nevski 693; tab. 49.
— Komarovii Roshev. 694; tab. 49.
— sibirica Trautv. 693; tab. 49.
Asperugo L. 19, 530.
— procumbens L. 530.
Asperula L. 23, 193.
— abchasica V. Krecz. 255.
— accrescens Klok. 249, 704.
— aemulans V. Krecz. 241, 700.
— affinis Boiss. et Huet 257.
— Albovii Manden. 207.
— alpina M. B. 253.
— aparine M. B. 275.
— arvensis L. 281.
— attenuata Klok. 235, 698.
— azurea Jaub. et Spach 283.
— balchanica Bobr. 213, 691; tab. 11.
— bidentata Klok. 233, 697.
— Biebersteinii V. Krecz. 247.
— caespitans Juz. 230.
— campanulata (Vill.) Klok. 166.
— caucasica Pobed. 215, 692; tab. 11.
— cimmerica V. Krecz. 246, 702.
— creticola Klok. 225, 694.
— cristata (Somm. et Levier) V. Krecz. 250.
— cynanchica L. 224; tab. 12.
— Danilewskiana Basin. 264.
— dasyantha Klok. 249, 704.
— diminuta Klok. 263, 707.
— dolichophylla Klok. 258, 706.
— exasperata V. Krecz. 227, 695.
— ferganica Pobed. 212, 690; tab. 11.
— galioides M. B. 267.
— glomerata (M. B.) Griseb. 208.
— gracilis C. A. Mey. 285.
— graniticola Klok. 239, 699.
— graveolens M. B. 262.
— hirsutiscula Pobed. 211, 690.
— humifusa (M. B.) Bess. 276.
— hypanica Klok. 235, 698.
— infracta Klok. 244, 701.
— insuavis Pobed. 278, 708.
— intersita Klok. 254, 705.
— karataviensis Pavl. 275; tab. 14.
— Kemulariae Manden. 259.
— Kryloviana Serg. 271.
— laevis Schischk. 280, 709; tab. 15.
— laevissima Klok. 264, 708; tab. 13.
— leiograveolens M. Pop. et Chrshan. 262.
— Lipskyana V. Krecz. 241.
— markothensis Klok. 240, 700.
— Maximoviczii Kom. 273.
— molluginoides (M. B.) Reichb. 205.
— octonaria Klok. 265; tab. 14.
— odorata L. 217.
— oppositifolia Rgl. et Schmalh. 278; tab. 15.
— pamirica Pobed. 212, 691.
— paniculata Bge. 271; tab. 14.
— pedicellata Klok. 247, 703.
— petraea V. Krecz. 228, 696.
— platygalium Maxim. 273.
— pontica Boiss. 256.
— Popovii Schischk. 280.
— praepilosa V. Krecz. 245, 702.
— praevestita Klok. 242, 701.
— propinqua Pobed. 216, 693.
— prostrata (Adam) C. Koch 206.
— rivalis Sibth. et Smith 274.
— rumelica Boiss. 236.
— savranica Klok. 261, 707; tab. 13.
— semiamicta Klok. 223, 694.
— setosa Jaub. et Spach 282.
— setulosa Boiss. 260.
— Stevenii V. Krecz. 232; tab. 12.
— supina M. B. 229; tab. 13.
— taurica Pacz. 284; tab. 11.
— tephrocarpa Czern. 226; tab. 12.
— tinctoria L. 272.
— Tranzshelii Klok. 231, 697.
— turcomanica Pobed. 214, 692.
— tyraica Bess. 267.
— vestita V. Krecz. 243.
— Woronowii V. Krecz. 259.
— xerotica Klok. 268.
Asphodeline Reichb. 4, 31.

— dendroides (Hoffm.) Woronow ex Grossh. 32; tab. 5.
— lutea (L.) Reichb. 33; tab. 3.
— taurica (Pall.) Kunth. 33; tab. 3.
— tenuiflora (C. Koch) Miscz. 34.
— tenuior (Fisch.) Ledeb. 34.
Asplenium L. 1, 61.
— adiantum-nigrum L. 70.
— anagrammoides Christ. 68; tab. 4.
— daghestanicum Christ. 69.
— germanicum Weis 64.
— Hermanni-Christi Fomin 67.
— Heufleri Reichardt 65.
— incisum Thunb. 66; tab. 4.
— pseudofontanum C. Koss. 68; tab. 4.
— pseudolanceolatum Fomin 69.
— ruta-muraria L. 66.
— samarcandense C. Koss. 67; tab. 4.
— Sarelii Hook. 67; tab. 4.
— septentrionale (L.) Hoffm. 64.
— trichomanes L. 65.
— viride Huds. 65.
— Woronowii Christ. 69.
Astenolobium Nevski 12, 233.
Aster L. 25, 77.
— ageratoides Turcz. 100.
— alpinus L. 105.
— amelloides Bess. 87.
— amellus L. 86.
— fallax Tamamsch. 109, 580; tab. 8.
— Fauriei Lévl. et Vaniot 93.
— Glehnii Fr. Schmidt 91.
— ibericus Stev. 87; tab. 6.
— Korshinskyi Tamamsch. 109, 578; tab. 8.
— laevis L. 84.
— luxurifolius Tamamsch. 100, 579; tab. 7.
— Maackii Rgl. 93.
— Novae-Angliae L. 86.
— Novi-Belgii L. 83.
— parviceps (Burg.) Mackenz. et Bush 84.
— pensauensis Tamamsch. 102, 580.
— Richardsonii Spreng. 96.
— salignus Willd. 84.
— see-burejensis Tamamsch. 101.
— serpentimontanus Tamamsch. 108.
— sibiricus L. 94.
— subintegerrimus (Trautv.) Ostenf. et Resvoll. 95.
— suputinicus Tamamsch. 99, 591.
— sutschanensis Kom. 103; tab. 7.
— tataricus L. f. 92.
— Tolmatschevii Tamamsch. 107.
Asteriscus Sch. Bip. 25, 508.
Asterocyticus Schur ex Fuss 11, 54.
Asterolinon Hoffmgg. et Link 18, 272.
— linum-stellatum (L.) Hoffmgg. et Link 272.
Asterothamnus Novopokr. 25, 124.
— fruticosus (C. Winkl.) Novopokr. 126; tab. 9.
— heteropappoides Novopokr. 125.
— poliifolius Novopokr. 126; tab. 9.
— Schischkinii Tamamsch. 129; tab. 9.
Astilbe Hamilt. 9, 135.
— chinense (Maxim.) Franch. et Savat. 135.
Astomatopsis Korov. 16, 203.
— galiocarpa Korov. 203.
Astragalus L. 12, 1.
— abbreviatus Kar. et Kir. 218.
— Abolinii M. Pop. 790.
— Abramovii Gontsch. 85, 869.
— acanthocarpus Boriss. 665.
— achtalensis Conr. et Freyn 476.
— ackerbergensis Freyn 632.
— acormosus Basil. 138; tab. 12.
— adpressipilosus Gontsch. 230.
— adsurgens Pall. 509.
— aduncus Willd. 495.
— adzharicus M. Pop. 501.
— aegobromus Boiss. et Hohen. 166.
— aemulans (Nevski) Gontsch. 815.
— aflatunensis B. Fedtsch. 86.
— agameticus Lipsky 399; tab. 26.
— aiwadzhi B. Fedtsch. 779.
— aksaicus Schischk. 47.
— aksuensis Bge. 30.
— aktauensis Gontsch. 220.
— alabugensis B. Fedtsch. 845.
— alaicus Freyn 227.
— alatavicus Kar. et Kir. 224.
— albanicus Grossh. 273, 886.
— Albertii Bge. 581.
— Albertoregelia C. Winkl. et B. Fedtsch. 135; tab. 16.
— albescens Boriss. 486.
— albicans Bong. 550.
— albicaulis DC. 784.
— Alexandri Char. 610.
— Alexeenkoanus B. Fedtsch. et Ivanov 360.
— Alexeenkoi Gontsch. 195.
— Alexeji Gontsch. 214.
— alitschuri O. Fedtsch. 543.
— alopecias Pall. 387; tab. 26.
— alopecurus Pall. 388.
— alpinus L. 43; tab. 3.
— altaicus Bge. 177.
— alyssoides Lam. 519.

Astragalus amabilis M. Pop. 687.
— amarus Pall. 549; tab. 37.
— ambigens M. Pop. 816.
— ammodendron Bge. 765; tab. 45.
— ammodytes Pall. 528; tab. 35.
— ammophilus Kar. et Kir. 313; tab. 21.
— ammotrophus Bge. 551.
— amygdalinus Bge. 122.
— andaulgensis B. Fedtsch. 193.
— Androssovianus Gontsch. 763.
— angarensis Turcz. ex Bge. 689.
— Angreni Lipsky 120.
— angustidens Freyn et Sint. 221.
— angustiflorus C. Koch 166.
— angustissimus Bge. 464.
— anisomerus Bge. 118.
— ankylotus Fisch. et Mey. 303.
— Antoninae Grig. 238.
— Aphanassjievii Gontsch. 187.
— apiculatus Gontsch. 194.
— applicatus Boriss. 475.
— arbuscula Pall. 667.
— arcuatus Kar. et Kir. 533.
— arenarius L. 456; tab. 31.
— arganaticus Bge. ex Rgl. et Herd. 541.
— arguricus Bge. 502; tab. 33.
— argutensis Bge. 457.
— argyroides G. Beck ex Stapf 682.
— arianus Gontsch. 112.
— arkalycensis Bge. 852.
— arnacantha M. B. 351.
— arnacanthoides Boriss. 352.
— arpilobus Kar. et Kir. 292; tab. 21.
— artvinensis M. Pop. 696.
— arvatensis Gontsch. 765.
— asephes Bge. 358.
— aschuturi B. Fedtsch. 811.
— asper Jacq. 443; tab. 30.
— Atlasovii Kom. 52.
— atripurpureus Boiss. 629.
— atrovinosus M. Pop. 204.
— aulieatensis M. Pop. 673.
— auratus Gontsch. 800.
— aureus Willd. 379; tab. 25.
— australis (L.) Lam. 66; tab. 5.
— austriacus L. 449.
— austrosibiricus Schischk. 510.
— aznabjurticus Grossh. 267.
— babatagi M. Pop. 131.
— bacaliensis Bge. 297; tab. 21.
— bachmarensis Grossh. 253.
— bactrianus Fisch. 359; tab. 24.
— badachschanicus Boriss. 26.
— badamensis M. Pop. 715.
— badghysi M. Pop. 764.
— baissunensis Lipsky 95.
— bakuensis Bge. 178.
— balchanensis Boriss. 185.
— balchaschensis Sumn. 217.
— baldschuanicus M. Pop. 724.
— Barinovii M. Pop. 716.
— Barrowianus Aitch. et Baker 221.
— Basineri Trautv. 96.
— Beckerianus Trautv. 243.
— Beketovii (Krassn.) B. Fedtsch. 72 tab. 5.
— bicolor Lam. 517.
— Biebersteinii Bge. 563.
— bischkendicus Gontsch. 139.
— Bobrovii B. Fedtsch. 233, 883; tab. 18.
— Borissianus Gontsch. 567.
— Bornmüllerianus B. Fedtsch. 237.
— Borodinii Krassn. 581; tab. 35.
— bossuensis M. Pop. 730.
— brachybotrys Bge. 457.
— brachycarpus M. B. 634.
— brachylobus DC. 705.
— brachymorphus Nikif. 304.
— brachypetalus Trautv. 259; tab. 29.
— brachypus Schrenk 774.
— brachyrachis M. Pop. 633.
— brachytropis (Stev.) C. A. Mey. 50.
— brevidens Freyn et Sint. 500.
— brevifolius Ledeb. 585.
— breviscapus B. Fedtsch. 822.
— bucharicus Rgl. 793; tab. 47.
— buchtarmensis Pall. 198.
— Bungeanus Boiss. 488.
— Bungei C. Winkl. et B. Fedtsch. 305; tab. 20.
— Butkovii M. Pop. 799.
— caespitosulus Gontsch. 576.
— calycinus M. B. 857.
— campylorrhynchus Fisch. et Mey. 290; tab. 22.
— campylosema Boiss. 631.
— campylotrichus Bge. 288; tab. 21
— cancellatus Bge. 496.
— candidissimus Ledeb. 554.
— Candolleanus Boiss. 601; tab. 39.
— captiosus Boriss. 491.
— caraganae Fisch. et Mey. 95.
— cartilagineus Gontsch. 171.
— caspicus M. B. 355.
— caucasicus Pall. 354; tab. 24.
— cerasocrenus Bge. 373; tab. 25.
— ceratoides M. B. 693.
— cernuiflorus Gontsch. 552.

— chaetodon Bge. 821; tab. 48.
— chaetolobus Bge. 692.
— chaeturus M. Pop. 810.
— charguschanus Freyn 219.
— chinensis L. f. 24.
— chionanthus M. Pop. 20.
— chionocalyx (Nevski) Boriss. 347; tab. 23
— chivensis Bge. 411; tab. 27.
— chlorodontus Bge. 200.
— chodshenticus B. Fedtsch. 775.
— Chomutovii B. Fedtsch. 593; tab. 35.
— chordorrhizus Fisch. ex Bge. 253.
— chorgossicus Lipsky, 718.
— chorinensis Bge. 72.
— chrysomallus Bge. 792.
— cicer L. 249.
— cinerascens M. Pop. 317.
— cinereus Willd. 637.
— circassicus Grossh. 487; tab. 33.
— cisdarvasicus Gontsch. 660.
— Clerceanus Iljin et Krasch. 450.
— coarctatus Trautv. 243.
— coelestis Boiss. 639.
— cognatus C. A. Mey. 770.
— coluteocarpus Boiss. 23; tab. 36.
— commixtus Bge. 302; tab. 20.
— compactus Willd. 374.
— compositus Pavl. 299.
— compressus Ledeb. 692.
— condensatus Ledeb. 375.
— confiniorum Boriss. 623.
— Conrathii Freyn 472.
— consanguineus Bong. et Mey. 452.
— consentaneus Boriss. 344.
— conspicuus Boriss. 492; tab. 33.
— contortuplicatus L. 281; tab. 20.
— corniculatus M. B. 698.
— cornubovis Lipsky 309.
— cornutus Pall. 704.
— corrugatus Bert. 294; tab. 21.
— Cottonianus Aitch. et Bak. 619.
— Cristophii Trautv. 395.
— cruciatus Link 310; tab. 21.
— curvipes Trautv. 222.
— cuscutae Bge. 628.
— cyri Fomin 612.
— cyrtobasis Bge. ex Boiss. 819.
— cysticalyx Ledeb. 834.
— cysticarpus Boriss. 15; tab. 1.
— cytisoides Bge. 740.
— czorochensis Char. 614.
— daghestanicus Grossh. 423.
— dahuricus (Pall.) DC. 285.
— danicus Retz. 256; tab. 34.
— darwasicus Basil. 16.
— dasyanthus Pall. 114.
— dasyglottis Fisch. 257.
— dealbatus Pall. 711.
— declinatus Willd. 183.
— Demetri Char. 610.
— dendroides Kar. et Kir. 838; tab. 49.
— densiflorus Kar. et Kir. 59.
— densissimus Boriss. 343.
— densus M. Pop. 131.
— denudatus Stev. 334.
— depauperatus Ledeb. 546.
— deserti M. Pop. 762.
— devius Boriss. 349.
— dianthoides Boriss. 731.
— dianthus Bge. 731.
— dictamnoides Gontsch. 31.
— dignus Boriss. 228.
— dilutus Bge. 850.
— dipelta Bge. 317; tab. 20.
— discessiflorus Gontsch. 441.
— djilgensis Franch. 508.
— dolichocarpus M. Pop. 745; tab. 44.
— dolichophyllus Pall. 589.
— dolichopodus Freyn 523.
— Drobovii M. Pop. et Vved. 291; tab. 21.
— dschangartensis Sumn. 466.
— dshimensis Gontsch. 23.
— dsharfi B. Fedtsch. 241.
— dsharkenticus M. Pop. 686.
— duanensis Saposhn. et Sumn. 670.
— dzhebrailicus Grossh. 518.
— elbrusensis Boiss. 520.
— elegans Bge. 270.
— ellypsoideus Ledeb. 850.
— elongatus Willd. 640.
— ephemeretorum Gontsch. 129.
— eremobius M. Pop. 737.
— eremospartoides Rgl. 468; tab. 32.
— erinaceus Fisch. et Mey. ex Fisch. 335.
— erioceras Fisch. et Mey. 536; tab. 44.
— erioceratiformis M. Pop. 540.
— erivanensis Bornm. et Woronow 591.
— Eugenii Grossh. 269, 884.
— euoplus Trautv. 244; tab. 18.
— exasperatus Basil. 232.
— excapus L. 213.
— excedens M. Pop. et Kult. 758.
— excelsior M. Pop. 806; tab. 48.
— exilis A. Korol. 679.
— eximius Bge. 389.
— fabaceus M. B. 167.
— falcatus Lam. 442.
— falcigerus M. Pop. 672.

Astragalus farctissimus Lipsky **215.**
— farctu, Bge. **217.**
— Fedorovii Takht. 603.
— Fedtschenkoanus Lipsky 663; tab. 43.
— ferganensis (M. Pop.) B. Fedtsch. ex A. Korol. 536.
— ferox Boriss. **376.**
— Fetissovii B. Fedtsch. 59.
— filicaulis Fisch. et Mey. 307.
— finitimus Bge. 396.
— fissuralis Alexeenko 716.
— flaccidus M. B. 251.
— flavidus M. Pop. 817.
— flexilispinus Boriss. 371.
— flexus Fisch. 180; tab. 14.
— floccosifolius Sumn. 129.
— fodinarum Boiss. et Noë ex Bge. 631.
— follicularis Pall. 857.
— fragrans Willd. 524, tab. 39.
— fraxinifolius DC. 92.
— Freynii Albov 250.
— Frickii Bge. 645.
— frigidus (L.) Bge. 33.
— fruticosus Pall. 706.
— Fuhsii Freyn et Sint. 426.
— fuliginosus G. Beck ex Stapf 624.
— fumosus Boriss. 494.
— galactites Pall. 585.
— galegiformis L. 25; tab. 36.
— gaudanensis B. Fedtsch. 186.
— Gebleri Fisch. ex Bong. 769.
— Georgii Gontsch. 747.
— gezeldarensis Grossh. 425.
— glabrescens Gontsch. 172; tab. 15.
— glaucophylloides Bornm. et Woronow 611.
— globiceps Bge. 399; tab. 26.
— glochideus Boriss. 471.
— glomeratus Ledeb. 584.
— glycyphylloides DC. 92.
— glycyphyllus L. 91; tab. 9.
— goktschaicus Grossh. 490.
— Gontsharovii Vass. 671. 887.
— grammocalyx Boiss. et Hohen. 424.
— grandiflorus Bge. 125; tab. 11.
— Grigorjevii B. Fedtsch. 34.
— haesitabundus Lipsky 697.
— hajastanus Grossh. 493.
— halicacabus Lam. 416.
— hamosus L. 781.
— harpocarpus W. Meff. 93; tab. 7.
— Helmii Fisch. 548.
— hemiphaca Kar. et Kir. 84; tab. 6.
— Henningii (Stev.) Boriss. 199.
— heptapotamicus Sumn. 547.
— heterodontus Boriss. 61.
— heteromorphus Boriss. 603.
— heterotrichus Gontsch. 797.
— Hilariae Boriss. 360.
— hissaricus Lipsky 640; tab. 40.
— Hohenackeri Boiss. 429; tab. 29.
— Holdichianus Aitch. et Bak. 621.
— holopterus Turcz. ex Bge. 452.
— holosericeus M. Pop. 121.
— humilis M. B. 590.
— hyalolepis Bge. 516.
— hypogaeus Ledeb. 580.
— hyrcanus Pall. 766.
— igniarius M. Pop. 760.
— iliensis Bge. 778.
— imbricatus Boriss. 365.
— imetensis Boriss. 60.
— inaequalifolius Basil. 861; tab. 50.
— incertus Ledeb. 514.
— indurescens Gontsch. 190; tab. 15.
— inflatus DC. 832.
— infractus Sumn. 539.
— innominatus Boriss. 544.
— inopinatus Boriss. 511.
— insidiosus Boriss. 351.
— insignis Gontsch. 103.
— intarrensis Franch. 462.
— intermedius Kar. et Kir. 835.
— intermixtus Litv. ex Boriss. 340.
— interpositus Boriss. 492; tab. 34.
— involutivus Sumn. 576.
— Ionae Palib. 713, 887.
— Irinae B. Fedtsch. 459.
— irisuensis Boriss. 566.
— ischnocarpus Gontsch. 664.
— ishigensis Maxim. ex Kom. 36.
— iskanderi Lipsky 707; tab. 42.
— ispahanicus Boiss. 13; tab. 36.
— isphairamicus B. Fedtsch. 567.
— ispirensis Boiss. 473.
— jagnobicus Lipsky 58; tab. 4.
— Jarmolenkoi Gontsch. 213; tab. 15.
— jodostachys Boiss. et Buhse 472.
— jolderensis B. Fedtsch. 622; tab. 39.
— juniperetorum Gontsch. 125.
— kabadianus Lipsky 661; tab. 42.
— kabristanicus Grossh. 270, 884.
— kadshorensis Bge. 499.
— kaghysmani Gontsch. 417.
— kamtschaticus (Kom.) Gontsch. 437.
— karabagensis Bge. 378.
— karabilicus M. Pop. 727.
— karakalensis Freyn et Sint. 340, tab. 23.
— karakugensis Bge. 770.

— karakuschensis Gontsch. 418.
— karataviensis Pavl. 723.
— karategini Gontsch. 135.
— Karelinianus M. Pop. 695.
— Karjaginii Boriss. 377.
— karkarensis M. Pop. 691.
— kaschkadarjensis Gontsch. 230.
— kazbeki Char. 616.
— kazymbeticus Saposhn. ex Sumn. 854.
— kelifi Lipsky 408.
— Kelleri M. Pop. 751; tab. 44.
— kendyrlyki M. Pop. 719.
— kenkolensis B. Fedtsch. 85.
— Kessleri Trautv. 725.
— Kikodzeanus Sosn. 564.
— kirghisorum Gontsch. 172.
— kisylkumi Boriss. 860.
— Klopotovskyi Sosn. 256.
— Knorringianus Boriss. 732.
— kokandensis Bge. 86.
— Komarovii Lipsky 709.
— kopalensis Lipsky 839; tab. 49.
— kopetdaghi Boriss. 185.
— Korolkovii Bge. 130; tab. 11.
— Kozlovskyi Grossh. 506.
— Krasnovii M. Pop. 849.
— Krassnovianus Gontsch. 742.
— Krauseanus Rgl. 802.
— Kronenburgii B. Fedtsch. 464.
— kubensis Grossh. 253, 871.
— Kudrjaschovii A. Korol. 659.
— kuhitangi (Nevski) Boriss. 347.
— kulabensis Lipsky 391; tab. 27.
— kurdaicus Saposhn. ex Sumn. 560.
— kurtschumensis Bge. 852.
— Kuschakevitschii B. Fedtsch. ex O. Fedtsch. 46; tab. 4.
— kuschkensis Boriss. 219.
— Kusnetzovii M. Pop. 206.
— kustanaicus M. Pop. 734.
— laceratus Lipsky 85.
— Lagowskyi Trautv. 380.
— laguroides Pall. 855.
— lagurus Willd. 434.
— lancifolius Gontsch. 678.
— lanuginosus Kar. et Kir. 127.
— lasiocalyx Gontsch. 798.
— lasioglottis Stev. 499; tab. 34.
— lasiopetalus Bge. 113; tab. 10.
— lasiophyllus Ledeb. 563; tab. 37.
— lasiosemius Boiss. 240.
— lasiostylus Fisch. 367.
— latifolius Lam. 602.
— Lehmannianus Bge. 407, tab. 27.
— leiophysa Bge. 410.
— lejosemius (Lipsky) M. Pop. 239.
— lepsensis Bge. 32.
— leptocaulis Ledeb. 446.
— leptopoides M. Pop. 703.
— leucocalyx M. Pop. 658.
— leucocladus Bge. 832.
— Levieri Freyn ex Somm. et Levier 527.
— Linczevskii Gontsch. 203.
— Lipschitzii Pavl. 729.
— Lipskyi M. Pop. 209.
— lithophilus Kar. et Kir. 19; tab. 12.
— Litvinovianus Gontsch. 121; tab. 10.
— Litvinovii Lipsky 404.
— longicuspis Bge. 630.
— longiflorus Pall. 178.
— longipetiolatus M. Pop. 772.
— longiramosus Boriss. 343.
— longistipitatus Boriss. 16.
— Lorinserianus Freyn 674.
— lunatus Pall. 505.
— lupulinus Pall. 856.
— luxurians Bge. 51.
— Mackeviczii Gontsch. 463.
— macranthoides Boriss. 368.
— macrobotrys Bge. 777.
— macroceras C. A. Mey. 688.
— macrocladus Bge. 777.
— macronyx Bge. 208; tab. 14.
— macropetalus C. A. Mey. 748.
— macrophysus Somm. et Levier 416.
— macropodium Lipsky 12; tab. 36.
— macropterus DC. 82.
— macropus Bge. 734.
— macrostachys DC. 265.
— macrotropis Bge. 686.
— macrourus Fisch. et Mey 268; tab. 19.
— maculatus Bge. 620.
— mailiensis B. Fedtsch. 447.
— Majevskianus Kryl. 835.
— managildensis B. Fedtsch. 447; tab 31.
— marguzaricus Lipsky 724.
— marinus Boriss. 512.
— Marschallianus Fisch. 336.
— Massagetovii B. Fedtsch. 22; tab. 17.
— Maximowiczii Trautv. 403; tab. 27.
— maverranagri M. Pop. 677.
— maximus Willd. 389; tab. 26.
— meanus Boriss. 358.
— medius Schrenk 783.
— megalanthus DC. 577.
— megalomerus Bge. 795; tab. 46.
— megalotropis C. A. Mey. ex Bge. 397.
— megricus Grossh. 271, 885.

Astragalus melanocladus Lipsky **831**.
— melanocomus M. Pop. **813**.
— melanostachys Benth. ex Bge. **54**.
— melilotoides Pall. 89; tab. 6.
— membranaceus (Fisch.) Bge. 38.
— mendax Freyn 200.
— meracus Boriss. 339.
— meschchedensis Bge. 338.
— mesites Boiss. et Buhse 415; tab. 28.
— meskheticus Manden. 505.
— Meyeri Boiss. 366.
— Michaelis Boriss. **703**.
— micracme Boiss. **427**.
— microcephalus Willd. 335.
— miniatus Bge. 458.
— mirabilis Lipsky **141**.
— mogoltavicus M. Pop. 392.
— Mokeevae M. Pop. 732.
— mollis M. B. 271.
— mongholicus Bge. 37.
— mongutensis Lipsky 45.
— monspessulanus L. 606.
— montis-aquilis Grossh. 612.
— mucidus Bge. 128.
— mugodsharicus Bge. **710**.
— multicaulis Ledeb. 81.
— multifoliolatus Boriss. 337.
— munitus Boriss. 350.
— myriophyllus Bge. 224.
— namanganicus M. Pop. 641; tab. 41.
— Nathaliae W. Meff. 825.
— Neireichianus Freyn et Sint. 684.
— nematodes Bge. ex Boiss. 467; tab. 32.
— neo-Lipskyanus M. Pop. 657.
— nephtonensis Freyn 209.
— neurophyllus Franch. 796.
— Nevskii Gontsch. 228.
— Nicolai Boriss. **119**.
— nigrescens M. Pop. **763**.
— nigriceps M. Pop. 666.
— nigrimontanus M. Pop. **809**.
— nigrocalyx Slob. 238; tab. **17**.
— Nikitinae B. Fedtsch. **746**, **888**.
— Ninae Gontsch. 303.
— nivalis Kar. et Kir. 824.
— nobilis Bge. ex B. Fedtsch. 867; tab. 50.
— nuciferus Bge. 101.
— nucleosus M. Pop. 750.
— nuratavicus Boriss. 368.
— nuratensis M. Pop. **138**.
— ochranthus Gontsch. 130.
— odoratus Lam. 441.
— oeroilanicus M. Pop. 126.
— olchonensis Gontsch. 78.
— Oldenburgii B. Fedtsch. 757; tab. **45**.
— oleifolius DC. 381.
— Olgae Bge. 21.
— oltensis Grossh. 355.
— onobrychioides M. B. 498.
— onobrychis L. 484.
— ophiocarpus Benth. ex Boiss. 285; tab. 21.
— orbiculatus Ledeb. 113.
— ordubadensis Grossh. 623.
— oreades C. A. Mey. 250.
— ornithopodioides Lam. 476.
— ornithorrhynchus M. Pop. **749**.
— oroboides Hornem. 48; tab. 3.
— orthanthoides Boriss. 825.
— ortholobiformis Sumn. 689.
— ortholobus Bge. 690.
— Ovczinnikovii Boriss. **14**.
— Overinii Bge. 516.
— oxyglottis Stev. 308; tab. 21.
— oxypterus Boriss. **12**.
— pachyrrhizus M. Pop. 562.
— paliurus Boriss. 353.
— pallescens M. B. 735.
— paradoxus Bge. 231.
— paraglycyphyllus Boissieu 92, 870.
— patentivillosus Gontsch. 229.
— paucijudus C. A. Mey. **771**; tab. 45.
— pauper Bge. **141**.
— pauperiformis B. Fedtsch. **142**.
— peduncularis Royle 438; tab. 300.
— pedunculosus M. Pop. 814.
— pendulinus M. Pop. et B. Fedtsch. 725
— perembelicus Grossh. 489.
— perrarus Boriss. 489.
— persicus Fisch. et Mey. 433; tab. **29**.
— Petkoffii B. Fedtsch. 448.
— petraeus Kar. et Kir. 532.
— petropylensis Bge. 851.
— Petunnikovii Litw. 736.
— physocarpus Ledeb. 558.
— physodes L. 559.
— piletocladus Freyn et Sint. 330. tab. 32.
— pischtovensis Gontsch. 702.
— platyphyllus Kar. et Kir. 638; tab. 41.
— plumatus Boriss. 369.
— plumbeus (Nevski) Gontsch. 868.
— podocarpus C. A. Mey. 266.
— podolobus Boiss. et Hohen. 759.
— Polakovii M. Pop. 547.
— poliotes Bge. 579.
— Politovii Kryl. 50.
— polyceras Kar. et Kir. 670.
— polygala Pall. 609; tab. 41.
— polyphyllus Bge. 184.

— polytimeticus M. Pop. **693.**
— polyzygus M. Pop. 818.
— ponticus Pall. 387.
— Popovii Pavl. 565.
— Prilipkoanus Grossh. 613.
— projecturus Sumn. 579.
— prominens Boriss. 357; tab. 24.
— promontoriorum Gontsch. 741.
— propinquus Schischk. 38; tab. 2.
— protractus Boriss. 270; 871.
— proximus Boriss. 348.
— pseudoamygdalinus M. Pop. 122.
— pseudanthylloides Gontsch. 418.
— pseudoaustralis Fisch. et Mey. 70.
— pseudobrachytropis Gontsch. 49.
— pseudocancellatus Grossh. 497.
— pseudocytisoides M. Pop. 740.
— pseudoeremophysa M. Pop. **174; tab. 13.**
— pseudomegalomerus M. Pop. 791.
— pseudonobilis M. Pop. 867.
— pseudorhacodes Gontsch. **790.**
— pseudoscoparius Gontsch. **462; tab. 32.**
— pseudotataricus Boriss. 701
— pseudoutriger Grossh. 197.
— psilophus Schrenk 727.
— pskemensis M. Pop. 714.
— psoraloides Lam. 518.
— pterocephalus Bge. 370.
— puberulus Ledeb. 78; tab. 6.
— pubiflorus DC. 117.
— pulcher Korov. 861.
— pulposus M. Pop. 179.
— pulvinatus Bge. 341.
— punctatus Bge. 630.
— pycnanthus Boriss. 371.
— pycnolobus Bge. 713.
— quisqualis Bge. 102; tab. 9.
— Raddei Basil. 419; tab. 28.
— rariflorus Ledeb. 737.
— rarissimus M. Pop. 542.
— Rawlinsianus Aitch. et Baker 267.
— reduncus Pall. 535.
— refractus C. A. Mey. 627; tab. 40.
— Regelii Trautv. 398.
— retamocarpus Boiss. et Hohen. ex Boiss. 97; tab. 7.
— reticulatus M. B. 293.
— Reverdattoanus Sumn. 698.
— rhacodes Bge. 789; tab. 47.
— robustus Bge. 635; tab. 40.
— Rollovii Grossh. 604.
— roschanicus B. Fedtsch. 240.
— roseus Ledeb. 568.
— rostratus C. A. Mey. 605; tab. 39.
— rotundus Gontsch. 207.
— rubellus Gontsch. 180.
— rubens B. Fedtsch. et Ivanova 358.
— rubrigalli M. Pop. 540.
— rubrivenosus Gontsch. 797.
— rubromarginatus Czernjak. 126.
— rumpens W. Meff. 810.
— rupifragiformis M. Pop. 572; tab. 38.
— rupifragus Pall. 575.
— Ruprechtii Bge. 515.
— rytidocarpus Ledeb. 83.
— rytilobus Bge. 308.
— Rzaevii Gross. 272, 885.
— sabuletorum Ledeb. 859.
— saccocalyx Schrenk ex Fisch. et Mey. 820.
— sachalinensis Bge. 76.
— Sachokianus Sosn. 683.
— saganlugensis Trautv. 254.
— sahendi Buhse 242.
— salareticus Boriss. 590.
— salatavicus Bge. 617.
— salicetorum Kom. 43.
— salsugineus Kar. et Kir. 582.
— sangesuricus Boriss. 245; tab. 18.
— sanguinolentus M. B. 615, tab 41.
— saralensis Gontsch. 34.
— saratagius Bge. 553.
— sarbasnensis B. Fedtsch. 507.
— sarchanensis Gontsch. 70.
— sarytavicus M. Pop. 194.
— satteotoichus Gontsch. 717.
— scaberrimus Bge. 586.
— scabrisetus Bong. 592; tab. 38.
— schachdarinus Lipsky 466.
— schachimardanus Basil. 866.
— schahrudensis Bge 395.
— schanginianus Pall. 169.
— Schelichovii Turcz. 436.
— Schelkovnikovii Grossh. 470; tab. 31.
— schemachensis Karjag. 168.
— Scheremetevianus B. Fedtsch. 15; tab. 1.
— Schischkinianus Gontsch. 272.
— Schischkinii Grossh. 521.
— schizopterus Boiss. 614.
— Schmalhausenii Bge. 298; tab. 20.
— Schrenkianus Fisch. et Mey. 862.
— schugnanicus B. Fedtsch. 57; tab. 4.
— schuschaensis Grossh. 517.
— schutensis Gontsch. 186.
— scleropodius Ledeb. 836.
— scleróxylon Bge. 673; tab. 43.
— scoparius Schrenk 461.
— secundus DC. 32.
— Semenovii Bge. 465.

Astragalus semideserti Gontsch 833.
— sericeocanus Gontsch. 39.
— sericeopuberulus Boriss. 195.
— sericopetalus Trautv. 413.
— sesamoides Boiss. 306.
— setosulus Gontsch. 260.
— sevangensis Grossh. 487.
— Severzovii Bge. 170.
— Sieversianus Pall. 101; tab. 8.
— sinaicus Boiss. 311.
— sisyrodites Bge. 120.
— Skorniakovii B. Fedtsch. 561; tab. 37.
— sogdianus Bge. 719.
— sogotensis Lipsky 701.
— Sommieri Freyn 361.
— spartioides Kar. et Kir. 779.
— speciosissimus Pavl. 728.
— sphaerocalyx Ledeb. 425.
— sphaerocephalus Stev. 397.
— sphaerocystis Bge. 854.
— sphaerophysa Kar. et Kir. 410.
— spinescens Bge. 233; tab. 16.
— spongocarpus W. Meff. 98.
— Sprygimii M. Pop. 665.
— Steinbergianus Sumn. 557.
— stenanthus Bge. 522.
— stenocarpus Gontsch. 745; tab. 44.
— stenoceras C. A. Mey. 694.
— stenoceroides Boriss. 685.
— stenocystis Bge. 805.
— Stevenianus DC. 474; tab. 31.
— stipulosus Boriss. 372; tab. 25.
— striatellus Pall. ex M. B. 287; tab. 22.
— strictifolius Boiss. 374.
— subalabugensis M. Pop. 846.
— subarcuatus M. Pop. 534.
— subauriculatus Gontsch. 583.
— subbijugus Ledeb. 541.
— subexcedens Gontsch. 680.
— subinduratus Gontsch. 188.
— subkahiricus Gontsch. 402.
— submaculatus Boriss. 618.
— subpolaris Boriss. et Schischk. 44.
— subrobustus Boriss. 636.
— subrosularis Gontsch. 187.
— subscaposus M. Pop. ex Boriss. 57.
— subschachimardanus M. Pop. 866.
— subspinescens M. Pop. 216.
— substenoceras Boriss. 695.
— substipitatus Gontsch. 205.
— subternatus Pavl. 726.
— subtrijugus M. Pop. 815.
— subulatus M. B. 699.
— subverticillatus Gontsch. 223.
— suidunensis Bge. 821.
— sulcatus L. 451.
— suluklensis Freyn et Sint. 269.
— sumbari M. Pop. 683.
— supinus Bge. 255.
— suprapilosus Gontsch. 559.
— surchanensis M. Pop. 776.
— surchobi Gontsch. 662.
— Syreitschikovii Pavl. 204.
— Szovitsii Fisch. et Mey. 428.
— Takhtadzhjanii Grossh. 273, 886.
— talassicus M. Pop. 189.
— talyschensis Bge. 184.
— tanaiticus C. Koch. 117.
— tanchassii Gontsch. 812.
— taschkendicus C. Bge. 637; tab. 41.
— Tatjanae Lincz. 402.
— tauczilikensis Golosk. 85, 869, 882.
— tauricus Pall. 455; tab. 31.
— tecti-mundi Freyn 31; tab. 2.
— teheranicus Boiss. 496.
— Tekutjevii Gontsch. 120.
— temirensis M. Pop. 668.
— tenuis Turcz. 90.
— tephrolobus Bge. 691.
— terekliensis Gontsch. 206.
— terrae-rubrae Butk. 382.
— testiculatus Pall. 578; tab. 38.
— tetrastichus Bge. 564.
— Theodori Grossh. 591.
— thlaspi Lipsky 301; tab. 20.
— tibetanus Benth. ex Bge. 258.
— Titovii Gontsch. 201.
— torrentum Bge. 167.
— trachycarpus Gontsch. 222; tab. 13.
— transcaspicus Freyn 760.
— transcaucasicus Boriss. 362.
— transhyrcanus M. Pop. 680.
— transiliensis Gontsch. 853.
— transoxanus Fisch. 356.
— Tranzschelii Boriss. 508.
— Trautvetteri Bge. 169.
— tribuloides Delile 312; tab. 21.
— trichocalyx Trautv. 391; tab. 26.
— trigonelloides Boiss. 473.
— trigonocarpus (Turcz.) Bge. 66.
— Troizkii Grossh. 486.
— tscharynensis M. Pop. 669.
— tschimganicus M. Pop. 137.
— tschuensis Bge. 71.
— Tulinovii B. Fedtsch. 60.
— tumescens M. Pop. 729.
— tupalangi Gontsch. 170.
— turbinatus Bge. 398.

— turcomanicus Bge. 761.
— Turczaninovii Kar. et Kir. 545.
— turkestanus Bge. ex Boiss. 96.
— turkmenorum Boriss. 342.
— tyttocarpus Gontsch. 837.
— ucrainicus M. Pop. et Klok. 700.
— ugamicus M. Pop. 708.
— ujalensis Gontsch. 289.
— uliginosus L. 435.
— umbellatus Bge. 35; tab. 1.
— unguiculatus Boriss. 333.
— unifoliolatus Bge. 773.
— unijugus Bge. 780.
— unilateralis Kar. et Kir. 506; tab. 33.
— uninodus M. Pop. et Vved. 297.
— uraniolimneus Boiss. 430.
— urgutinus Lipsky 659.
— ustiurtensis Bge. 494.
— utriger Pall. 196.
— vachanicus Boriss. et A. Korol. 314.
— vaginatus Pall. 65; tab. 5.
— vallicola Gontsch. 82.
— variegatus Franch. 656; tab. 43.
— varzobicus Gontsch. 202.
— vedicus Takht. 378.
— vegetior Gontsch. 139.
— velatus Trautv. 618.
— Veresczaginii Kryl. et Sumn. 828.
— versicolor Pall. 77.
— vesicarius L. 785.
— vicarius Lipsky 300; tab. 20.
— viciifolius DC. 252.
— villosissimus Bge. 775.
— violaceus Basil. 865.
— virens Pavl. 173.
— virgatus Pall. 733.
— virgeus Borris. 474.
— viridiflorus Boriss. 140.
— viridis Bge. 709.
— volgensis Bge. 198.
— Voronovianus Boriss. 365.
— vulpinus Willd. 390.
— Vvedenskyi M. Pop. 757.
— wachschi B. Fedtsch. 756.
— Willisii M. Pop. 420.
— Winckleri Trautv. 409.
— Woldemari Juz. 837, 888.
— xanthomeloides Korov. et M. Pop. 801; tab. 46.
— xanthotrichus Ledeb. 827; tab. 49.
— xanthoxiphidium Freyn et Sint. 681.
— xerophilus Ledeb. 521.
— xiphidioides Freyn et Sint. 681.
— xiphidium Bge. 662.
— xipholobus M. Pop. 750.
— zaissanensis Sumn. 858.
— zangelanus Grossh. 605.
— Zaprjagaevii Gontsch. 193.
— zerabulaki M. Pop. 818.
— Zingeri Korsh. 712.
— zuvanticus Grossh. 519.
Astrantia L. 16, 66.
— colchica Albov 70.
— major L. 69.
— maxima Pall. 67.
— pontica Albov 68.
— trifida Hoffm. 69; tab. 3.
Astrocodon Fed. 24, 374, 471.
— expansus (J. Rudolph) Fed. 378.
— Kruhseanus (Fisch. ex Rgl. et Til.) Fed. 377; tab. 22.
Astrodaucus Drude 16, 169.
— littoralis (M. B.) Drude 171.
— orientalis (L.) Drude 170.
— persicus (Boiss.) Drude 170.
Astrolobium Desv. 13, 247.
Asyneuma Griseb. et Schenk 24, 396.
— amplexicaule (Willd.) Hand.-Mazz. 406
— argutum (Rgl.) Bornm. 400.
— attenuatum (Franch.) Bornm. 412.
— baldshuanicum (O. Fedtsch.) Fed. 400.
— campanuloides (M. B.) Bornm. 405; tab. 24.
— canescens (Waldst. et Kit.) Griseb. et Schenk 411.
— cichoriforme (Boiss.) Bornm. 414.
— debile Fed. 402, 472.
— japonicum (Miq.) Briq. 406.
— lanceolatum (Willd.) Hand.-Mazz. 409.
— leianthum (Trautv.) Bornm. 416.
— lobelioides (Willd.) Hand-Mazz. 417.
— otites (Boiss.) Bornm. 419.
— pulchellum (Fisch. et Mey.) Bornm. 415; tab. 24.
— ramosum Pavl. 401.
— rigidum (Willd.) Grossh. 410.
— salignum (Waldst. et Kit. ex Bess.) Fed. 408.
— talyschense Fed. 407, 473.
— Trautvetteri (B. Fedstch.) Bornm. 413.
— urceolatum (Fomin) Fed. 416.
— Woronowii (Fomin) Bornm. 418.
Atalanthus D. Don 29, 241.
— acanthodes (Boiss.) Kirp. 241; tab. 15.
Atelanthera Hook. f. et Thoms. 8, 272.
— perpusilla Hook. f. et Thoms. 272.
Athamus Neck. 27, 77.
Athyrium Roth 1, 51.

Athyrium acrostichoides (Sw.) Diels 53; tab. 2.
— alpestre (Hoppe) Ryl. 57; tab. 2.
— austro-ussuriense (Kom.) Fomin 58.
— crenatum (Sommerf.) Rupr. 59.
— filix-femina (L.) Roth 53; tab. 2.
— Göringianum (Kuntze) S. Moore 59.
— monomachii Kom. 54.
— pterorachis Christ 57, tab. 2.
— rubripes Kom. 54.
— spinulosum (Maxim.) Milde 60; tab. 2.
— Wardii (Hook.) Makino 59.
— yokoscense (Franch. et Savat.) Christ 58; tab. 2.
Atractylodes DC. 27, 88.
— ovata (Thunb.) DC. 89.
Atragene L. 7, 307.
— macropetala Ledeb. 308, tab. 20.
— ochotensis Pall. 309.
— sibirica L. 308.
Atraphaxis L. 5, 501.
— angustifolia Jaub. et Spach 525.
— badghysi Kult. 524.
— canescens Bge. 508.
— caucasica (Hoffm.) Pavl. 516.
— compacta Ledeb. 509.
— decipiens Jaub. et Spach 523.
— frutescens (L.) Ewersm. 520.
— karataviensis Lipsch. et Pavl. 510.
— laetevirens (Ledeb.) Jaub. et Spach 514.
— Muschketovii Krassn. 513; tab. 29a.
— pungens (M. B.) Jaub. et Spach 517; tab. 29a.
— pyrifolia Bge. 518.
— replicata Lam. 506.
— seravschanica Pavl. 519.
— spinosa L. 505.
— teretifolia (M. Pop.) Kom. 526; tab. 29a.
— Tournefortii Jaub. et Spach 515.
— virgata (Rgl.) Krassn. 522.
Atrema DC. 16, 198.
Atriplex L. 6, 77.
— amblyostegia Turcz. 85.
— calotheca (Raf.) Fries 92; tab. 4.
— cana C. A. Mey. 98.
— centralasiatica Iljin 102; tab. 4.
— crassifolia C. A. Mey. 88.
— dimorphostegia Kar. et Kir. 93; tab. 4.
— fera (L.) Bge. 107; tab. 4.
— flabellum Bge. 105; tab. 4.
— Gmelinii C. A. Mey. 90; tab. 4.
— hastata L. 91.
— heterosperma Bge. 93; tab. 4.
— hortensis L. 85.
— laevis C. A. Mey. 87.
— litoralis L. 86.
— megalotheca M. Pop. 101, 873; tab. 4.
— moneta Bge. 106; tab. 4.
— nitens Schk. 84; tab. 4.
— nudicaulis Bogusl. 90; tab. 4.
— oblongifolia Waldst. et Kit. 89.
— ornata Iljin 97; tab. 4.
— pamirica Iljin 98; tab. 4.
— patens (Litv.) Iljin 87.
— patula L. 88.
— pedunculata L. 106; tab. 4.
— pungens Trautv. 96.
— rosea L. 94; tab. 4.
— schugnanica Iljin 97.
— sibirica L. 102; tab. 4.
— sphaeromorpha Iljin 94; tab. 1, 4.
— tatarica L. 95.
— thunbergiaefolia Boiss. 105; tab. 4.
— turcomanica Fisch. et Mey. 103.
— verrucifera M. B. 104.
Atropa L. 22, 71.
— belladonna L. 72.
— caucasica Kreyer 73.
— Komarovii Blin. et Schal. 74.
Atropis Rupr. 2, 460.
— alaskana (Scribn. et Merr.) V. Krecz. 480.
— angustata R. Br. 472; tab. 35.
— anisoclada V. Krecz. 487, 764; tab. 38.
— bulbosa Grossh. 490; tab. 37.
— chilochloa V. Krecz. 487, 764.
— choresmica V. Krecz. 479, 761; tab. 36.
— convoluta (Kunth) Griseb. 488; tab. 38.
— diffusa V. Krecz. 490, 766; tab. 38.
— distans (L.) Griseb. 484; tab. 36.
— dolicholepis V. Krecz. 488, 764; tab. 38.
— geniculata (Turcz.) V. Krecz. 471, 759.
— gigantea Grossh. 486; tab. 37.
— glauca (Rgl.) V. Krecz. 484, 762; tab. 36.
— Grossheimiana V. Krecz. 477, 761; tab. 35.
— Hackeliana V. Krecz. 484, 762; tab. 36.
— Hauptiana (Trin.) V. Krecz. 485, 763; tab. 36.
— humilis (Litv.) V. Krecz. 473, 759; tab. 35.
— iliensis V. Krecz. 485, 763; tab. 36.
— jenisseiensis Roshev. 471; tab. 38.
— kamtschatica (Holmb.) V. Krecz. 479; tab. 36, tab. 37.
— kurilensis Takeda 473.
— laeviuscula V. Krecz. 483, 762; tab. 37.
— macranthera V. Krecz. 471, 759; tab. 35.

— macropus V. Krecz. 489, 765; tab. 38.
— maritima (Huds.) Richt. 470; tab. 35.
— pamirica (Roshev.) V. Krecz. 474, 760; tab. 35.
— pauciramea (Hack.) V. Krecz. 477, 760; tab. 36.
— paupercula (Holmb.) V. Krecz. 480; tab. 37.
— phryganodes (Trin.) V. Krecz. 470.
— poecilantha (C. Koch) V. Krecz. 472.
— pulvinata (Fries) V. Krecz. 478, 761; tab. 36.
— Roshevitsiana Schischk. 493.
— sclerodes V. Krecz. 488, 765.
— sevangensis (Grossh.) V. Krecz. 486, 763; tab. 37.
— sibirica (Holmb.) V. Krecz. 479; tab. 36.
— subspicata V. Krecz. 474, 750; tab. 35.
— suecica Holmb. 478; tab. 36.
— tenella Lange 483; tab. 37.
— tenuiflora (Turcz.) Griseb. 493; tab. 38.
— tenuissima (Litv.) V. Krecz. 489, 765; tab. 38.
Aucklanda Falc 27, 361.
Aucuba Thunb. 17, 347.
— japonica Thunb. 347.
Audibertia Benth. 21, 623.
Aulacospermum Ledeb. 16, 238.
— anomalum Ledeb. 241; tab. 15.
— darvasicum (Lipsky) Schischk. 243; tab. 15.
— isetense (Spreng.) Schischk. 242.
— simplex Rupr. 243; tab. 15.
— tianschanicum (Korov.) C. Norm. 245.
— turkestanicum (Franch.) Schischk. 244.
Aulonix Raf. 11, 83.
Aurelia Cass. 25, 30.
Aurinia Desv. 8, 348.
Avena L. 2, 259.
— barbata Pott ex Link 262; tab. 19.
— Bruhnsiana Grun. 261; tab. 19.
— clauda Durand 260; tab. 19.
— cultiformis Malz. 268; tab. 20.
— fatua L. 267; tab. 20.
— Ludoviciana Durand 269; tab. 20.
— macrantha (Hack.) Malz. 266; tab. 20.
— meridionalis Malz. 266; tab. 19.
— nodipilosa Malz. 265; tab. 19.
— nuda L. 262; tab. 19.
— pilosa M. B. 261; tab. 19.
— praegravis (Krause) Roshev. 268; tab. 20.
— sativa L. 267; tab. 20.
— septentrionalis Malz. 265; tab. 19.
— sterilis L. 269; tab. 20.
— strigosa Schreb. 261; tab. 19.
— trichophylla C. Koch. 269; tab. 20.
— Wiestii Steud. 265; tab. 19.
Avenastrum Jessen 2, 270.
— adzharicum (Albov) Roshev. 274.
— armeniacum (Schischk.) Roshev. 274; tab. 21.
— asiaticum Roshev. 275; tab. 21.
— compactum (Boiss. et Heldr.) Roshev. 281; tab. 21.
— dahuricum (Kom.) Roshev. 275; tab. 21.
— desertorum (Less.) Podp. 276; tab. 21.
— Fedtschenkoi (Hack.) Roshev. 280; tab. 21.
— hissaricum Roshev. 279; tab. 21.
— Krylovii Pavl. 279.
— mongolicum Roshev. 281; tab. 21.
— pratense (L.) Jessen 273; tab. 21.
— pubescens (Huds.) Jessen 275; tab. 21.
— Schellianum (Hack.) Roshev. 274; tab. 21.
— tianschuanicum Roshev. 280; tab. 21.
Avenella Bluff et Fingerh. 2, 244.
Axyris L. 6, 112.
— amaranthoides L. 113; tab. 1, 2.
— caucasica (Somm. et Levier) Lipsky 115.
— hybrida L. 113.
— prostrata L. 114.
— sphaerosperma Fisch. et Mey. 114; tab. 2.
Azalea L. 18, 57.
Baccharis L. 25, 289.
— halimifolia L. 289.
Balansaea Boiss. et Reut. 16, 94.
Baldingera Gaertn. 2, 54.
Ballota L. 21, 187.
— borealis Schweigg. 190.
— grisea Pojark. 191, 651.
— nigra L. 188.
Balsamita Mill. 26, 184, 196.
Baphorhiza Link 19, 231, 232.
Barbarea Beckm. 8, 130.
— arcuata Reichb. 133.
— grandiflora N. Busch 134.
— minor C. Koch 134.
— orthoceras Ledeb. 132.
— plantaginea DC. 134.
— stricta Andrz. 131.
— vulgaris R. Br. 132.
Barbeya Albov 27, 75.
Bardana Hill 27, 93.
Barkhausenia Hoppe 29, 595, 687.
Barkhausia Moench 29, 594, 687.
Barnardia Lindl. 4, 372.
Bartsia L. 22, 657.
— alpina L. 657.

Bassia All. 6, 124.
Batrachium S. F. Gray 7, 335.
— carinatum Schur 344.
— dichotomum Schmalh. 349; tab. 22.
— divaricatum (Schrank) Schur 340; tab. 21, 22.
— eradicatum (Laest.) Fries 339; tab. 21. 22.
— foeniculaceum (Gilib.) V. Krecz. 338, tab. 21, 22.
— Gilibertii V. Krecz. 345; tab. 22.
— Kauffmannii (Clerc) V. Krecz. 343; tab. 21, 22.
— Langei F. Schultz ex Nym. 350.
— marinum (Arrhen et Fries) Fries 349.
— mongolicum (Kryl.) V. Krecz. 350: tab. 22.
— pachycaulon Nevski 346.
— Rionii (Lag.) Nym. 346; tab. 21.
— trichophyllum (Chaix) van den Bossche 344; tab. 21, 22.
— triphyllum (Wallr.) Dum. 343.
Beccabunga Fourr. 22, 329, 468.
Beckeria Bernh. 2, 342.
Beckmannia Host. 2, 287.
— eruciformis (L.) Host. 288; tab. 22.
— syzigachne (Steud.) Fern. 288; tab. 22.
Behen Moench 6, 595.
Belamcanda Adans. 4, 576.
— chinensis (L.) Leman 577; tab. 31.
Belenia Decne. 22, 103.
Belladonna Adans. 22, 71.
Bellardia All. 22, 658.
— trixago (L.) All. 659.
Bellardiachloa Chiov. 2, 426.
— polychroa (Trautv.) Roshev. 426; tab. 32.
Bellevalia Lapeyr. 4, 395.
— acutifolia (Boiss.) Deloney 405; tab. 24.
— albana Woronow 397; tab. 23.
— araxina Woronow 401; tab. 23.
— atroviolacea Rgl. 397; tab. 23.
— Aucheri (Baker) Losinsk. 401.
— Fominii Woronow 402; tab. 23.
— Lipskyi (Miscz.) Wulf 402; tab. 23.
— longistyla (Miscz.) Grossh. 401.
— lutea Bordz. 402.
— montana (C. Koch) Boiss. 403.
— pycnantha (C. Koch) Losinsk. 404.
— sarmatica (Pall.) Woronow 397; tab. 23.
— Saviczii Woronow 403; tab. 23.
— speciosa Woronow 404; tab. 23.
— turkestanica Franch. 396; tab. 23.
— Wilhelmsii (Stev.) Woronow 403; tab. 23.
— zygomorpha Woronow 398; tab. 23.
Bellia Bub. 16, 94.
Bellis L. 25, 53.
— hyrcanica Woronow 53.
— perennis L. 54; tab. 2.
Belowia Moq. 6, 182.
Bennetia S. F. Gray 27, 361.
Benthamia Lindl. 17, 323.
Benthamidia Spach 17, 320.
Berberis L. 7, 553.
— amurensis Rupr. 557.
— crataegina DC. 558.
— densiflora Boiss. et Buhse 559.
— emarginata Willd. 560.
— heteropoda Schrenk 555.
— integerrima Bge. 558; tab. 37.
— kaschgarica Rupr. 555.
— nummularia Bge. 559.
— oblonga (Rgl.) Schneid. 556.
— orientalis Schneid. 557.
— Poiretii Schneid. 557.
— sibirica Pall. 554; tab. 37.
— tragacanthoides DC. 560.
— turcomanica Kar. 559.
— vulgaris L. 556.
Bergenia Moench 9, 136.
— crassifolia (L.) Fritsch 137.
— pacifica Kom. 137.
Bergia L. 15, 259.
— aquatica Roxb. 260.
— ammannioides Roxb. 260.
Berinia Brignol. 29, 594.
Berteroa DC. 8, 338.
— ascendens C. Koch 339.
— incana (L.) DC. 339.
— spathulata (Steph.) C. A. Mey. 339.
Berula Hoffm. ex Bess. 16, 458, 463, 466.
— erecta (Huds.) Cov. 466.
— orientalis Woronow 467.
Beta L. 6, 33.
— lomatogona Fisch. et Mey. 39; tab. 2.
— macrorrhiza Stev. 39; tab. 2.
— perennis (L.) Freyn 34.
— trigyna Waldst. et Kit. 40; tab. 2.
— vulgaris L. 35; tab. 2.
Betonica L. 21, 237.
— abchasica (Bornm.) Chinth. 240.
— foliosa Rupr. 241.
— grandiflora Willd. 239.
— nivea Stev. 238.
— officinalis L. 240.
— orientalis L. 242.
— ossetica (Bornm.) Chinth. 239.
Betula L. 5, 269.

— ajanensis Kom. 294.
— alajica Litv. 304.
— baicalensis Sukacz. 300.
— Cajanderi Sukacz. 294.
— costata Trautv. 275.
— dahurica Pall. 290; tab. 14.
— Ermanii Cham. 276; tab. 13.
— exilis Sukacz. 282.
— fruticosa Pall. 286; tab. 14.
— Gmelinii Bge. 289.
— grandifolia Litv. 294; tab. 14.
— humilis Schrank 285.
— irkutensis Sukacz. 300.
— japonica Sieb. 293.
— Kelleriana Sukacz. 302.
— kirghisorum Sav.-Ryczg. 301.
— Korshinskyi Litv. 305.
— Kusmisscheffii (Rgl.) Sukacz. 299.
— mandshurica (Rgl.) Nakai 293.
— Medwedewii Rgl. 274; tab. 13.
— megrelica Sosn. 275.
— microphylla Bge. 301; tab. 15.
— Middendorfii Trautv. et Mey. 284.
— nana L. 281; tab. 13.
— ovalifolia Rupr. 289; tab. 14.
— pamirica Litv. 305.
— platyphylla Sukacz. 292; tab. 14.
— Prochorowii Kuzen. et Litv. 279.
— procurva Litv. 304; tab. 15.
— pubescens Ehrh. 295; tab. 15.
— Raddeana Trautv. 280.
— Rezniczenkoana (Litv.) Schischk. 302.
— rotundifolia Spach 283; tab. 13.
— Saposhnikovii Sukacz. 303.
— Schmidtii Rgl. 280; tab. 13.
— schugnanica (B. Fedtsch.) Litv. 304; tab. 15.
— tianschanica Rupr. 301.
— tortuosa Ledeb. 299; tab. 15.
— turkestanica Litv. 303.
— verrucosa Ehrh. 291; tab. 14.
Betulaster Spach 5, 269.
Biasoletia Bertol. 16, 93.
Bicuccula Adans. 7, 648.
Bidens L. 25, 551.
— bipinnata L. 561.
— cernua L. 555; tab. 33.
— frondosa L. 559.
— kamtschatica Vass. 556, 593.
— Maximoviczia Oett. 559.
— parviflora Willd. 560; tab. 33.
— radiata Thuill. 558.
— tripartita L. 557.
Biebersteinia Steph. 14, 74.
— multifida DC. 75.
— odora Steph. 74.
Bielzia Schur 28, 370, 377.
Bienertia Bge. 6, 199.
— cycloptera Bge. 200; tab. 10.
Bifolium Gaertn. 4, 453.
Bifora Hoffm. 16, 188.
— radians M. B. 202; tab. 17.
— testiculata (L.) DC. 201; tab. 17.
Biformis Spreng. 16, 198.
Bilderdykia Dum. 5, 693.
Bilegnum Brand 19, 586.
— Bungei (Boiss.) Brand 587; tab. 19.
Billotia Sch. Bip. 29, 595, 687.
Biota D. Don 1, 192.
— orientalis Endl. 192; tab. 7.
Biotia Maxim. 25, 121.
Bistorta Tourn. 5, 672.
Bitteria Boern. 3, 358.
Blackstonia Huds. 18, 535.
— perfoliata (L.) Huds. 536; tab. 27.
Blechnum L. 1, 71.
— spicant With. 71.
Blitum L. 6, 48.
Blysmus Panz. 3, 58.
— compressus (L.) Link 58; tab. 5.
— rufus (Huds.) Link 58; tab. 5.
Boehmeria Jacq. 5, 396.
— nivea (L.) Gaud. 396.
Boissiera Hochst. ex Steud. 2, 583.
— pumilio (Trin.) Hack. 584; tab. 42.
Boltonia L'Her. 25, 57.
— Lautureana Deb. 57; tab. 3.
Bombycilaena (DC.) Smoljan. 25, 301.
— erecta (L.) Smoljan. 302; tab. 18.
Bongardia C. A. Mey. 7, 551.
— chrysogonum (L.) Boiss. 552; tab. 37.
Bootia Neck. 6, 861.
Boraginoides Moench 19, 690.
Borago L. 19, 293.
— officinalis L. 293.
Borodinia N. Busch 8, 199.
— baicalensis N. Busch 200.
Borszczowia Bge. 6, 199.
— aralocaspica Bge. 199; tab. 10.
Boschniakia C. A. Mey. 23, 113.
— rossica (Cham. et Schlecht.) B. Fedtsch. 114; tab. 6.
Bothriospermum Bge. 19, 236.
— tenellum (Hornem.) Fisch. et Mey. 236.
Botrocaryum (Koehne) Pojark. 17, 330.
— controversum (Hemsl.) Pojark. 330; tab. 25.
Botryadenia Fisch. et Mey. 25, 52.

Botryanthus Kunth 4, 416.
Botrychium Sw. 1, 94.
— boreale (Franch.) Milde 97.
— lanceolatum Angstr. 98; tab. 4.
— lunaria (L.) Sw. 98.
— multifidum (J. G. Gmel.) Rupr. 99; tab. 4.
— ramosum Aschers. 98; tab. 4.
— robustum (Rupr.) Und. 99; tab. 3.
— simplex Hitchc. 98; tab. 4.
— virginianum (L.) Sw. 100.
Botrydium Spach 6, 46.
Botryostege Stapf 18, 26.
— bracteata (Maxim.) Stapf 26; tab. 3.
Botrys Fourr. 20, 39.
Brachanthemum DC. 26, 390.
— Baranovii (Krasch. et Poljak.) Krasch. 394.
— fruticulosum (Ledeb.) DC. 396.
— kasakhorum Krasch. 397.
— kirghisorum Krasch. 396.
— Titovii Krasch. 395.
Brachiaria (Trin.) Griseb. 2, 30.
— eruciformis (Sibth.) Griseb. 30; tab. 2.
Brachyactis Ledeb. 25, 188.
— ciliata Ledeb. 189; tab. 17.
Brachybotrys Maxim. 19, 259.
— paridiformis Maxim. ex Oliv. 259.
Brachycodon Fed. 24, 341, 468.
— fastigiatus (Duf.) Fed. 342; tab. 21.
Brachyderea Cass. 29, 595, 630.
Brachylepis C. A. Mey. 6, 287.
Brachypodium Beauv. 2, 593.
— pinnatum (L.) Beauv. 594; tab. 44.
— rupestre (Host) Roem. et Schult. 595; tab. 44.
— silvaticum (Huds.) Beauv. 594; tab. 44.
— villosum Drob. 596.
Bradshawia F. Muell. 22, 529.
Brasenia Schreb. 7, 4.
— Schreberi J. F. Gmel. 5; tab. 1.
Brassica L. 8, 459.
— campestris L. 462.
— cretica Lam. 461.
— elongata Ehrh. 464.
— juncea (L.) Czern. 465.
— napus L. 461.
— nigra Koch 466.
— oleracea L. 460.
— rapa L. 463.
— Tournefortii Gouan 466.
Braya Sternb. et Hoppe 8, 69.
— aenea Bge. 73.
— angustifolia (N. Busch) Vass. 73, 636.
— brachycarpa Vass. 72, 636.
— oxycarpa Hook. f. et Thoms. 74.
— pamirica (Korsh.) O. Fedtsch. 71.
— purpurascens (R. Br.) Bge. 74.
— rosea (Turcz.) Bge. 72; tab. 19.
— Scharnhorstii Rgl. et Schmalh. 72.
— siliquosa Bge. 71.
Breea Less. 28, 52, 209.
Bresia Moq. 6, 191.
Breviglandium Dulac 18, 252.
Bromus L. 2, 554.
— adjaricus Somm. et Levier 564.
— anatolicus Boiss. et Heldr. 578.
— angrenicus Drob. 565.
— arvensis L. 576.
— Benekenii (Lange) Trimen 558.
— Biebersteinii Roem. et Schult. 562.
— briziformis Fisch. et Mey. 577; tab. 42
— cappadocicus Boiss. et Bal. 568.
— commutatus Schrad. 579.
— Danthoniae Trin. 582; tab. 42.
— gracillimus Bge. 574; tab. 42.
— gracilis M. Pop. 581.
— inermis Leyss. 558.
— irkutensis Kom. 559; tab. 43.
— japonicus Thunb. 578; tab. 42.
— kopetdagensis Drob. 567; tab. 43.
— Korotkyi Drob. 560.
— macrostachys Desf. 582.
— madritensis L. 571.
— mollis L. 579.
— ornans Kom. 560; tab. 43.
— oxyodon Schrenk 581; tab. 42.
— pamiricus Drob. 566.
— Paulsenii Hack. ex Pauls. 565.
— Popovii Drob. 580.
— racemosus L. 579.
— Richardsonii Link 562; tab. 43.
— rigens L. 572; tab. 42.
— riparius Rehm. 566.
— rubens L. 572.
— scoparius L. 581.
— secalinus L. 576; tab. 42.
— sericeus Drob. 573; tab. 42.
— Severtzovii Rgl. 580.
— sibiricus Drob. 561; tab. 43.
— squarrosus L. 577; tab. 42.
— sterilis L. 571; tab. 42.
— tectorum L. 573.
— tomentellus Boiss. 567; tab. 43.
— tomentosus Trin. 563.
— turkestanicus Drob. 565.
— tyttholepis Nevski 563; tab. 43
— variegatus M. B. 564.

— vogulicus Socz. 561.
Broussonetia L'Hér. 5, 379.
— papyrifera L'Hér. 379.
Bruckenthalia Reichb. 18, 92.
— spiculifolia (Salisb.) Reichb. 92.
Brunella Moench 20, 494.
Brunnera Stev. 19, 294.
— macrophylla (M. B.) Johnst. 296.
— sibirica Stev. 295; tab. 15.
Bryanthus Stell. 18, 65.
— Gmelinii D. Don 66.
Brylkinia Fr. Schmidt 2, 354.
— caudata (Thunb.) Fr. Schmidt 354; tab. 27.
Bryomorpha Kar. et Kir. 6, 545.
Bryonia L. 24, 101.
— alba L. 102.
— dioica Jacq. 104.
— lappifolia Vass. 109, 458; tab. 8.
— monoica Aitch. et Hemsl. 104.
— transoxana Vass. 106, 458; tab. 8.
Bryza L. 2, 354.
— elatior Sibth. et Smith 356; tab. 27.
— Marcowiczii Woronow 356; tab. 27.
— maxima L. 355, tab. 27.
— media L. 357; tab. 27.
— minor L. 356; tab. 27.
— spicata Sibth. et Smith 355; tab. 27.
Buchingera Boiss. et Hohen. 8, 368.
— axillaris Boiss. 368.
Buda Adans. 6, 556.
Buffonia L. 6, 475.
— macrocarpa Ser. 476; tab. 26.
— Oliveriana Ser. 479; tab. 26.
— parviflora Griseb. 476; tab. 26.
Buglossum Gaertn. 19, 306.
Bugula Benth. 20, 30.
Bulbocapnos Bernh. 7, 651.
Bulbocastanum Adans. 16, 401.
Bulbocodium L. 4, 20.
— versicolor (Ker.-Gawl.) Spreng. 20; tab. 2.
Bulboschoenus Palla 3, 55.
— affinis (Roth) Drob. 57; tab. 3.
— compactus (Hoffm.) Drob. 57; tab. 3.
— maritimus (L.) Palla 56; tab. 3.
Bulbostylis Kunth 3, 96.
— capillaris (L.) Nees 97; tab. 8.
— tenerrima (Fisch. et Mey.) Palla 97.
— Woronowii Palla 97; tab. 8.
Bumalda Thunb. 14, 578.
Bungea C. A. Mey. 22, 797.
— trifida (Vahl) C. A. Mey. 798.
— vesiculifera (Herd.) Schischk. 799.
Bunias 8, 235.
— cochlearioides Murr. 236; tab. 20.
— orientalis L. 235; tab. 20.
Bunium L. 16, 396.
— angreni Korov. 404.
— badghysi Korov. 410.
— Bourgaei (Boiss.) Freyn et Sint. 409.
— Capusii (Franch.) Korov. 401.
— chaerophylloides (Rgl. et Schmalh.) Drude 400.
— cylindricum (Boiss. et Hohen.) Drude 411.
— elegans (Fenzl.) Freyn 402.
— ferulaceum Sibth. et Smith 407.
— gypsaceum Korov. 398.
— hissaricum Korov. 406.
— intermedium Korov. 405.
— kuhitangi Nevski 409.
— longipes Freyn 408.
— paucifolium DC. 403.
— persicum (Boiss.) B. Fedtsch. 404.
— scabrellum Korov. 407, 597.
— seravschanicus Korov. 405.
— vaginatum Korov. 399.
Buphtalmum L. 25, 510.
Bupleurum L. 16, 275.
— abchasicum Manden. 294.
— aenigma K.-Pol. 348; tab. 23.
— affine Sadl. 338; tab. 22.
— asperuloides Heldr. 343; tab. 23.
— aureum Fisch. 295.
— badachschanicum Lincz. 318, 596.
— bicaule Helm. 322.
— Boissieri Post 339; tab. 22.
— brachiatum C. Koch 340; tab. 22.
— commutatum Boiss. et Bal. 337; tab. 22.
— czimganicum Lincz. 316; 595.
— densiflorum Rupr. 303; tab. 20.
— exaltatum M. B. 324; tab. 21.
— falcatum L. 310.
— fruticosum L. 333.
— Gerardii All. 334; tab. 22.
— glaucum Rob. et Cast. 346; tab. 23.
— gulczense O. et B. Fedtsch.
— Komarovianum Lincz. 319.
— Koso-Poljanskyi Grossh. 331.
— Krylovianum Schischk. ex Kryl. 315.
— lancifolium Hornem. 286; tab. 20.
— longiinvolucratum Kryl. 299.
— longiradiatum Turcz. 287; tab. 21.
— Marschallianum C. A. Mey. 345; tab. 23.
— Martjanovii Kryl. 329.
— multinerve DC. 297; tab. 20.
— Nordmannianum Ledeb. 303.

Bupleurum pauciradiatum Fenzl ex Boiss. 341; tab. 22.
— polymorphum Albov 313.
— polyphyllum Ledeb. 312.
— pusillum Kryl. 323.
— Rischavii Albov 290.
— rotundifolium L. 283; tab. 20.
— sachalinense Fr. Schmidt 289.
— scorzonerifolium Willd. 320; tab. 21.
— sibiricum Vest 308.
— Sosnovskyi Manden. 314.
— tenuissimum L. 344; tab. 23.
— tianschanicum Freyn 330; tab. 31.
— triradiatum Adam ex Hoffm. 301; tab. 20.
— Wittmannii Stev. 285; tab. 20.
— Woronowii Manden. 327.
Burgsdorffia Moench 20, 253.
Butia Becc. 3, 469.
— capitata (Mart.) Becc. 470.
Butomus L. 1, 292.
— junceus Turcz. 293; tab. 10.
— umbellatus L. 292.
Buxus L. 14, 504.
— colchica Pojark. 505; tab. 27.
— hyrcana Pojark. 509; tab. 27.
— sempervirens L. 509.

Cacalia L. 26, 683.
— auriculata DC. 694.
— hastata L. 687.
— kamtschatica (Maxim.) Kudo 695.
— Komaroviana (Pojark.) Pojark. 691.
— praetermissa (Pojark.) Pojark. 692.
— robusta Tolm. 689.
Caccinia Savi 19, 685.
— crassifolia (Vent.) C. Koch 686.
— dubia Bge. 688; tab. 40.
Cachrys L. 16, 254.
— alpina M. B. 157.
— Herderi Rgl. 258.
— macrocarpa Ledeb. 257; tab. 19.
— odontalgica Pall. 259.
— pubescens (Pall.) Schischk. 259.
Cacnotus Raf. 25, 191.
Caesalpinia L. 11, 22.
— Gilliesii Wall. 22.
Cajanus DC. 13, 533.
— indicus Spreng. 533.
Cakile Mill. 8, 473.
— maritima Scop. 473
Calamagrostis Adans. 2, 189.
— agrostidiformis Roshev. 201.
— alajica Litv. 211; tab. 15.
— aleutica Trin. 222; tab. 16.
— Alexeenkoana Litv. 212, 749; tab. 15.
— alopecuroides Roshev. 202; tab. 14.
— Andrejewii Litv. 229; tab. 16.
— angustifolia Kom. 214 tab. 15.
— anthoxanthoides (Munro) Rgl. 209; tab. 14.
— arundinacea (L.) Roth 222; tab. 16.
— Balansae Boiss 197; tab. 14.
— brachytricha Steud. 227; tab. 16.
— Bungeana V. Petrov 218; tab. 15.
— caucasica Trin. 230; tab. 16.
— chalybaea (Laest.) Fries 221; tab. 16.
— czukczorum Socz. 218, 749; tab. 15.
— deschampsioides Trin. 219; tab. 15.
— dubia Bge. 195; tab. 14.
— elata Blytt 210; tab. 15.
— emodensis Griseb. 202.
— epigeios (L.) Roth 194; tab. 14.
— flexuosa Rupr. 209; tab. 15.
— fusca Kom. 215; tab. 15.
— gigantea Roshev. 195; tab. 14.
— glauca (M. B.) Trin. 196; tab. 14.
— grandiflora Hack. 206; tab. 14.
— groenlandica (Schrad.) Kunth 216; tab. 15.
— hissarica Nevski 197, 747; tab. 14.
— Holmii Lange 217; tab. 15.
— iberica (Griseb.) Litv. 230; tab. 16.
— kolymaensis Kom. 219; tab. 15.
— Korotkyi Litv. 228; tab. 16.
— Korshinskyi Litv. 211; tab. 15.
— laguroides Rgl. 209.
— lanceolata Roth 203; tab. 14.
— Langsdorffii (Link) Trin. 213; tab. 15.
— lapponica (Wahl.) Hartm. 219; tab. 16.
— macilenta Griseb. 205; tab. 14.
— monticola V. Petrov 228; tab. 16.
— neglecta (Ehrh.) Beauv. 215; tab. 15.
— obtusata Trin. 220; tab. 16.
— olympica Boiss. 201; tab. 14.
— paradoxa Lipsky 204; tab. 14.
— Pavlovii Roshev. 221; tab. 16.
— persica Boiss. 204; tab. 14.
— Poplawskiae Roshev. 211; tab. 15.
— pseudophragmites (Hall. f.) Koel. 196; tab. 14.
— purpurascens R. Br. 228; tab. 16.
— Ruprechtii Nevski 198; 748.
— sachalinensis Fr. Schmidt 221; tab. 16.
— schugnanica Litv. 212; tab. 15.
— Steinbergii Roshev. 217; tab. 15.
— teberdensis Litv. 214; tab. 15.
— tianschanica Rupr. 205; tab. 14.
— trichantha Schischk. 198, 748; tab. 14.

— Turczaninowii Litv. 227; tab. 16.
— turkestanica Hack. 202; tab. 14.
— uralensis Litv. 213; tab. 15.
— vilnensis Bess. 210; tab. 15.
— wiluica Litv. 229; tab. 16.
Calamintha Lam. 21, 429.
— debilis (Bge.) Benth. 430.
— grandiflora (L.) Moench 435.
— nepeta (L.) Savi 435; tab. 23.
— officinalis Moench 434.
Calatea Cass. 25, 138.
Calcitrapa Adans. 28, 370, 570, 573.
Caldesia Parl. 1, 283.
— parnassifolia (Bassi) Parl. 283; tab. 14.
Calendula L. 26, 857.
— arvensis L. 859.
— gracilis DC. 859.
— karakalensis Vass. 860, 895.
— officinalis L. 860.
— persica C. A. Mey. 858.
Calepina Adans. 8, 491.
— irregularis (Asso) Thell. 491; tab. 20.
Calestania K.-Pol. 17, 168, 201.
Calimeris DC. 25, 110, 122.
Calla L. 3, 481.
— palustris L. 482.
Callianthemum C. A. Mey. 7, 54.
— alatavicum Freyn 56.
— angustifolium Witas. 55; tab. 6.
— isopyroides (DC.) Witas. 56; tab. 6.
— sajanense (Rgl.) Witas. 55; tab. 6.
Callicephalus C. A. Mey. 28, 340.
— nitens (M. B.) C. A. Mey. 342; tab. 21.
Calligonum Endl. 5, 583.
Calligonum L. 5, 527.
— acanthopterum Borszcz. 554; tab. 32.
— affine T. Pop. 546; tab. 31.
— alatiforme Pavl. 544; tab. 31.
— alatum Litv. 539; tab. 30.
— Androssowii Litv. 560; tab. 33.
— aphyllum (Pall.) Guerke 538; tab. 30.
— aralense Borszcz. 544; tab. 31.
— arborescens Litv. 592; tab. 37.
— bakuense Litv. 568; tab. 34.
— batiola Litv. 559; tab. 33.
— Borszczowii Litv. 550; tab. 31.
— Bubyri B. Fedtsch. ex Pavl. 557; tab. 33.
— calcareum Pavl. 553; tab. 32.
— caput-Medusae Schrenk 593; tab. 37.
— cartilagineum Pavl. 562; tab. 34.
— colubrinum Borszcz. 573; tab. 35.
— cordatum Korov. ex Pavl. 579; tab. 35.
— coriaceum Pavl. 561; tab. 34.
— crispum Bge. 552; tab. 32.
— cristatum Pavl. 557; tab. 33.
— densum Borszcz. 572; tab. 34.
— dissectum T. Pop. 546; tab. 31.
— Dubjanskyi Litv. 557; tab. 33.
— elatum Litv. 587; tab. 37.
— erinaceum Borszcz. 573; tab. 36.
— eriopodum Bge. 585; tab. 36.
— Eugenii-Korovinii Pavl. 558; tab. 33.
— ferganense Pavl. 592; tab. 37.
— flavidum Bge. 549; tab. 31.
— gracile Litv. 543; tab. 31.
— griseum Korov. ex Pavl. 588; tab. 37.
— humile Litv. 543; tab. 30.
— involutum Pavl. 558; tab. 33.
— junceum (Fisch. et Mey.) Litv. 594; tab. 37.
— kzyl-kumi Pavl. 579; tab. 36.
— lanciculatum Pavl. 559; tab. 33.
— leucocladum (Schrenk) Bge. 545; tab. 31.
— Lipskyi Litv. 561; tab. 33.
— Litwinowii Drob. 588; tab. 37.
— macrocarpum Borszcz. 572; tab. 34.
— membranaceum• (Borszcz.) Litv. 553; tab. 32.
— microcarpum Borszcz. 586; tab. 37.
— molle Litv. 580; tab. 36.
— Muravljanskyi Pavl. 574; tab. 34.
— murex Bge. 591; tab. 37.
— obtusum Litv. 560; tab. 33.
— orthotrichum Pavl. 585; tab. 36.
— Paletzkianum Litv. 580; tab. 36.
— patens Litv. 561; tab. 33.
— pellucidum Pavl. 587; tab. 37.
— Petunnikowii Litv. 568; tab. 34.
— physopterum Pavl. 559; tab. 33.
— platyacanthum Borszcz. 574; tab. 35.
— plicatum Pavl. 550; tab. 32.
— polygonoides L. 568; tab. 34.
— pulcherrimum Korov. ex Pavl. 577; tab. 35.
— quadraepterum Korov. 540; tab. 30.
— rigidum Litv. 540; tab. 30.
— rotula Borszcz. 577; tab. 34.
— rubicundum Bge. 549; tab. 31.
— Russanovii Pavl. 552; tab. 32.
— santoanum Korov. 571; tab. 34.
— setosum Litv. 578; tab. 35.
— spinosissimum Pavl. 553; tab. 32.
— squarrosum Pavl. 573; tab. 36.
— tenue Pavl. 543; tab. 30.
— tetrapterum Jaub. et Spach 551; tab. 32.
— triste Litv. 591; tab. 37.
— turbineum Pavl. 551; tab. 32.
— turkestanicum (Korov.) Pavl. 586; tab. 37.

Calligonum undulatum Litv. 550; tab. 32.
Calliopea D. Don 29, 595, 615.
Callipeltis Stev. 23, 417.
— cucullaris (Jusl.) Rothm. 418; tab. 15.
Calliphysa Fisch. et Mey. 5, 594.
Callisace Fisch. 17, 11, 23.
Callistemma Cass. 25, 73.
Callistephus Cass. 25, 73.
— chinensis (L.) Nees 74.
Callitriche L. 14, 495.
— autumnalis L. 502; tab. 26.
— elegans V. Petrov 497; tab. 26.
— fallax V. Petrov 498; tab. 26.
— pedunculata DC. 502; tab. 26.
— polymorpha Lönnr. 498; tab. 26.
— stagnalis Scop. 501; tab. 26.
— subanceps V. Petrov 501; tab. 26.
— verna L. 496; tab. 26.
Calloscordum Herb. 4, 279.
Calluna Salisb. 18, 88.
— vulgaris (L.) Hill 88.
Calocapnos Spach 7, 685.
Calomecon Spach 7, 616.
Calophaca Fisch. 11, 368.
— grandiflora Rgl. 372.
— Howenii Schrenk 371.
— sericea B. Fedtsch. ex Boriss. 375.
— tianschanica (B. Fedtsch.) Boriss. 370; tab. 25.
— wolgarica (L. f.) Fisch. 369.
Caltha L. 7, 35.
— arctica R. Br. 37.
— caespitosa Schipcz. 37.
— fistulosa Schipcz. 38; tab. 4.
— membranacea (Turcz.) Schipcz. 41; tab. 4.
— natans Pall. 36.
— palustris L. 37.
— polypetala Hochst. 42.
Calycomorphum C. Presl 11, 258.
Calydermos Ruiz et Pav. 22, 115.
Calypso Salisb. 4, 604.
— bulbosa (L.) Reichb. f. 604; tab. 37.
Calyptrostigma Trautv. et Mey. ex C. A. Mey. 23, 582.
Calysphyrum Bge. 23, 575.
Calystegia R. Br. 19, 33.
— hederacea Wall. 36.
— pellita (Ledeb.) G. Don 36; tab. 2.
— sepium (L.) R. Br. 34.
— silvestris (Willd.) Roem. et Schult. 35; tab. 2.
— soldanella (L.) R. Br. 34; tab. 2.
Calvella Moq. 6, 182.
Camelina Crantz 8, 596.
— albiflora Kotschy 598.
— caucasica (Sinsk.) Vass. 601, 652.
— glabrata (DC.) Fritsch ex N. Zing. 600.
— laxa C. A. Mey. 597.
— linicola Schimp. et Spenn. 601.
— microcarpa Andrz. 599.
— pilosa N. Zing. 600.
— silvestris Wallr. 598.
Camirium Rumph. 14, 303.
Campanula L. 24, 133.
— abietina Griseb. et Schenk 309; tab. 18.
— Achverdovii Char. 242.
— Alberti Trautv. 311; tab. 18.
— Albovii Kolak. 224.
— aldanensis Fed. et Karav. 280, 466.
— alliariifolia Willd. 215.
— alpigena C. Koch 264.
— alpina Jacq. 186; tab. 9.
— altaica Ledeb. 312; tab. 18.
— andina Rupr. 286; tab. 16.
— Annae Kolak. 222.
— anomala Fomin 274.
— ardonensis Rupr. 275; tab. 14.
— argunensis Rupr. 268.
— armazica Char. 263.
— Aucheri A. DC. 261.
— Autraniana Albov 237; tab. 12.
— Bayerniana Rupr. 244; tab. 12.
— Beauverdiana Fomin 316.
— bellidifolia Adam 272.
— besenginica Fomin 277.
— betulifolia C. Koch 241.
— Biebersteiniana Roem. et Schult. 259; tab. 15.
— bononiensis L. 197.
— brassicifolia Somm. et Levier 181.
— Brotheri Somm. et Levier 213.
— calcarata Somm. et Levier 232.
— calcarea (Alb.) Char. 228.
— Capusii (Franch.) Fed. 289.
— carpatica Jacq. 319; tab. 19.
— caucasica M. B. 167.
— cephalotes Nakai 199, 462; tab. 11.
— cervicaria L. 210.
— Chamissonis Fed. 279; tab. 15.
— Charadzae Grossh. 173.
— Charkeviczii Fed. 174, 459.
— Choziatowskyi Fomin 244.
— ciliata Stev. 256; tab. 14.
— circassica Fomin 274.
— collina M. B. 222; tab. 12.
— cordifolia C. Koch 197.
— crispa Lam. 162.

— daghestanica Fomin 169.
— darialica Char. 180.
— dasyantha M. B. 277; tab. 15.
— dolomitica E. Busch 214; tab. 12.
— Doluchanovii Char. 270.
— Dzaaku Albov 257; tab. 14.
— dzyschrica Kolak. 233.
— elatior (Fomin) Grossh. 175.
— elegantissima Grossh. 246.
— engurensis Char. 238.
— erinus L. 189.
— Eugeniae Fed. 290, 446.
— evolvulacea Royle 254.
— farinosa (Rochel) Andrz. ex Bess. 204.
— Fedorovii Char. 167.
— finitima Fomin 239.
— Fominii Grossh. 271; tab. 14.
— Fonderwisii Alb. 225.
— Giesekiana Vest ex Roem. et Schult. 295; tab. 17.
— glomerata L. 200.
— Grossheimii Char. 196.
— hemschinica C. Koch 310.
— hieracioides Kolak. 302.
— Hohenackeri Fisch. et Mey. 179; tab. 10.
— hypopolia Trautv. 291; tab. 17.
— imeretina Rupr. 184; tab. 10.
— incanescens Boiss. 250; tab. 13.
— Irinae Kuthath. 227.
— Jadvigae Kolak. 233.
— kachetica Z. Kantsch. 250; tab. 13.
— Kantschavelii Zagareli 247.
— karakuschensis Grossh. 287; tab. 16.
— Kemulariae Fomin 243.
— Ketzkhovelii Sosn. 255.
— Kirpicznikovii Fed. 217.
— Kladniana (Schur) Witas. 296.
— kluchorica Kolak. 224.
— Kolakovskyi Char. 234.
— Kolenatiana C. A. Mey. ex Rupr. 242.
— Komarovii Maleev 168; tab. 10.
— kryophila Rupr. 276.
— lactiflora M. B. 302.
— Lambertiana A. DC. 306.
— Langsdorffiana Fisch. ex Trautv. 292.
— lasiocarpa Cham. 322; tab. 19.
— latifolia L. 191.
— Ledebouriana Trautv. 281.
— Lehmanniana Bge. 288; tab. 16.
— Leskovii Fed. 217.
— letschchumensis Kem.-Nath. 221.
— longistyla Fomin 170.
— lyrata Lam. 162.
— macrochlamys Boiss. et Hult 211.
— macrostachya Waldst. et Kit. ex Willd. 209.
— Makaschvilii E. Busch 218.
— Maleevii Fed. 207, 462.
— Massalskyi Fomin 255.
— medium L. 163.
— megrelica Mand. et Kuthath. 192.
— Meyeriana Rupr. 269.
— Minsteriana Grossh. 282.
— mirabilis Albov 165; tab. 9.
— napuligera Schur 298.
— oblongifolia (C. Koch) Char. 205.
— ochroleuca Kem.-Nath. 216.
— odontosepala Boiss. 193; tab. 17.
— ossetica M. B. 240.
— Panjutinii Kolak. 207.
— patula L. 308.
— persicifolia L. 307.
— petrophila Rupr. 285; tab. 16.
— polymorpha Witas. 297.
— pontica Alb. 303.
— propinqua Fisch. et Mey. 190; tab. 10.
— punctata Lam. 185; tab. 9.
— radchensis Char. 262.
— Raddeana Trautv. 240.
— radula Fisch. et Tchich. 248; tab. 13.
— rapunculoides L. 194.
— rapunculus L. 305.
— rotundifolia L. 294.
— Ruprechtii Boiss. 267.
— sarmatica Ker-Gawl. 228.
— saxifraga M. B. 267.
— Schelkownikowii Grossh. 182, 460.
— Schischkinii Kolak. et Sachok. 170.
— schistosa Kolak. 223.
— sclerotricha Boiss. 213.
— sibirica L. 177.
— Siegizmundii Fed. 230, 464.
— Sommieri Char. 231.
— Sosnowskyi Char. 273.
— sphaerocarpa Kolak. 225.
— Stevenii M. B. 316.
— stricta L. 212.
— suanetica Rupr. 237.
— subcapitata M. Pop. 205.
— symphytifolia (Albov) Kolak. 206.
— Takhtadzhianii Fed. 245.
— Talievii Juz. 176.
— taurica Juz. 178.
— trachelium L. 193.
— Trautvetteri Grossh. 208, 463; tab. 11.
— tridentata Schreb. 258.
— Turczaninovii Fed. 304; tab. 18.
— uniflora L. 320; tab. 19.

Campanula Vaidae Pénzes 311.
— wolgensis Smirn. 317.
— Woronovii Char. 230.
Campanulastra Nym. 24, 166.
Campanulastrum Small 24, 301.
Campelia Kunth 2, 245.
Campella Link 2, 245.
Camphorosma L. 6, 116.
— annuum Pall. 118; tab. 5.
— Lessingii Litv. 118; tab. 5.
— monspeliacum L. 117; tab. 5.
— soongoricum Bge. 119.
Campsis Lour. 23, 1.
— radicans (L.) Seem. 2.
Camptaparine K. Schmidt 23, 304.
Camptocarpus C. Koch 19, 231.
Camptosorus Link 1, 61.
— sibiricus Rupr. 61.
Canavalia Adans. 13, 532.
— gladiata DC. 532.
Cancrinia Kar. et Kir. 26, 293.
— Botschantzevii (S. Kovalevsk.) Tzvel. 309.
— chrysocephala Kar. et Kir. 314; tab. 13.
— discoidea (Ledeb.) Poljak. 313.
— ferganensis (S. Kovalevsk.) Tzvel. 304.
— Goloskokovii (Poljak.) Tzvel. 301.
— karatavica Tzvel. 306.
— mucronata (Rgl. et Schmalh.) Tzvel. 302.
— Nevskii Tzvel. 309.
— pamiralaica (S. Kovalevsk.) Tzvel. 307.
— paropamisica (Krasch.) Tzvel. 310.
— Pjataeviae (S. Kovalevsk.) Tzvel. 306.
— santoana (Krasch., M. Pop. et Vved.) Poljak. 311.
— setacea (Rgl. et Schmalh.) Tzvel. 304.
— submarginata (S. Kovalevsk.) Tzvel. 305.
— subsimilis (Rech. f.) Tzvel. 302.
— tadshikorum (Kudr.) Tzvel. 299.
— tianschanica (Krasch.) Tzvel. 315.
— urgutensis (M. Pop.) Tzvel. 308.
Cancriniella Tzvel. 26, 292, 876.
— Krascheninnikovii (N. Rubtz.) Tzvel. 293; tab. 13.
Cannabis L. 5, 383.
— ruderalis Janisch. 384.
— sativa L. 383.
Capnogorium Bernh. 7, 685.
Capnoiodes Adans. 7, 649.
Capparis L. 8, 1.
— Rosanoviana B. Fedtsch. 3; tab. 1.
— spinosa L. 2.
Caprifolium Adans. 23, 569.
Caprifolium Zinn 23, 467.
Capsella Medic. 8, 603.
— bursa-pastoris (L.) Medic. 604.
— hyrcana Grossh. 605.
— orientalis Klok. 605.
Capsicum L. 22, 57.
— annuum L. 57.
Caragana Lam. 11, 327.
— acanthophylla Kom. 361; tab. 24.
— alaica Pojark. 349, 999; tab. 23.
— altaica (Kom.) Pojark. 345, 397.
— arborescens Lam. 362.
— aurantiaca Koehne 348; tab. 23.
— balchaschensis (Kom.) Pojark. 342; tab. 23.
— Bongardiana (Fisch. et Mey.) Pojark. 357, 400.
— Bungei Ledeb. 366; tab. 24.
— Camilli-Schneideri Kom. 335.
— dasyphylla Pojark. 350, 400; tab. 24.
— frutescens DC. 335.
— frutex (L.) C. Koch 333.
— fruticosa (Pall.) Bess. 363.
— grandiflora (M. B.) DC. 338.
— hololeuca Bge. ex Kom. 353.
— jubata (Pall.) Poir. 359.
— kirghisorum Pojark. 342, 396.
— laeta Kom. 336.
— laetevirens Pojark. 360, 402; tab. 24.
— leucophloea Pojark. 347, 399; tab. 24.
— microphylla (Pall.) Lam. 367.
— mollis (DC.) Bess. 334.
— pleiophylla (Rgl.) Pojark. 357, 401; tab. 24.
— Prainii Schneid. 365.
— pruinosa Kom. 352.
— pumila Pojark. 346, 398.
— pygmaea (L.) DC. 343.
— scythica (Kom.) Pojark. 341, 396.
— spinosa (L.) DC. 351.
— stenophylla Pojark. 344, 397.
— tragacanthoides (Pall.) Poir. 354.
— turfanensis (Krassn.) Kom. 337.
— turkestanica Kom. 364.
— ussuriensis (Rgl.) Pojark. 331, 395; tab. 23.
Cardamine L. 8, 153.
— amara L. 163.
— bellidifolia L. 168.
— dentata Schult. 166.
— flexuosa With. 159.
— graeca L. 169.
— hirsuta L. 159.
— hyperborea O. E. Schulz 167.
— impatiens L. 158.

— lazica Boiss. et Bal. 161.
— leucantha (Tausch) O. E. Schulz 157.
— lyrata Bge. 162.
— macrophylla Willd. 156.
— minuta Willd. 168.
— parviflora L. 161.
— pectinata Pall. 158; tab. 8.
— pedata Rgl. et Til. 169.
— pratensis L. 166.
— prorepens Fisch. 163.
— purpurea Cham. et Schlecht. 169.
— Regeliana Miq. 160.
— Sachokiana N. Busch 165, 640.
— Seidlitziana Albov 164.
— tenera S. G. Gmel. ex C. A. Mey. 162.
— uliginosa M. B. 165.
— umbellata Greene 160.
— Victoris N. Busch 167; tab. 8.
— yezoensis Maxim. 157.
Cardaria Desv. 8, 504.
Cardiaca Lam. 21, 145.
Carduus L. 28, 4.
— acanthocephalus C. A. Mey. 13.
— acanthoides L. 24.
— adpressus C. A. Mey. 29.
— arabicus Jacq. 34.
— atropatanicus Sosn. ex Grossh. 20.
— Beckerianus Tamamsch. 33; tab. 3.
— cinereus M. B. 34; tab. 3.
— coloratus Tamamsch. 12.
— crispus L. 23.
— furiosus Tamamsch. 17, 599; tab. 1.
— glaucus Baumg. 28.
— hajastanicus Tamamsch. 16, 599; tab. 1.
— hamulosus Ehrh. 22.
— hystrix C. A. Mey. 14; tab. 2.
— Kerneri Simk. 31.
— multijugus C. Koch 30.
— Navaschinii Bordz. 25.
— nervosus C. Koch 38.
— Nikitinii Tamamsch. 37, 600; tab. 3.
— nutans L. 10.
— onopordioides Fisch. ex M. B. 18; tab. 1.
— personata (L.) Jacq. 27; tab. 2.
— poliochrus Trautv. 16.
— pseudocollinus (Schmalh.) Klok. 21.
— pycnocephalus L. 32.
— Schischkinii Tamamsch. 12.
— seminudus M. B. 31.
— songoricus Tamamsch. 11.
— stenocephalus Tamamsch. 27.
— transcaspicus Gaud. 21; tab. 2.
— uncinatus M. B. 26; tab. 2.
Careum Adans. 16, 385.
Carex L. 3, 111.
— aa Kom. 185.
— abbreviata Prescott ex Boott 464.
— accrescens Ohwi 136; tab. 10.
— acrifolia V. Krecz. 192, 591.
— acuta L. 207, tab. 15.
— acutiformis Ehrh. 400; tab. 22.
— adelostoma V. Krecz. 275, 603; tab. 16.
— aequivoca V. Krecz. 257, 600.
— alajica Litv. 355.
— alba Scop. 373; tab. 19.
— Alexeenkoana Litv. 338.
— algida Turcz. 345, 612; tab. 20.
— altaica Gorodk. 227.
— alticola Popl. ex Sukatsch. 382.
— amblyorhyncha V. Krecz. 184, 595; tab. 12.
— amgunensis Fr. Schmidt 319.
— amphilogos C. Koch 461.
— amurensis Kük. 457; tab. 24.
— aneurocarpa V. Krecz. 369, 614.
— angarae Steud. 270; tab. 16.
— angustior Mackenz. ex Rydb. 153; tab. 11.
— anisoneura V. Krecz. 410; tab. 22.
— anthoxanthea C. Presl 295; tab. 18.
— aomorensis Franch. 299.
— aperta Boott 224.
— appendiculata (Trautv.) Kük. 214
— appropinquata Schum. 158; tab. 10.
— approximata Bell. ex All. 313.
— aquatilis Wahl. 229.
— arakamensis Clarke 265.
— arcatica Meinsh. 225.
— argunensis Turcz. ex Bess. 382.
— arenaria L. 133; tab. 10.
— arenicola Fr. Schmidt 156; tab. 11.
— argyroglochin Hornem. 162.
— Arnellii Christ ex Scheutz 424.
— arrhyncha Franch. 174.
— aspratilis V. Krecz. 398, 618; tab. 21.
— astracanica Willd. ex Kunth 464.
— atrata L. 252; tab. 17.
— atrofusca Schkuhr 282.
— Augustinowiczii Meinsh. ex Korsh. 272.
— behringensis Clarke 259.
— bicolor Bell. ex All. 264.
— blepharicarpa Franch. 333; tab. 19.
— bonanzensis Britt. ex Britt. et Rydb. 175; tab. 12.
— Bordzilowskii V. Krecz. 378, 615.
— bostrichostigma Maxim. 334.
— brevicollis DC. ex Lam. et DC. 350.
— brizoides Jusl. ex L. 136; tab. 10.

Carex Brotherorum Christ 460.
— brunnescens (Pers.) Poir. 179; tab. 12.
— bucharica Kük. 305.
— Buekii Wimm. 213; tab. 15.
— burjatorum Korot. 461.
— Buxbaumii Wahl. 276; tab. 16.
— caespitosa L. 217; tab. 15.
— callitrichos V. Krecz. 364, 613.
— campylorhina V. Krecz. 342, 612.
— canescens L. 176; tab. 12.
— capillaris L. 428.
— capitata L. 186; tab. 14.
— capitellata Boiss. et Bal. ex Boiss. 301.
— capituliformis Meinsh. ex Maxim. 300.
— capricornis Meinsh. ex Maxim. 459; tab. 24.
— caucasica Stev. 252.
— chalcodeta V. Krecz. 279, 604.
— Chamissonis Meinsh. 430.
— chinganensis Litv. 320.
— chlorostachys Stev. 428.
— chordorrhiza Ehrh. ex L. fil. 194; tab. 13.
— chosenica Ohwi 298.
— cilicica Boiss. 293.
— circinata C. A. Mey. 295; tab. 18.
— coarcta Boott 128; tab. 10.
— colchica J. Gay 134; tab. 10.
— compacta Lam. 151; tab. 11.
— congesta C. A. Mey. 464.
— conspissata V. Krecz. 326, 611.
— coriophora Fisch. et Mey. ex Kunth 282; tab. 17.
— cryptocarpa C. A. Mey. 238.
— cuprina (Sander) Nendtv. 155; tab. 11.
— curaica Kunth 137; tab. 10.
— cuspidata Host 402.
— cyperoides Murr. 163, tab. 13.
— czarwakensis Litv. 395; tab. 21.
— dacica Heuff. 207.
— dahurica Kük. 301.
— Davalliana Smith 165; tab. 14.
— decaulescens V. Krecz. 255, 599.
— depauperata Good. 352; tab. 20.
— descendens Kük. 215; tab. 15.
— diandra Schrank 157; tab. 10.
— diastena V. Krecz. 172, 594; tab. 12.
— dichora Freyn 447; tab. 23.
— dichroandra V. Krecz. 210, 596.
— digitata L. 360; tab. 19.
— diluta M. B. 394; tab. 21.
— dioica L. 164; tab. 14.
— diplasiocarpa V. Krecz. 135, 590; tab. 10.
— discolor Nyl. 236.
— dispalata Boott ex A. Gray 399.
— disperma Dew. 185; tab. 12.
— distans L. 397; tab. 21.
— disticha Huds. 129; tab. 10.
— diversicolor Crantz 402.
— divisa Huds. 128.
— divulsa Stokes ex With. 156; tab. 11.
— dolichocarpa C. A. Mey. ex Kom. 458, 623.
— drymophila Turcz. 456; tab. 24.
— duriuscula C. A. Mey. 140; tab. 10.
— duriusculiformis V. Krecz. 142, 591.
— elata Bell. ex All. 215.
— eleusinoides Turcz. ex Bess. 231.
— elongata L. 171; tab. 12.
— emasculata V. Krecz. 277, 604; tab. 16.
— enervis C. A. Mey. 192; tab. 13.
— ensifolia Turcz. ex Bess. 227.
— erawinensis Korot. 212.
— eremopyroides V. Krecz. 384, 617.
— ericetorum Poll. 312; tab. 19.
— eriophora Fisch. ex Steud. 464.
— eriophylla (Kük.) Kom. 455.
— euxina Marc. 321.
— extensa Good. 396.
— falcata Turcz. 346; tab. 20.
— fedia Nees ex Wight 417.
— Fedtschenkoana Kük. 414; tab. 22.
— flava L. 387; tab. 21.
— flavella V. Krecz. 388, 617; tab. 21.
— foliosissima Fr. Schmidt 356.
— forticula Franch. et Sav. 421.
— Fujitae Kudo 246.
— fuscocuprea (Kük.) V. Krecz. 244, 599.
— fusco-vaginata Kük. 211.
— glacialis Mackenzie 374.
— glareosa Wahl. 182; tab. 12.
— glaucaeformis Meinsh. 404.
— Glehnii Fr. Schmidt 350.
— globularis L. 317.
— Gmelinii Hook. et Arn. 277.
— Gorodkovii V. Krecz. 287, 606.
— gracilis Curt. 210; tab. 15.
— Grahamii Boott 446; tab. 23.
— Griffithii Boott 281; tab. 17.
— Grioletii Roem. ex Schkuhr 304.
— Grossheimii V. Krecz. 130, 589.
— gynocrates Wormsk. 165; tab. 14.
— haematostachys Lévl. et Vaniot 415; tab. 22.
— hakkodensis Franch. 295; tab. 18.
— Hallerana Asso 353.
— Halleri Gunn. 265; tab. 16.
— Hancockiana Maxim. 271.

— heleonastes Ehrh. ex L. 184; tab. 12.
— Hepburnii Boott 187.
— hirta L. 455; tab. 24.
— holostoma Drej. 264.
— hordeistichos Vill. 383.
— Hostiana DC. 397.
— Huetiana Boiss. 328.
— humilis Leyss. 365.
— hylaea V. Krecz. 179, 594; tab. 12.
— hypaneura V. Krecz. 424, 621.
— hyperborea Drej. 226.
— hypochlora Freyn 330.
— Iljinii V. Krecz. 137, 590.
— inflata Huds. 442; tab. 23.
— infuscata Nees ex Wight 270.
— inumbrata V. Krecz. 218; tab. 15.
— irrigua Wahl. 245.
— jacutica V. Krecz. 441, 621; tab. 23.
— jaluensis Kom. 420.
— Jonkowskii Gorodk. 300.
— juncella Fries 208.
— jungenda Steud. 462.
— Kabanovii V. Krecz. 166, 593.
— kamtschatica Gorodk. 226.
— Karelinii Meinsh. 395; tab. 21.
— Karoi Freyn 430.
— kattegatensis Fries 236.
— kenkolensis Litv. 337.
— kirganica Kom. 411; tab. 22.
— Kirilowii Turcz. 370.
— Knorringiae Kük. 212.
— kobomugi Ohwi 124; tab. 11.
— koraginensis Meinsh. 258.
— koreana Kom. 432.
— Korshinskyi Kom. 375.
— Koshewnikowii Litw. 337.
— Kotschyana Boiss. et Hohen. 229.
— Ktausipali Meinsh. 286; tab. 18.
— laeviculmis Meinsh. 145; tab. 12.
— lanceolata Boott 362; tab. 19.
— lapponica O. F. Lang 179.
— lasiocarpa Ehrh. 416.
— latifrons V. Krecz. 425.
— latisquamea Kom. 450; tab. 20.
— laxa Wahl. 243.
— Ledebouriana C. A. Mey. ex Trev. 431.
— leiorhyncha C. A. Mey. 144; tab. 11.
— lenaeensis Kük. 429.
— lepidocarpa Tausch 389; tab. 21.
— leporina L. 161; tab. 11.
— leucochlora Bge. 323.
— limosa L. 244.
— lineolata Cham. ex Meinsh. 223, 598.
— lithophilla Turcz. 130.
— Litwinowii Kük. 338.
— livida Wahl. 343.
— loliacea L. 173; tab. 12.
— longirostrata C. A. Mey. 354.
— Lumnitzeri Rouy ex Rouy et Fouc. 154.
— Maackii Maxim. 161; tab. 11.
— Mackenziei V. Krecz. 183; tab. 12.
— macrocephala Willd. ex Spreng. 124; tab. 11.
— macrochaeta C. A. Mey. 259; tab. 17.
— macrogyna Turcz. ex Bess. 293; tab. 18.
— macrostigmatica Kük. 271.
— macroura Meinsh. 367; tab. 19.
— mandshurica Meinsh. 418.
— marina Dew. 182.
— maritima Gunn. 189; tab. 14.
— Medwedewii Leskov 256.
— Meinshauseniana V. Krecz. 287, 606; tab. 18.
— melanantha C. A. Mey. 273; tab. 16.
— melananthaeformis Litv. 274; tab. 16.
— melanocarpa Cham. ex Trautv. 312.
— melanocephala Turcz. ex Bess. 266; tab. 16.
— melanostachya M. B. ex Willd. 412; tab. 22.
— melanostoma Fisch. 261, 602.
— membranacea Hook. 449; tab. 23.
— Meyeriana Kunth 262.
— Michelii Host 354.
— microglochin Wahl. 302.
— micropoda C. A. Mey. 297.
— micropodioides V. Krecz. 298, 608.
— microtricha Franch. 327.
— Middendorffii Fr. Schmidt 238.
— mimula V. Krecz. 266, 603; tab. 16.
— mingrelica Kük. 352; tab. 20.
— minuta Franch. 220.
— minutiscabra Kük. ex B. Fedtsch. 380, 616.
— misandra R. Br. 292; tab. 18.
— mollissima Christ ex Scheutz 433; tab. 23.
— montana L. 310; tab. 19.
— mucronata Bell. ex All. 336.
— muricata L. 152; tab. 11.
— musartiana Kük. ex B. Fedtsch. 261, 602.
— nanella Ohwi 364.
— nemurensis Franch. 172.
— nervata Franch. et Sav. 326; tab. 19.
— neurocarpa Maxim. 144; tab. 11.
— nigricans C. A. Mey. 298.
— nikolskensis Kom. 346; tab. 20.
— nitida Host 377.
— Novograblenovii Kom. 432

Carex obtusata Liljebl. 381.
— occultans (Franch.) V. Krecz. 417, 620.
— Oederi Retz. 390; tab. 21.
— oligantha Steud. 269; tab. 16.
— oligostachys Meinsh. ex Maxim. 351; tab. 20.
— Oliveri W. Beck. 280; tab. 17.
— omskiana Meinsh. 216.
— ontakensis Franch. et Savat. 299.
— orbicularis Boott. 224.
— oreophila C. A. Mey. 186; tab. 14.
— orthostachys C. A. Mey. 452; tab. 24.
— oxyandra (Franch. et Sav.) Kudo 319; tab. 19.
— oxyleuca V. Krecz. 284, 605; tab. 17.
— pachystachya Cham. ex Steud. 161; tab. 11.
— pachystylis J. Gay 197; tab. 13.
— pallescens L. 278.
— pamirensis Clarke ex B. Fedtsch. 443; tab. 23.
— panicea L. 344; tab. 20.
— paniculata Jusl. ex L. 159; tab. 10.
— paralia V. Krecz. 237.
— parallela Laest. ex Sommerf. 167; tab. 14.
— parva Nees 303.
— pauciflora Lightf. 302.
— pauxilla V. Krecz. 260, 601.
— pedata L. 361.
— pediformis C. A. Mey. 368; tab. 19.
— pedunculifera Kom. 240.
— peiktusani Kom. 272.
— pendula Huds. 420.
— pentastachys Fisch. ex Steud. 464.
— perfusca V. Krecz. 256, 600; tab. 17.
— philocrena V. Krecz. 393, 618.
— phyllostachys C. A. Mey. 358.
— physodes M. B. 194; tab. 13.
— pilosa Scop. 342; tab. 20.
— pilulifera L. 318; tab. 19.
— planiculmis Kom. 434.
— platyrhyncha Franch. et Savat. 412.
— plumbea Wahl. 462.
— polyphylla Kar. et Kir. 155; tab. 11.
— pontica Albov 286.
— praecox Schreb. 135; tab. 10.
— pribylovensis Macoun ex Jord. 181.
— prionocarpa Franch. 240.
— procerula V. Krecz. 449, 622; tab. 23.
— prolixa Fries 209.
— pseudo-curaica Fr. Schmidt 138; tab. 10.
— pseudo-cyperus L. 460; tab. 24.
— pseudo-foetida Kük. 191; tab. 13.
— pseudo-lanceolata V. Krecz. 363, 613.
— pseudo-loliacea Fr. Schmidt 173; tab. 12.
— pulchella Loennr. 393; tab. 21.
— pulicaris L. 198; tab. 14.
— pumila Thunb. 408; tab. 22.
— Putjatinii Kom. 263.
— pycnostachya Kar. et Kir. 138.
— quadriflora (Kük.) Ohwi 361.
— Raddei Kük. 457; tab. 24.
— Ramenskii Kom. 235.
— rariflora Wahl. 242.
— Redowskiana C. A. Mey. 166; tab. 14.
— Regeliana Kük. ex Litw. 292; tab. 18.
— relaxa V. Krecz. 379, 616.
— remota L. 145; tab. 12.
— remotiuscula Wahl. 146; tab. 12.
— reptabunda (Trautv.) V. Krecz. 190; tab. 13.
— retorta (Fries) V. Krecz. 219, 598; tab. 15.
— reventa V. Krecz. 367, 614.
— rhizina Blytt ex Lindbl. 366.
— rhynchophysa C. A. Mey. 438; tab. 23.
— Riabushinskii Kom. 239.
— rigescens (Franch.) V. Krecz. 142; 592.
— rigidioides Gorodk. 228.
— riparia Curt. 409; tab. 22.
— ripariaeformis Litv. 413; tab. 22.
— rotundata Wahl. 443; tab. 23.
— rubra Levl. et Vaniot 218.
— rugulosa Kük. 410; tab. 22.
— rupestris Bell. ex All. 381.
— ruthenica V. Krecz. 325, 610; tab. 19.
— sabulosa Turcz. ex Bess. 275; tab. 16.
— sabynensis Less. ex Kunth 329; tab. 19.
— sachalinensis Fr. Schmidt 332; tab. 19.
— sadoensis Franch. 422.
— sajanensis V. Krecz. 190; tab. 13.
— sareptana V. Krecz. 213, 597.
— saxatilis L. 448; tab. 23.
— scabricuspis V. Krecz. 324; 609.
— scabrifolia Steud. 408, tab. 22.
— scabrinervia Franch. 258.
— Schmidtii Meinsh. 223.
— scirpoidea Michx. 303.
— secalina Willd. ex Wahl 384.
— Sedakowii C. A. Mey. ex Meinsh. 427.
— seiskoensis Freyn 463.
— semiplena Kük. 214.
— setina (Christ) V. Krecz. 192; tab. 14.
— siderosticta Hance 357.
— Siegertiana Uechtr. 451; tab. 24.
— sikokkiana Franch. et Savat. 331.
— silvatica Huds. 423.
— similegena V. Krecz. 193, 596; tab. 13.

— Slobodovii V. Krecz. 191, 595.
— Smirnovii V. Krecz. 411, 619; tab. 22.
— Soczavaena Gorodk. 219.
— songorica Kar. et Kir. 414; tab. 22.
— sordida Cham. ex Van Heurck et Muell. 462.
— soyaeensis Kük. 273.
— spaniocarpa Steud. 375.
— spicata Huds.. 153; tab. 11.
— stans Drej. 230.
— stenocarpa Turcz. ex Bess, 291, 607; tab. 18.
— stenolepis Less. 443.
— stenophylloides V. Krecz. 141, 592; tab. 10.
— stilbophaea V. Krecz. 283, 605; tab. 17.
— stipata Muehl. ex Willd. 152; tab. 11.
— strigosa Huds. 419.
— stygia Fries 242.
— stylosa C. A. Mey. 262.
— subebracteata (Kük.) Ohwi 330.
— sublivida Norrl. 464.
— subspathacea Worsmk. ex Hornem. 235.
— subumbellata Meinsh. 322.
— suifunensis Kom. 239.
— Sukaczovii V. Krecz. 415, 620; tab. 22.
— supina Willd. ex Wahl. 376.
— sutschanensis Kom. 368.
— Szovitsii V. Krecz. 159, 593; tab. 10.
— taldycola Meinsh. 228.
— tasorum Kom. 333.
— tenuiflora Wahl. 174; tab. 12.
— tomentosa L. 318.
— traiziscana Fr. Schmidt 172; tab. 12.
— Trautvetteriana Kom. 331.
— tripartita All. 181; tab. 12.
— tristis M. B. 288; tab. 18.
— tuminensis Kom. 234.
— turkestanica Rgl. 377.
— uda Maxim. 300.
— ulobasis V. Krecz. 311, 608.
— umbrosa Host 328.
— ungurensis Litv. 379.
— uralensis C. B. Clarke 141; tab. 10.
— ursina Dew. 183; tab. 12.
— ussuriensis Kom. 373.
— utriculata Boott 444.
— Uzoni Kom. 230.
— vaginata Tausch 344; tab. 20.
— Vanheurckii Muell. 314.
— verna Chaix ex Vill. 323.
— vesicaria L. 445; tab. 23.
— vesicata Meinsh. 446; tab. 23.
— viridula Michx. 389.
— vitilis Fries 180; tab. 12.
— vulpina L. 150; tab. 11.
— wiluica Meinsh. 208.
— xyphium Kom. 349.
Carlina L. 27, 77.
— acaulis L. 86; tab. 6.
— Biebersteinii Bernh. ex Hornem. 82.
— cirsioides Klok. 87.
— lanata L. 83.
— onopordifolia Bess. ex Szafer, Kulcz. et B. Pawl. 84.
— vulgaris L. 79.
Carlowitzia Moench 27, 77.
Carolinella Hemsl. 18, 133.
Caropodium Stapf et Wettst. 16, 123.
— armenum (Bordz.) Schischk. 123; tab. 8.
— platycarpum (Boiss. et Hausskn.) Schischk. 124.
Caroselinum Griseb. 17, 168, 194.
Carota Rupr. 17, 288.
Carpesium L. 25, 498.
— abrotanoides L. 505; tab. 31.
— cernuum L. 501.
— eximium C. Winkl. 502.
— triste Maxim. 503.
Carpinus L. 5, 254.
— betulus L. 258.
— cordata Blume 256.
— grosseserrata H. Winkl. 261.
— hybrida H. Winkl. 261.
— orientalis Mill. 257.
— oxycarpa H. Winkl. 261; tab. 12.
— schuschaensis H. Winkl. 257; tab. 12.
Carpoceras Boiss. 8, 591.
— brevistylum N. Busch 592; tab. 32.
— ceratocarpum (Pall.) N. Busch 592.
— hastulatum (Stev.) Boiss. 591; tab. 32.
— stenocarpum Boiss. 592; tab. 32.
Carthamus L. 28, 581.
— glaucus M. B. 584.
— gypsicola Iljin 585.
— lanatus L. 583.
— oxyacanthus M. B. 584.
— tinctorius L. 586.
Carthusianastrum Will. 6, 812.
Carthusianum Will. 6, 813.
Carum Baill. 16, 422, 451.
Carum L. 16, 385.
— alpinum (M. B.) Benth. 388.
— atrosanguineum Kar. et Kir. 395; tab. 26.
— buriaticum Turcz. 395; tab. 26.
— carvi L. 386.
— caucasicum (M. B.) Boiss. 393; tab. 26
— Grossheimii Schischk. 394.

Carum Komarovii Karjag. 392.
— meifolium (M. B.) Boiss. 394.
— porphyrocoleon (Freyn et Sint.) Woronow 388.
— saxicolum Albov 391.
Carvi Bernh. 16, 560.
Carvi Mill. 16, 385.
Carvifolia Vill. 16, 560.
Caryophyllastrum Will. 6, 822.
Cassandra D. Don 18, 75.
Casselia Dum. 19, 238.
Cassia L. 11, 20.
— marylandica L. 20.
Cassiope D. Don 18, 66.
— ericoides (Pall.) D. Don 69.
— lycopodioides (Pall.) D. Don 67.
— Redowskii (Cham. et Schlecht.) G. Don 68.
— tetragona (L.) D. Don 68.
Castanea Mill. 5, 320.
— sativa Mill. 321; tab. 19.
Castilleja L. f. 22, 530.
— arctica Kryl. et Serg. 532.
— pallida (L.) Kunth 531.
Castillejoa Post et Kuntze 22, 530.
Catabrosa Beauv. 2, 445.
— aquatica (L.) Beauv. 445.
— Capusii Franch. 446; tab. 33.
Catalpa Scop. 23, 9.
— bignonioides Walt. 10.
— ovata G. Don 11.
— speciosa Warder ex Engelm. 10.
Cataria Adans. 20, 286.
Catharanthus G. Don 18, 646.
Cathartolinum Reichb. 14, 127.
Catonia Moench 29, 594, 627.
Caucalis L. 16, 172.
— Bischoffii K.-Pol. 173.
— lappula (Web.) Grande 172; tab. 13.
Caulinia Willd. 1, 273.
Caulophyllum Rich. 7, 552.
— robustum Maxim. 552; tab. 37.
Cavotuber Rupr. 7, 658.
Cedrus Link 1, 59.
— deodara Laws. 159.
Celasia Cass. 29, 55.
Celastrus L. 14, 574.
— flagellaris Rupr. 576.
— orbiculata Thunb. 575.
— strigillosus Nakai 576.
Celosia L. 6, 355.
— argentea L. 355, tab. 18.
— cristata L. 356; tab. 18.
Celsia L. 22, 170.
— heterophylla Desf. 171.
— nudicaulis (Wydl.) B. Fedtsch. 172.
— orientalis L. 171.
— Suworowiana C. Koch 173.
Celtis L. 5, 373.
— caucasica Willd. 374; tab. 20.
— glabrata Stev. 375; tab. 20.
Cenekia Opiz 24, 133, 301.
Cenolophium Koch 16, 582.
— Fischeri (Spreng.) Koch 583.
Centaurea L. 28, 370.
— abbreviata (C. Koch) Hand.-Mazz. 453.
— abchasica (Albov) Sosn. 425.
— abnormis Czer. 455.
— acmophylla Boiss. 411.
— Adamii Willd. 573.
— adjarica Albov 492.
— adpressa Ledeb. 507.
— aemulans Klok. 533, 617.
— aggregata Fisch. et Mey. 537.
— alaica Iljin 382.
— albida C. Koch 536.
— Albovii Sosn. 424.
— Alexandri Bordz. 559.
— alutacea Dobrocz. 454.
— amblyolepis Ledeb. 479.
— Androssovii Iljin 382.
— Angelescui (Prod.) Czer. 406.
— apiculata Ledeb. 506.
— appendicata Klok. 545.
— arenaria M. B. 522.
— armena Boiss. 569.
— arpensis (Czer.) Czer. 481; tab. 24.
— avarica Tzvel. 484.
— Barbeyi (Albov) Sosn. 426.
— behen L. 561.
— Belangeriana (DC.) Stapf 578; tab. 24.
— bella Trautv. 489.
— Besseriana DC. 526.
— bogadensis Woronow 490.
— borysthenica Grun. 523.
— breviceps Iljin 548.
— calcitrapa L. 575.
— carbonata Klok. 465.
— carduiformis DC. 509.
— carpatica (Porc.) Porc. 451.
— carthalinica (Sosn.) Sosn. 439.
— caspica Grossh. 534.
— cheiranthifolia Willd. 399.
— circassica (Albov) Sosn. 425.
— ciscaucasica Sosn. 428, 610.
— colchica (Sosn.) Sosn. 438.
— cyanus L. 416.
— czerkessica Dobrocz. et Kotov 405.

— daghestanica (Lipsky) Czer. 479.
— dealbata Willd. 434.
— declinata M. B. 430.
— depressa M. B. 415.
— diffusa Lam. 532.
— Dimitriewiae Sosn. 491.
— donetzica Klok. 547.
— Dubjanskii Iljin 550.
— Edmondi Sosn. 432.
— erivanensis (Lipsky) Bordz. 475.
— exsurgens Sosn. 435.
— Fischeri Willd. 396.
— fuscimarginata (C. Koch) Juz. 408.
— georgica Klok. 557, 619.
— Gerberi Stev. 550.
— Glehnii Trautv. 509.
— Gontscharovii Iljin 382.
— Grossheimii Sosn. 568.
— Gulissashvilii Dumb. 531.
— hajastana Tzvel. 383.
— holophylla Socz. et Lipat. 426, 610.
— Huetii Boiss. 410; tab. 23.
— hymenolepis Trautv. 433.
— hyrcanica Bornm. 460.
— iberica Trev. 574.
— Iljinii Czerniak. 419.
— integrifolia Tausch 504.
— jacea L. 444.
— karabaghensis (Sosn.) Sosn. 437.
— kasakorum Iljin 383; tab. 22.
— kobstanica Tzvel. 483.
— koktebelica Klok. 528, 617.
— Kolakovskyi Sosn. 429, 611.
— konkae Klok. 545.
— kopetdaghensis Iljin 420.
— Kotschyana Heuff. 502; tab. 25.
— kubanica Klok. 518, 615.
— Kultiassovii Iljin 384.
— lasiopoda M. Pop. et Kult. 385.
— latiloba Klok. 557, 620.
— Lavrenkoana Klok. 526.
— leucophylla M. B. 429.
— leuzeoides (Jaub. et Sp.) Walp. 487.
— Majorovii Dumb. 520.
— Maleevii (Sosn.) Sosn. 428.
— maramarosiensis (Jáv.) Czer. 392.
— margarita-alba Klok. 544.
— margaritacea Ten. 543.
— maris-nigri Sosn. 431.
— Marschalliana Spreng. 466.
— meskhetica Sosn. 439, 612.
— Meyeriana Tzvel. 473.
— micranthos I. F. Gmel. 519.
— Modesti Fed. 381.
— mollis Waldst. et Kit. 393.
— Nathadzeae Sosn. 490.
— nigriceps Dobrocz. 452.
— nigrifimbria (C. Koch) Sosn. 394; tab. 23.
— novorossica Klok. 556, 619.
— odessana Prod. 523.
— oltensis Sosn. 570.
— orientalis L. 501.
— ossethica Sosn. 505, 614.
— ovina Pall. 530.
— Paczoskii Kotov ex Klok. 549.
— pambakensis (Sosn.) Sosn. 437.
— pannonica (Heuff.) Hayek 446.
— pauciloba Trautv. 427.
— Pecho Albov 493.
— phaeopappoides Bordz. 486.
— phrygia L. 449.
— phyllopoda Iljin 385.
— pineticola Iljin 548.
— polypodiifolia Boiss. 563; tab. 26.
— proto-Gerberi Klok. 546.
— proto-margaritacea Klok. 543.
— pseudoleucolepis Kleop. 542.
— pseudomaculosa Dobrocz. 518.
— pseudophrygia C. A. Mey. 456.
— pseudoscabiosa Boiss. et Buhse 508.
— Razdorskyi Karjag. 386, 607.
— reflexa Lam. 499.
— rhenana Boreau 517.
— rhizantha C. A. Mey. 567; tab. 27.
— rhizanthoides Tzvel. 568, 620.
— rubriflora Illar. 498.
— Ruprechtii (Boiss.) Czer. 480.
— ruthenica Lam. 380.
— salicifolia M. B. 458.
— salonitana Vis. 498.
— salviifolia (Boiss.) Sosn. 432.
— Sarandinakiae Illar. 555.
— savranica Klok. 524.
— scabiosa L. 503.
— Schelkovnikovii Sosn. 412.
— schistosa Sosn. 433, 611.
— semijusta Juz. 551.
— Sergii Klok. 471, 614.
— sevanensis Sosn. 436.
— sibirica L. 469.
— simplicicaulis Boiss. et Huet 491.
— Sintenisiana Gand. 510.
— solstitialis L. 571.
— somchetica (Sosn.) Sosn. 438.
— Sophiae Klok. 521, 615.
— Sosnovskyi Grossh. 500; tab. 25.
— squarrosa Willd. 535.

Centaurea stenolepis A. Kerner 458.
— stereophylla Bess. 511.
— sterilis Stev. 553.
— Steveniana Klok. 527, 616.
— Stevenii M. B. 564.
— stricta Waldst. et Kit. 402.
— substituta Czer. 448, 612.
— sumensis Kalen. 467.
— Szovitsiana Boiss. 562.
— Talievii Kleop. 386.
— tanaitica Klok. 404.
— taochia (Sosn.) Sosn. 435.
— ternopoliensis Dobrocz. 401.
— transcaucasica Sosn. ex Grossh. 558.
— trichocephala M. B. 462.
— trinervia Steph. 483; tab. 22.
— Troitzkyi (Sosn.) Sosn. 427.
— tuapsensis Sosn. 431.
— turgaica Klok. 470, 613.
— turkestanica Franch. 387.
— Vankovii Klok. 552, 618.
— vicina Lipsky 530.
— Willdenowii Czer. 398, 608.
— Woronowii Bornm. ex Sosn. 413.
— xanthocephala (DC.) Sosn. 476.
— xanthocephaloides Tzvel. 476.
— zangezuri (Sosn.) Sosn. 436.
— zuvandica (Sosn.) Sosn. 435.
Centaurium Cass. 28, 370, 377.
Centaurium Gilib. 18, 527.
— Meyeri (Bge.) Druce 529.
— pulchellum (Swartz) Druce 528.
— spicatum (L.) Fritsch 535; tab. 27.
— tenuiflorum (Hoffmgg. et Link) Fritsch 530.
— uliginosum (Waldst. et Kit.) G. Beck 534.
— umbellatum Gilib. 533.
— vulgare Raf. 533; tab. 27.
Centella L. 16, 59.
— asiatica (L.) Urb. 60.
Centipeda Lour. 26, 421.
— orbicularis Lour. 422.
Centranthus DC. 23, 640.
Centunculus L. 18, 278.
— minimus L. 278.
Cephalanthera Rich. 4, 630.
— caucasica Kraenzl. 633.
— epipactoides Fisch. 631; tab. 38.
— grandiflora (L.) Bab. 632; tab. 38.
— longibracteata Blume 631; tab. 38.
— longifolia (Huds.) Fritsch 632; tab. 38.
— rubra (L.) Rich. 634; tab. 38.
Cephalaria Schrad. 24, 25.
— armena Grossh. 47.
— armeniaca Bordz. 36.
— aristata C. Koch 46; tab. 3.
— brevipalea (Somm. et Levier) Litv. 33.
— calcarea Albov 34.
— coriacea (Willd.) Steud. 39; tab. 3.
— dagestanica Bobr. 38.
— Demetrii Bobr. 39; tab. 3.
— gigantea (Ledeb.) Bobr. 29.
— Grossheimii Bobr. 33.
— Kotschyi Boiss. et Hohen. 32.
— Litvinovii Bobr. 31.
— media Litv. 43.
— microdonta Bobr. 45; tab. 3.
— nachiczevanica Bobr. 35, 457.
— pilosa Boiss. et Huet 37.
— procera Fisch. et Avé-Lall. 32.
— sublanata (Bornm.) Szabo 37.
— syriaca (L.) Schrad. 47.
— Tchihatchewii Boiss. 35.
— transsylvanica (L.) Schrad. 45; tab. 3.
— uralensis (Murr.) Schrad. 40.
— velutina Bobr. 44.
Cephalonoplos Neck. 28, 52, 209.
Cephalorrhizum M. Pop. et Korov. 18, 405.
— oopodum M. Pop. et Korov. 406; tab. 20.
— turcomanicum M. Pop. 407, 746; tab. 20.
Cephalorrhynchus Boiss. 29, 338.
— Kirpicznikovii Grossh. 343; tab. 20.
— Kossinskyi (Krasch.) Kirp. 347; tab. 20.
— polycladus (Boiss.) Kirp. 350; tab. 20.
— soongoricus (Rgl.) S. Kovalevsk 346; tab. 20.
— subplumosus S. Kovalevsk. 348; tab. 20.
— Takhtadzhianii (Sosn.) Kirp. 349; tab. 20.
— talyschensis Kirp. 342; tab. 20.
— tuberosus (Stev.) Schchian 340; tab. 20.
Cephaloxys Spreng. ex Desv. 3, 558.
Ceramanthe Dum. 22, 229.
Ceramiocephalum Sch. Bip. 29, 595.
Ceramocarpus Wittst. 16, 184.
Cerastium L. 6, 430.
— Alexeenkoanum Schischk. 451.
— alpinum L. 458.
— anomalum Waldst. et Kit. 436; tab. 22.
— argenteum M. B. 464; tab. 24.
— armeniacum Gren. 443.
— arvense L. 460.
— Beeringianum Cham. et Schlecht. 456; tab. 24.
— Bialynickii Tolm. 459.
— Biebersteinii DC. 463.
— caespitosum Gilib. 455.

— cerastioides Britt. 435; tab. 22.
— chlorifolium Fisch. et Mey. 446.
— dagestanicum Schischk. 465, 883.
— dahuricum Fisch. 444.
— dentatum Moeschl 451.
— dichotomum L. 447; tab. 24.
— falcatum Bge. 439; tab. 22.
— Fischerianum Ser. 457.
— furcatum Cham. et Schlecht. 454.
— glomeratum Thuill. 450.
— glutinosum Fries 452.
— Gorodkovianum Schischk. 457; 883.
— hemschinicum Schischk. 445.
— holosteum Fisch. ex Hornem. 443; tab. 23.
— inflatum Link 447; tab. 23.
— Kasbek Parrot 441.
— lithospermifolium Fisch. 441.
— longifolium Willd. 448.
— maximum L. 440; tab. 22.
— Meyerianum Rupr. 453.
— microspermum C. A. Mey. 442; tab. 23.
— multiflorum C. A. Mey. 442.
— nemorale M. B. 444; tab. 23.
— oreades Schischk. 445.
— pauciflorum Stev. ex Ser. 439.
— perfoliatum L. 446; tab. 23.
— polymorphum Rupr. 459; tab. 24.
— ponticum Albov 465, tab. 24.
— purpurascens Adam 453.
— pusillum Ser. 458; tab. 24.
— Regelii Ostenf. 459.
— ruderale M. B. 449.
— Schmalhausenii Pacz. 452.
— semidecandrum L. 450.
— Sosnowskyi Schischk. 464.
— Stevenii Schischk. 466.
— Szowitsii Boiss. 465.
— tauricum Spreng. 449.
— tianschanicum Schischk. 456.
— undulatifolium Somm. et Levier 440; tab. 23.
Cerasus Juss. 10, 547.
— alaica Pojark. 573.
— amygdaliflora Nevski 570.
— araxina Pojark. 567; tab. 36.
— austera (L.) Roem. 561.
— avium (L.) Moench 556.
— collina Lej. et Court. 561.
— erythrocarpa Nevski 569.
— fruticosa (Pall.) Woronow 558; tab. 36.
— glandulifolia (Rupr. et Maxim.) Kom. 555; tab. 35.
— glandulosa (Thunb.) Lois. 562; tab. 35.
— incana (Pall.) Spach 568.
— Jacquemontii (Hook. f.) Buser 572.
— mahaleb (L.) Mill. 551.
— Maximowiczii (Rupr.) Kom. 550; tab. 35.
— microcarpa (C. A. Mey.) Boiss. 563: tab. 36.
— pseudoprostrata Pojark. 568.
— sachalinensis (Fr. Schmidt) Kom. et Aliss. 552.
— tianschanica Pojark. 573.
— turcomanica Pojark. 574.
— verrucosa (Franch.) Nevski 571.
— vulgaris Mill. 559.
Ceratocarpus L. 6, 110.
— arenarius L. 110.
— turkestanicus Sav.-Rycz. 111, 874: tab. 2.
Ceratocephalus Moench 7, 324.
— falcatus Pers. 324.
— orthoceras DC. 324.
Ceratochloa Beauv. 2, 583.
— unioloides (Willd.) Beauv. 583.
Ceratonia L. 11, 19.
— siliqua L. 19.
Ceratophyllum L. 7, 15.
— demersum L. 15; tab. 2.
— Komarovii Kuzen. 16, 721; tab. 2.
— Kossinskyi Kuzen. 19, 721; tab. 2
— oryzetorum Kom. 16; tab. 2.
— submersum L. 19; tab. 2.
— tanaiticum Sapjeg. 19; tab. 2.
Cercis L. 11, 15.
— Griffithii Boiss. 16; tab. 1.
— siliquastrum L. 16; tab. 1.
Cerefolium Haller 16, 125.
Cerinthe L. 19, 268.
— alpina Kit. 268.
— major L. 270.
— minor L. 269.
Cervaria Gaertn. 17, 168, 188.
Ceterach Willd. 1, 71.
— officinarum Willd. 71.
Chaenorrhinum Lange 22, 226.
— Klokovii Kotov 228.
— rytidospermum (Fisch. et Mey.) Kuprian. 228.
— spicatum Korov. 228.
— viscidum (Moench) Simk. 227.
Chaerefolium Haller 16, 125.
Chaerophyllum L. 16, 94.
— angelicifolium M. B. 110; tab. 6.
— aromaticum L. 99.
— astrantiae Boiss. et Bal. 104.
— Bobrovii Schischk. 115, 591.
— Borodinii Albov 107; tab. 6.

Chaerophyllum bulbosum L. 113.
— caucasicum (Fisch.) Schischk. 114.
— cicutaria Vill. 100.
— confusum Woronow 111; tab. 6.
— crinitum Boiss. 112.
— humile Stev. 101; tab. 7.
— khorossanicum Czerniak. 108, 590.
— kiapazi Woronow 102, 590; tab. 7.
— macrospermum (Willd.) Fisch. et Mey. ex Hohen. 109.
— maculatum Willd. 99.
— Meyeri Boiss. et Buhse 111.
— Prescottii DC. 116.
— roseum M. B. 102.
— rubellum Albov 104; tab. 7.
— temuloides Boiss. 112.
— temulum L. 108.
Chaetolimon (Bge.) Lincz. 18, 372.
— limbatum Lincz. 375.
— setiferum (Bge.) Lincz. 374; tab. 19.
— sogdianum Lincz. 373.
Chaiturus Willd. 21, 144.
— marrubiastrum (L.) Spenn. 144.
Chalcanthus Boiss. 8, 499.
— tuberosus Kom. 499.
Chamaeclema Moench 20, 437.
Chamaedaphne Mitchell 23, 169.
Chamaedaphne Moench 18, 75.
— calyculata (L.) Moench 76.
Chamaedryos Koch 22, 430.
Chamaedrys S. F. Gray 20, 39.
Chamaegeron Schrenk 25, 118.
— Bungei (Boiss.) Botsch. 119.
— oligocephalus Schrenk 120; tab. 15.
Chamaemelum Mill. 26, 66.
— nobile (L.) All. 67.
Chamaemelum Vis. 26, 157.
Chamaenerium Adans. 15, 621.
— angustifolium (L.) Scop. 622; tab. 27.
— angustissimum (Hausskn.) Sosn. 624.
— caucasicum (Hausskn.) Sosn. 625; tab. 27.
— colchicum (Albov) Steinb. 626.
— halimifolium Salisb. 627.
— latifolium (L.) Th. Fries et Lange 626; tab. 27.
— Stevenii (Boiss.) Sosn. 624.
Chamaeorchis Rich. 4, 644.
— alpinus (L.) Rich. 644.
Chamaepericlymenum Graebn. 17, 322.
— canadense (L.) Graebn. 327; tab. 25.
— suecicum (L.) Graebn. 325; tab. 25.
— unalaschkense (Ledeb.) Rydb. 328; tab. 25.
Chamaepeuce DC. 28, 219.
Chamaepitys Link 20, 17.
Chamaerhodos Bge. 10, 233.
— altaica (Laxm.) Bge. 235.
— erecta (L.) Bge. 238; tab. 17.
— grandiflora (Pall.) Ledeb. 234.
— sabulosa Bge. 237.
— songarica Juz. 239, 615.
— trifida Ledeb. 236; tab. 17.
Chamaerops L. 3, 475.
— humilis L. 475; tab. 25.
Chamaesaracha Franch. et Savat. 22, 60.
Chamaesciadium C. A. Mey. 16, 421.
— acaule (M. B.) Boiss. 421; tab. 30.
Chamaespartium Adans. 11, 68.
Chamaesphacos Schrenk 21, 243.
— ilicifolius Schrenk 244.
Chamaesyce S. F. Gray 14, 482.
Chamitea A. Kerner 5, 24.
Chamomilla DC. 26, 148.
Chardinia Desf. 27, 69.
— macrocarpa C. Koch 71.
— orientalis (L.) Kuntze 70; tab. 5.
Chartocalyx Rgl. 21, 182, 183.
Chartolepis Cass. 28, 335.
— Biebersteinii Jaub. et Sp. 339.
— glastifolia (L.) Cass. 336; tab. 19.
— intermedia Boiss. 337.
— pterocaula (Trautv.) Czer. 338.
Chartoloma Bge. 8, 223.
— platycarpum Bge. 223; tab. 13.
Chasmone E. Mey. 11, 51.
Cheilanthes Sw. 1, 74.
— argentea (S. G. Gmel.) Kuntze 75.
— Kuhnii Milde 76; tab. 4.
— persica (Bory) Mett. ex Kuhn 76.
— pteridioides (Reichb.) C. Chr. 76.
Cheirolepis Boiss. 28, 356.
— persica Boiss. 358; tab. 19.
Chelidonium L. 7, 582.
— majus L. 582; tab. 40.
Chenopodina Moq. 6, 177, 182, 191.
Chenopodium L. 6, 41.
— acuminatum Willd. 56; tab. 3.
— album L. 61; tab. 3.
— ambrosioides L. 45.
— aristatum L. 47; tab. 3.
— Berlandieri Moq. 70.
— Bonus-Henricus L. 52.
— botrys L. 46; tab. 3.
— bryoniaefolium Bge. 66; tab. 3.
— capitatum (L.) Aschers. 49.
— chenopodioides (L.) Aell. 50.
— foetidum Schrad. 47.

— foliosum (Moench) Aschers. 48; tab. 3.
— frutescens C. A. Mey. 60; tab. 3.
— giganteum D. Don 65.
— glaucum L. 52.
— hircinum Schrad. 72.
— hybridum L. 55; tab. 3.
— jenissejense Aell. et Iljin 69, 873.
— Klinggraeffii (Abromeit) Aell. 70; tab. 3.
— Korshinskyi Litw. 49.
— murale L. 55; tab. 3.
— opulifolium Schrad. 63.
— pamiricum Iljin 70, 873.
— polyspermum L. 54; tab. 3.
— pratericola Rydb. 65.
— prostratum Bge. ex Herd. 69.
— quinoa Willd. 72.
— rubrum L. 53.
— serotinum L. 71; tab. 3.
— Sosnowskyi Kapell. 66.
— strictum Roth 64.
— urbicum L. 59; tab. 3.
— Vachelii Hook. et Arn. 59.
— viride L. 68; tab. 3.
— vulvaria L. 67.
Chesneya Lindl. 11, 376.
— crassipes Boriss. 380.
— elegans Fomin 382.
— ferganensis Korsh. 382; tab. 25.
— hissarica Boriss. 381.
— kopetdaghensis Boriss. 380, 402; tab. 25.
— Linczevskyi Boriss. 378.
— ternata (Korsh.) M. Pop. 378.
— tribuloides Nevski 383.
— turkestanica Franch. 379.
Chilochloa Beauv. 2, 129.
Chimaphila Pursch 18, 17.
— japonica Miq. 18.
— umbellata (L.) Nutt. 17.
Chlora L. 18, 535.
Chloranthus Sw. 5, 21.
— japonicus Sieb. 21.
Chloris Sw. 2, 286.
— villosa (Desf.) Pers. 286; tab. 22.
— virgata Sw. 286; tab. 22.
Chlorocyperus Rikli 3, 17.
Chomutowia B. Fedtsch. 18, 327.
Chondrilla L. 29, 560.
— acantholepis Boiss. 569.
— ambigua Fisch. ex Kar. et Kir. 580.
— aspera (Schrad. ex Willd.) Poir. 571.
— Bosseana Iljin 585.
— brevirostris Fisch. et Mey. 568.
— canescens Kar. et Kir. 572; tab. 27.
— gibborostris M. Pop. 573.
— graminea M. B. 566.
— hispida (Pall.) Poir. 586.
— juncea L. 565; tab. 27.
— Kusnezovii Iljin 580.
— laticoronata Leonova 576, 754; tab. 28.
— latifolia M. B. 570; tab. 27.
— lejosperma Kar. et Kir. 574; tab. 28.
— macra Iljin 584.
— macrocarpa Leonova 581, 754; tab. 29.
— maracandica Bge. 578.
— mujunkumensis Iljin et Igolk. 584, 755.
— ornata Iljin 579.
— pauciflora Ledeb. 582.
— phaeocephala Rupr. 577; tab. 28.
— piptocoma Fisch. et Mey. 574.
— Rouillieri Kar. et Kir. 576.
Chondrosea Haw. 9, 186.
Chorisis DC. 29, 393.
— repens (L.) DC. 394; tab. 24.
Chorisma D. Don 29, 393.
Chorispora DC. 8, 310.
— Bungeana Fisch. et Mey. 315.
— elegans Cambess. 315.
— Greigii Rgl. 312.
— iberica (M. B.) DC. 316.
— macropoda Trautv. 312; tab. 21.
— sibirica (L.) DC. 317.
— songorica Schrank 311.
— tenella (Pall.) DC. 316.
Chosenia Nakai 5, 23.
— macrolepis (Turcz.) Kom. 23.
Christolea Cambess. 8, 328.
— crassifolia Cambess. 329; tab. 15.
— flabellata (Rgl.) N. Busch 330; tab. 15.
— linearis N. Busch 331, 636; tab. 15.
— maidantalica (M. Pop. et P. Baran.) N. Busch 330.
— pamirica Korsh. 329.
— parryoides (Cham.) N. Busch 331; tab. 15.
Christophoriana Moench 7, 79.
Chromatolepis Dulac 27, 77.
Chromogalia Boiss. 23, 354.
Chrozophora Neck. 14, 288.
— gracilis Fisch. et Mey. 290.
— hierosolymitana Spreng. 291; tab. 17.
— obliqua (Vahl) Juss. ex Spreng. 292; tab. 17.
— sabulosa Kar. et Kir. 289; tab. 17.
— tinctoria (L.) Juss. 290; tab. 17.
Chrysanthemum L. 26, 131.
— carinatum Schousb. 133.
— coronarium L. 134.
— Roxburghii Cass. 135.

Chrysanthemum segetum L. 136.
Chryseis Cass. 28, 323.
Chrysochamelia (Fenzl) Boiss. 8, 85.
— draboides Woronow 85.
Chrysocyathus Falc. 7, 528.
Chrysopogon Trin. 2, 15.
— gryllus (L.) Trin. 15; tab. 1.
Chrysosplenium L. 9, 200.
— alternifolium L. 203.
— baicalense Maxim. 215.
— beringianum Rose 204.
— dubium J. Gay 214.
— filipes Kom. 206.
— flagelliferum Fr. Schmidt. 205; tab. 12.
— kamtschaticum Fisch. 210.
— Komarovii Losinsk. 206, 488.
— nudicaule Bge. 202; tab. 12.
— ovalifolium M. B. ex Bge. 207.
— peltatum Turcz. 202.
— pilosum Maxim. 209.
— ramosum Maxim. 208; tab. 12.
— rimosum Kom. 214.
— Sedakowii Turcz. 207; tab. 12.
— sinicum Maxim. 210
— tetrandrum (Lund) Th. Fries 204.
— thianschanicum Krassn. 208; tab. 12.
— trachyspermum Maxim. 213.
— Wrightii Franch. et Savat. 204.
Chusua Nevski 4, 670, 753.
— secunda Nevski 670; tab. 42.
Chymsydia Albov 17, 37.
— agasylloides Albov 37; tab. 4.
Cicer L. 13, 386.
— anatolicum Alef. 391.
— arietinum L. 388; tab. 21.
— baldshuanicum (M. Pop.) Lincz. 395; tab. 21.
— ervoides (Sieb.) Fenzl 390.
— Fedtschenkoi Lincz. 399; tab. 21.
— flexuosum Lipsky 392; tab. 21.
— Jacquemontii Jaub. et Spach 400; tab. 22.
— kopetdaghense Lincz. 405; tab. 22.
— Korshinskyi Lincz. 396.
— macracanthum M. Pop. 404; tab. 22.
— minutum Boiss. et Hohen. 391.
— mogoltavicum (M. Pop.) A. Korol. 397.
— pungens Boiss. 403; tab. 22.
— songoricum Steph. ex DC. 397; tab. 21.
Cicerbita Wallr. 29, 351.
— alpina (L.) Wallr. 355; tab. 21.
— azurea (Ledeb.) Beauverd 358; tab. 21.
— Bourgaei (Boiss.) Beauverd 366; tab. 21.
— deltoidea (M. B.) Beauverd 368; tab. 22.
— Kovalevskiana Kirp. 372; tab. 22.
— macrophylla (Willd.) Wallr. 361; tab. 21.
— prenanthoides (M. B.) Beauverd 365; tab. 21.
— racemosa (Willd.) Beauverd 363; tab. 21.
— rosea (M. Pop. et Vved.) S. Kovalevsk. 370; tab. 22.
— sevanensis Kirp. 362; tab. 21.
— tianschanica (Rgl. et Schmalh.) Beauverd 355; tab. 21.
— uralensis (Rouy) Beauverd 360; tab. 21.
— zeravschanica M. Pop. ex S. Kovalevsk. 369; tab. 22.
Cicerbita Wallr. emend. Beauverd 29, 351.
Cichorium L. 29, 14.
— endivia L. 19.
— glandulosum Boiss. et Huet 18.
— intybus L. 16.
— pumilum Jacq. 18.
Cicuta L. 16, 375.
— virosa L. 376.
Cicutaria Lam. 16, 376.
Cicutaria Moench 16, 349.
Ciliaria Haw. 9, 180.
Cimicifuga L. 7, 82.
— dahurica (Turcz.) Maxim. 83.
— europaea Schipcz. 85, 721.
— foetida L. 84.
— heracleifolia Kom. 83.
— simplex Wormsk. 84.
Cinna L. 2, 169.
— latifolia (Trev.) Griseb. 169; tab. 12.
Circaea L. 15, 633.
— alpina L. 633; tab. 31.
— cordata Royle 637; tab. 31.
— intermedia Ehrh. 634; tab. 31.
— lutetiana L. 635.
— mollis Sieb. et Zucc. 636.
— quadrisulcata (Maxim.) Franch. et Savat. 636.
Cirsium Mill. 28, 51.
— abkhasicum (Petrak) Grossh. 86.
— acaule (L.) Scop. 205.
— adjaricum Somm. et Lev. 79.
— aduncum Fisch. et Mey. ex DC. 97; tab. 5.
— aggregatum Ledeb. 69.
— alatum (S. G. Gmel.) Bobr. 192.
— Alberti Rgl. et Schmalh. 145.
— Albowianum Somm. et Levier 189.
— anatolicum (Petrak) Grossh. 102.
— apiculatum DC. 190.
— arachnoideum (M. B.) M. B. 110.
— argillosum V. Petrov ex Char. 123; tab. 6.

— armenum DC. 200.
— arvense (L.) Scop. 213.
— badakhschanicum Char. 144, 604.
— balkharicum Char. 117.
— Biebersteinii Char. 170.
— Bornmülleri Sint. ex Bornm. 99.
— bracteosum DC. 93.
— brevipapposum Tschern. 204.
— Buschianum Char. 84, 602.
— canum (L.) All. 169.
— caput-medusae Somm. et Levier 88.
— caucasicum (Adam) Petrak 78; tab. 4.
— cephalotes Boiss. 114.
— chlorocomos Somm. et Lev. 76.
— ciliatum (Murr.) Moench 130.
— congestum Fisch. et Mey. ex DC. 90.
— coryletorum Kom. 174; tab. 7.
— cosmelii (Adam) Fisch. ex Hohen. 107.
— czerkessicum Char. 72, 602.
— daghestanicum Char. 119.
— dealbatum M. B. 168.
— depilatum Boiss. et Bal. 185.
— echinus (M. B.) Hand.-Mazz. 147.
— elbrusense Somm. et Lev. 188.
— elodes M. B. 191.
— erisithales (Jacq.) Scop. 163.
— erythrolepis C. Koch 81.
— esculentum (Sievers) C. A. Mey. 201.
— euxinum Char. 110.
— Fominii Petrak 87.
— Frickii Fisch. et Mey. 204.
— Gagnidzei Char. 71, 600.
— glabrifolium (C. Winkl.) O. et B. Fedtsch. 143; tab. 8.
— helenioides (L.) Hill 166.
— heterophyllum (L.) Hill 165.
— hydrophiloides Char. 186.
— hygrophilum Boiss. 187.
— hypoleucum DC. 154.
— imereticum Boiss. 74.
— incanum (S. G. Gmel.) Fisch. ex M. B. 211.
— kamtschaticum Ledeb. ex DC. 179.
— Kemulariae Char. 75.
— Ketzkhovelii Char. 120.
— Komarovii Schischk. 178.
— Kusnetzowianum Somm. et Levier 150.
— lamyroides Tamamsch. 138.
— laniflorum (M. B.) M. B. 111.
— lappaceum (M. B.) M. B. 101; tab. 5.
— longiflorum Char. 71, 601.
— Maackii Maxim. 157.
— macrobotrys (C. Koch) Boiss. 106.
— macrocephalum C. A. Mey. 122.
— megricum Char. 104, 603.
— oblongifolium C. Koch 153.
— obvallatum (M. B.) M. B. 198.
— ochrolepideum Juz. 212.
— oleraceum (L.) Scop. 161.
— osseticum (Adam) Petrak 83.
— palustre (L.) Scop. 183.
— pannonicum (L. f.) Link 171.
— pectinellum A. Gray 182.
— pendulum Fisch. 208.
— polonicum (Petrak) Iljin 127.
— pseudolappaceum Char. 125.
— pugnax Somm. et Levier 116.
— rhabdotolepis Petrak 197.
— rhizocephalum C. A. Mey. 203.
— rigidum DC. 96.
— rivulare (Jacq.) All. 162.
— sairamense (C. Winkl.) O. et B. Fedtsch. 142.
— schantarense Trautv. et Mey. 158.
— Schelkownikowii Petrak 105.
— Schischkinii Serg. 177.
— Semenovii Rgl. et Schmalh. 140.
— serratuloides (L.) Hill 176.
— serrulatum (M. B.) M. B. 128.
— setosum (Willd.) M. B. 210.
— Sieversii (Fisch. et Mey.) Petrak 146; tab. 8.
— simplex C. A. Mey. 199.
— sinuatum (Trautv.) Boiss. 137; tab. 7.
— sorocephalum Fisch. et Mey. 91.
— Sosnowskyi Char. 85; tab. 6.
— strigosum (M. B.) M. B. 94.
— subinerme Fisch. et Mey. 195.
— sublaniflorum Soják 112.
— svaneticum Somm. et Lev. 193.
— sychnosanthum Petrak 152.
— Szowitsii (C. Koch) Boiss. 132.
— tauricum Soják 114.
— tindaicum Char. 121.
— tomentosum C. A. Mey. 118; tab. 4.
— tricholoma Fisch. et Mey. 82.
— turkestanicum (Rgl.) Petrak 124.
— ukranicum Bess. 129.
— uliginosum (M. B.) M. B. 194.
— Vlassovianum Fisch. 173.
— vulgare (Savi) Ten. 135.
— Waldsteinii Rouy 153.
— Weyrichii Maxim. 181; tab. 9.
Cistanche Hoffmgg. et Link 23, 21.
— ambigua (Bge.) G. Beck 25; tab. 3.
— fissa (C. A. Mey.) G. Beck 26; tab. 3.
— flava (C. A. M.) Korsh. 23.

Cistanche mongolica G. Beck 24.
— Ridgewayana Aitch. et Hemsl. 27; tab. 3.
— salsa (C. A. Mey.) G. Beck. 24.
— tubulosa (Schenk) Wight 23; tab. 3.
Cistus L. 15, 328.
— salvifolius L. 329.
— tauricus (J. et C.) Presl 328; tab. 17.
Cithareloma Bge. 8, 333.
— Lehmannii Bge. 333; tab. 21.
— vernum Bge. 334.
Citrullus Forsk. 24, 111.
— colocynthus (L.) Schrad. 113.
— vulgaris Schrad. 111.
Citrus L. 14, 234.
— limon Burm. 237.
— unshiu Marc. 237; tab. 14.
Cladium Schrad. 3, 101.
— mariscus (L.) R. Br. 102; tab. 5.
Cladochaeta DC. 25, 430.
— candidissima (M. B.) DC. 431; tab. 26.
— caspica Sosn. ex Grossh. 431.
Clausia Korn.-Tr. 8, 251.
— aprica (Steph.) Korn.-Tr. 252.
— hispida (Rgl.) Lipsky 254.
— kasakorum Pavl. 253.
— mollissima Lipsky 254.
— Olgae (Rgl. et Schmalh.) Lipsky 255.
— papillosa Vass. 255, 641; tab. 17.
— serawschanica (Rgl. et Schmalh.) Lipsky 255.
— turkestanica Lipsky 253.
Claytonia Gronov. 6, 377.
— acutifolia Pall. ex Willd. 378.
— arctica Adam 381.
— Eschscholtzii Cham. 382; tab. 19
— Joanneana Roem. et Schult. 381.
— sarmentosa C. A. Mey. 382.
— sibirica L. 383.
— tuberosa Pall. ex Willd. 378; tab. 19.
— Vassilievii Kuzen. 383, 880.
Clematis L. 7, 310.
— aethusifolia Turcz. 315; tab. 20.
— asplenifolia Schrenk 317; tab. 20.
— brevicaudata DC. 320; tab. 20.
— flammula L. 319.
— fusca Turcz. 312; tab. 20.
— glauca Willd. 321.
— hexapetala Pall. 318.
— integrifolia L. 312; tab. 20.
— ispahanica Boiss. 317.
— manschurica Rupr. 319; tab. 20.
— orientalis L. 322; tab. 20.
— pseudoflammula Schmalh. ex Lipsky 319; tab. 20.
— recta L. 318.
— serratifolia Rehd. 321.
— songarica Bge. 316; tab. 20.
— tangutica (Maxim.) Korsh. 322.
— vitalba L. 310; tab. 20.
— viticella L. 316.
Clementsia Rose 9, 29.
Cleome L. 8, 3.
— chorossanica Bge. 9; tab. 1.
— coluteoides Boiss. 13.
— Gordjaginii M. Pop. 5.
— Lipskyi M. Pep. 12.
— Noeana Boiss. 10; tab. 1.
— ornithopodioides L. 9.
— quinquenervia DC. 10.
— Raddeana Trautv. 14.
— rostrata Bobr. 12, 633; tab. 1.
— tomentella M. Pop. 11.
— turkmena Bobr. 6, 633.
Clerodendron L. 19, 699.
— foetidum Bge. 700.
Clerodendrum L. 19, 699.
Clinelymus (Griseb.) Nevski 2, 690.
— dahuricus (Turcz.) Nevski 691; tab. 49.
— excelsus (Turcz.) Nevski 692; tab. 49.
— nutans (Griseb.) Nevski 691; tab. 49.
— sibiricus (L.) Nevski 690; tab. 49.
Clinopodium L. 21, 436.
— chinense (Benth.) Kuntze 440.
— integerrimum Boriss. 438; tab. 26.
— sachalinense (Fr. Schmidt) Koidz. 440.
— umbrosum (M. B.) C. Koch 439; tab. 26.
— vulgare L. 437.
Clintonia Raf. 4, 448.
— udens Trautv. et Mey. 448; tab. 27.
Clypeola L. 8, 364.
— dichotoma Boiss. 367; tab. 22.
— echinata DC. 367.
— elegans Boiss. et Huet 366.
— jonthlaspi L. 365; tab. 22.
— microcarpa Boiss. 366.
— Raddeana Alb. 366.
Cnicus L. 28, 52, 209.
Cnicus L. 28, 587.
— benedictus L. 587.
Cnidium Guss. 16, 549.
— ajanense (Rgl. et Til.) Drude 554; tab. 34.
— cnidiifolium (Turcz.) Schischk. 552; tab. 34, 36.

— dahuricum (Jacq.) Turcz. 550; tab. 15, 36.
— dubium (Schkuhr) Thell. 552.
— Grossheimii Manden. 558.
— Monnieri (L.) Guss. 559.
— multicaule (Turcz.) Ledeb. 551; tab. 34.
— orientale Boiss. 557.
— pauciradiatum Somm. et Lev. 558.
— salinum Turcz. 553.
Cobresia Willd. 3, 105.
— Bellardii (All.) Degl. 109; tab. 9.
— calillifolia (Decne.) Clarke 109; tab. 9.
— filifolia (Turcz.) Meinsh. 109.
— humilis (C. A. Mey.) Serg. 111; tab. 9.
— paniculata Meinsh. 111; tab. 9.
— Royleana (Nees) W. Beck. 110; tab. 9.
— schoenoides (C. A. Mey.) Steud. 106; tab. 9.
— simpliciuscula (Wahl.) Maack 110; tab. 9.
Cochlearia L. 8, 568.
— arctica Schlecht. 569.
— groenlandica L. 570.
— lenensis Ad. 569.
— oblongifolia DC. 569.
Cochlidiospermum Reichb. 22, 329, 392.
Cocos Mart. 469.
Codonocephalum Fenzl 25, 432.
— paecockianum Aitch. et Hemsl. 432.
Codonopsis Wall. 24, 433.
— clematidea (Schrenk) Clarke 437; tab. 26.
— lanceolata (Sieb. et Zucc.) Benth. et Hook. 435.
— pilosula (Franch.) Nannfeldt 434.
— ussuriensis (Rupr. et Maxim.) Hemsl. 436; tab. 26.
Coeloglossum Hartm. 4, 647.
— viride (L.) Hartm. 647.
Coelopleurum Ledeb. 17, 33.
— Gmelinii (DC.) Ledeb. 33; tab. 8.
Colchicum Tourn. 4, 23.
Colchicum L. 4, 23.
— autumnale L. 30.
— Biebersteinii Rouy 25.
— laetum Stev. 29.
— luteum Baker 27.
— nivale Boiss. et Huet ex Stef. 26; tab. 2.
— Regelii Stef. 27.
— serpentinum Miscz. 25; tab. 2.
— speciosum Stev. 29.
— Szovitsii Fisch. et Mey. 26; tab. 2.
— umbrosum Stev. 28; tab. 2.
— zangezurum Grossh. 26.
Coleanthus Seidl. 2, 160.
— subtilis (Tratt.) Seidl. 160; tab. 12.
Coleostephus Cass. 26, 146.
— myconis (L.) Cass. 147.
Collomia Nutt. 19, 94.
— linearis Nutt. 94.
Colpodium Trin. 2, 434.
— altaicum Trin. 436.
— araraticum Woronow 438.
— chrysanthum Woronow 439.
— colchicus (Albov) Woronow 437; tab. 33.
— fibrosum Trautv. 438.
— humile (M. B.) Griseb. 442; tab. 33.
— lakium Woronow 441.
— leucolepis Nevski 437.
— ornatum Nevski 441, 757.
— parviflorum Boiss. et Buhse 445; tab. 33.
— ponticum (Bal.) Woronow 440.
— Vahlianum (Liebm.) Nevski 436; tab. 35.
— variegatum Boiss. 439.
— versicolor (Stev.) Schmalh. 440; tab. 33.
Coluria R. Br. 10, 241.
— geoides (Pall.) Ledeb. 241; tab. 18.
Colutea L. 11, 315.
— acutifolia Shap. 318, 394.
— arborescens L. 318.
— armena Boiss. et Huet 321; tab. 22.
— Buhsei (Boiss.) Shap. 320; tab. 22.
— canescens Shap. 312, 394.
— cilicica Boiss. et Bal. 319; tab. 22.
— gracilis Freyn et Sint. ex Freyn 321; tab. 22.
— hybrida Shap. 322, 395.
— Jarmolenkoi Shap. 320, 394; tab. 22.
— Komarovii Takht. 323.
— kopetdaghensis B. Fedtsch. 320.
— orientalis Mill. 317; tab. 22.
— Paulsenii Freyn et Sint. ex Freyn 322; tab. 22.
Coluteocarpus Boiss. 8, 572.
— vesicaria (L.) Holmb. 572.
Comaropsis Rich. 10, 239.
Comarum L. 10, 73.
— palustre L. 74; tab. 7.
— Salesovianum (Steph.) Aschers. Graebn. 77.
Commelina L. 3, 498.
— communis L. 501; tab. 26.
Comperia C. Koch 4, 676.
— taurica C. Koch 676; tab. 43.
Condylocarpus Hoffm. 17, 276.
Coniogramma Fée 1, 73.
— fraxinea (D. Don) Diels 73.

Conioselinum Fisch. 17, 1.
— boreale Schischk. 6, 351.
— kamtschaticum Rupr. 4; tab. 8.
— latifolium Rupr. 3; tab. 8.
— longifolium Turcz. 5; tab. 1.
— pinnatifolium (Korov.) Schischk. 9.
— schugnanicum B. Fedtsch. 4.
— vaginatum (Spreng.) Thell. 2; tab. 8.
— victoris Schischk. 6, 351; tab. 8.
Conium L. 16, 225.
— maculatum L. 225.
Conringia Adans. 8, 495.
Conringia (Heist.) Pers. 8, 495.
— austriaca (Jacq.) Sweet 496.
— orientalis (L.) Andrz. 496.
— perfoliata (C. A. Mey.) N. Busch 497.
— persica Boiss. 497.
— planisiliqua Fisch. et Mey. 497.
Consolida S. F. Gray 7, 99.
Convallaria L. 4, 467.
— majalis L. 467; tab. 29.
Convolvulus L. 19, 6.
— Ammanii Desr. 26.
— arvensis L. 29.
— ascabadensis Bornm. et Sint. ex Bornm. 20.
— Calvertii Boiss. 23.
— campanulatus Zapr. 13.
— cantabrica L. 22.
— commutatus Boiss. 23.
— divaricatus Rgl. et Schmalh. 18.
— erinaceus Ledeb. 16; tab. 1.
— fruticosus Pall. 14.
— Gortschakovii Schrenk 14.
— hamadae (Vved.) V. Petrov 17.
— hirsutus Stev. 28; tab. 1.
— holosericeus M. B. 25.
— Korolkovii Rgl. et Schmalh. 18.
— Krauseanus Rgl. et Schmalh. 19.
— lineatus L. 26.
— lycioides Boiss. 10.
— Michelsonii V. Petrov 19.
— Olgae Rgl. et Schmalh. 13.
— persicus L. 27; tab. 1.
— pilosellifolius Desr. 22.
— pseudocantabrica Schrenk 20.
— pseudoscammonia C. Koch 28.
— scammonia L. 29.
— sericocephalus Juz. 24.
— subhirsutus Rgl. et Schmalh. 21.
— subsericeus Schrenk 17.
— tauricus (Bornm.) Juz. 24.
— tragacanthoides Turcz. 15.
Conyza L. 25, 191.
Conyza Less. 25, 191, 239.
Conyza Spreng. 25, 185.
Conyzanthus Tamamsch. 25, 185, 583.
— graminifolius (Spreng.) Tamamsch. 186; tab. 4.
— squamatus (Spreng.) Tamamsch. 186.
Conyzella Rupr. 25, 191.
Coptis Salisb. 7, 78.
— trifolia (L.) Salisb. 78.
Corallorhiza Hall. 4, 608.
— trifida Chatel. 608; tab. 37.
Corchorus L. 15, 2.
— capsularis L. 3.
— olitorius L. 3.
Coreopsis L. 25, 546.
— grandiflora Hoog. 547.
— tinctoria Nutt. 547.
Coriandrum L. 16, 184.
— sativum L. 185.
Corion Hoffmgg. et Link 16, 198.
Corispermum L. 6, 135.
— algidum Iljin 139.
— altaicum Iljin 148; tab. 6.
— aralo-caspicum Iljin 144.
— canescens Kit. 140.
— chinganicum Iljin 147.
— confertum Bge. 157.
— crassifolium Turcz. 143.
— declinatum Steph. ex Stev. 147; tab. 7.
— Dutreuilii Iljin 154; tab. 6.
— elongatum Bge. 156; tab. 6.
— erosum Iljin 150; tab. 6.
— filifolium C. A. Mey. 146.
— gelidum Iljin 156, 875.
— heptapotamicum Iljin 152.
— Hilariae Iljin 154, 874.
— hyssopifolium L. 148; tab. 6.
— Komarovii Iljin 143.
— Korovinii Iljin 155; tab. 7.
— Krylovii Iljin 150.
— laxiflorum Schrenk 145; tab. 6.
— Lehmannianum Bge. 152; tab. 6.
— macrocarpum Bge. 157; tab. 6.
— Marschallii Stev. 138; tab. 6.
— mongolicum Iljin 151; tab. 6.
— nitidum Kit. 145; tab. 6.
— orientale Lam. 150; tab. 6.
— pamiricum Iljin 153; tab. 6.
— papillosum (Kuntze) Iljin 155; tab. 7.
— piliferum Iljin 151; tab. 7.
— Redowskii Fisch. 149; tab. 7.
— sibiricum Iljin 149; tab. 6.
— squarrosum L. 140; tab. 6.
— tibeticum Iljin 153.

— ulopterum Fenzl 158; tab. 7.
Cormus Spach 9, 374.
Cornella Rydb. 17, 324.
Corniola Adans. 11, 54.
Cornulaca Delile 6, 350.
— Korshinskyi Litv. 350; tab. 14.
Cornus L. 17, 317.
— mas L. 317; tab. 24.
Corona Fisch. et J. Grah. 4, 302.
Coronaria L. 6, 699.
— coriacea (Moench) Schischk. 699.
— flos-cuculi (L.) A. Br. 700.
Coronilla L. 13, 247.
— Balansae Boiss. 251.
— coronata L. 249.
— cretica L. 253.
— emeroides Boiss. et Sprun. 248.
— latifolia Jav. 251.
— orientalis Mill. 250.
— parviflora Willd. 254.
— scorpioides (L.) Koch 254.
— varia L. 252.
Coronopus (Rupr.) Gaertn. 8, 537.
— didymus (L.) Smith 538.
— procumbens Gilib. 538.
Coroxylon Moq. 6, 248.
Coroxylon Thunb. 6, 248.
Cortaderia Stapf 2, 302.
— dioica (Spreng.) Speg. 302; tab. 23.
Cortusa L. 18, 242.
— altaica Losinsk. 246.
— amurensis Fed. 246, 731.
— Brotheri Pax ex Lipsky 245.
— Matthioli L. 248.
— pekinensis (A. Rich.) Losinsk. 244; tab. 10.
— sibirica Andrz. 247.
— turkestanica Losinsk. 245; tab. 10.
Corvisartia Mérat 25, 433, 440.
Corydalis Medic. 7, 649.
— Aitchisonii M. Pop. 658, 750.
— Alexeenkoana N. Busch 674.
— alpestris C. A. Mey. 676.
— ambigua Cham. et Schlecht. 666.
— angustifolia (M. B.) DC. 673.
— arctica M. Pop. 680, 751; tab. 43.
— bracteata (Steph.) Pers. 674.
— bucharica M. Pop. 693, 752.
— Buschii Nakai 680.
— capnoides (L.) Pers. 702.
— caucasica DC. 664; tab. 43.
— cava (Mill.) Schweigg. et Korte 659; tab. 43.
— chionophila Czerniak. 654.
— conorhiza Ledeb. 675; tab. 43.
— darwasica Rgl. ex Prain 653.
— Emanueli C. A. Mey. 679.
— Erdelii Zucc. 655; tab. 43.
— Fedtschenkoana Rgl. 698.
— fimbrillifera Korsh. 691.
— fumariifolia Maxim. 669.
— gigantea Trautv. et Mey. 683.
— glaucescens Rgl. 663.
— Gortschakovii Schrenk 688.
— Halleri Willd. 671.
— impatiens (Pall.) Fisch. 705.
— inconspicua Bge. 699.
— intermedia (L.) Merat 665.
— kaschgarica Rupr. 695.
— Ledebouriana Kar. et Kir. 655.
— macrantha (Rgl.) M. Pop. 682, 751; tab. 44.
— macrocalyx Litv. 697; tab. 44.
— macrocentra Rgl. 652.
— Marschalliana (Pall.) Pers. 659; tab. 43.
— Nevski M. Pop. 658, 750; tab. 43.
— nobilis (L.) Pers. 687.
— nudicaulis Rgl. 660.
— ochotensis Turcz. 703; tab. 44.
— Paczoskii N. Busch 672.
— paeoniifolia (Steph.) Pers. 684.
— pallida Pers. 700.
— pallidiflora (Rupr.) N. Busch 679.
— paniculigera Rgl. et Schmalh. 695; tab. 44.
— pauciflora (Steph.) Pers. 678; tab. 43.
— persica Cham. et Schlecht. 654; tab. 43.
— Popovii Nevski 656; tab. 43.
— pseudoadunca M. Pop. 696, 753; tab. 44.
— pseudoalpestris M. Pop. 677, 750.
— pseudostricta M. Pop. 696, 752.
— Raddeana Rgl. 704; tab. 44.
— Redowskii Fedde 683.
— remota Fisch. ex Maxim. 670.
— repens Mandl et Muehld. 666.
— Schanginii (Pall.) B. Fedtsch. 663.
— Schelesnowiana Rgl. et Schmalh. 694.
— Semenovii Rgl. 687.
— sempervirens (L.) Pers. 701.
— Severzovii Rgl. 657.
— sibirica (L. f.) Pers. 704.
— stricta Steph. 692.
— tenella Kar. et Kir. 700; tab. 44.
— transalaica M. Pop. 693, 751.
Corylus L. 5, 262.
— avellana L. 264.
— brevituba Kom. 268.
— colchica Albov 267; tab. 12.

Corylus colurna L. 263; tab. 12.
— heterophylla Fisch. ex Bess. 265.
— manshurica Maxim. 268.
— maxima Mill. 266; tab. 12.
— pontica C. Koch 266; tab. 12.
Corynephorus Beauv. 2, 240.
— articulatus (Desf.) Beauv. 241.
— canescens (Bernh.) Beauv. 241; tab. 18.
Cosmea Willd. 25, 561.
Cosmos Cav. 25, 561.
— bipinnatus Cav. 562.
Cota J. Gay ex Guss. 26, 9, 39.
Cotinus Adans. 14, 526.
— coggygria Scop. 526.
Cotoneaster Medic. 9, 319.
— insignis Pojark. 330; tab. 19.
— integerrima Medic. 323.
— lucida Schlecht. 323; tab. 19.
— melanocarpa Lodd. 320.
— multiflora Bge. 329.
— oligantha Pojark. 326.
— racemiflora (Desf.) C. Koch. 331.
— saxatilis Pojark. 333.
— taurica Pojark. 332; tab. 19.
— uniflora Bge. 324; tab. 19.
Cotula L. 26, 418.
— coronopifolia L. 421; tab. 6.
Cousinia Cass. 27, 108.
— abbreviata Tscherneva 347.
— Abolinii Kult. ex Tsherneva 218.
— acrodroma Tscherneva 348; tab. 28.
— adenophora Juz. 337.
— affinis Schrenk 153, tab. 8.
— agelocephala Tscherneva 216.
— alaica Juz. ex Tscherneva 254.
— alata Schrenk 163.
— Alberti Rgl. et Schmalh. 239.
— Albertoregelia C. Winkl. 168.
— albiflora (Bornm. et Sint.) Bornm. 147; tab. 8.
— allolepis Tscherneva et Vved. 343.
— alpestris Bornm. ex Juz. 282.
— alpina Bge. 284.
— ambigens Juz. 209.
— amoena C. Winkl. 228.
— Androssovii Juz. 217.
— angreni Juz. 199.
— angusticeps Juz. 275; tab. 10.
— anomala Franch. 182; tab. 13.
— Antonowii C. Winkl. 158.
— apiculata Tscherneva 174.
— arachnoidea Fisch. et Mey. 316.
— araxena Takht. 192.
— arctioides Schrenk 177; tab. 12.
— armena Takht. 189; tab. 22.
— astracanica (Spreng.) Tamamsch. 152.
— aurea C. Winkl. 244.
— Badghysi Kult. 144.
— Baranovii Juz. ex Tscherneva 255.
— Batalinii C. Winkl. 295.
— Beckeri Trautv. 265.
— Bobrovii Juz. 169.
— Bonvalotii Franch. 354.
— Botschantzevii Juz. ex Tscherneva 267; tab. 23.
— brachyptera DC. 192.
— Bungeana Rgl. et Schmalh. 272.
— buphtalmoides Rgl. 287.
— Butkovii Tscherneva et Vved. 304.
— caespitosa C. Winkl. 356.
— calca Juz. 295.
— campyloraphis Tscherneva 228.
— candicans Juz. 230.
— carduncelloidea Rgl. et Schmalh. 285.
— centauroides Fisch. et Mey.
— ceratophora Kult. 355.
— chaetocephala Kult. 151.
— chlorantha Kult. 247; tab. 27.
— chlorocephala C. A. Mey. 188.
— chrysantha Kult. 340.
— coerulea Kult. ex Tscherneva 235.
— congesta Bge. 216.
— coronata Franch. 279.
— corymbosa C. Winkl. 235.
— cryptadena Juz. 157.
— cynaroides (M. B.) C. A. Mey. 262.
— daralaghezica Takht. 189.
— darwasica C. Winkl. 309.
— decurrentifolia Juz. ex Tscherneva 257; tab. 20.
— dichotoma Bge. 269.
— dichromata Kult. 214; tab. 15.
— Dimoana Kult. 309.
— dissecta Kar. et Kir. 160; tab. 9.
— dissectifolia Kult. 257.
— divaricata C. Winkl. 310.
— dolichoclada Juz. 162.
— dolicholepis Schrenk 197.
— dolichophylla Kult. 219.
— dshisakensis Kult. 300.
— dubia M. Pop. 327.
— egens Juz. 215.
— egregia Juz. 218; tab. 14.
— erectispina Tscherneva 269.
— eriotricha Juz. 352.
— erivanensis Bornm. 193.
— eryngioides Boiss. 186; tab. 14.
— Eugenii Kult. 275.

— fallax C. Winkl. 255; tab. 19.
— fascicularis Juz. 184.
— Fedorovii Takht. 190.
— Fedtschenkoana Bornm. 224.
— ferganensis Bornm. 320.
— ferruginea Kult. 308.
— Fetissowii C. Winkl. 293; tab. 23.
— finitima Juz. 207.
— Franchetii C. Winkl. 221.
— Freynii Bornm. et Sint. 265.
— glabriseta Kult. 292.
— glandulosa Kult. 158.
— glaphyrocephala Juz. ex Tscherneva 334.
— glochidiata Kult. 175.
— Gnezdilloi Tscherneva 233.
— Gomolitzkii Juz. ex Tscherneva 238.
— Gontscharowii Juz. 170.
— grandiflora Kult. 249; tab. 18.
— Grigoriewii Juz. 290.
— grisea Kult. 252.
— gulczensis Kult. 297.
— haesitabunda Juz. 247; tab. 19
— hamadae Juz. 155.
— hastifolia C. Winkl. 326; tab. 26.
— Hilariae Kult. 288.
— Hohenackeri Fisch. et Mey. 195.
— hoplophylla Tscherneva 234.
— horrescens Juz. 240.
— horridula Juz. 212.
— hypopolia Bornm. et Sint. 148.
— hystrix C. A. Mey. 276; tab. 22.
— Iljinii Takht. 194.
— integrifolia Franch. 310.
— jassyensis C. Winkl. 332.
— Juzepczukii Tscherneva 203.
— karatavica Rgl. et Schmalh. 246.
— kazachorum Juz. ex Tscherneva 212.
— Knorringiae Bornm. 289.
— kokanica Rgl. et Schmalh. 328.
— Komarowii (Kuntze) C. Winkl. 250.
— Korolkovii Rgl. et Schmalh. 248.
— Korshinskyi C. Winkl. 225; tab. 16.
— Krauseana Rgl. et Schmalh. 324.
— Kuekenthalii Bornm. 256.
— laetevirens C. Winkl. 352.
— lanata C. Winkl. 335.
— laniceps Juz. 302.
— lappacea Schrenk 180.
— lasiosiphon Juz. 259; tab. 21.
— leiocephala (Rgl.) Juz. 314.
— leptacantha (Bornm.) Juz. 184.
— leptacma Tscherneva 227.
— leptocampyla Bornm. 283.
— leptocephala Fisch. et Mey. 146.
— leptoclada Kult. 234.
— leptocladoides Tscherneva 232.
— leucantha Bornm. et Sint. 252.
— Linczewskii Juz. 145.
— Litvinovii Kult. ex Juz. 227.
— Lomakinii C. Winkl. 194.
— lyrata Bge. 263; tab. 21.
— macilenta C. Winkl. 226.
— macrocephala C. A. Mey. 263.
— macroptera C. A. Mey. 277.
— magnifica Juz. 334; tab. 27.
— maracandica Juz. 213.
— Margaritae Kult. 330.
— margiana Juz. 187.
— medians Juz. 245.
— microcarpa Boiss. 317.
— microcephala C. A. Mey. 196.
— mindshelkensis B. Fedtsch. 237.
— Minkwitziae Bornm. 200.
— minuta Boiss. 324.
— mogoltavica Tscherneva et Vved. 303.
— mollis Schrenk 185.
— mucida Kult. 150.
— mulgediifolia Bornm. 282.
— multiloba DC. 350.
— murgabica Tscherneva 274.
— neglecta Juz. 256.
— Newesskiana C. Winkl. 344.
— Ninae Juz. 318.
— Olgae Rgl. et Schmalh. 166; tab. 9.
— omphalodes Tscherneva 326; tab. 26.
— onopordioides Ledeb. 260.
— oopoda Juz. 170.
— oreodoxa Bornm. et Sint. 174; tab. 11.
— oreoxerophila Kult. 150.
— orientalis (Adam) C. Koch 196.
— orthacantha Tscherneva 273.
— ortholepis Juz. ex Tscherneva 319.
— outichaschensis Franch. 285.
— Ovczinnikovii Tscherneva 296; tab. 24.
— oxiana Tscherneva 163.
— oxytoma Rech. f. 296.
— pannosa C. Winkl. 220.
— pannosiformis Tscherneva 222.
— patentispina Tscherneva 270.
— pauciramosa Kult. 315.
— peduncularis Juz. ex Tscherneva 297.
— pentacantha Rgl. et Schmalh. 239; tab. 17.
— pentacanthoides Juz. ex Tscherneva 241.
— perovskiensis (Bornm.) Juz. 161; tab. 9.
— platylepis Schrenk ex Fisch. et Mey. 318.
— platystegia Tscherneva 214.
— podophylla Tscherneva 232; tab. 16.

Cousinia polycephala Rupr. 206.
— polytimetica Tscherneva 270.
— praestans Tscherneva et Vved. 305.
— princeps Franch. 300.
— proxima Juz. 229.
— psammophila Kult. 164; tab. 10
— pseudarctium Bornm. 179.
— pseudoaffinis Kult. 155.
— pseudodshisakensis Tscherneva et Vved. 299; tab. 24.
— pseudolanata M. Pop. ex Tscherneva 336.
— pseudomollis C. Winkl. 183; tab. 13.
— pterolepida Kult. 225.
— pulchella Bge. 314.
— pulchra C. Winkl. 280.
— pungens Juz. 205.
— purpurea C. A. Mey. 264.
— pusilla C. Winkl. 273.
— pygmaea C. Winkl. 167; tab. 10.
— Raddeana C. Winkl. 176.
— radians Bge. 278.
— ramulosa Rech. f. 348.
— rava C. Winkl. 290.
— refracta Juz. 245.
— Regelii C. Winkl. 316.
— resinosa Juz. 210; tab. 15.
— rhodantha Kult. 307.
— rigida Kult. 236; tab. 18.
— rosea Kult. 306.
— rotundifolia C. Winkl. 307.
— rubiginosa Kult. 349.
— scabrida Juz. 199.
— Schischkinii Juz. 294.
— schistoptera Juz. 165.
— Schmalhausenii C. Winkl. 246.
— Schtschurowskiana Rgl. et Schmalh. 298.
— schugnanica Juz. 346.
— sclerantha Kult. 205.
— sclerophylla Juz. 173.
— semidecurrens C. Winkl. 222.
— semilacera Juz. 342.
— serawschanica C. Winkl. 355.
— Severtzovii Rgl. 313; tab. 25.
— simulatrix C. Winkl. 322.
— Smirnowii Trautv. 266.
— sogdiana Bornm. 161.
— sororia Juz. 346.
— speciosa C. Winkl. 294.
— Spiridonovii Juz. 208.
— splendida C. Winkl. 337; tab. 28.
— sporadocephala Juz. 207.
— Sprygínii Kult. 268.
— Stahliana Bornm. et Gauba 150.
— stellaris Bornm. 287.
— stenophylla Kult. 251.
— stephanophora C. Winkl. 340.
— stricta Tscherneva 172; tab. 11.
— strobilocephala Tscherneva et Vved. 305.
— subappendiculata Kult. 320.
— subcandicans Tscherneva 230.
— submutica Franch. 325.
— sylvicola Bge. 272.
— syrdariensis Kult. 202.
— talassica (Kult.) Juz. 154.
— Tamarae Juz. 326.
— tedshenica Tscherneva 274.
— tenella Fisch. et Mey. 333.
— tenuisecta Juz. 202.
— tianschanica Kult. 303.
— tomentella C. Winkl. 179.
— trachyphylla Juz. 339.
— transiliensis Juz. 329.
— transoxana Tscherneva 208.
— triceps Kult. 292.
— trichopora Kult. 345.
— triflora Schrenk 242; tab. 17.
— turcomanica C. Winkl. 145.
— turkestanica (Rgl.) Juz. 203.
— turkmenorum Bornm. 165.
— ulotoma Bornm. 258; tab. 20.
— umblicata Juz. 204.
— umbrosa Bge. 178; tab. 12.
— Vavilovii Kult. 249.
— verticillaris Bge. 353.
— vicaria Kult. 198.
— Vvedenskyi Tscherneva 323.
— Waldheimiana Bornm. 330.
— xanthina Bornm. 338.
— xanthiocephala Tscherneva 156.
Cousiniopsis Nevski 27, 57.
— atractyloides (C. Winkl.) Nevski 58; tab. 4.
Covola Briq. 21, 359.
Craccina Stev. 12, 444.
Craepalia Aschers. et Graebn. 2, 546.
Crambe L. 8, 474.
— aculeolata (N. Busch) Czerniak. 488, 651.
— armena N. Busch 486.
— aspera M. B. 480.
— cordifolia Stev. 477.
— edentula Fisch. et Mey. 486.
— gibberosa Rupr. 480.
— Gordjaginii Spryg. et M. Pop. 489.
— grandiflora DC. 485.
— juncea M. B. 487.
— koktebelica (Junge) N. Busch 487.
— Kotschyana Boiss. 477; tab. 27.

— Litwinowii Grossh. 483.
— maritima L. 478; tab. 27.
— orientalis L. 489.
— pinnatifida R. Br. 479.
— schugnanica Korsh. 490.
— Steveniana Rupr. 485.
— tataria Sebeók 484.
Craniospermum Lehm. 19, 531.
— canescens DC. 534.
— echioides (Schrenk) Bge. 535.
— subfloccosum Kryl. 534.
— subvillosum Lehm. 533.
Crassina Scepin 26, 532.
Crataegus L. 9, 416.
— altaica Lange 424; tab. 28.
— ambigua C. A. Mey. 443; tab. 29.
— armena Pojark. 467, 509.
— atrosanguinea Pojark. 448, 504.
— Beckeriana Pojark. 453, 505.
— caucasica C. Koch 447.
— chlorosarca Maxim. 429; tab. 28.
— dahurica Koehne ex Schneid. 423; tab. 28.
— dipyrena Pojark. 462, 508.
— dsungarica Zab. 461.
— eriantha Pojark. 439, 500.
— kyrtostyla Fingerh. 450; tab. 30.
— Maximowiczii Schneid. 428.
— Meyeri Pojark. 438, 500; tab. 29.
— microphylla C. Koch 454; tab. 30.
— monogyna Jacq. 459; tab. 30.
— orientalis Pall. ex M. B. 433; tab. 29.
— oxyacantha L. 437; tab. 30.
— pentagyna Waldst. et Kit. 430.
— pinnatifida Bge. 421; tab. 28.
— pontica C. Koch 435; tab. 29.
— pseudoambigua Pojark. 466, 509.
— pseudoazarolus M. Pop. 465.
— pseudoheterophylla Pojark. 456, 506; tab. 30.
— pseudomelanocarpa M. Pop. 432, 498.
— remotilobata Raik. ex M. Pop. 428.
— sanguinea Pall. 422; tab. 28.
— Schraderiana Ledeb. 464.
— songorica C. Koch 449; tab. 29.
— sphaenophylla Pojark. 441, 502.
— Stevenii Pojark. 453, 505; tab. 30.
— Szovitsii Pojark. 434, 499.
— taurica Pojark. 440, 501.
— tianschanica Pojark. 461, 507.
— transcaspica Poiark. 447, 503; tab. 29.
— turcomanica Pojark. 457, 507; tab. 30.
— turkestanica Pojark. 451, 504.
— ucrainica Pojark. 444, 502.
— volgensis Pojark. 444, 502.
— zangezura Pojark. 463, 508.
Crawfurdia Wall. 18, 536.
— japonica Sieb. et Zucc. 537.
— volubilis (Maxim.) Makino 537.
Crepidium Tausch 29, 595, 697.
Crepinia Reichb. 29, 699, 708.
Crepis auct. 29, 699.
Crepis L. 29, 594.
— alaica Krasch. 653.
— Alikeri Tamamsch. 646, 756.
— alpina L. 688.
— astrachanica Stev. 671, 759.
— biennis L. 637.
— Bungei Ledeb. 665.
— burejensis Fr. Schmidt 618.
— capillaris (L.) Wallr. 682.
— caucasica C. A. Mey. 612.
— caucasigena Czer. 663, 758; tab. 32.
— chrysantha (Ledeb.) Turcz. 617.
— ciliata C. Koch 636.
— conyzifolia (Gouan) Dalla Torre 629.
— corniculata Rgl. et Schmalh. 650; tab. 31.
— crocea (Lam.) Babc. 648.
— czuensis Serg. 621.
— darvazica Krasch. 632.
— flexuosa (Ledeb.) Clarke 651.
— foetida L. 689.
— glabra Boiss. 613.
— Gmelinii (L.) Tausch 678.
— hokkaidoensis Babc. 615.
— karakuschensis Czer. 622, 755.
— Karelinii M. Pop. et Schischk. 656, 757; tab. 31.
— khorassanica Boiss. 642.
— Kotschyana Boiss. 693.
— lactea Lipsch. 658; tab. 31.
— lyrata (L.) Froel. 623.
— Marschallii (C. A. Mey.) F. Schultz 696; tab. 33.
— micrantha Czer. 684.
— mollis (Jacq.) Aschers. 624.
— multicaulis Ledeb. 679.
— nana Richardson 659.
— nigrescens Pohle 672.
— oreades Schrenk 647.
— paludosa (L.) Moench 610.
— pannonica (Jacq.) C. Koch 639.
— polytricha (Ledeb.) Turcz. 620.
— pontica C. A. Mey. 627; tab. 32
— praemorsa (L.) Tausch 661.
— pulchra L. 674.
— ramosissima Urv. 669.
— rhoeadifolia M. B. 690.

Crepis setosa Hall. f. 698.
— sibirica L. 608.
— sogdiana (Krasch.) Czer. 657; tab. 30.
— sonchifolia (M. B.) C. A. Mey. 634.
— tectorum L. 667.
— trichocephala (Krasch.) V. V. Nikit. 692; tab. 33.
— turcomanica Krasch. 643.
— Willdenowii Czer. 645.
— willemetioides Boiss. 626.
Cressa L. 19, 2.
— cretica L. 2.
Crinita Moench 25, 173.
Crinitaria Cass. 25, 173.
Critamus Bess. 16, 382.
Critesion Raf. 2, 721.
— jubatum (L.) Nevski 721; tab. 50.
Crithmum L. 16, 468.
— maritimum L. 468.
Crithodium Nevski 2, 677.
Crocus L. 4, 499.
— Adamii J. Gay 509.
— alatavicus Rgl. et Semen. 505.
— artvinensis (Philipp.) Grossh. 510.
— aureus Sibth. et Smith 505.
— Autranii Albov 502.
— caspicus Fisch. et Mey. 505.
— Heuffelianus Herb. 503; tab. 31.
— karsianus Fomin 502.
— Korolkowii Rgl. et Maw 506; tab. 31.
— Michelsonii B. Fedtsch. 506, 749.
— Pallasii Goldb. 503; tab. 31.
— Roopiae Woronow 510.
— Scharojanii Rupr. 502.
— speciosus M. B. 510; tab. 31.
— susianus Ker-Gawl. 504.
— Suworowianus C. Koch 501.
— tauricus (Trautv.) Puring 509.
— vallicola Herb. 501; tab. 31.
— variegatus Hoppe et Hornsch. 504; tab. 31.
Crossophyllum Spach 15, 203.
Crossostephium auct. 26, 315.
Crotalaria L. 11, 47.
— juncea L. 47.
Crucianella L. 23, 177.
— angustifolia L. 187.
— baldshuanica Krasch. 185; tab. 10.
— bucharica B. Fedtsch. 179.
— chlorostachys Fisch. et Mey. 186; tab. 10.
— divaricata Korov. 184; tab. 10.
— exasperata Fisch. et Mey. 185; tab. 10.
— filifolia Rgl. et C. Winkl. 183; tab. 9.
— gilanica Trin. 189.
— latifolia L. 188.
— sabulosa Korov. et Krasch. 180, 688.
— Schischkinii Lincz. 178, 687; tab. 9.
— Sintenisii Bornm. 191; tab. 10.
— suaveolens C. A. Mey. 190; tab. 10.
Crupina Cass. 28, 252.
— crupinastrum (G. Moris) Vis. 253.
— vulgaris Cass. 254.
Crypsis Ait. 2, 121.
— aculeata (L.) Ait. 122, tab. 8.
— alopecuroides (Pill. et Mitt.) Schrad. 125; tab. 8.
— Borszczowii Rgl. 126; tab. 8.
— schoenoides (L.) Lam. 125; tab. 8.
— turkestanica Eig. 125; tab. 8.
Cryptoceras Schott 7, 651.
Cryptocodon Fed. 24, 422, 474.
— monocephalus (Trautv.) Fed. 425; tab. 25.
Cryptodiscus Schrenk 16, 260.
— ammophilus Bge. 260.
— arenarius Schischk. 263, 594.
— cachroides Schrenk 262; tab. 19.
— didymus (Rgl.) Korov. 261; tab. 30.
Cryptogramma R. Br. 1, 77.
— acrostichoides R. Br. 78; tab. 2.
— crispa (L.) R. Br. 77.
— Raddeana Fomin 78; tab. 3.
— Stelleri Prantl 78.
Cryptospora Kar. et Kir. 8, 284.
— falcata Kar. et Kir. 285, tab. 17.
Cryptotaenia DC. 16, 380.
— Flahaultii (Woronow) K.-Pol. 380.
Ctenadena Prokh. 14, 476.
Cucubalus L. 6, 729.
— baccifer L. 729; tab. 45.
Cucumis L. 24, 114.
— sativus L. 114.
Cucurbita L. 24, 120.
— maxima Duchartre 122.
— moschata Duchartre 122.
— pepo L. 121.
Cuminia J. F. Gmel. 16, 369.
Cuminum L. 16, 369.
— cyminum L. 369.
Cummin Hill 16, 369.
Cupressus L. 1, 194.
— sempervirens L. 194; tab. 7.
Cuscuta L. 19, 38.
— alba J. et C. Presl 44.
— appoximata Bab. 49.
— araratica Butk. 55, 703.
— australis R. Br. 60.
— babylonica Auch. ex Choisy 43.
— basarabica Buia 62.

— brevistyla H. Br. ex A. Rich. 55.
— bucharica Palib. 70.
— callinema Butk. 50.
— campestris Yunck. 61.
— chinensis Lam. 59.
— colorans Maxim. 74.
— convallariiflora Pavl. 76.
— cupulata Engelm. 49.
— Engelmanii Korsh. 72.
— epilinum Weihe 54.
— epithymum Murr. 45.
— europaea L. 52.
— ferganensis Butk. 51.
— gigantea Griff. 75.
— Gronovii Willd. ex Roem. 64.
— indica (Engelm.) Petr. 52; tab. 3.
— japonica Choisy 73; tab. 3.
— karatavica Pavl. 63.
— kotschyana Boiss. 57.
— Lehmanniana Bge. 69.
— lophosepala Butk. 68.
— lupuliformis Krock. 71.
— monogyna Vahl **67**.
— pamirica Butk. 75.
— pedicellata Ledeb. 56.
— Stapfiana Palib. 58.
— stenocalycina Palib. 58.
— suaveolens Ser. 63.
— tianschanica Palib. 69.
— trifolii Bab. 46; tab. 3.
Cutandia Willk. 2, 339.
— memphitica (Spreng.) Richt. 340; tab. 27.
Cuviera Koel. 2, 720.
— europea (L.) Koel. 720; tab. 50.
Cyanus Juss. 28, 370, 387.
Cyathidium Lindl. 27, 361.
Cyathocephalum Nakai 26, 788.
Cycas L. 1, 129.
— revoluta Thunb. 129.
Cyclachaena Fresen. 25, 515.
— xanthifolia (Nutt.) Fresen. 515.
Cyclamen L. 18, 279.
— abchasicum (Medw.) Kolak. 288; tab. 15.
— adsharicum Pobed. 288.
— circassicum Pobed. 285; tab. 15.
— coum Mill. 286; tab. 15.
— elegans Boiss. et Buhse 289; tab. 15.
— parviflorum Pobed. 285; tab. 15.
— ponticum (Albov) Pobed. 281; tab. 15.
— verum Sw. 287.
Cyclophorus Desv. 1, 86.
— lingua (Thunb.) Desv. 86.
Cydonia Mill. 9, 334.
— oblonga Mill. 334; tab. 27.
Cyclactis Raf. 10, 12.
Cylindrocarpa Rgl. 24, 426.
— Sewerzowii Rgl. 426; tab. 25.
Cymatocarpus O. E. Schulz 8, 80.
— Grossheimii N. Busch 81.
— heterophyllus (M. Pop.) N. Busch 82.
— pilosissimus (Trautv.) O. E. Schulz 81.
Cymbalaria Medic. 22, 175.
— muralis Gaertn., Mey. et Scherb. 175.
Cymbaria L. 22, 801.
Cymbaria L. 22, 799.
— dahurica L. 800.
Cymbocarpum DC. 17, 57.
— anethoides DC. 58.
— erythraeum (DC.) Boiss. 58.
— Wiedemannii Boiss. 62; tab. 2.
Cymbochasma (Endl.) Klok. et Zoz 22, 801.
— borysthenica (Pall.) Klok. et Zoz 802.
Cymbolaena Smoljan. 25, 305.
— longifolia (Boiss. et Reut.) Smoljan. 306; tab. 18.
Cymboseris Boiss. 29, 595, 673.
Cyminon St. Lager 16, 369.
Cyminum Hill 16, 369.
Cynanchica Fourr. 23, 218.
Cynanchum L. 18, 714.
— acutum L. 715.
— Maximoviczii Pobed. **717**; tab. 39.
— sibiricum Willd. 716.
— stellatum Pobed. **717**, **754**; tab. 39.
Cynara L. 28, 225.
— scolymus L. 226.
Cynocrambe Gaertn. 6, 372.
Cynoctonum F. Mey. 18, 709.
— purpureum (Pall.) Pobed. 709; tab. 39.
Cynodon Rich. 2, 285.
— dactylon (L.) Pers. 285; tab. 22.
Cynoglossum L. 19, 657.
— divaricatum Steph. 670.
— holosericeum Stev. 672; tab. 39.
— montanum L. 668.
— nebrodense Guss. 667.
— officinale L. 671.
— pictum Ait. 664.
— seravschanicum (B. Fedtsch.) M. Pop. 661; tab. 38.
— tianschanicum M. Pop. 663.
— viridiflorum Pall. ex Lehm. 662.
Cynomorium L. 15, 671.
— songaricum Rupr. 671.
Cynosurus L. 2, 362.
— cristatus L. 363; tab. 27.
— echinatus L. 363; tab. 27.
— elegans Desf. 364; tab. 27.

Cynoxylon Raf. 17, 319.
— capitata (Wall.) Nakai 323; tab. 25.
— florida (L.) Raf. 320.
Cyperus L. 3, 13.
— amuricus Maxim. 21.
— aureus Ten. 21.
— badius Desf. 20.
— difformis L. 16.
— esculentus L. 21.
— flavidus Retz. 17.
— fuscus L. 15; tab. 1.
— glaber L. 17.
— glomeratus L. 18.
— haspan L. 16.
— iria L. 18.
— longus L. 19; tab. 1.
— rotundus L. 20.
— soongoricus Kar. et Kir. 15.
— truncatus Turcz. 18.
Cyphomattia Boiss. 19, 565, 573.
Cypripedium L. 4, 595.
— calceolus L. 598.
— guttatum Sw. 596; tab. 37.
— macranthum Sw. 598; tab. 37.
— Yatabeanum Makino 597; tab. 37.
Cyssopetalum Turcz. 16, 526, 536.
Cystidospermum Prokh. 14, 480.
Cystopteris Bernh. 1, 24.
— Dickieana Sims 25; tab. 1.
— fragilis (L.) Bernh. 24; tab. 1.
— montana (Lam.) Bernh. 25.
— regia (L.) C. Presl 25, tab. 1.
— sudetica A. Br. et Milde 26.
Cytherea Salisb. 4, 604.
Cytinus L. 5, 441.
— rubra (Fourr.) Kom. 442; tab. 23.
Cytisus L. 11, 75.
— aggregatus Schur 92.
— albus Hacq. 90; tab. 6.
— austriacus L. 86; tab. 6.
— Blockii V. Krecz. 86.
— borysthenicus Grun. 78; tab. 6.
— hirsutissimus C. Koch 83.
— Lindemannii V. Krecz. 79.
— Litwinowii V. Krecz. 89.
— Paczoskii V. Krecz. 82.
— podolicus Blocki 90.
— polytrichus M. B. 85.
— ratisbonensis Schaeff. 79.
— Rochelii Wierzb. 91.
— ruthenicus Fisch. ex Bess. 82.
— Skrobiszewskii Pacz. 92.
— Wulffii V. Krecz. 84.
— Zingeri (Nenuk.) V. Krecz. 80; tab. 6.
Cyttarium Peterm. 25, 381, 404.
Czernaevia Turcz. 17, 11, 26.

Dactylis L. 2, 361.
— glomerata L. 361; tab. 27.
— Woronowii Ovcz. 362, 752.
Dactylorhiza Neck. 4, 697.
Dahlia Cav. 25, 548.
— pinnata Cav. 548.
Damasonium Juss. 1, 285.
— alisma Mill. 285; tab. 14.
— constrictum Juz. 286; tab. 14.
Danaa All. 16, 223.
— denaensis (B. Fedtsch.) Schischk. 224, 593; tab. 15.
— nudicaulis (M. B.) Grossh. 223; tab. 16.
Danaë Medic. 4, 442.
— racemosa (L.) Moench 443; tab. 26.
Danthonia DC. 2, 283.
— calycina (Vill.) Reichb. 284.
— Forskahlei (Vahl) Trin. 284.
— intermedia Vasey 284.
Daphne L. 15, 483.
— Albowiana Woronow 488; tab. 23.
— altaica Pall. 490.
— angustifolia C. Koch 494.
— axilliflora (Keissl.) Pobed. 492; tab. 23.
— baksanica Pobed. 497; tab. 23.
— Blagayana Freyer 499.
— caucasica Pall. 491; tab. 23.
— circassica Woronow 498; tab. 23.
— cneorum L. 499
— glomerata Lam. 489.
— Julia K.-Pol. 500.
— kamtschatica Maxim. 493.
— laureola L. 488.
— mezereum L. 486.
— pontica L. 487; tab. 23.
— pseudosericea Pobed. 498.
— Sophia Kalen. 491.
— transcaucasica Pobed. 497; tab. 23.
Dargeria Decne. 22, 526.
Dasiphora Raf. 10, 68.
— davurica (Nestl.) Kom. et Aliss. 72.
— dryadanthoides Juz. 72, 608.
— fruticosa (L.) Rydb. 69; tab. 7.
— mandshurica (Maxim.) Juz. 73.
— parvifolia (Fisch.) Juz. 71; tab. 7.
— phyllocalyx Juz. 70, 607.
Dasyanthus Bub. 25, 381.
Dasyloma DC. 16, 526, 536.
Dasyspermum Neck. 16, 93, 153.
Datisca L. 15, 480.
— cannabina L. 480.

Datura L. 22, 109.
— innoxia Mill. 114.
— metel L. 112.
— stramonium L. 109.
— tatula L. 111.
Daucalis Pomel 16, 153, 163.
Daucus L. 17, 288.
— carota L. 288.
— sativus (Hoffm.) Roehl. 291.
Decaneurum Sch. Bip. 26, 137.
Deeringia Kuntze 16, 380.
Delia Dum. 6, 562.
Delphinium L. 7, 99.
— aemulans Nevski 161, 725.
— Ajacis L. 104.
— albomarginatum Simonova 178.
— altaicum Nevski 161, 726.
— arcuatum N. Busch 139.
— barbatum Bge. 102.
— Batalinii Huth 169.
— biternatum Huth 182; tab. 11.
— brachycentrum Ledeb. 163.
— bracteosum Somm. et Lev. 138.
— Brunonianum Royle 123; tab. 9.
— bucharicum M. Pop. 170.
— Buschianum Grossh. 143.
— camptocarpum Fisch. et Mey. 106; tab. 8.
— caucasicum C. A. Mey. 126.
— cheilanthum Fisch. ex DC. 162; tab. 10.
— confusum M. Pop. 136, 723.
— consolida L. 102; tab. 8.
— corymbosum Rgl. 155.
— crassifolium Schrad. ex Spreng. 154; tab. 10.
— crispulum Rupr. 143.
— cryophilum Nevski 146, 724.
— cuneatum Stev. ex DC. 148; tab. 10.
— cyananthum Nevski 160, 725.
— dasyanthum Kar. et Kir. 132; tab. 9.
— dasycarpum Stev. ex DC. 127.
— dictyocarpum DC. 159; tab. 10.
— divaricatum Ledeb. 104; tab. 8.
— elatum L. 147; tab. 10.
— flexuosum M. B. 141.
— foetidum Lomak. 125.
— Freynii Conr. 179; tab. 11.
— grandiflorum L. 164.
— Hohenackeri Boiss. 101; tab. 8.
— iliense Huth 155; tab. 9.
— inconspicuum Serg. 145.
— inopinatum Nevski 173; 727.
— karategini Korsh. 170.
— Knorringianum B. Fedtsch. 165.
— Korshinskyanum Nevski 153, 724.
— Lacostei Danguy 124; tab. 9.
— laxiflorum DC. 158.
— laxiusculum (Boiss.) Rouy 175.
— leiocarpum Huth 178.
— leptocarpum Nevski 110; tab. 8.
— Lipskyi Korsh. 167.
— longipedunculatum Rgl. et Schmalh. 167.
— Maackianum Rgl. 156.
— Mariae N. Busch 144.
— megalanthum Nevski 129, 722.
— mirabile Serg. 131.
— ochotense Nevski 146, 724; tab. 10.
— ochroleucum Stev. ex DC. 181.
— oreophilum Huth 136; tab. 9.
— orientale J. Gay 105.
— osseticum N. Busch 140.
— Pallasii Nevski 177, 727.
— paniculatum Host 103; tab. 8.
— paradoxum Bge. 113.
— persicum Boiss. 111.
— Poltoratzkii Rupr. 135; tab. 9.
— propinquum Nevski 124, 722.
— puniceum Pall. 176.
— pyramidatum Albov 142.
— quercetorum Boiss. et Hausskn. 175.
— retropilosum (Huth) Sambuk 152; tab. 10.
— rugulosum Boiss. 112; tab. 8.
— Ruprechtii Nevski 138, 723.
— sauricum Schischk. 130.
— Schmalhausenii Albov 179.
— semibarbatum Bienert ex Boiss. 182; tab. 11.
— semiclavatum Nevski 168, 726.
— songoricum (Kar. et Kir.) Nevski 109.
— speciosum M. B. 128.
— Stocksianum Boiss. 111, tab. 8.
— Szovitsianum Boiss. 180.
— ternatum Huth 166.
— triste Fisch. ex DC. 157; tab. 10.
— turkmenum Lipsky 174.
— uralense Nevski 159, 725.
— villosum Stev. ex DC. 152.
Dematra Raf. 14, 476.
Dendranthema (DC.) Des Moul. 26, 364.
— arcticum (L.) Tzvel. 386.
— erubescens (Stapf) Tzvel. 374.
— Hultenii (A. et'D. Löve) Tzvel. 387; tab. 16.
— integrifolium (Richards.) Tzvel. 388; tab. 16.
— kurilense Tzvel. 385.
— littorale (Maekawa) Tzvel. 384.
— Maximowiczii (Kom.) Tzvel. 379.

Dendranthema mongolicum (Ling) Tzvel. 378.
— naktongense (Nakai) Tzvel. 375.
— sichotense Tzvel. 380, 879; tab. 16.
— sinuatum (Ledeb.) Tzvel. 370.
— Weyrichii (Maxim.) Tzvel. 383.
— Zawadskii (Herbich) Tzvel. 376.
Dendrobenthamia Hutch. 17, 323.
Dendrostellera Van-Tiegh. 15, 511.
— arenaria Pobed. 512, 689.
— linearifolia Pobed. 512, 689.
— macrorhachis Pobed. 513, 690; tab. 24.
— Olgae Pobed. 514, 691.
— stachyoides (Schrenk) Van-Tiegh. 511.
— turkmenorum Pobed. 514, 691.
Dentaria L. 8, 144.
— bipinnata C. A. Mey. 151; tab. 8.
— bulbifera L. 147; tab. 7.
— glandulosa Waldst. et Kit. 148; tab. 7.
— microphylla Willd. 151; tab. 8.
— quinquefolia M. B. 148; tab. 7.
— sibirica (O. E. Schulz) N. Busch 151.
— tenuifolia Ledeb. 152; tab. 8.
Dentidia Lour. 21, 630.
Derderia Jaub. et Spach 27, 643.
Deriandra Cambess. 6, 545.
Deringa Adans. 16, 380.
Derouetia Boiss. et Bal. 29, 699, 703.
Deschampsia Beauv. 2, 243.
— alpina (L.) Roem. et Schult. 248.
— arctica (Spreng.) Schischk. 248.
— beringensis Hult. 251; tab. 18.
— borealis (Trautv.) Roshev. 246, 750.
— bottnica (Wahl.) Trin. 247; tab. 18.
— caespitosa (L.) Beauv. 245; tab. 18.
— flexuosa (L.) Trin. 244; tab. 18.
— koelerioides Rgl. 251; tab. 18.
— media (Gouan) Roem. et Schult. 245.
— obensis Roshev. 247.
— pamirica Roshev. 252; 750.
— Sukatschewii (Popl.) Roshev. 246; tab. 18.
— Turczaninowii Litv. 246; tab. 18.
Descurainia Webb et Berth. 8, 82.
— Kochii (Petri) O. E. Schulz 84.
— Sophia (L.) Schur 83.
— sophioides (Fisch.) O. E. Schulz 84.
Descurea Guett 8, 82.
Desmiograstis Boern. 3, 149.
Desmodium Desv. 13, 374.
— manschuricum (Maxim.) Nakai 374.
— Oldhami Oliv. 375.
Deutzia Thunb. 9, 223.
— amurensis (Rgl.) Airy-Show 224; tab. 1[2]
— glabrata Kom. 224.
Deyeuxia Clar. 2, 202.
Dianthus L. 6, 803.
— acantholimonoides Schischk. 837; tab. 53.
— acicularis Fisch. ex Ledeb. 849; tab. 54.
— amurensis Jacq. 827.
— Andronakii Woronow 841.
— Andrzejowskianus (Zapal.) Kulcz. 813; tab. 51.
— arenarius L. 848.
— armeria L. 812; tab. 51.
— azkurensis Sosn. 845.
— barbatus L. 821.
— bicolor Adam 829; tab. 52.
— Borbasii Vand. 820; tab. 51.
— calocephalus Boiss. 819; tab. 51.
— campestris M. B. 830; tab. 52.
— canescens C. Koch 854.
— capitatus Balb. 813, tab. 51.
— carbonatus Klok. 830.
— carthusianorum L. 814; tab. 51.
— caryophyllus L. 847.
— chinensis L. 823.
— crinitus Smith 850; tab. 54.
— crossopetalus Fenzl 850.
— cyri Fisch. et Mey. 860; tab. 54.
— deltoides L. 838.
— discolor Smith 825.
— elatus Ledeb. 833.
— Eugeniae Kleop. 834.
— eu-ponticus Zapal. 818; tab. 51.
— Fischeri Spreng. 824.
— floribundus Boiss. 845.
— fragrans Adam 846.
— Grossheimii Schischk. 838.
— guttatus M. B. 829; tab. 52.
— Hoeltzeri C. Winkl. 856.
— humilis Willd. ex Ledeb. 837; tab. 53.
— hypanicus Andrz. 833.
— imereticus (Rupr.) Schischk. 827.
— inamoensis Schischk. 842, 897.
— karataviensis Pavl. 852, 898.
— kirghizicus Schischk. 837, 896.
— kubanensis Schischk. 828; tab. 52.
— Kuschakewiczii Rgl. et Schmalh. 851.
— Kusnezovii Marc. 826.
— lanceolatus Stev. ex Reichb. 839; tab. 53.
— leptopetalus Willd. 840· tab. 52.
— libanotis Labill. 848.
— macronyx Fenzl 855.
— Marschallii Schischk. 841; tab. 53
— membranaceus Borb. 817; tab. 51.
— multicaulis Boiss. et Huet 846.
— orientalis Adam 854.

— pallidiflorus Ser. 835.
— petraeus M. B. 845.
— polymorphus M. B. 820; tab. 51.
— pratensis M. B. 828; tab. 52.
— Preobrashenskii Klok. 834.
— pseudoarmeria M. B. 812; tab. 51.
— Raddeanus Vierh. 823.
— ramosissimus Pall. 835.
— recognitus Schischk. 851.
— repens Willd. 822.
— rigidus M. B. 836; tab. 53.
— Rogowiczii Kleop. 817.
— Ruprechtii Schischk. 818, 896.
— schemachensis Schischk. 840.
— Semenovii (Rgl. et Herd.) Vierh. 825; tab. 52.
— seravschanicus Schischk. 852, 898.
— soongoricus Schischk. 853, 899.
— squarrosus M. B. 849.
— stenocalyx (Trautv.) Juz. 859; tab. 54.
— subulosus Freyn et Conr. 814; tab. 51.
— superbus L. 856; tab. 54.
— tabrisianus Bienert ex Boiss. 855.
— talyschensis Boiss. et Buhse 847.
— tetralepis Nevski 854, 829.
— tianschanicus Schischk. 851, 897.
— transcaucasicus Schischk. 819.
— turcomanicus Schischk. 853, 899.
— turkestanicus Preobr. 827.
— uralensis Korsh. 836; tab. 53.
— versicolor Fisch. 824.
Diapensia L. 18, 105.
— lapponica L. 105.
— obovata (Fr. Schmidt) Nakai 106.
Diaphane Salisb. 4, 554.
Diarrhena Beauv. 2, 352.
— japonica Franch. et Savat. 353; tab. 27.
— manshurica Maxim. 352; tab. 27.
Diarthron Turcz. 15, 500.
— linifolium Turcz. 501.
— vesiculosum (Fisch. et Mey.) C. A. Mey. 501.
Diatropa Dum. 16, 283.
Diaxulon Raf. 11, 78.
Dicentra Bernh. 7, 648.
— peregrina (Rudolphi) Fedde 648; tab. 45.
Dichodon Bartl. 6, 435.
Dichoglottis Fisch. et Mey. 6, 772.
Dichostylis Beauv. 3, 59.
— hamulosa (M. B.) Nees 63; tab. 5.
— Micheliana (L.) Nees 59; tab. 5.
— pygmaea (Rottb.) Nees 60, tab. 5.
Dichrocephala L'Hér. 25, 51.
— bicolor (Roth) Schlecht. 51; tab. 2.
Diclytra Borckh. 7, 648.
Diclytra DC. 7, 648.
Dictamnus L. 14, 227.
— angustifolius G. Don 230.
— caucasicus Fisch. ex Grossh. 231.
— dasycarpus Turcz. 229.
— gymnostylus Stev. 231; tab. 11.
— tadshikorum Vved. 229.
Didymopelta Taub. 12, 317.
Didymophysa Boiss. 8, 573.
— Aucheri Boiss. 573.
— Fedtschenkoana Rgl. 574.
Dielsiocharis O. E. Schulz 8, 56.
Digitalis L. 22, 514.
— ciliata Trautv. 521.
— ferruginea L. 522.
— grandiflora Mill. 520.
— lanata Ehrh. 525.
— nervosa Steud. et Hochst. 524.
— purpurea L. 518.
— Schischkinii Ivanina 523.
Digitaria Heist. ex Adans. 2, 27.
— ciliaris (Retz.) Koel. 30; tab. 2.
— horizontalis Willd. 29.
— linearis (Krock.) Crép. 28; tab. 2.
— longiflora Pers. 29.
— sanguinalis (L.) Scop. 29; tab. 2.
Digraphis Trin. 2, 54.
— arundinacea (L.) Trin. 55; tab. 4.
Dilophia Thoms. 8, 574.
— salsa Thoms. 575.
Dimorphanthus Miq. 16, 25.
Diolotus Tausch 11, 51.
Dionysia Fenzl 18, 208.
— hissarica Lipsky 213; tab. 11.
— involucrata Zapr. 214; tab. 12.
— Kossinskyi Czerniak. 210; tab. 12.
— tapetodes Bge. 209; tab. 12.
Dioscorea L. 4, 495.
— batatus Decne. 496; tab. 29.
— caucasica Lipsky 495; tab. 29.
— polystachya Turcz. 497; tab. 29.
Diospyros L. 18, 475.
— kaki L. f. 477; tab. 24.
— lotus L. 481; tab. 24.
— virginiana L. 482; tab. 24.
Diotis Desf. 26, 127.
Dipelta Rgl. et Schmalh. 12, 317.
Diphryllum Raf. 4, 611.
Diphylleia Rich. 7, 541.
— Grayi Fr. Schmidt 541; tab. 37.
Diplachne Beauv. 2, 309.
— bulgarica (Bornm.) Roshev. 310, 751; tab. 23.

Diplachne serotina (L.) Link 310; tab. 23.
— sinensis Hance 311; tab. 23.
— songorica Roshev. 311, 752; tab. 23.
— squarrosa (Trin.) Maxim. 310; tab. 23.
Diploloma Schrenk 19, 531.
Diplopappus DC. 25, 110.
Diplophyllum Lehm. 22, 329, 392, 412.
Diplotaxis DC. 8, 456.
— cretacea Kotov 458.
— muralis (L.) DC. 458.
— tenuifolia (L.) DC. 457.
— viminea (L.) DC. 457.
Dipsacus L. 24, 19.
— azureus Schrenk 23; tab. 2.
— fullonum L. 21.
— Gmelinii M. B. 23; tab. 2.
— laciniatus L. 21.
— pilosus L. 24; tab. 1.
— sativus (L.) Honck. 22.
— strigosus Willd. 25.
Diptera Borkh. 9, 198.
Diptychocarpus Trautv. 8, 309.
— strictus (Fisch.) Trautv. 310; tab. 17.
Disarenum Labill. 2, 59.
Discrneston Jaub. et Spach 17, 155.
Disperma J. F. Gmel. 23, 169.
Disporum Salisb. 4, 454.
— viridescens B. Fedtsch. 454.
Distegia Raf. 23, 479.
Ditrita Raf. 14, 482.
Dodartia L. 22, 318.
— orientalis L. 319.
Dodecatheon L. 18, 290.
— frigidum Cham. et Schlecht. 291.
Doellingeria Nees 25, 121.
— scabra (Thunb.) Nees 121; tab. 3.
Donia R. Br. 5, 444.
Dontostemon Andrz. 8, 306.
— dentatus Ledeb. 307.
— hispidus Maxim. 307.
— integrifolius (L.) C. A. Mey. 308.
— micranthus C. A. Mey. 308.
— pectinatus (DC.) Ledeb. 309.
— perennis C. A. Mey. 308.
Dopatrium Buch.-Ham. 22, 320.
— junceum (Roxb.) Buch.-Ham. 320.
Dorema D. Don 17, 155.
— Aitchisonii Korov. 158.
— glabrum Fisch. et Mey. 163.
— gummiferum (Jaub. et Spach) K. Korol. 162.
— hyrcanum K.-Pol. 163.
— karataviense Korov. 161.
— microcarpum Korov. 164.
— namanganicum K. Korol. 165.
— pruinosum K. Korol. 162.
— sabulosum Litv. 161; tab. 17.
Doronicum L. 26, 669.
— altaicum Pall. 672.
— austriacum Jacq. 676.
— bargusinense Serg. 673.
— carpaticum (Griseb. et Schenk) Nym. 675.
— Clusii (All.) Tausch 674.
— longifolium Griseb. et Schenk 680.
— macrophyllum Fisch. 678.
— oblongifolium DC. 681.
— orientale Hoffm. 677.
— pardalianches L. 679.
— Schischkinii Serg. 673.
— turkestanicum Cavill. 681.
Dornera Heuff. 3, 296.
Dortmannia Neck. 24, 451.
Dorycnium L. 11, 282.
— graecum (L.) Ser. 282; tab. 18.
— intermedium Ledeb. 283; tab. 18.
Draba L. 8, 371.
— alajica Litv. 412.
— Alberti Rgl. et Schmalh. 407.
— alpina L. 394; tab. 24.
— altaica (C. A. Mey.) Bge. 423; tab. 25.
— alticola Kom. 452; tab. 26.
— aradani N. Busch 404.
— araratica Rupr. 388.
— Arseniewii (B. Fedtsch.) Gilg 410.
— Aucheri Boiss. 453.
— baicalensis Tolm. 421, 650; tab. 25.
— barbata Pohle 386.
— Behringii Tolm. 393, 649; tab. 23.
— borealis DC. 439.
— bruniifolia Stev. 381; tab. 23.
— bryoides DC. 380.
— camtschatica Andrz. 417.
— cardaminiflora Kom. 436; tab. 25.
— Chamissonis G. Don 418; tab. 24.
— cinerea Adam 419; tab. 25.
— cuspidata M. B. 378; tab. 23.
— darwasica Lipsky 410; tab. 24.
— diversifolia Boiss. 382.
— Elisabethae N. Busch 444.
— eriopoda Turcz. 446.
— Eschscholtzii Pohle ex N. Busch 396; tab. 24.
— Fedtschenkoi (Pohle) Gilg 407.
— fladnizensis Wulf 403.
— glacialis Adam 396; tab. 24.
— globifera Ledeb. 381.
— heterocoma Fenzl 381.
— hirta L. 430.

— hispida Willd. 441; tab. 26.
— hissarica Lipsky 409; tab. 24.
— Huetii Boiss. 449.
— hyperborea (L.) Desv. 440.
— imeretica Rupr. 413.
— incana L. 435.
— incompta Stev. 444.
— juvenilis Kom. 429.
— Kjellmanii Lid ex Ekman 394; tab. 24.
— Korshinsky (O. Fedtsch.) Pohle 389; tab. 23.
— kurilensis (Turcz.) Fr. Schmidt 438.
— Kuznetzovii (Turcz.) Hayek 424.
— lactea Adam 403.
— lanceolata Royle 433.
— lasiophylla Royle 431; tab. 25.
— Lipskyi Tolm. 409, 650; tab. 24.
— longisiliqua Schmalh. 415; tab. 24.
— macrocarpa Adam 395.
— magna (N. Busch) Tolm. 442; tab. 26.
— melanopus Kom. 453; tab. 26.
— micropetala Hook. 390.
— mollissima Stev. 414; tab. 24.
— mongolica Turcz. 433.
— muralis L. 451.
— nemorosa L. 451; tab. 26.
— nivalis Liljebl. 416.
— norvegica Gunn. 422.
— oblongata R. Br. 390; tab. 23.
— ochroleuca Bge. 399; tab. 24.
— odudiana Lipsky 410.
— Olgae Rgl. et Schmalh. 400.
— oreades Schrenk 393; tab. 24.
— ossetica (Rupr.) Somm. et Lev. 414.
— pamirica (O. Fedtsch.) Pohle 389.
— parviflora O. E. Schulz 423.
— parvisiliquosa Tolm. 420.
— physocarpa Kom. 408.
— pilosa DC. 385; tab. 23.
— Pohlei Tolm. 395.
— polytricha Ledeb. 388; tab. 23.
— primuloides Turcz. 400.
— Prozorovskii Tolm. 429.
— pseudopilosa Pohle 402.
— pygmaea Turcz. 402.
— sachalinensis Fr. Schmidt 438; tab. 25.
— scabra C. A. Mey. 380.
— sibirica (Pall.) Thell. 445.
— siliquosa M. B. 405.
— stenocarpa Hook. f. et Thoms. 450.
— stenopetala Trautv. 386.
— stylaris J. Gay 434.
— subamplexicaulis C. A. Mey. 427.
— subcapitata Simm. 387.
— subglabra (Rupr.) Tolm. 405.
— subsecunda Somm. et Levier 443.
— supranivalis Rupr. 406.
— talassica Pohle 411.
— tibetica Hook. f. et Thoms. 412.
— Turczaninovii Pohle et N. Busch 418; tab. 24.
— ussuriensis Pohle 437.
Dracocephalum L. 20, 439.
— argunense Fisch. ex Link 471.
— austriacum L. 472.
— bipinnatum Rupr. 464; tab. 26.
— botryoides Stev. 448.
— Bungeanum Schischk. ex Serg. 446.
— discolor Bge. 447.
— diversifolium Rupr. 466; tab. 27.
— foetidum Bge. 462.
— formosum Gontsch. 452.
— fragile Turcz. ex Benth. 454.
— fruticulosum Steph. 456.
— grandiflorum L. 451.
— heterophyllum Benth. 465.
— imberbe Bge. 453.
— integrifolium Bge. 457.
— Komarovii Lipsky 469.
— Krylovii Lipsky 456.
— moldavica L. 463.
— multicaule Montbr. et Auch. 470.
— multicolor Kom. 455.
— nodulosum Rupr. 460; tab. 27.
— nutans L. 458.
— oblongifolium Rgl. 461.
— origanoides Steph. ex Willd. 445.
— palmatum Steph. ex Willd. 445; tab. 27.
— Paulsenii Briq. 448.
— peregrinum L. 464.
— pinnatum L. 444; tab. 26.
— Ruyschiana L. 472.
— scrobiculatum Rgl. 461; tab. 26.
— spinulosum M. Pop. 466.
— stamineum Kar. et Kir. 473.
— Stellerianum Hilteb. 454.
— subcapitatum (Kuntze) Lipsky 470.
— thymiflorum L. 459.
Drepanocaryum Pojark. 20, 228, 516.
— Sewerzowii (Rgl.) Pojark. 228; tab. 14.
Drepanophyllum Hoffm. 16, 458.
Drepanophyllum Wib. 16, 382.
Drosanthe Spach 15, 203.
Drosera L. 9, 2.
— anglica Huds. 5, tab. 1.
— intermedia Hayne 6; tab. 1.
— obovata Mert. et Koch 6; tab. 1.
— rotundifolia L. 2; tab. 1.

Drosocarpium Fourr. 15, 203.
Dryadanthe Endl. 10, 228.
— tetrandra (Bge.) Juz. 229; tab. 17.
Dryas L. 10, 264.
— ajanensis Juz. 275.
— caucasica Juz. 274; tab. 20.
— Chamissonis Spreng. 278; tab. 20.
— crenulata Juz. 277; tab. 20.
— grandis Juz. 265; tab. 20.
— Henricae Juz. 616.
— octopetala L. 269; tab. 20.
— oxyodonta Juz. 273; tab. 20.
— punctata Juz. 267; tab. 20.
— Tschonoskii Juz. 276; tab. 20.
— viscosa Juz. 266; tab. 20.
Drymocallis Fourr. 10, 93.
Drymocodon Fourr. 24, 133, 191.
Dryopteris Adans. 1, 30.
— Alexeenkoana Fomin 41.
— amurensis Christ 42.
— austriaca (Jacq.) Woynar 41.
— barbellata Fomin 37; tab. 2.
— Buschiana Fomin 35.
— continentalis V. Petrov 43.
— cristata (L.) A. Gray 39.
— filix-mas (L.) Schott 36.
— fragrans (L.) Schott 38.
— kamtschatica Kom. 35.
— Komarovii C. Koss. 39.
— laeta (Kom.) C. Chr. 42.
— Linnaeana C. Chr. 43.
— mediterranea Fomin 35.
— oreades Fomin 37.
— oreopteris (Ehrh.) Maxon 34.
— phegopteris (L.) C. Chr. 44.
— Raddeana Fomin 39.
— rigida (Hoffm.) Und. 38.
— Robertiana (Hoffm.) C. Chr. 44.
— sichotensis Kom. 37.
— spinulosa (Muell.) Kuntze 40.
— thelypteris (L.) A. Gray 33.
— wladiwostokensis B. Fedtsch. 42.
Duchesnea Smith 10, 67.
— indica Andr. 67; tab. 6.
Dufresnia DC. 23, 642.
Dulcamara Moench 22, 10.
Dupontia R. Br. 2, 432.
— Fischeri R. Br. 432; tab. 33.
Dysophylla Blume 21, 637.
— Jatabeana Makino 637.
Ebulum Garcke 23, 426.
Eckballium A. Rich. 24, 109.
— elaterium (L.) A. Rich. 110.
Echenais Cass. 28, 52, 139.
Echinacea Moench 25, 540.
— purpurea (L.) Moench 541.
Echinanthus Neck. 27, 2.
Echinaria Desf. 2, 298.
— capitata (L.) Desf. 298; tab. 23.
Echinochloa Beauv. 2, 31.
— caudata Roshev. 35.
— colona (L.) Beauv. 33.
— crus-galli (L.) Roem. et Schult. 32; tab. 2.
— frumentacea (Roxb.) Link 32.
— macrocarpa Vasing. 34, 739.
— oryzicola Vasing. 33.
— spiralis Vasing. 34, 739.
Echinocystis Torr. et Gray 24, 124.
— echinata (Muhl.) Vass. 125.
Echinopanax Decne. et Planch. 16, 17.
— elatum Nakai 18; tab. 2.
Echinophora L. 16, 88.
— trichophylla Smith 89; tab. 5.
— Sibthorpiana Guss. 90; tab. 5.
Echinops L. 27, 2.
— abstersibilis Iljin 46.
— albicaulis Kar. et Kir. 18.
— babatagensis Tscherneva 24.
— brevipenicillatus Tscherneva 12.
— chantavicus Trautv. 22.
— colchicus Sosn. ex Kem.-Nath. 43.
— connatus C. Koch 27.
— Conrathii Freyn 34.
— dagestanicus Iljin 41.
— dasyanthus Rgl. et Schmalh. 21.
— dissectus Kitag. 45.
— Dubjanskyi Iljin 19.
— erevanensis Mulk. 41.
— exaltatus Schrad. 47.
— Fedtschenkoi Iljin 25.
— foliosus Somm. et Lev. 35.
— galaticus Freyn 43.
— Gmelinii Turcz. 52.
— humilis M. B. 50; tab. 3.
— Iljinii Mulk. 37.
— integrifolius Kar. et Kir. 48; tab. 3.
— kafirniganus Bobr. 18, 714.
— karabachensis Mulk. 38.
— karatavicus Rgl. et Schmalh. 23.
— Knorringianus Iljin 53.
— latifolius Tausch 44.
— leiopolyceras Bornm. 14.
— leucographus Bge. 14.
— Lipskyi Iljin 16; tab. 1.
— macrophyllus Boiss. et Hausskn. 28; tab. 1.
— maracandicus Bge. 11.

— Meyeri (DC.) Iljin 32.
— multicaulis Nevski 17.
— nanus Bge. 52.
— obliquilobus Iljin 23.
— opacifolius Iljin ex Grossh. 26.
— orientalis Trautv. 25.
— ossicus C. Koch 44.
— polyacanthus Iljin 34.
— polygraphus Tscherneva 11.
— praetermissus Nevski 15.
— pubisquameus Iljin 20.
— pungens Trautv. 33; tab. 2.
— ritro L. 30.
— ritrodes Bge. 16.
— saissanicus (B. Keller) Bobr. 51.
— sevanensis Mulk. 38.
— spraerocephalus L. 40.
— spiniger Iljin ex Bobr. 32; tab. 2.
— subglaber Schrenk 20.
— Szovitsii Fisch. et Mey. ex DC. 36.
— tjanschanicus Bobr. 22, 714; tab. 1.
— Tournefortii Ledeb. ex Trautv. 39.
— transcaspicus Bornm. 10.
— transcaucasicus Iljin 36.
— tricholepis Schrenk 46.
— tschimganicus B. Fedtsch. 50.
— viridifolius Iljin 42.
Echinopsilon Moq. 6, 124.
— divaricatum Kar. et Kir. 126; tab. 5.
— hirsutum (L.) Moq. 127; tab. 5.
— hyssopifolium (Pall.) Moq. 125; tab. 5.
— sedoides (Pall.) Moq. 126; tab. 5.
Echinopsus St.-Lag. 27, 2.
Echinopus Adans. 27, 2.
Echinospermum Sw. ex Lehm. 19, 403.
Echium L. 19, 271.
— amoenum Fisch. et Mey. 274.
— italicum L. 277.
— plantagineum L. 275.
— rubrum Jacq. 273.
— vulgare L. 276.
Eclipta L. 25, 537.
— alba (L.) Hasskarl. 538; tab. 34.
— prostrata L. 537; tab. 34.
Edrajanthus A. DC. 24, 442.
— Owerinianus Rupr. 443; tab. 27.
Edritria Raf. 3, 335.
Edwardsia Salisb. 11, 25
Eichwaldia Ledeb. 15, 277.
Elaeagnus L. 15, 518.
— angustifolia L. 522; tab. 25.
— argentea Pursh 521.
— orientalis L. 524.
Elaeagrus Pall. 15, 518.
Elaeochytris Fenzl 17, 62.
Elatine L. 15, 261.
— alsinastrum L. 262; tab. 13.
— ambigua Wight 270.
— callitrichoides (Rupr.) Kauffm. 269.
— hexandra DC. 264.
— hungarica Moesz. 266.
— hydropiper L. 265.
— orthosperma Dueben 264.
— triandra Schkuhr 270; tab. 13.
Elatinoides (Chav.) Wettst. 22, 176.
Elephantina Bertol. 22, 686.
Elephas Adans. 22, 686.
Eleusine Gaertn. 2, 291.
— indica (L.) Gaertn. 291; tab. 22.
Eleutherococcus Maxim. 16, 20.
— senticosus (Rupr. et Maxim.) Maxim. 20; tab. 2.
Eleutherospermum C. Koch 16, 236.
— cicutarium (M. B.) Boiss. 236; tab. 15, 18a.
— lazicum Boiss. et Bal. 237.
Elisanthe Reichb. 6, 704.
Elisma Buchenau 1, 284.
— natans (L.) Buchenau 284.
Elodea Juss. 15, 203.
Elodea Rich. 1, 294.
— canadensis Rich. 295; tab. 10.
Elsholzia Willd. 21, 634.
— densa Benth. 636.
— Patrinii (Lepech.) Garcke 635.
— serotina Kom. 636.
Elwendia Boiss. 16, 398.
Elymus L. 2, 694.
— arenarius L. 695; tab. 49.
— giganteus Vahl 696; tab. 49.
— mollis Trin. 695; tab. 49.
— villosissimus Scribn. 696; tab. 49.
Elytrigia Desv. 2, 627.
Eminium (Blume) Schott 3, 488.
— Alberti (Rgl.) Engl. 489.
— Lehmannii (Bge.) Kuntze 488; tab. 26.
Emmenospermum Clarke ex Hook. 22, 555.
Empedoclea Raf. 20, 253.
Empetrum L. 14, 511.
— androgynum V. Vassil. 515.
— arcticum V. Vassil. 516.
— hermaphroditum (Lange) Hagerup 515.
— Kardakovii V. Vassil. 517.
— kurilense V. Vassil. 514, 744.
— nigrum L. 512.
— sibiricum V. Vassil. 513.
Endocellion Turcz. ex Herd. 26, 653.
Endogonia Turcz. ex Lindl. 19, 260.

Endoptera DC. 29, 595.
Enemion Raf. 7, 73.
— Raddeanum Rgl. 73; tab. 5.
Enula Neck. 25, 433, 443.
Ephedra L. 1, 195.
— ciliata C. A. Mey. 198; tab. 9.
— distachya L. 201.
— equisetina Bge. 203.
— intermedia Schrenk ex C. A. Mey. 198.
— Fedtschenkoi Pauls. 202; tab. 9.
— lomatolepis Schrenk 197; tab. 9.
— monosperma C. A. Mey. 202; tab. 9.
— procera Fisch. et Mey. 203.
— strobilacea Bge. 196; tab. 9.
Ephemerum Reichb. 18, 263.
Ephippianthus Reichb. f. 4, 607.
— Schmidtii Reichb. f. 607; tab. 37.
Epigaea L. 18, 77.
— gaultherioides (Boiss. et Bal.) Takht. 77, tab. 3.
Epilasia (Bge.) Benth. et Hook. f. 29, 111.
— acrolasia (Bge.) Clarke 112.
— hemilasia (Bge.) Clarke 114.
— mirabilis Lipsch. 112.
Epilobium L. 15, 571.
— adenocaulon Hausskn. 602.
— adnatum Griseb. 586; tab. 29.
— algidum M. B. 604; tab. 29.
— almaatense Steinb. 590.
— alpestre (Jacq.) Krock. 618; tab. 30.
— alpinum L. 607.
— alsinifolium Vill. 609.
— amurense Hausskn. 594.
— anatolicum Hausskn. 598.
— angulatum Kom. 597.
— arcticum Sam. 619; tab. 30.
— behringianum Hausskn. 610.
— bifarium Kom. 604; tab. 29.
— Bongardii Hausskn. 611.
— calycinum Hausskn. 596.
— cephalostigma Hausskn. 595.
— collinum C. C. Gmel. 582.
— confusum Hausskn. 606.
— consimile Hausskn. 590.
— cylindrostigma Kom. 596.
— davuricum Fisch. 617.
— frigidum Hausskn. 594.
— gemmascens C. A. Mey. 605.
— glandulosum Lehm. 603.
— hirsutum L. 578; tab. 28.
— Hornemannii Reichb. 612.
— japonicum Hausskn. 601.
— lactiflorum Hausskn. 608; tab. 29.
— Lamyi F. Schultz 587.
— lanceolatum Seb. et Mauri 585; tab. 28.
— minutiflorum Hausskn. 620.
— modestum Hausskn. 620.
— montanum L. 581; tab. 28.
— nervosum Boiss. et Buhse 589.
— nudicarpum Kom. 597.
— nutans F. W. Schmidt 614.
— obscurum (Schreb.) Roth 588.
— palustre L. 613.
— parviflorum (Schreb.) DC. 580; tab. 28.
— ponticum Hausskn. 593.
— prionophyllum Hausskn. 598.
— roseum (Schreb.) Pers. 588.
— rupicola Pavl. 593.
— sertulatum Hausskn. 607; tab. 30.
— subalgidum Hausskn. 605.
— subnivale M. Pop. ex Pavl. 592.
— tenue Kom. 597.
— thermophilum Pauls. 591.
— tianschanicum Pavl. 591.
— tundrarum Sam. 618; tab. 30.
— uralense Rupr. 612.
Epimedium L. 7, 542.
— circinnatocucullatum Sosn. 543.
— colchicum (Boiss.) Trautv. 543.
— koreanum Nakai 544.
— pinnatum Fisch. 543.
— pubigerum (DC.) Morr. et Decne. 544.
Epipactis Adans. 4, 620.
— latifolia (L.) All. 624; tab. 39.
— microphylla (Ehrh.) Sw. 629; tab. 39.
— palustris (L.) Crantz 623; tab. 39.
— papillosa Franch. et Sav. 625; tab. 39.
— Royleana Lindl. 622.
— rubiginosa Crantz 626; tab. 39.
— Thunbergii A. Gray 622; tab. 38.
— veratrifolia Boiss. et Hohen. 623.
Epipogon S. G. Gmel. 4, 635.
— aphyllum (F. W. Schmidt) Sw. 635; tab. 39.
Epitrachys C. Koch 28, 52, 69.
Epurga Fourr. 14, 479.
Equisetum L. 1, 101.
— arvense L. 103; tab. 5.
— heleocharis Ehrh. 108.
— hiemale L. 110, tab. 5.
— Komarovii Iljin 110; tab. 5.
— litorale Kühl. 109; tab. 5.
— majus Gars. 104; tab. 5.
— palustre L. 108; tab. 5.
— pratense Ehrh. 104; tab. 5.
— ramosissimum Desf. 109; tab. 5.
— scirpoides Michx. 111; tab. 5.
— silvaticum L. 107; tab. 5.

— trachyodon A. Br. 111, tab. 5.
— variegatum Schleich. 111; tab. 5.
Eragrostis Host 2, 313.
— aegyptica (Willd.) Delile 315; tab. 24.
— arundinacea (L.) Roshev. 319; tab. 24.
— Kossinskyi Roshev. 314; tab. 24.
— megastachya (Koel.) Link 316; tab. 24.
— minor Host 315; tab. 24.
— nigra Nees ex Steud. 314; tab. 24.
— pilosa (L.) Beauv. 315; tab. 24.
— Starosselskyi Grossh. 319; tab. 24.
— suaveolens Beck. ex Claus 316; tab. 24.
Eranthis Salisb. 7, 60.
— longistipitata Rgl. 60; tab. 6.
— sibirica DC. 61; tab. 6.
— stellata Maxim. 61; tab. 6.
Erechtites Raf. 26; 682.
— valerianifolia (Wolf) DC. 683.
Eremanthe Spach 15, 203, 211.
Eremodaucus Bge. 16, 250.
— Lehmannii Bge. 250; tab. 13.
Eremogone Fenzl 6, 520.
Eremopappus Takht. 28, 367.
Eremopoa Roshev. 2, 429, 756.
— altaica (Trin.) Roshev. 431; tab. 32.
— bellula (Rgl.) Roshev. 431; tab. 32.
— multiradiata (Trautv.) Roshev. 430; tab. 32.
— oxyglumis (Boiss.) Roshev. 430, 756.
— persica (Trin.) Roshev. 430; tab. 32.
— soongorica (Schrenk) Roshev. 431; tab. 32.
Eremopyrum Jaub. et Spach 2, 661.
— Buonapartis (Spreng.) Nevski 663; tab. 47a.
— distans (C. Koch) Nevski 665; tab. 47a.
— hirsutum (Bertol.) Nevski 663.
— orientale (L.) Jaub. et Spach 664; tab. 47a.
— triticeum (Gaertn.) Nevski 662; tab. 47a.
Eremosparton Fisch. et Mey. 11, 310.
— aphyllum (Pall.) Fisch. et Mey. 310; tab. 21.
— flaccidum Litv. 311; tab. 21.
— songoricum (Litv.) Vass. 311; tab. 21.
Eremostachys Bge. 21, 1.
— affinis Schrenk 54.
— Alberti Rgl. 13.
— angreni M. Pop. 17; tab. 1.
— aralensis Bge. 50.
— arctiifolia M. Pop. 9.
— baissunensis M. Pop. 35.
— baldschuanica Rgl. 37.
— Boissieriana Rgl. 47.
— canescens M. Pop. 20.
— cephalariifolia M. Pop. 30.
— cordifolia Rgl. 9.
— czuiliensis Golosk. 28.
— desertorum Rgl. 49.
— dschambulensis Knorr. 29, 645.
— ebracteolata M. Pop. 17.
— eriocalyx Rgl. 18.
— Fetissovii Rgl. 10.
— fulgens Bge. 36.
— glanduligera M. Pop. 35.
— gymnocalyx Schrenk 22.
— gypsacea M. Pop. 20.
— hajastanica Sosn. 26.
— hissarica Rgl. 13.
— iberica Vis. 25.
— iliensis Rgl. 22.
— jakka aghi M. Pop. 34.
— Kaufmanniana Rgl. 40.
— Korovinii M. Pop. 11.
— labiosa Bge. 39.
— labiosiformis (M. Pop.) Knorr. 26.
— laciniata (L.) Bge. 24.
— Lehmanniana Bge. 12.
— mogianica M. Pop. 52.
— moluccelloides Bge. 55.
— nuda Rgl. 23.
— paniculata Rgl. 42.
— pectinata M. Pop. 21.
— phlomoides Bge. 49.
— Popovii Gontsch. 37; tab. 2.
— pulchra M. Pop. 38; tab. 1.
— Regeliana Aitch. et Hemsl. 47; tab. 3.
— rotata Schrenk 56.
— schugnanica (M. Pop.) Knorr. 33.
— serawschanica Rgl. 33.
— speciosa Rupr. 27.
— subspicata M. Pop. 44.
— tadschikistanica B. Fedtsch. 43; tab. 3.
— tianschanica M. Pop. 51.
— transoxana Bge. 41.
— tuberosa (Pall.) Bge. 53.
— uniflora Rgl. 19.
— Zenaidae M. Pop. 14; tab. 1.
Eremurus M. B. 4, 37.
— Alberti Rgl. 48; tab. 4.
— altaicus (Pall.) Stev. 41; tab. 4.
— anisopteris (Kar. et Kir.) Rgl. 50; tab. 4
— bucharicus Rgl. 52.
— Capusii Franch. 51.
— comosus O. Fedtsch. 43; tab. 4.
— Hilariae M. Pop. et Vved. 49.
— inderiensis (MB.) Rgl. 43.
— Kaufmannii Rgl. 47; tab. 4.
— kopetdaghensis M. Pop. 50; tab. 4.

Eremurus Korovinii B. Fedtsch. 48, 733; tab. 4.
— Korshinskii O. Fedtsch. 32; tab. 4.
— lactiflorus O. Fedtsch. 49; tab. 4.
— luteus Baker 51.
— Olgae Rgl. 44; tab. 4.
— Regelii Vved. 40; tab. 4.
— robustus Rgl. 47; tab. 4.
— Saprjagajevii B. Fedtsch. 51, 734.
— sogdianus (Rgl.) Benth. et Hook. f. 42.
— spectabilis M. B. 39; tab. 4.
— stenophyllus (Boiss. et Buhse) Baker 43.
— tauricus Stev. 40.
— turkestanicus Rgl. 41.
Erianthera Benth. 21, 140.
— anomala Juz. 140.
Erianthus Rich. 2, 9.
— purpurascens Anderss. 9; tab. 1.
Erica L. 18, 89.
— arborea L. 90.
— tetralix L. 90.
Erigeron L. 25, 191.
— acer L. 246.
— Alexeenkoi Krasch. 271.
— allochrous Botsch. 208.
— alpinus L. 252.
— altaicus M. Pop. 203.
— andryaloides (DC.) Benth. 274.
— annuus (L.) Pers. 244; tab. 15.
— aurantiacus Rgl. 209.
— azureus Rgl. ex M. Pop. 207.
— badachschanicus Botsch. 263, 586.
— baicalensis Botsch. 250, 584.
— bellidiformis M. Pop. 260.
— biramosus Botsch. 279, 588.
— bonariensis L. 240.
— borealis (Vierh.) Simm. 253; tab. 16.
— brachyspermus Botsch. 269.
— cabulicus (Boiss.) Botsch. 277; tab. 17.
— canadensis L. 239; tab. 15.
— caucasicus Stev. 227.
— crispus Pourr. 286.
— dolichostylus Botsch. 278, 588.
— elongatus Ledeb. 251; tab. 16.
— eriocalyx (Ledeb.) Vierh. 228; tab. 14.
— eriocephalus J. Vahl. 232.
— flaccidus (Bge.) Botsch. 217.
— heterochaeta (Benth. ex Clarke) Botsch. 218; tab. 13.
— hissaricus Botsch. 262, 586.
— humilis J. Grah. 233; tab. 14.
— kamtschaticus DC. 248.
— karatavicus Pavl. 272.
— khorossanicus Boiss. 280; tab. 17.
— Komarovii Botsch. 213.
— koraginensis (Kom.) Botsch. 212.
— Krylovii Serg. 254.
— lachnocephalus Botsch. 230.
— leioreodes M. Pop. 201.
— leucophyllus (Bge.) Boiss. 272.
— lonchophyllus Hook. 234; tab. 14.
— nigromontanus Boiss. et Buhse 282.
— Oharai (Nakai) Botsch. 216; tab. 13.
— Olgae Rgl. et Schmalh. 270.
— oreades (Schrenk) Fisch. et Mey. 236.
— orientalis Boiss. 261.
— pallidus M. Pop. 225.
— peregrinus (Pursh) Greene 200.
— petiolatis Vierh. 219.
— petroiketes Rech. f. 231.
— plurifolius Botsch. 223, 583.
— podolicus Bess. 249.
— polymorphus Scop. 224.
— Poncinsii (Franch.) Botsch. 275.
— Popovii Botsch. 235.
— primuloides M. Pop. 268.
— pseudelongatus Botsch. 264, 587.
— pseuderigeron (Bge.) M. Pop. 267.
— pseuderiocephalus M. Pop. 237; tab. 14.
— pseudoseravschanicus Botsch. 258, 585.
— sachalinensis Botsch. 266.
— schalbusi Vierh. 225.
— Schmalhausenii M. Pop. 264.
— seravschanicus M. Pop. 202.
— silenifolius (Turcz.) Botsch. 212.
— sogdianus M. Pop. 226.
— strigosus Muhl. ex Willd. 245.
— Thunbergii A. Gray 214.
— tianschanicus Botsch. 259.
— trimorphopsis Botsch. 238, 584.
— umbrosus (Kar. et Kir.) Boiss. 285.
— uniflorus L. 227.
— venustus Botsch. 210; tab. 13.
— vicarius Botsch. 204.
— violaceus M. Pop. 257.
Erinia Noulet 24, 133, 189.
Eriobotrya Lindl. 9, 407.
— japonica Lindl. 407; tab. 27.
Eriocaulon L. 3, 494.
— chinorossicum Kom. 497; tab. 27.
— decemflorum Maxim. 496; tab. 27.
— robustius (Maxim.) Makino 497; tab. 27.
— Sieboldianum Sieb. et Zucc. 495; tab. 27.
— ussuriense Koern. 496; tab. 27.
Eriochloa H. B. K. 2, 26.
— succincta (Trin.) Kunth 27; tab. 2.
— villosa (Thunb.) Kunth 27; tab. 2.
Eriolepis Cass. 28, 52, 69, 89.

Eriophorum L. 3, 26.
— angustifolium Roth 29; tab. 2.
— brachyantherum Trautv. 34; tab. 2.
— callitrix Cham. ex C. A. Mey. 35; tab. 2.
— Chamissonis C. A. Mey. ex Ledeb. 36; tab. 2.
— coreanum Palla 30; tab. 2.
— gracile Koch 30; tab. 2.
— humile Turcz. 33; tab. 2.
— japonicum Maxim. 28.
— latifolium Hoppe 29; tab. 2.
— russeolum Fries 35; tab. 2.
— Scheuchzeri Hoppe 36; tab. 2.
— vaginatum L. 33; tab. 2.
Eriostemon Less. 27, 361.
Eriosynaphe DC. 17, 144.
— longifolia (Fisch.) DC. 147.
Eritrichium Schrad. 19, 482.
— aretioides (Cham.) DC. 519; tab. 25.
— caucasicum (Albov) Grossh. 514; tab. 24.
— Czekanowskii Trautv. 520.
— dubium O. Fedtsch. 495.
— Fetissovii Rgl. 494.
— incanum (Turcz.) DC. 509.
— jacuticum M. Pop. 511, 712; tab. 24.
— jenisseense Turcz. 503.
— latifolium Kar. et Kir. 489.
— kamtschaticum Kom. 511.
— Maackii Maxim. 506.
— mandshuricum M. Pop. 505, 711.
— nanum (Vill.) Schrad. 516.
— pamiricum B. Fedtsch. 490.
— pectinatum (Pall.) DC. 508.
— pseudolatifolium M. Pop. 494; tab. 19.
— pseudostrictum M. Pop. 496, 708.
— pulviniforme M. Pop. 504, 710.
— rupestre (Pall.) Bge. 501.
— sachalinense M. Pop. 499, 709.
— sericeum (Lehm.) DC. 512; tab. 25.
— sichotense M. Pop. 510, 171.
— sub-Jacquemontii M. Pop. 497, 709.
— turkestanicum Franch. 498.
— villosum (Ledeb.) Bge. 517.
Ermania Cham. 8, 328.
Erodium L'Hér. 14, 63.
— anthemidifolium M. B. 68.
— armenum Woronow 68.
— Beketovii Schmalh. 67.
— chium (Burm. f.) Willd. 69.
— ciconium (L.) Ait. 71.
— cicutarium (L.) L'Hér. 71.
— fumarioides Stev. 67.
— Hoefftianum C. A. Mey. 70.
— Litvinovii Woronow 65.
— malacoides (L.) Willd. 69.
— oxyrrhynchum M. B. 64.
— ruthenicum M. B. 68.
— schemachense Grossh. 67.
— Stephanianum Willd. 65.
— Stevenii M. B. 66.
— strigosum Kar. 70.
— tataricum Willd. 66.
— tibetanum Edgew. 65.
Erophila DC. 8, 454.
— Krockeri Andrz. 456.
— minima C. A. Mey. 455.
— praecox (Stev.) DC. 455.
— verna (L.) Bess. 456.
Eruca Adans. 8, 469.
— sativa Lam. 469.
Ervum L. 13, 410.
Eryngium L. 16, 73.
— balchanicum Bobr. 79, 589; tab. 4.
— Biebersteinianum Nevski 86.
— Bungei Boiss. 83.
— campestre L. 78.
— giganteum M. B. 76.
— incognitum Pavl. 80; tab. 4.
— karatavicum Iljin 84; tab. 4.
— macrocalyx Schrenk 80; tab. 4.
— maritimum L. 87.
— mirandum Bobr. 84, 589; tab. 4.
— nigromontanum Boiss. 78.
— Noeanum Boiss. 76.
— planum L. 85.
— Wanaturi Woronow 87; tab. 4.
Erysimum L. 8, 92.
— altaicum C. A. Mey. 106; tab. 6.
— argyrocarpum N. Busch 108, 638.
— artvinense N. Busch 114.
— aurantiacum (Bge.) Maxim. 107.
— aureum M. B. 98.
— babataghi Korsh. 119.
— badghysi (Korsh.) Lipsky 123.
— brachycarpum Boiss. 115.
— brevistylum Somm. et Levier 108.
— callicarpum Lipsky 102.
— canescens Roth 118.
— caspicum N. Busch 112, 639.
— caucasicum Trautv. 114.
— chazarjurti N. Busch 112, 637.
— cheiranthoides L. 108.
— collinum (M. B.) Andrz. 111.
— contractum Somm. et Lev. 105.
— crassipes Fisch. et Mey. 120.
— cretaceum (Rupr.) Schmalh. 113.
— croceum M. Pop. 101.
— cuspidatum (M. B.) DC. 127.

Erysimum Czernjajevii N. Busch 115.
— epikeimenum N. Busch 126, 637.
— Franchetii N. Busch 125, 639.
— gaudanense Litv. 111.
— gelidum Bge. 109; tab. 5.
— humillimum (Ledeb.) N. Busch 106, 638.
— ibericum (Adam) DC. 101; tab. 5.
— inense N. Busch 119.
— ischnostylum Freyn et Sint. 110; tab. 6.
— krynkense Lavr. 113.
— lazistanicum (Rupr.) Lipsky 105.
— leptophyllum (M. B.) Andrz. 114; tab. 5.
— leptostylum DC. 110.
— leucanthemum (Steph.) B. Fedtsch. 116; tab. 6.
— lilacinum Steinb. 124.
— Marschallianum Andrz. 100.
— Meyerianum (Rupr.) N. Busch 107.
— Pallasii (Pursh) Fern. 126.
— pannonicum Crantz 99.
— passgalense Boiss. 117.
— persicum Boiss. 118; tab. 5.
— pulchellum (Willd.) J. Gay 102; tab. 5.
— repandum L. 116; tab. 6.
— sisymbrioides C. A. Mey. 124.
— strictisiliquum N. Busch 126, 638.
— substrigosum (Rupr.) N. Busch 112.
— sylvaticum M. B. 99.
— Szovitsianum Boiss. 123.
— violascens M. Pop. 125.
— vitellinum M. Pop. 120.
Erythraea Renealme ex Borkh. 18, 527.
Erythrochaete Sieb. et Zucc. 26, 788, 798.
Erythronium L. 4, 364.
— caucasicum Woronow 365; tab. 20.
— sibiricum (Fisch. et Mey.) Kryl. 365; tab. 20.
Eshenbachia Moench 25, 91.
Esquirolia Lév. 18, 518.
Estellata Schur 19, 194.
Esula Haw. 14, 309.
Eubotryoides (Nakai) Hara 18, 72.
— Grayana (Maxim.) Hara 73; tab. 3.
Eucalligonum Bge. ex Boiss. 5, 562.
Eucalyptus L'Her. 15, 555.
— camaldulensis Dehnh. 560.
— cinerea F. Muell. ex Benth. 564.
— citriodora Hook. 560.
— globulus Labill. 562.
— gummifera (Gaertn.) Hochr. 559.
— Gunnii Hook. f. 561.
— Macarthuri Deane et Maid. 563.
— pauciflora Sieb. 564.
— salicifolia (Soland.) Cav. 565.
— urnigera Hook. f. 561.
— viminalis Labill. 563.
Euchroma Nutt. 22, 530.
Euclidium R. Br. 8, 323.
— syriacum (L.) R. Br. 323; tab. 13.
Eucommia Oliv. 9, 273.
— ulmoides Oliv. 273.
Eufragia Griseb. 22, 641.
Euilus Stev. 12, 594.
Eukrania Raf. 17, 317, 324.
Eulophia R. Br. 4, 609.
— turkestanica (Litw.) Schlechter 610.
Eumoltkea Wettst. 19, 234.
Eumoltkia Guerke 19, 234.
Eunomia DC. 8, 590.
— rotundifolia C. A. Mey. 590; tab. 32.
Euonymus L. 14, 548.
— europaea L. 553; tab. 30.
— japonica L. 552.
— Koopmannii Lauche 565; tab. 30.
— Maackii Rupr. 555; tab. 30.
— nana M. B. 561; tab. 30.
— pauciflora Maxim. 560.
— sacrosancta Koidz. 557; tab. 30.
— Semenovii Rgl. et Herd. 559.
— Sieboldiana Blume 556.
— velutina (C. A. Mey.) Fisch. et Mey. 557.
— verrucosa Scop. 559; tab. 30.
Eupatorium L. 25, 18.
— cannabinum L. 19.
— Glehnii Fr. Schmidt ex Trautv. 21.
— Lindleyanum DC. 20.
Euphorbia L. 14, 304.
— acuminata Lam. 465.
— agraria M. B. 430; tab. 22.
— alaica Prokh. 388.
— alatavica Boiss. 373; tab. 20.
— aleppica L. 460.
— alpina C. A. Mey. 370.
— altaica C. A. Mey. ex Ledeb. 381.
— amygdaloides L. 453.
— andrachnoides Schrenk 418.
— angulata Jacq. 381.
— anisopetala Prokh. 492; tab. 25.
— aristata Schmalh. 361.
— armena Prokh. 411, 741.
— arvalis Boiss. et Heldr. 472; tab. 23.
— aserbajdzhanica Bordz. 470; tab. 23.
— astrachanica C. A. Mey. ex Claus 440; tab. 22.
Aucheri Boiss. 456.
— aulacosperma Boiss. 464; tab. 23.
— biglandulosa Desf. 405.

— blepharophylla C. A. Mey. ex Ledeb. 345.
— Boissieriana (Woronov) Prokh. 445.
— Borodinii Sambuk 427.
— Borszczowii Prokh. 425, 742; tab. 22.
— buchtormensis C. A. Mey. ex Ledeb. 371.
— Buhsei Boiss. 419.
— Bungei Boiss. 386.
— Buschiana Grossh. 413.
— canescens L. 491; tab. 25.
— carniolia Jacq. 365.
— carpatica Woloszcz. 357.
— chamaesyce L. 488; tab. 25.
— cheirolepis Fisch. et Mey. 481; tab. 24.
— condylocarpa M. B. 377.
— coniosperma Boiss. et Buhse 369; tab. 20.
— consanguinea Schrenk 471; tab. 23.
— cyparissias L. 439.
— cyrtophylla Prokh. 441.
— deltobracteata Prokh. 457.
— densa Schrenk 473; tab. 23.
— densiuscula M. Pop. 475.
— denticulata Lam. 412.
— discolor Lede . 423.
— dulcis L. 379.
— eriophora Boiss. 382.
— esula L. 419.
— Eugeniae Prokh. 356, 735; tab. 19.
— exigua L. 460; tab. 23.
— falcata L. 464; tab. 23.
— ferganensis B. Fedtsch. 346.
— Forskalli J. Gay 493; tab. 24.
— Franchetii B. Fedtsch. 467; tab. 23.
— glaberrima C. Koch 454.
— glareosa Pall. ex M. B. 402; tab. 21.
— glomerulans Prokh. 429.
— Gmelinii Steud. 422.
— Goldei Prokh. 404, 738.
— graeca Boiss. et Sprun. 461; tab. 23.
— Grossheimii Prokh. 391.
— guntensis Prokh. 442.
— helioscopia L. 383.
— humifusa Willd. 487; tab. 24.
— humilis C. A. Mey. ex Ledeb. 397.
— iberica Boiss. 435.
— inderiensis Less. ex Kar. et Kir. 468.
— indica Lam. 486.
— irgisensis Litv. 417.
— ispahanica Boiss. 390.
— jaxartica Prokh. 446.
— Karoi Freyn 424.
— Komaroviana Prokh. 343, 734; tab. 18.
— kopetdaghi Prokh. 398.
— Kudrjaschevii (Pazij) Prokh. 352.
— lamprocarpa Prokh. 362.
— lanata Sieber ex Spreng. 476.
— lathyris L. 479.
— latifolia C. A. Mey. 426.
— Ledebouri Boiss. 462.
— leptocaula Boiss. 443.
— Lipskyi Prokh. 347; tab. 18.
— lucida Waldst. et Kit. 434.
— lucorum Rupr. ex Maxim. 374.
— macrocarpa Boiss. et Buhse 350.
— macroceras Fisch. et Mey. 455.
— macroclada Boiss. 400.
— macrorrhiza C. A. Mey. ex Ledeb. 371.
— maculata L. 494; tab. 24.
— mandshurica Maxim. 433; tab. 22.
— Marschalliana Boiss. 410; tab. 21.
— microcarpa Prokh. 420.
— microsphaera Boiss. 368.
— mongolica Prokh. 342.
— monocyathium Prokh. 338.
— monostyla Prokh. 406, 739; tab. 21.
— mucronulata Prokh. 351.
— myrsinites L. 408; tab. 21.
— Normannii Schmalh. ex Lipsky 466; tab. 23.
— nutans Lag. 485; tab. 24.
— oblongifolia C. Koch 455.
— orientalis L. 352.
— pachyrrhiza Kar. et Kir. 378.
— Pallasii Turcz. 343; tab. 18.
— palustris L. 355; tab. 19.
— pamirica Prokh. 438.
— paralias L. 391.
— peplis L. 486.
— peplus L. 463; tab. 23.
— petrophila C. A. Mey. 392; tab. 21.
— pilosa L. 363.
— platyphylla L. 367.
— poecilophylla Prokh. 428.
— polychroma A. Kerner 364.
— polytimetica Prokh. 458.
— pontica Prokh. 408, 740; tab. 21.
— Prokhanovii M. Pop. 340.
— pseudagraria Smirn. 436.
— pubescens Vahl 372.
— rapulum Kar. et Kir. 344.
— rosularis Al. Theod. 339.
— rupestris C. A. Mey. ex Ledeb. 341.
— salicifolia Host 429.
— sareptana Beck. 425; tab. 22.
— Savaryi Kiss. 452.
— schugnanica B. Fedtsch. 387.
— sclerocyathium Korov. et M. Pop. 336
— scripta Somm. et Levier 348.
— Seguieriana Neck. 395.

Euphorbia semivillosa Prokh. 360; tab. 19.
— serawschanica Rgl. 338.
— Sewerzowii Herd. 433.
— Sieboldiana Morr. et Decne. 451.
— sogdiana M. Pop. 399.
— soongorica Boiss. 362.
— sororia Schrenk 470; tab. 23.
— spinidens Bornm. ex Prokh. 407; tab. 21.
— squamosa Willd. 349; tab. 20.
— stepposa Zoz 401, 738.
— stricta L. 366.
— subcordata C. A. Mey. ex Ledeb. 448.
— subtilis Prokh. 421.
— Szovitsii Fisch. et Mey. 474; tab. 23.
— talastavica Prokh. 379; tab. 20.
— tauricola Prokh. 358, 736; tab. 19.
— tianshanica Prokh. 339.
— tibetica Boiss. 385.
— transoxana Prokh. 350; tab. 19.
— Tranzschelii Prokh. 386.
— tridonta Prokh. 469.
— tshuiensis (Prokh.) Serg. 440.
— turcomanica Boiss. 492; tab. 25.
— Turczaninowii Kar. et Kir. 472.
— turkestanica Rgl. 388.
— undulata M. B. 414.
— uralensis Fisch. ex Link 437.
— villosa Waldst. et Kit. 359.
— virgata Waldst. et Kit. 443.
— volgensis Krysht. 403.
— Wittmannii Boiss. 377; tab. 20.
— Woronowii Grossh. 409.
— zhiguliensis Prokh. 447.
Euphrasia L. 22, 557.
— adenocaulon Juz. 623; tab. 30.
— Alboffii Chabert 619; tab. 30.
— altaica Serg. 611.
— amblyodonta Juz. 613; tab. 29.
— amurensis Freyn 628; tab. 31.
— bajankolica Juz. 594.
— bakurianica Juz. 638.
— brevipila Burn. et Gremli 582.
— caucasica Juz. 587.
— condensata Jord. 579.
— cyclophyla Juz. 595.
— daghestanica Juz. 625; tab. 30.
— drosophylla Juz. 608; tab. 29.
— Fedschenkoana Wettst. ex B. Fedtsch. 590; tab. 28.
— fennica Kihlm. 634.
— frigida Pugsl. 605; tab. 29.
— georgica Kem.-Nath. 577.
— glabrescens (Wettst.) Wiinst. 602; tab. 28.
— Grossheimii Kem.-Nath. 607.
— hirtella Jord. 635; tab. 31.
— huperborea Jorgens. 591.
— Irenae Juz. 575.
— jacutica Juz. 579.
— Juzepczukii Deniss. 614.
— Kemulariae Juz. 620.
— Kerneri Wettst. 615.
— Krassnowii Juz. 599.
— macrocalyx Juz. 576.
— macrodonta Juz. 619.
— Maximowiczii Wettst. 568; tab. 27.
— micrantha Reichb. 603.
— mollis Ledeb. ex Benth. 611; tab. 29.
— montana Jord. 633.
— Murbeckii Wettst. 586.
— onegensis Cajand. 635.
— ossica Juz. 624.
— parviflora Schag. 600; tab. 28.
— pectinata Ten. 577.
— peduncularis Juz. 617.
— petiolaris Wettst. 621; tab. 30.
— picta Wimm. 616.
— pseudomollis Juz. 612.
— Regelii Wettst. 589; tab. 28.
— Reuteri Wettst. 581.
— Rostkoviana Hayne 623.
— saamica Juz. 592.
— salisburgensis Funck 638.
— schugnanica Juz. 618.
— sevanensis Juz. 625; tab. 30.
— sibirica Serg. 574.
— Sosnowskyi Kem.-Nath. 637.
— subpolaris Juz. 593.
— svanica Kem.-Nath. 588.
— Syreitschikovii Govor. 574.
— tatarica Fisch. 570; tab. 27.
— Tatrae Wettst. 606.
— taurica Ganesch. 627; tab. 30.
— tenuis (Brenn.) Wettst. 587.
— Townsendiana Freyn ex Wettst. 578.
— Tranzszelii Juz. 596.
— Uechtritziana Jung. et Engl. 601.
— ussuriensis Juz. 569.
— Woronowii Juz. 626.
Eurotia Adans. 6, 108.
— ceratoides (L.) C. A. Mey. 108; tab. 1, 2.
— Ewersmanniana Stschegl. 109.
Eurotia L. 6, 423.
Euryale Salisb. 7, 5.
— ferox Salisb. 6.
Euryangium Kauffm. 17, 62.
Eusalsola Fenzl 6, 224, 235, 236.
Eutrema R. Br. 8, 33.

— cordifolium Turcz. 35.
— Edwardsii R. Br. 34.
— integrifolium (DC.) Bge. 35.
— pseudocordifolium M. Pop. 35, 634.
Evax Gaertn. 25, 307.
— anatolica Boiss. et Heldr. 308; tab. 19.
— arenaria Smoljan. 310; tab. 19.
— contracta Boiss. 309.
— filaginoides Kar. et Kir. 313; tab. 19.
— micropodioides Willk. 313.
Ewersmannia Bge. 13, 258.
— subspinosa (Fisch.) B. Fedtsch. 258.
Exochorda Lindl. 9, 316.
— Alberti Rgl. 317; tab. 19.
— tianschanica Gontsch. 317.

Faba Adans. 13, 474.
Fabago Adans. 14, 158.
Facelos Raf. 3, 167.
Fagopyrum Gaertn. 5, 702.
— sagittatum Gilib. 702.
— suffruticosum Fr. Schmidt 703.
— tataricum (L.) Gaertn. 703.
Fagotriticum L. 5, 702.
Fagus L. 5, 354.
— orientalis Lipsky 355; tab. 19.
— silvatica L. 356; tab. 19.
Falcaria Bernh. 16, 382.
— falcarioides (Bornm. et Wolff) Wolff 384; tab. 27.
— sioides (Wib.) Aschers. 383.
Falcata J. F. Gmel. 13, 526.
— japonica (Oliv.) Kom. 527.
Faldermannia Trautv. 21, 381, 407.
Farfara Gilib. 26, 641.
Farnesia Gasp. 11, 11.
Farsetia Turra 8, 332.
— spathulata Kar. et Kir. 332.
Fatsia Decne. et Planch. 16, 2.
— japonica (Thunb.) Decne. et Planch. 3.
Fauria Franch. 18, 641.
— crista-galli (Menz.) Makino 641.
Fedczenkiella Gontsch. 20, 473.
Fedia Adans. 23, 587.
Fedia Gaertn. 23, 642.
Ferula L. 17, 62.
— Aitchisonii K.-Pol. 123; tab. 13.
— akitschkensis B. Fedtsch. ex K.-Pol. 135.
— angreni Korov. 115.
— assa-foetida L. 73.
— badhysi Korov. 126.
— badrakema K.-Pol. 90.
— canescens Ledeb. 92.
— caspica M. B. 141; tab. 15.
— caucasica Korov. 140.
— ceratophylla Rgl. et Schmalh. 107.
— clematidifolia K.-Pol. 131.
— conocaula Korov. 83.
— dissecta Ledeb. 113.
— diversivittata Rgl. et Schmalh. 83.
— dshaudshamyr Korov. 141.
— dshizakensis Korov. 111.
— equisetacea K.-Pol. 100.
— eremophila Korov. 130.
— Fedtschenkoana K.-Pol. 101; tab. 13.
— ferganensis Lipsky 113.
— ferulaeoides (Steud.) Korov. 139; tab. 14, 17.
— foetidissima Rgl. et Schmalh. 89.
— foliosa Lipsky 94.
— gigantea B. Fedtsch. 84.
— glaberrima Korov. 75.
— gracilis Ledeb. 137; tab. 11.
— Grigorjevii B. Fedtsch. 102.
— gumosa Boiss. 90.
— gypsacea Korov. 124.
— iliensis Krasn. 84.
— inflata Korov. 85.
— involucrata Korov. 123.
— Jaeschkeana Vatke 86.
— karatavica Rgl. et Schmalh. 125.
— karataviensis (Rgl. et Schmalh.) Korov. 137.
— karategina Lipsky 107.
— Kaschkarovii Korov. 132.
— kelifi Korov. 74.
— Kelleri K. Pol. 95.
— kokanica Rgl. et Schmalh. 93.
— kopetdagensis Korov. 115.
— Korshinskyi Korov. 131.
— Koso-Poljanskyi Korov. 100.
— Krylovii Korov. 91.
— kuhistanica Korov. 89.
— lapidosa Korov. 112.
— latifolia Korov. 85.
— latiloba Korov. 120.
— Lehmannii Boiss. 77; tab. 10.
— leiophylla (K.-Pol.) Korov. 96.
— leucographa Korov. 122.
— Linczevskii Korov. 99.
— lingulata Korov. 119.
— Lipskyi Korov. 101.
— Litwinoviana K.-Pol. 81.
— microcarpa Korov. 110.
— microloba Boiss. 77.
— Minkwitziae Korov. 103.
— mogoltavica Lipsky 78.
— mollis Korov. 120.

Ferula moschata (Reinsch) K.-Pol. 121.
— Nevskii Korov. 99.
— nuda Spreng 108; tab. 22.
— oopoda (Boiss. et Buhse) Boiss. 129.
— orientalis L. 116.
— ovina Boiss. 110.
— pachyphylla Korov. 104.
— pallida Korov. 114.
— penninervis Rgl. et Schmalh. 132.
— persica Willd. 76; tab. 10.
— plurivittata Korov. 75.
— Potaninii Korov. ex Pavl. 108; tab. 12.
— prangifolia Korov. 103.
— primaeva Korov. 75.
— pseudooreoselinum (Rgl. et Schmalh.) K.-Pol. 121.
— rigidula DC. 109.
— rubroarenosa Korov. 112.
— samarcandica Korov. 95.
— schair Borszcz. 126.
— Schtschurowskiana Rgl. et Schmalh. 138; tab. 17.
— songorica Pall. ex Schult. 136.
— stylosa Korov. 111.
— subtilis Korov. 96.
— Syreitschikovii K.-Pol. 92.
— Szovitsiana DC. 81; tab. 9.
— tatarica Fisch. 130.
— tenuisecta Korov. 114.
— tersakensis Korov. 76; tab. 9.
— teterrima Kar. et Kir. 91; tab. 11.
— transitora Korov. 135.
— tschimganica Lipsky 104.
— tuberifera Korov. 94.
— ugamica Korov. 124.
— vicaria Korov. 138.
— xeromorpha Korov. 119.
Ferulago Koch 17, 149.
— campestris (Bess.) Grec. 150.
— daghestanica Schischk. 151.
— latiloba Schischk. 152.
— setifolia C. Koch 153.
— silvatica (Bess.) Reichb. 154; tab. 16.
— taurica Schischk. 152; tab. 16.
— turcomanica Schischk. 154.
Festuca L. 2, 497.
— alatavica (Hack.) Roshev. 528.
— altaica Trin. 528.
— amblyodes V. Krecz. et Bobr. 529, 771; tab. 39.
— amethystina L. 515.
— arenaria Osbeck ex Retz. 520; tab. 40.
— arundinacea Schreb. 531; tab. 39.
— aucta V. Krecz. et Bobr. 518, 767.
— auriculata Drob. 511.
— baicalensis (Griseb.) V. Krecz. et Bobr. 518, 768.
— Beckeri Hack. 508.
— Bornmuelleri (Hack.) V. Krecz. et Bobr. 510.
— brevifolia R. Br. 514; tab. 40.
— chalcophaea V. Krecz. et Bobr. 513, 768; tab. 40.
— coelestis (St.-Yves) V. Krecz. et Bobr. 514, 740; tab. 40.
— cretacea Czern. 527; tab. 39.
— cryophila V. Krecz. et Bobr. 519, 767; tab. 40.
— dahurica (St.-Yves) V. Krecz. et Bobr. 517, 771; tab. 40.
— djimilensis Boiss. et Bal. 524.
— duriuscula L. 507.
— egena V. Krecz. et Bobr. 527, 769; tab. 40.
— eriantha Honda et Tatewaki 523; tab. 40.
— extremiorientalis Ohwi 534; tab. 40.
— Fenas Lag. 532.
— Ganeschinii Drob. 509.
— gigantea (L.) Vill. 534; tab. 40.
— heterophylla Lam. 515; tab. 40.
— jacutica Drob. 516.
— Kirilovii Steud. 524; tab. 40.
— kolymensis Drob. 511.
— Kryloviana Reverd. 512; tab. 40.
— lenensis Drob. 513; tab. 40.
— longiaristata Somm. et Lev. 511.
— mollissima V. Krecz. et Bobr. 512, 770
— montana M. B. 533.
— orientalis Kerner 531; tab. 39.
— ovina L. 503; tab. 40.
— polesica Zapal. 508.
— pratensis Huds. 530; tab. 39.
— pseudo-sulcata Drob. 504.
— pseudovina Hack. 508; tab. 40.
— rubra L. 517; tab. 40.
— Ruprechtii (Boiss.) V. Krecz. et Bobr. 507, 769.
— silvatica (Poll.) Vill. 533; tab. 40.
— sulcata Hack. 509; tab. 40.
— supina Schur 504; tab. 40.
— taurica Kerner ex Hack. 510.
— tianschanica Roshev. 526, 772; tab. 39.
— tristis Kryl. et Ivanitzk. 527; tab. 39.
— vaginata Waldst. et Kit. 516.
— varia Haenke 526; tab. 39.
— venusta St.-Yves. 527.
Fibigia Medic. 8, 336.
— clypeata (L.) Medic. 336.

— eriocarpa Boiss. 337.
— macroptera Boiss. 337; tab. 22.
— suffruticosa (Vent.) Sweet 337; tab. 22.
Ficaria Dill. 7, 332.
— calthifolia Reichb. 333.
— fascicularis C. Koch 334.
— ficarioides (Bory et Chaub.) Halacsy 335.
— verna Huds. 333.
Ficus L. 5, 380.
— carica L. 380; tab. 21.
Fiebera Opiz. 16, 93.
Filaginella Opiz. 25, 381.
Filago L. 25, 342.
Filago L. 25, 314.
— arvensis L. 322; tab. 20.
— Bornmuelleri Hausskn. 318.
— eriocephala Guss. 317; tab. 20.
— gallica L. 325.
— germanica L. 316.
— lagopus Parl. 323.
— minima (Smith) Pers. 324.
— montana L. 323.
— spathulata J. et C. Presl 321; tab. 20.
Filifolium Kitam. 26, 416.
— sibiricum (L.) Kitam. 417; tab. 19.
Filipendula Adans. 10, 279.
— angustiloba (Turcz.) Maxim. 283; tab. 21.
— denudata (J. et C. Presl) Fritsch 285.
— glabra Nakai ex Kom. et Aliss. 281.
— hexapetala Gilib. 286; tab. 21.
— intermedia (Glehn) Juz. 284.
— kamtschatica (Pall.) Maxim. 281; tab. 21.
— palmata (Pall.) Maxim. 282; tab. 21.
— purpurea Maxim. 282.
— stepposa Juz. 286, 617.
— ulmaria (L.) Maxim. 284; tab. 21.
Fimbriatum Will. 6, 848.
Fimbristylis Vahl 3, 90.
— annua (All.) Roem. et Schult. 93.
— dichotoma (L.) Vahl 92; tab. 8.
— ferruginea (L.) Vahl 94.
— leiocarpa Maxim. 93; tab. 8.
— ochotensis (Meinsh.) Kom. 94; tab. 8.
— quinquangularis (Vahl) Kunth 96; tab. 8.
— Sieberiana Kunth 95.
— squarrosa Vahl 95; tab. 8.
— verrucifera (Maxim.) Makino 92; tab. 8.
Fleischeria Hochst. et Steud. ex Endl. 29, 55.
Floerkea Spreng. 24, 343.
Flourensia Cambess. 6, 545.
Flox Adans. 6, 692.
Flüggea Willd. 14, 284.
— ussuriensis Pojark. 284, 734; tab. 16.
Foeniculum Mill. 16, 541.
— vulgare Mill. 542.
Fornicium Cass. 28, 308, 311.
Fortunella Swingle 14, 238.
— japonica (Thunb.) Swingle 241; tab. 14.
Fragaria L. 10, 58.
— ananassa Duchartre 63.
— bucharica Losinsk. 63; tab. 6.
— campestris Stev. 62.
— chiloensis Duchartre 64.
— moschata Duchartre 60; tab. 6.
— orientalis Losinsk. 61; tab. 6.
— vesca L. 59; tab. 6.
— virginiana Duchartre 64.
— viridis Duchartre 61; tab. 6.
Fragariastra Th. Wolf 10, 91.
Francoeuria Cass. 25, 487.
Frangula Mill. 14, 641.
— alnus Mill. 642.
— grandifolia (Fisch. et Mey.) Grub. 643; tab. 35.
— rupestris (Scop.) Schur 644; tab. 35.
Frankenia L. 15, 272.
— bucharica Basil. 275.
— hirsuta L. 274; tab. 25.
— pulverulenta L. 273.
Fraxinus L. 18, 485.
— americana L. 491.
— angustifolia Vahl 500; tab. 25.
— coriariifolia Scheele 496; tab. 25.
— excelsior L. 495; tab. 25.
— lanceolata Borkh. 490.
— mandschurica Rupr. 492; tab. 25.
— ornus L. 488.
— oxycarpa Willd. 499; tab. 25.
— Pallisae Willem. ex M. Pallis 497.
— pennsylvanica Marsh. 491.
— Pojarkoviana V. Vassil. 501, 750; tab. 25.
— raibocarpa Rgl. 489; tab. 25.
— rhynchophylla Hance 489; tab. 25.
— sogdiana Bge. 502; tab. 25.
— syriaca Boiss. 498; tab. 25.
Freirea Gaud. 5, 401.
Fritillaria L. 4, 302.
— armena Boiss. 311.
— caucasica Adam 311.
— dagana Turcz. 319.
— Eduardii Rgl. 317.
— ferganensis Losinsk. 315, 740; tab. 19.
— grandiflora Grossh. 308.
— Grossheimiana Losinsk. 313, 739; tab. 19.

Fritillaria kamtschatcensis (L.) Fisch. ex Hook. 318; tab. 19.
— Kotschyana Herb. 307.
— kurdica Boiss. et Noë 312.
— latifolia Willd. 308.
— lutea Mill. 307; tab. 19.
— Maximowiczii Freyn 320; tab. 19.
— meleagris L. 306.
— meleagroides Patr. ex Schult. 306.
— Michailovskyi Fomin 312; tab. 19.
— montana Hoppe 305.
— Olgae Vved. 314.
— orientalis Adam 311.
— pallidiflora Schrenk 304.
— Raddeana Rgl. 317; tab. 19.
— Regelii Losinsk. 314, 739.
— ruthenica Wikstr. 305; tab. 19.
— ussuriensis Maxim. 315.
— verticillata Willd. 316.
— Walujewii Rgl. 313.
Frolovia DC. 27, 361, 384.
Froriepia C. Koch 16, 368.
— subpinnata (Ledeb.) Baill. 368; tab. 30.
Fuernrohria C. Koch 16, 191.
— setifolia C. Koch 192; tab. 15, 17.
Fumana Spach 15, 348.
— procumbens (Dunal) Gren. et Godr. 349.
— viscidula (Stev.) Juz. 348, 681; tab. 17.
Fumaria L. 7, 706.
— asepala Boiss. 715.
— capreolata L. 708.
— Kralikii Jord. 709.
— micrantha Lag. 710.
— officinalis L. 713; tab. 45.
— parviflora Lam. 714.
— pikermiana Boiss. 708.
— rostellata Knaf 709.
— Schleicheri Soy.-Willem. 713.
— Thuretii Boiss. 708.
— Vaillantii Loisel. 715; tab. 45.
Fumariola Korsh. 7, 716.
— turkestanica Korsh. 716.
Funkia Spreng. 4, 54.
— lancifolia Spreng. 55; tab. 5.
— ovata Spreng. 54; tab. 5.

Gaertnera Retz. 24, 449.
Gagea Salisb. 4, 61.
— afghanica Terr. 107.
— Alberti Rgl. 103; tab. 8.
— Alexeenkoana Miscz. 99; tab. 8.
— altaica Schischk. et Sumn. 107; tab. 8.
— anisanthos C. Koch 87, 737; tab. 7.
— anisopoda M. Pop. 103, 737; tab. 9.
— arvensis Dum. 89.
— Bergii Litv. 95; tab. 8.
— bulbifera (Pall.) Roem. et Schult. 108; tab. 8.
— Callieri Pasch. 93.
— capillifolia Vved. 110; tab. 9.
— Capusii Terr. 81; tab. 7.
— Caroli-Kochii Grossh. 99, 736; tab. 44.
— Chanae Grossh. 82.
— chlorantha (M. B.) Roem. et Schult. 103; tab. 8.
— Chomutoviae Pasch. 102; tab. 8.
— commutata C. Koch 100; tab. 8.
— confusa Terr. 75; tab. 6.
— divaricata Rgl. 95; tab. 8.
— dshungarica Rgl. 75; tab. 6.
— dubia Terr. 90; tab. 7.
— emarginata Kar. et Kir. 88; tab. 7.
— erubescens Bess. 79; tab. 6.
— Fedtschenkoana Pasch. 81; tab. 6.
— filiformis (Ledeb.) Kunth 76.
— fistulosa (Ram.) Ker-Gawl. 83.
— gageoides (Zucc.) Vved. 110; tab. 9.
— Germainiae Grossh. 82; tab. 7.
— glacialis C. Koch 87; tab. 7.
— graminifolia Vved. 94; tab. 7.
— Granatellii Parl. 90.
— granulosa Turcz. 71; tab. 6.
— Heldreichii Terr. 93; tab. 7.
— Helenae Grossh. 80; tab. 6.
— hiensis Pasch. 72; tab. 6.
— hissarica Lipsky 95; tab. 8.
— improvisa Grossh. 102, 737; tab. 44.
— Joannis Grossh. 70, 734; tab. 6, 44.
— kopetdagensis Vved. 77; tab. 6.
— Korshinskyi Grossh. 77, 735; tab. 6.
— longiscapa Grossh. 79, 735.
— lutea (L.) Ker-Gawl. 78; tab. 6.
— minima (L.) Ker-Gawl. 70; tab. 6.
— minuta Grossh. 76, 734; tab. 6.
— minutiflora Rgl. 111; tab. 8.
— mirabilis Grossh. 88, 736; tab. 7.
— Olgae Rgl. 104; tab. 9.
— ova Stapf 109; tab. 9.
— Paczoskii (Zapal.) Grossh. 91; tab. 7.
— pamirica Grossh. 108, 738; tab. 6.
— parva Vved. 82.
— pauciflora Turcz. 111; tab. 9.
— Popovii Vved. 77; tab. 7.
— pratensis (Pers.) Roem. et Schult. 91.
— provisa Pasch. 112; tab. 9.
— pseudoerubescens Pasch. 75; tab. 6.
— pseudoreticulata Vved. 100; tab. 8.
— pusilla (Schmidt) Roem. et Schult. 80.

— reticulata (Pall.) Roem. et Schult. 101; tab. 8.
— rigida Boiss. et Sprun. 94; tab. 7.
— sacculifera Rgl. 78; tab. 6.
— samojedorum Grossh. 84, 736; tab. 7.
— spathacea (Hayne) Salisb. 70; tab. 6.
— stipitata Merkcl. ex Bge. 109; tab. 9.
— sulfurea Miscz. 87.
— Szovitsii (Lang.) Bess. 92; tab. 7.
— taurica Stev. 99; tab. 8.
— tenera Pasch. 109; tab. 9.
— tenuifolia (Boiss.) Fomin 101; tab. 8.
— tenuissima Miscz. 89; tab. 7.
— transversalis Stev. 92.
— triquetra Vved. 83; tab. 7.
— ucrainica Klok. 96; tab. 8.
— Vvedenskyi Grossh. 107, 737; tab. 6.
Gaillonia A. Rich. 23, 170.
— asperuliformis Lincz. 175, 689.
— Bruguierei A. Rich. 171.
— bucharica B. Fedtsch. et Desjat. 170; tab. 9.
— Szovitsii DC. 171; tab. 9.
Galagania Lipsky 16, 418.
Galanthus L. 4, 476.
— alpinus Sosn. 479.
— caucasicus (Baker) Grossh. 478.
— latifolius Rupr. 477.
— nivalis L. 478.
— plicatus M. B. 479; tab. 30.
— transcaucasicus Fomin 478.
— Woronowii Losinsk. 479, 749.
Galatella Cass. 25, 138.
— altaica Tzvel. 144, 582.
— angustissima (Tausch) Novopokr. 167.
— biflora (L.) Nees 151.
— chromopappus Novopokr. 149.
— coriacea Novopokr. 156.
— dahurica DC. 143; tab. 12.
— divaricata (Fisch. ex M. B.) Novopokr. 171.
— dracunculoides (Lam.) Nees 162.
— eldarica Kem.-Nath. 160.
— fastigiiformis Novopokr. 157.
— Hauptii (Ledeb.) Lindl. 166.
— hissarica Novopokr. 158.
— Litvinovii Novopokr. 158.
— macrosciadia Gand. 145.
— Pastuchovii (Kem.-Nath.) Tzvel. 159.
— polygaloides Novopokr. 165.
— punctata (Waldst. et Kit.) Nees 154.
— Regelii Tzvel. 153.
— saxatilis (M. Pop. ex Novopokr.) Novopokr. 163.
— scoparia (Kar. et Kir.) Novopokr. 172; tab. 11.
— Sosnovskyana Kem.-Nath. 161.
— tianschanica Novopokr. 164; tab. 12.
— trinervifolia (Less.) Novopokr. 152.
— Villosula Novopokr. 150.
Galatella (Cass.) Benth. et Hook. 25, 138
Galbanum D. Don 17, 278.
Galearia C. Presl 11, 226.
Galega L. 11, 303.
— officinalis L. 303; tab. 20.
— orientalis Lam. 304.
Galeobdolon Adans. 21, 138.
— luteum Huds. 139.
Galeopsis L. 21, 111.
— bifida Boenn. 119; tab. 7.
— ladanum L. 112; tab. 7.
— pubescens Bess. 115.
— speciosa Mill. 116.
— tetrahit L. 117; tab. 7.
Galeorchis Rydb. 4, 669.
— cyclochilus (Franch. et Savat.) Nevski 669; tab. 42.
Galinsoga Ruiz et Pav. 25, 563.
— parviflora Cav. 563; tab. 34.
— quadriradiata Ruiz et Pav. 563.
Galinsogaea Zucc. 25, 563.
Galium L. 23, 287.
— achurense Grossh. 363.
— amblyophyllum Schrenk ex Fisch. et Mey. 351; tab. 19.
— amurense Pobed. 350, 716.
— anfractum Somm. et Lev. 360.
— anisophyllum Vill. 338.
— aparine L. 306.
— apsheronicum Pobed. 313, 710.
— articulatum Lam. 345; tab. 19.
— atropatanum Grossh. 369.
— aureum Vis. 362.
— baicalense Pobed. 336, 714; tab. 18.
— boreale L. 349.
— brachyphyllum Roem. et Schult. 380 tab. 20.
— Braunii Zelen. 322.
— bullatum Lipsky 377; tab. 20.
— calcareum (Alb.) Pobed. 372, 719.
— caspicum Stev. 375.
— chloroleucum Fisch. et Mey. 310.
— cordatum Roem. et Schult. 324.
— coriaceum Bge. 379; tab. 21.
— coronatum Sibth. et Sm. 322.
— cruciata (L.) Scop. 319.
— Czerepanovii Pobed. 379, 720.
— dahuricum Turcz. 341.

Galium Decaisnei Boiss. 311.
— densiflorum Ledeb. 364.
— elbrussicum Pobed. 319, **711**.
— eldaricum Grossh. 311; tab. 17.
— elongatum J. et C. Presl 333.
— erectum Huds. 371.
— fistulosum Somm. et Levier 359.
— Grossheimii Pobed. 378, 720.
— grusinum Trautv. 375.
— hercynicum Weig. 336.
— hyrcanicum C. A. Mey. 376.
— Juzepczukii Pobed. 372, **719**.
— kamtschaticum Steller ex Schult. 330.
— karakulense Pobed. 332, **712**; tab. 18.
— kiapazi Manden. 340; tab. 17.
— kopetdaghense Pobed. 323, **711**; tab. 16.
— Krylovii Iljin 318; tab. 16.
— krymense Pobed. 334, **713**.
— Kutzingii Boiss. et Buhse 369; tab. 20.
— lacteum (Maxim.) Pobed. 363, **718**.
— leiophyllum Boiss. et Hohen. 377.
— Linczevskyi Pobed. 305, **710**.
— majmechense Bordz. 361; tab. 20.
— mite Boiss. et Hohen. 376.
— mollugo L. 370.
— mugodsharicum Pobed. 352, **717**.
— nupercreatum M. Pop. 313; tab. 21.
— palustre L. 331.
— pamiro-alaicum Pobed. 361, **718**.
— paradoxum Maxim. 326.
— pedemontanum (Bell.) All. 325.
— Pojarkoviae Pobed. 345, **715**.
— polonicum Blocki 374.
— pseudopolycarpon Somm. et Levier 325.
— pumilum Murr. 338.
— rubioides L. 347.
— Ruprechtii Pobed. 335, **713**; tab. 18.
— ruthenicum Willd. 367.
— saturejifolium Trev. 332; tab. 18.
— saurense Litv. 368; tab. 21.
— scabrum L. 329.
— Schischkinii Pobed. 318, **711**.
— Schultesii Vest 373.
— septentrionale Roem. et Schult. 353.
— sevanense Pobed. 324, **712**.
— soongoricum Schrenk 299; tab. 16.
— Sosnowskyi Manden. 321; tab. 17.
— spurium L. 306.
— subuliferum Somm. et Levier 358; tab. 20.
— tauricum (Willd.) Roem. et Schult. 320; tab. 16.
— tenuissimum M. B. 308; tab. 17.
— tianschanicum M. Pop. 337; tab. 17.
— transcaucasicum Stapf 309.
— trichophorum Kar. et Kir. 308; tab. 17.
— tricorne Stokes 304.
— trifidum L. 334.
— trifloriforme Kom. 303.
— triflorum Michx. 300.
— turkestanicum Pobed. 351, **717**.
— uliginosum L. 340.
— ussuriense Pobed. 348, **716**.
— valantioides M. B. 354; tab. 19.
— Vartanii Grossh. 339.
— Vassilczenkoi Pobed. 342, **714**.
— vernum Scop. 314.
— verticillatum Danth. 312.
— verum L. 357.
— volgense Pobed. 348, **715**; tab. 19.
— Wirtgenii F. Schultz 358.
Gallotropis G. Don 11, 303.
Gamanthus Bge. 6, 346.
— ferganicus Iljin 349.
— gamocarpus (Moq.) Bge. 346.
— kelifi Korov. 348; ta . 16.
— pilosus (Pall.) Bge. 348; tab. 16.
Gamochaeta Wedd. 25, 397.
Garhadiolus Jaub. et Spach 29, 231.
— angulosus Jaub. et Spach 232.
— papposus Boiss. et Buhse 231.
Gastridium Beauv. 2, 188.
— lendigerum (L.) Gaud. 188; tab. 12.
Gastrocotyle Bge. 19, 315.
— hispida (Forsk.) Bge. 316.
Gastrodia R. Br. 4, 636.
— elata Blume 636; tab. 39.
Gastrolychnis Reichb. 6, 714.
Gastrolychnis Rupr. 6, 714.
Gatyona Cass. 29, 595.
Gaudinia Beauv. 2, 283.
— fragilis (L.) Beauv. 283.
Gaudinopsis Eig 2, 258.
— macra (M. B.) Eig 258; tab. 18.
Gaultheria L. 18, 78.
— Miqueliana Takeda 78.
Gaya Gaud. 16, 579.
Genersichia Heuff. 3, 380.
Genista L. 11, 54.
— albida Willd. 57; tab. 4.
— angustifolia Schischk. 59, 389.
— armenica Spach 59.
— artwinensis Schischk. 65, 390.
— compacta Schischk. 58, 389.
— depressa M. B. 60.
— flagellaris Somm. et Levier 60.
— germanica L. 57; tab. 4.
— glaberrima Novopokr. 64.
— humifusa L. 63; tab. 4.

— Lipskii Novopokr. et Schischk. 63.
— mingrelica Albov 64.
— patula M. B. 67.
— pilosa L. 68.
— scythica Pacz. 58.
— suanica Schischk. ex Grossh. 68, 390.
— tanaitica P. Smirn. 67; tab. 4.
— tetragona Bess. 64; tab. 4.
— tinctoria L. 65.
— transcaucasica Schischk. 67.
Genistoides Moench 11, 54.
Gentiana L. 18, 538.
— acuta Michx. 607.
— algida Pall. 559.
— angulosa M. B. 585.
— aquatica L. 579.
— arctica Grossh. 588.
— asclepiadea L. 543.
— atrata Bge. 610.
— aurea L. 611.
— auriculata Pall. 601; tab. 31.
— azurea Bge. 613.
— axillaris (F. M. Schmidt) Murb. 606.
— axilliflora Lévl. et Vaniot 549.
— baltica Murb. 599.
— barbata Froel. 595.
— Biebersteinii Bge. 601.
— blepharophora Bordz. 592.
— campestris L. 600.
— carpaticola Borb. 603.
— caucasica M. B. 603; tab. 31.
— ciliata L. 591.
— Clusii Perr. et Song. 583.
— cordifolia C. Koch 551.
— cruciata L. 568.
— Dechyana Somm. et Lév. 619.
— dahurica Fisch. 566.
— decumbens L. f. 562.
— detonsa Rottb. 596.
— dichotoma Pall. 617.
— djimilensis C. Koch 575.
— Doluchanovii Grossh. 592.
— dschungarica Rgl. 550.
— excisa C. Presl 583.
— falcata Turcz. 617; tab. 32.
— Fetissowii Rgl. et Winkl. 567.
— Fischeri Smirn. 549.
— frigida Haenke 560.
— gelida M. B. 555; tab. 28.
— glauca Pall. 561.
— grandiflora Laxm. 572.
— Grossheimii Doluch. 552.
— Karelinii Griseb. 578.
— Kaufmanniana Rgl. et Schmalh. 565.
— Kolakovskyi Doluch. 552.
— Komarovii Grossh. 596.
— Krylovii Grossh. 585.
— kurilensis Grossh. 576, 750; tab. 30.
— laciniata Kit. ex Kanitz 572.
— lagodechiana (Kusn.) Grossh. 553.
— leucomelaena Maxim. 578.
— lingulata Agardh 605.
— Lipskyi Kusn. 604.
— lutea L. 540.
— macrophylla Pall. 567; tab. 29.
— Makinoi Kusn. 548.
— Marcowiczii Kusn. 604.
— nipponica Maxim. 576; tab. 30.
— nivalis L. 588.
— nutans Bge. 577; tab. 30.
— Olgae Rgl. et Schmalh. 564; tab. 29.
— oschtenica (Kusn.) Woronow 586.
— Olivieri Griseb. 569.
— Overinii (Kusn.) Grossh. 555.
— paludicola Koidz. 608.
— pamirica Grossh. 612, 751; tab. 32.
— paradoxa Albov 554.
— phlogifolia Schott et Kotschy 569.
— plebeja Cham. et Schlecht. 607.
— pneumonanthe L. 547.
— pontica Soltok 587.
— praecox A. Kerner 602.
— propinqua Richards. 609.
— prostrata Haenke 576.
— pseudoaquatica Kusn. 580.
— pulmonaris Turcz. 614.
— punctata L. 540.
— pygmaea Rgl. et Schmalh. 619; tab. 32.
— riparia Kar. et Kir. 581.
— Romanzovii Ledeb. ex Bge. 559.
— scabra Bge. 544; tab. 28.
— schistocalyx C. Koch 543; tab. 28.
— septemfida Pall. 550.
— sibirica (Kusn.) Grossh. 612.
— squarrosa Ledeb. 581.
— tenella Rottb. 618.
— tianschanica Rupr. 563.
— triflora Pall. 547.
— turkestanorum Gand. 610.
— uliginosa Willd. 605.
— umbellata M. B. 611.
— utriculosa L. 589.
— verna L. 587.
— Vvedenskyi Grossh. 597, 751.
— Walujewii Rgl. et Schmalh. 565.
— Zollingeri Fawcett 582.
Gentianopsis Ma 18, 590.
Georgina Willd. 25, 548.

Geracium Reichb. 29, 595, 606, 622, 661.
Geranium L. 14, 2.
— affine Ledeb. 33.
— albanicum M. B. 49; tab. 3.
— albiflorum Ledeb. 27.
— bifolium Patrin 30.
— bohemicum L. 59; tab. 5.
— Charlesii (Aitch. et Hemsl.) Vved. ex Nevski 62; tab. 3.
— collinum Steph. ex Willd. 43; tab. 3.
— columbinum L. 12; tab. 5.
— dahuricum DC. 11.
— depilatum (Somm. et Lev.) Grossh. 52.
— dissectum L. 14; tab. 5.
— divaricatum Ehrh. 57; tab. 5.
— erianthum DC. 23; tab. 4.
— eriostemon Fisch. 22; tab. 2, 4.
— ferganense Bobr. 43, 713; tab. 3.
— finitimum Woronow 34.
— gracile Ledeb. 46; tab. 2.
— gymnocaulon DC. 18; tab. 1.
— ibericum Cav. 19; tab. 1.
— Kotschyi Boiss. 62.
— linearilobum DC. 61; tab. 3.
— lucidum L. 36; tab. 5.
— macrorrhizum L. 34.
— Maximowiczii Rgl. et Maak 42.
— Meeboldii Briq. 45.
— molle L. 55; tab. 5.
— montanum Habl. 20; tab. 1.
— pallens M. B. 50.
— palustre L. 40.
— phaeum L. 39; tab. 2.
— platypetalum Fisch. et Mey. ex Hohen. 20.
— pratense L. 31.
— pseudosibiricum J. Mayer 29.
— psilostemon Ledeb. 18.
— pusillum Burm. f. 52; tab. 5.
— pyrenaicum Burm. f. 51; tab. 3.
— rectum Trautv. 30; tab. 2.
— Renardii Trautv. 17; tab. 1.
— Robertianum L. 35; tab. 5.
— rotundifolium L. 56; tab. 5.
— Ruprechtii Woronow 33.
— sanguineum L. 10.
— saxatile Kar. et Kir. 44; tab. 3.
— Schrenkianum Trautv. ex A. Beck. 13.
— sibiricum L. 57.
— Sieboldii Maxim. 21; tab. 4.
— silvaticum L. 24.
— soboliferum Kom. 12; tab. 4.
— Sophiae Fed. 40.
— subcaulescens L'Her. 10.
— tauricum Rupr. 50; tab. 3.
— transbaicalicum Serg. 32; tab. 2.
— transversale (Kar. et Kir.) Vved. 61.
— tuberosum L. 60; tab. 3.
— Vlassovianum Fisch. ex Link 41; tab. 2, 4.
— Wilfordii Maxim. 49; tab. 4.
Gerbera auct. 28, 592.
Gerbera Cass. 28, 589.
— Jamesonii Bolus 590.
Geropogon L. 29, 197.
— hybridus (L.) Sch. Bip. 197; tab. 8.
Geum L. 10, 251.
— aleppicum Jacq. 254; tab. 19.
— coccineum Sibth. et Smith 253.
— Fauriei Lèvl. 256.
— glabricaule Juz. 255, 615.
— latilobum Somm. et Lev. 259.
— macrophyllum Willd. 259; tab. 19.
— rivale L. 252; tab. 19.
— urbanum L. 260; tab. 19.
Gifola Cass. 25, 314, 316.
Gigachilon Seidl. 2, 688.
Gingidium Forst. 17, 11.
Giraldia Baroni 27, 88.
Girardinia Gaud. 5, 395.
— cuspidata Wedd. 395.
Girgensohnia Bge. 6, 277.
— diptera Bge. 281.
— minima Korov. 278.
— oppositiflora (Pall.) Fenzl 277; tab. 14.
Gladiolus L. 4, 578.
— atroviolaceus Boiss. 587; tab. 36.
— communis L. 582; tab. 36.
— halophilus Boiss. et Heldr. 588.
— imbricatus L. 583; tab. 36.
— Kotschyanus Boiss. 584.
— palustris Gaud. 583.
— segetum Ker.-Gawl. 580; tab. 36.
— tenuiflorus C. Koch 582; tab. 36.
— turkmenorum Czerniak. 581, 750; tab. 36.
Glaucium Adans. 7, 585.
— bracteatum M. Pop. 595; tab. 40.
— corniculatum (L.) Curt. 588; tab. 40.
— elegans Fisch. et Mey. 594.
— fimbrilligerum Boiss. 593.
— flavum Crantz 592.
— grandiflorum Boiss. et Huet. 591.
— insigne M. Pop. 587.
— leiocarpum Boiss. 593.
— oxylobum Boiss. et Buhse 591.
— squamigerum Kar. et Kir. 594; tab. 40.
Glaux L. 18, 273.
— maritima L. 274.

Glebionis Cass. 26, 131, 134.
Glechoma L. 20, 437.
— hederacea L. 437.
— hirsuta Waldst. et Kit. 438.
— longituba (Nakai) Kuprian. 439.
Gleditschia L. 11, 21.
— caspia Desf. 21.
— triacanthos L. 22.
Glehnia Fr. Schmidt 17, 42.
— litoralis Fr. Schmidt 42; tab. 4.
Glinus Loefl. ex L. 6, 375.
— lotoides L. 375.
Globocarpus Car. 16, 526.
Globularia L. 23, 128.
— aphyllanthes Graebn. 129.
— trichosantha Fisch. et Mey. 130.
Glossanthis Poljak. 26, 273, 277.
Glossocomia D. Don 24, 433.
Glyceria R. Br. 2, 449.
— alnasteretum Kom. 457; tab. 34.
— aquatica (L.) Wahl. 458; tab. 34.
— arundinacea (M. B.) Kunth 459; tab. 34.
— caspica Trin. 454; tab. 34.
— fluitans (L.) R. Br. 451; tab. 34.
— kamtschatica Kom. 459, 758; tab. 34.
— leptorhiza (Maxim.) Kom. 452; tab. 34.
— lithuanica (Gorski) Lindm. 453; tab. 34.
— natans Kom. 453; tab. 34.
— nemoralis Uechtr. et Koern. 453; tab. 34.
— orientalis Kom. 454; tab. 34.
— plicata Fries 452; tab. 34.
— spiculosa (Fr. Schmidt) Roshev. 457; tab. 34.
— triflora (Korsh.) Kom. 459, 758; tab. 34.
— turcomanica Kom. 452, 757.
— ussuriensis Kom. 460, 758; tab. 34.
Glycine L. 13, 527.
— gracilis Skvorts. 528.
— hispida (Moench) Maxim. 529.
— ussuriensis Rgl. et Maack 528.
Glycyrrhiza L. 13, 230.
— aspera Pall. 234.
— bucharica Rgl. 231.
— echinata L. 238.
— glabra L. 233.
— Gontscharovii Masl. 232.
— Korshinskyi Grig. 237.
— kulabensis Masl. 233.
— laxissima Vass. 235.
— macedonica Boiss. 238.
— pallidiflora Maxim. 239.
— uralensis Fisch. 236.
— zaissanica Serg. 235.
Glycyphylla Stev. 12, 91.
Gnaphalium L. 25, 342, 384, 404.
Gnaphalium L. 25, 381.
— baicalense Kirp. 386.
— caucasicum Somm. et Lev. 401; tab. 24.
— kasachstanicum Kirp. 391.
— luteo-album L. 395.
— mandshuricum Kirp. 385.
— norvegicum Gunn. 399.
— rossicum Kirp. 389; tab. 24.
— sibiricum Kirp. 390.
— supinum L. 402.
— sylvaticum L. 397.
— Tranzschelii Kirp. 384; tab. 24.
— uliginosum L. 391.
Goldbachia DC. 8, 236.
— laevigata (M. B.) DC. 238; tab. 20.
— papulosa Vass. 237, 643; tab. 20.
— reticulata (Kuntze) Vass. 241, 644; tab. 20.
— verrucosa Kom. 237; tab. 20.
Golenkinianthe K.-Pol. 16, 94, 109.
Golia Adans. 18, 249.
Colowninia Maxim. 18, 536.
Gomphocarpus R. Br. 18, 668.
— fruticosus (L.) R. Br. 668; tab. 36.
Gomphopetalum Turcz. 17, 10, 11, 21.
Gomphrena L. 6, 369.
— globosa L. 369; tab. 18.
Goniolimon Boiss. 18, 382.
— Besserianum (Schult.) Kusn. 403.
— callicomum (C. A. Mey.) Boiss. 394.
— cuspidatum Gamajun. 396.
— dschungaricum (Rgl.) O. et B. Fedtsch. 390.
— elatum (Fisch.) Boiss. 395.
— eximium (Schrenk) Boiss. 391.
— graminifolium (Ait.) Boiss. 404.
— orthocladum Rupr. 392.
— rubellum (S. G. Gmel.) Klok. 401.
— Severzovii Herd. 386.
— speciosum (L.) Boiss. 388; tab. 20.
— tataricum (L.) Boiss. 397; tab. 20.
Gontscharovia Boriss. 21, 628.
— Popovii (B. Fedtsch. et Gontsch.) Boriss. 629; tab. 22.
Goodyera R. Br. 4, 639.
— repens (L.) R. Br. 639; tab. 39.
Gossypium L. 15, 170.
— albescens Raf. 179.
— Figarei Tod. 180.
— frutescens Lasteyr. 182.
— herbaceum L. 175.

8*

Gossypium hirsutum L. 178.
— Jumelianum (Tod.) Prokh. 183.
— Zaitzevii Prokh. 176.
Graellsia Boiss. 8, 201.
— saxifragifolia (DC.) Boiss. 202.
Grammarthron Cass. 26, 669, 671.
Grammopetalum C. A. Mey. ex Claus 16, 349.
Grammosciadium DC. 16, 120.
— daucoides DC. 120; tab. 15.
Gratiola L. 22, 321.
— japonica Miq. 323.
— officinalis L. 322.
Gregoria Duby 18, 208.
Grindelia Willd. 25, 30.
— squarrosa (Pursh) Dun. 30.
Groenlandia J. Gay 1, 261.
— densa (L.) Fourr. 261; tab. 12.
Grossheimia Sosn. et Takht. 28, 332.
— macrocephala (Muss.-Puschk.) Sosn. et Takht. 333.
— ossica (C. Koch) Sosn. et Takht. 334.
Grossularia Mill. 9, 267.
— acicularis (Smith) Spach 269; tab. 15.
— burejensis (Fr. Schmidt) Berger 270.
— reticulata (L.) Mill. 268.
Gueldenstaedtia Fisch. 11, 384.
— monophylla Fisch. 385; tab. 25.
— pauciflora Fisch. 386.
Guizotia Cass. 25, 545.
— abyssinica (L. f.) Cass. 545.
Gundelia L. 26, 862.
— Tournefortii L. 862.
Gymnadenia R. Br. 4, 667.
— conopsea (L.) R. Br. 668; tab. 40.
— odoratissima (L.) Rich. 667.
Gymnandra Pall. 22, 500.
Gymnaster Kitam. 25, 58.
— Savatieri (Makino) Kitam. 58; tab. 4
Gymnocline Cass. 26, 184, 193, 330.
Gypsophila L. 6, 731.
— acutifolia Fisch. 757.
— albida Schischk. 753.
— alsinoides Bge. 773; tab. 48.
— altissima L. 750.
— anatolica Boiss. et Heldr. 761.
— Antoninae Schischk. 744.
— aretioides Boiss. 744.
— aulieatensis B. Fedtsch. 757.
— bicolor Freyn et Sint. 749.
— brachypetala Trautv. 770.
— bucharica B. Fedtsch. 771.
— capitata M. B. 745; tab. 46.
— capituliflora Rupr. 768; tab. 48.
— cappadocia Boiss. et Bal. 747.
— cephalotes (Schrenk) Will. 752; tab. 47.
— dahurica Turcz. ex Fenzl 768.
— desertorum (Bge.) Fenzl 740; tab. 46.
— dichotoma Bess. 748.
— diffusa Fisch. et Mey. 753; tab. 47.
— elegans M. B. 764.
— eriocalyx Boiss. 757.
— fastigiata L. 747; tab. 47.
— Fedtschenkoana Schischk. 770; tab. 48.
— filipes (Boiss.) Schischk. 775.
— floribunda (Kar. et Kir.) Turcz. ex Ledeb. 775.
— glandulosa (Boiss.) Walp. 738.
— glauca Stev. 759.
— globulosa Stev. ex Bess. 746; tab. 46.
— glomerata Pall. 746; tab. 46.
— herniarioides Boiss. 743; tab. 46.
— heteropoda Freyen 767.
— hispida Boiss. 770.
— imbricata Rupr. 745; tab. 46.
— intricata Franch. 771.
— Krascheninnikovii Schischk. 751.
— linearifolia (Fisch. et Mey.) Boiss. 772.
— Lipskyi Schischk. 769; tab. 48.
— Litwinowii K.-Pol. 751.
— Meyeri Rupr. 759.
— microphylla (Schrenk) Fenzl 740; tab. 46.
— muralis L. 774.
— pacifica Kom. 768.
— paniculata L. 749; tab. 47.
— Patrinii Ser. 767.
— picta Boiss. 776.
— Popovii Preobr. 772.
— porrigens (L.) Fenzl 776.
— Preobrashenskyi Czernjak. 769.
— robusta Grossh. 761.
— Sambukii Schischk. 763; 892.
— scorzonerifolia Ser. 760.
— sericea (Ser.) Fenzl 739.
— silenoides Rupr. 764.
— simulatrix Bornm. et Woronow 752.
— spathulifolia (Fisch. et Mey.) Fenzl 773.
— stepposa Klok. 774.
— Steupii Schischk. 762.
— Stevenii Fisch. ex Bess. 758.
— Szowitsii Fisch. et Mey. 754.
— tenuifolia M. B. 762.
— tianschanica M. Pop. et Schischk. 750, 891.
— trichotoma Wend. 759.
— turkestanica Schischk. 743.
— ucrainica Kleop. 748, 891.

— uralensis Less. 763.
— violacea (Ledeb.) Fenzl 739.
— virgata Boiss. 754; tab. 47.
— yorae Woronow 761; tab. 48.
Gyrostachys Pers. 4, 637.
Gytonanthus Raf. 23, 587.

Habenaria Willd. 4, 641.
— dianthoides Nevski 642, 751; tab. 40.
— linearifolia Maxim. 642; tab. 40.
Hablitzia M. B. 6, 32.
— tamnoides M. B. 32; tab. 2.
Hackelia Opiz 19, 479.
— deflexa (Wahl.) Opiz 480.
— thymifolia (DC.) M. Pop. 481.
Haenselera Lag. 16, 223.
Hagenia Moench 6, 776.
Hahnia Medic. 9, 387.
Halanthium C. Koch 6, 340.
— kulpianum (C. Koch) Bge. 344.
— Lipskyi Pauls. 345.
— rariflorum C. Koch 343; tab. 16.
— roseum (Trautv.) Iljin 345.
Halenia Borkh. 18, 638.
— corniculata (L.) Cornaz 638.
— elliptica D. Don 639; tab. 34.
Halerpestes Green 7, 329.
— ruthenica (Jacq.) Ovcz. 331; tab. 34.
— salsuginosa (Pall.) Green 330; tab. 34.
Halianthus Fries 6, 516.
Halimocnemis C. A. Mey. 6, 329.
— Beresinii Iljin 334.
— glaberrima Iljin 335.
— Karelinii Moq. 331; tab. 17.
— lasiantha Iljin 336; tab. 17.
— latifolia Iljin 338; tab. 17.
— longifolia Bge. 333; tab. 17.
— macranthera Bge. 336.
— mollissima Bge. 334.
— sclerosperma (Pall.) C. A. Mey. 332; tab. 17.
— Smirnowii Bge. 337; tab. 17.
— villosa Kar. et Kir. 333; tab. 17.
Halimodendron Fisch. 11, 323.
— halodendron (Pall.) Voss 323; tab. 20.
Halimus Wallr. 6, 106.
Halocharis Moq. 6, 327.
— hispida (C. A. Mey.) Bge. 327; tab. 14.
— lachnantha Korov. 329, 880.
— turcomanica Iljin 328.
Halocnemum M. B. 6, 170.
— strobilaceum (Pall.) M. B. 171; tab. 8.
Halodendron DC. 11, 323.
Halogeton C. A. Mey. 6, 350.
— arachnoideus Moq. 352.
— glomeratus (M. B.) C. A. Mey. 351; tab. 14.
— tibeticus Bge. 352.
Halopeplis Bge. 6, 168.
— pygmaea (Pall.) Bge. 169; tab. 8.
Haloscias Fries 16, 566, 568.
Halostachys C. A. Mey. 6, 169.
— caspica (Pall.) C. A. Mey. 169; tab. 8.
Halotis Bge. 6, 338.
— pilosa (Moq.) Iljin 339; tab. 17.
Haloxylon Bge. 6, 310.
— ammodendron (C. A. Mey.) Bge. 313.
— aphyllum (Minkw.) Iljin 312; tab. 10.
— persicum Bge. ex Boiss. 311; tab. 10.
Handelia Heimerl 26, 124.
— trichophylla (Schrenk) Heimerl 125.
Hansenia Turcz. 16, 566.
Hapalostephium D. Don 29, 595, 606, 627.
Haplophyllum A. Juss. 14, 200.
— acutifolium (DC.) G. Don 226.
— affine (Aitch. et Hemsl.) Korov. 212.
— Alberti-Regelii Korov. 221.
— Bourgaei Boiss. 211.
— bucharicum Litv. 211.
— Bungei Trautv. 222.
— ciliatum Griseb. 208.
— ciscaucasicum Grossh. et Vved. 218.
— dauricum (L.) G. Don 14, 205.
— dshungaricum Rubtz. 216.
— dubium Korov. 220.
— Eugenii-Korovinii Pavl. 213.
— ferganicum Vved. 223.
— foliosum Vved. 216.
— Kowalenskyi Stschegl. 212.
— latifolium Kar. et Kir. 225.
— leptomerum Lincz. et Vved. 217, 731; tab. 12.
— monadelphum Afan. 215.
— multicaule Vved. 213, 731.
— obtusifolium Ledeb. 214.
— pedicellatum Bge. 14, 206.
— perforatum (M. B.) Kar. et Kir. 226.
— Popovii Korov. 222; tab. 12.
— ramosissimum Vved. 214.
— robustum Bge. 223.
— Schelkovnikovii Grossh. 219.
— suaveolens (DC.) G. Don 14; 207.
— tenue Boiss. 217.
— tenuisectum Lincz. et Vved. 216, 732.
— versicolor Fisch. et Mey. 224.
— villosum (M. B.) G. Don 218; tab. 12.
— Vvedenskyi Nevski 220.
Haplotaxis Endl. 27, 361.

Happophaestum S. F. Gray 28, 573.
Harmsiella Briq. 21, 182.
Harrimanella Cov. 18, 70.
— hypnoides (L.) Cov. 70.
— Stelleriana (Pall.) Cov. 71.
Harrisonia Neck. 27, 59.
Hasselquistia L. 17, 276.
Hasteola Raf. 26, 683.
Haynaldia Schur 2, 665.
— villosa (L.) Schur 665; tab. 47a.
Hecatonia Lour. 7, 351, 407.
Hedera L. 16, 3.
— caucasigena Pojark. 587; tab. 1.
— colchica C. Koch 5.
— chrysocarpa Walsh 8.
— helix L. 10.
— Pastuchovii Woronow 7; tab. 1.
— taurica Carr. 16.
Hedinia Ostenf. 8, 547.
— tibetica (Thoms.) Ostenf. 547.
Hedraianthus Griseb. 24, 442.
Hedranthus Rupr. 24, 442.
Hedyphylla Stev. 12, 91.
Hedypnois Schreb. 29, 229.
— cretica (L.) Willd. 230; tab. 1.
— persica M. B. 230.
Hedysarum L. 13, 259.
— alaicum B. Fedtsch. 294.
— alpinum L. 279.
— amankutanicum B. Fedtsch. 292.
— arcticum B. Fedtsch. 269.
— argenteum M. B. 310.
— armenum Boiss. 277.
— atropatanum Bge. ex Boiss. 285.
— auriculatum Eastw. 275.
— austrosibiricum B. Fedtsch. 271.
— babatagicum Korotk. 317.
— baicalense B. Fedtsch. 312.
— balchanense Boriss. 316.
— baldshuanicum B. Fedtsch. 294.
— Bordzilovskyi Grossh. 310.
— Branthii Trautv. 280; tab. 14.
— bucharicum B. Fedtsch. 318.
— candidum M. B. 308.
— caucasicum M. B. 277.
— cephalotes Franchet 313.
— chaitocarpum Rgl. et Schmalh. 294.
— connatum B. Fedtsch. 277.
— consanguineum DC. 270.
— cretaceum Fisch. 301.
— daghestanicum Rupr. ex Boiss. 311.
— dahuricum Turcz. 290.
— dasycarpum Turcz. 288.
— denticulatum Rgl. et Schmalh. 305.
— Drobovii Korotk. 288.
— elegans Boiss. et Huet 310.
— Fedtschenkoanum Rgl. et Schmalh. 307; tab. 14.
— ferganense Korsh. 312.
— flavescens Rgl. et Schmalh. 278.
— flexuosum L. 268.
— formosum Fisch. et Mey. 281.
— fruticosum Pall. 266.
— Gmelinii Ledeb. 289.
— grandiflorum Pall. 309.
— hedysaroides (L.) Schinz et Thell. 272.
— ibericum M. B. 287.
— iliense B. Fedtsch. 308; tab. 15.
— inundatum Turcz. 270.
— iomuticum B. Fedtsch. 287; tab. 15.
— jaxarticum M. Pop. 303.
— karataviense B. Fedtsch. 297.
— kirghisorum B. Fedtsch. 274.
— kopetdaghi Boriss. 286.
— Korshinskyanum B. Fedtsch. 282; tab. 15.
— Krassnovii B. Fedtsch. 267.
— Krylovii Sumn. 314.
— Kudrjaschevii Korotk. 303.
— kuhitangi Boriss. 306.
— Lehmannianum Bge. 304.
— Lipskyi B. Fedtsch. 285.
— macranthum Freyn et Sint. 317; tab. 14.
— magnificum Kudr. 282.
— microphyllum Turcz. 315.
— minussinense B. Fedtsch. 302.
— mogianicum B. Fedtsch. 293.
— monophyllum Boriss. 306.
— montanum B. Fedtsch. 292.
— neglectum Ledeb. 273.
— nuratense M. Pop. 317.
— Olgae B. Fedtsch. 318; tab. 14.
— Poncinsii Franch. 314.
— pskemense M. Pop. 295.
— Razoumovianum Fisch. 302.
— sachalinense B. Fedtsch. 276.
— santalaschi B. Fedtsch. 296.
— Schischkinii Sumn. 314.
— scoparium Fisch. et Mey. 267; tab. 15.
— Semenovii Rgl. et Herd. 278.
— sericeum M. B. 311.
— setigerum Turcz. 313.
— Severzovii Bge. 295.
— songoricum Bong. 291.
— splendens Fisch. 308.
— subglabrum (Kar. et Kir.) B. Fedtsch. 291.
— taschkendicum M. Pop. 290.

— tauricum L. 298.
— tenuifolium B. Fedtsch. 293.
— truncatum Eastw. 275.
— turkestanicum Rgl. et Schmalh. 315.
— Turkeviczii B. Fedtsch. 297; tab. 14.
— ucrainicum Kaschm. 301.
— ussuriense I. Schischk. et Kom. 280.
— varium Willd. 286.
— vegetius B. Fedtsch. 316.
— vicioides Turcz. 280.
Hegemone Bge. 7, 53.
— lilacina Bge. ex Ledeb. 53.
Hegemone Prantl 7, 53.
Heldreichia Boiss. 8, 593.
— longifolia Boiss. 594.
Heleocharis R. Br. 3, 63.
— acicularis (L.) Roem. et Schult. 70; tab. 6.
— afflata Steud. 88; tab. 7.
— argyrolepidoides Zinserl. 79, 584; tab. 6.
— argyrolepis Kjerulff ex Bge. 79; tab. 6.
— atropurpurea (Retz.) Kunth 72; tab. 6.
— carniolica C. Koch 89.
— crassa Fisch. et Mey. ex Beck. 77, 582; tab. 6.
— ecarinata Zinserl. 78, 583; tab. 6.
— equisetiformis (Meinsh.) B. Fedtsch. 80; tab. 6.
— eupalustris Lindb. f. 76, 582; tab. 6.
— euuniglumis Zinserl. 82, 584; tab. 6.
— fennica Palla 87, 587; tab. 7.
— globularis Zinserl. 78, 582; tab. 6.
— intersita Zinserl. 76, 581; tab. 6.
— kamtschatica (C. A. Mey.) Kom. 81; tab. 6.
— kasakstanica Zinserl. 78, 583; tab. 6.
— Klingei (Meinsh.) B. Fedtsch. 82, 585; tab. 7.
— Komarovii Zinserl. 81, 588; tab. 6.
— Korshinskyana Zinserl. 88, 588; tab. 7.
— leptostylopodiata Zinserl. 75, 581; tab. 6
— mamillata Lindb. f. 75; tab. 6.
— margaritacea Hult. 89; tab. 7.
— Maximoviczii Zinserl. 88, 588; tab. 7.
— meridionalis Zinserl. 69, 580; tab. 6.
— multiseta Zinserl. 83, 586; tab. 7.
— oxylepis (Meinsh.) B. Fedtsch. 83, 586; tab. 7.
— ovata (Roth) Roem. et Schult. 71; tab. 6, 7.
— parvula (Roem. et Schult.) Link 68; tab. 6, 7.
— paucidentata Zinserl. 87, 588; tab. 7.
— pauciflora (Lightf.) Link 69; tab. 6.
— petasata (Maxim.) Zinserl. 90, 589; tab. 7.
— sachalinensis (Meinsh.) Kom. 80; tab. 6.
— scythica Zinserl. 84, 586; tab. 7.
— septentrionalis Zinserl. 83, 586; tab. 7.
— Svensonii Zinserl. 71, 580; tab. 6.
— transcaucasica Zinserl. 81, 585; tab. 6.
— turcomanica Zinserl. 79, 583; tab. 6.
— ussuriensis Zinserl. 75, 581; tab. 6.
Heleochloa turkestanica Litv. 2, 743.
Helianthemum Adans. 15, 330.
— alpestre (Jacq.) DC. 343.
— Buschii (Palib.) Juz. et Pozd. 344.
— canum (L). Baumg. 345.
— ciscaucasicum Juz. et Pozd. 346, 680.
— dagestanicum Rupr. 332.
— georgicum Juz. et Pozd. 347, 680.
— grandiflorum (Scop.) Lam. et DC. 334; tab. 17.
— hirsutum (Thuill.) Mérat 333.
— lasiocarpum Willk. 337.
— ledifolium (L.) Mill. 336.
— nummularium (L.) Dunal 334.
— orientale (Grosser) Juz. et Pozd. 344.
— rupifragum A. Kerner 340.
— salicifolium (L.) Mill. 338.
— soongoricum Schrenk 338.
— Stevenii Rupr. ex Juz. et Pozd. 346, 679; tab. 17.
— tomentosum (Scop.) Spreng. 335.
Helianthus L. 25, 541.
— annuus L. 542.
— atrirubens L. 544.
— tuberosus L. 544.
Helichrysum Mill. 25, 404.
— araxinum Takht. ex Kirp. 424.
— arenarium (L.) Moench 410.
— armenium (Fisch. et Mey.) DC. 422.
— aurantiacum Boiss. et Huet ex Boiss. 429.
— callichrysum (Fisch. et Mey.) DC. 424; tab. 26.
— graveolens (M. B.) Sweet 421.
— kopetdagense Kirp., 415.
— maracandicum M. Pop. ex Kirp. 420.
— Mussae Nevski 417; tab. 25.
— nuratavicum Krasch. 417.
— Pallasii (Spreng.) Ledeb. 428; tab. 26.
— plicatum (Fisch. et Mey.) DC. 419.
— polylepis Bordz. ex Grossh. 427.
— polyphyllum Ledeb. 418.
— thianschanicum Rgl. 415; tab. 25.
— undulatum Ledeb. 412.
Helicilla Moq. 6, 177.
Heliopsis Pers. 25, 535.

Heliopsis scabra Dun. 535.
Heliotropium L. 19, 120.
— acutiflorum Kar. et Kir. 150.
— arguzioides Kar. et Kir. 131.
— biannulatiforme M. Pop. 133.
— bucharicum B. Fedtsch. 132.
— chorassanicum Bge. 137.
— dasycarpum Ledeb. 125.
— ellipticum Ledeb. 147; tab. 8.
— europaeum L. 145.
— Fedtschenkoanum M. Pop. 136; tab. 7.
— grande M. Pop. 126; tab. 8.
— Kowalenskyi Stchegl. 135.
— Litvinovii M. Pop. 149; tab. 8.
— micranthum (Pall.) Bge. 151.
— Olgae Bge. 138; tab. 8.
— parvulum M. Pop. 150.
— seravschanicum M. Pop. 140; tab. 8.
— styligerum Trautv. 144.
— suaveolens M. B. 140; tab. 8.
— supinum L. 153.
— Szovitsii (Stev.) Bge. 134; tab. 8.
— transoxanum Bge. 126.
— turcomanicum M. Pop. et Korov. 128; tab. 8.
Helleborine Hill 4, 620.
Helleborus L. 7, 57.
— abchasicus A. Br. 59; tab. 6.
— caucasicus A. Br. 58; tab. 6.
— guttatus A. Br. 59.
— purpurascens Waldst. et Kit. 58.
Helminthia Juss. 29, 227.
— echioides (L.) Gaertn. 228; tab. 12.
Helodium Dum. 16, 372.
Helosciadium Koch 16, 372.
— nodiflorum (L.) Koch 372; tab. 30.
Hemarthria R. Br. 2, 12.
— fasciculata (Lam.) Kunth 12.
— japonica (Hack.) Roshev. 13; tab. 1.
Hemerocallis L. 4, 55.
— Dumortieri Morr. 60.
— flava L. 59; tab. 5.
— fulva L. 56.
— Middendorfii Trautv. et Mey. 59.
— minor Mill. 60.
Hemipappus C. Koch 26, 359.
— canus C. Koch 360.
Hemisphace Benth. 21, 359.
Hepatica Mill. 7, 282.
— Falconeri (Thoms.) Juz. 284.
— nobilis Gars. 283; tab. 17.
Heracleum L. 17, 223.
— aconitifolium Woronow 233.
— Albovii Manden. ex Grossh. 255.
— antasiaticum Manden. ex Grossh. 249.
— apiifolium Boiss. 256.
— asperum M. B. 237.
— barbatum Ledeb. 238.
— calcareum Albov 240.
— carpaticum Porc. 234.
— caspicum DC. 259.
— chorodanum (Hoffm.) DC. 252.
— Clausii Ledeb. 259.
— colchicum Lipsky 239.
— cuneiforme DC. 259.
— cyclocarpum C. Koch 232.
— dissectum Ledeb. 238.
— dulce Fisch. 236.
— grandiflorum Stev. ex M. B. 250.
— Grossheimii Manden. ex Grossh. 243.
— jugatum Boiss. 259.
— Lehmannianum Bge. 246.
— Leskovii Grossh. 249.
— ligusticifolium M. B. 257.
— Mantegazzianum Somm. et Lévier 242.
— Olgae Rgl. et Schmalh. 257.
— osseticum Manden. 241.
— palmatum Baumg. 234.
— pastinacifolium C. Koch 251.
— ponticum (Lipsky) Schischk. ex Grossh. 235.
— pubescens M. B. 242.
— roseum Stev. 253.
— scabrum Albov 247.
— Schelkovnikovii Woronow 254.
— sibiricum L. 231.
— Sommieri Manden. ex Grossh. 245.
— Sosnowskyi Manden. 244.
— sphondylium L. 236.
— Stevenii Manden. 248.
— trachyloma Fisch. et Mey. 246.
— transcaucasicum Manden. ex Grossh. 253.
— transiliense (Rgl. et Herd.) O. et B. Fedtsch. 258.
— Wilhelmsii Fisch. et Avé-Lall. 244.
Herminium R. Br. 4, 643.
— monorchis (L.) R. Br. 643.
Herniaria L. 6, 566.
— caucasica Rupr. 567; tab. 34.
— glabra L. 567; tab. 34.
— hirsuta L. 571; tab. 34.
— incana Lam. 568; tab. 34.
— polygama J. Gay 571; tab. 34.
Hesiodia Moench 20, 253, 268.
Hesperis L. 8, 242.
— armena Boiss. 244.
— bicuspidata (Willd.) Poir. 243.
— karsiana N. Busch 247.

— matronalis L. 244.
— Meyeriana (Trautv.) N. Busch 246.
— pendula DC. 251.
— persica Boiss. 248.
— sibirica L. 246.
— silvestris Crantz 245.
— Steveniana DC. 246.
— tristis L. 248.
— violacea Boiss. 244.
— Voronovii N. Busch 247; tab. 14.
Heteracia Fisch. et Mey. 29, 591.
— Szovitsii Fisch. et Mey. 591.
Heteranthelium Hochst. 2, 689.
— piliferum Hochst. 689; tab. 47a.
Heterocaryum DC. 19, 537.
— echinophorum (Pall.) Brand 544.
— laevigatum (Kar. et Kir.) DC. 539; tab. 26.
— macrocarpum Zak. 545.
— oligacanthum (Boiss.) Bornm. 546.
— rigidum DC. 540; tab. 26.
— Szovitsianum (Fisch. et Mey.) DC. 546; tab. 26.
Heterochaeta DC. 25, 191.
Heterochroa Bge. 6, 738.
Heteroderis (Bge.) Boiss. 29, 587.
— leucocephala (Bge.) Leonova 590.
— pusilla (Boiss.) Boiss. 588.
Heterolophus Cass. 28, 370, 463.
Heteropappus Less. 25, 61.
— Alberti (Rgl.) Novopokr. et Tamamsch. 67.
— altaicus (Willd.) Novopokr. 66.
— canescens (Nees) Novopokr. 67.
— distortus (Turcz.) Tamamsch. 66; tab. 5.
— elisabethinus Tamamsch. 70, 578.
— hispidus (Thunb.) Less. 69; tab. 5.
— medius (Kryl.) Tamamsch. 68.
— Meyendorfii (Rgl. et Maack) Kom. et Aliss. 72.
— noneifolius Tamamsch. 71, 578.
— pinetorum Kom. 69.
— saxomarinus Kom. 70.
— tataricus (Lindl.) Tamamsch. 71.
Heterotrichum M. B. 27, 361.
Hexakistra Hook. f. 14, 269.
Hibiscus L. 15, 151.
— cannabinus L. 153.
— esculentus L. 165.
— planifolius Sweet 168.
— ponticus Rupr. 157; tab. 8.
— rosa-sinensis L. 156.
— sabdariffa L. 155.
— syriacus L. 152.
— tiliaceus L. 152.
— trionum L. 159.
Hieracioides auct. 29, 699, 703.
Hieracioides Rupr. 29, 595.
Hieracium L. 30, 1.
— abakurae Schelk. et Zahn 651.
— abastumanense Juxip 314.
— accline Norrl. 581.
— achalzichiense Juxip 113.
— acrifolium Dahlst. 96.
— acrocomum Naeg. et Peter 580; tab. 40.
— acrogymnon Malme 389.
— acroleucoides Dahlst. 249.
— acroleucum Stenstr. 239.
— acrosciadium Naeg. et Peter 512.
— acrothyrsum Naeg. et Peter 524.
— acroxanthum Sosn. et Zahn 53.
— acuminatifolium Litv. et Zahn 259.
— acutangulum Kozl. et Zahn 48.
— acutisquamum Naeg. et Peter 669.
— aczelmanicum Schischk. et Serg. 224
— adelum Juxip 351.
— adenoactis Juxip 320.
— adenobrachion Litv. et Zahn 22.
— adjarianum Peter 496.
— adspersum Norrl. 166.
— adunans Norrl. 251.
— aeruginascens Norrl. 663.
— agnostum Juxip 383.
— agronesaeum Juxip 230.
— Akinfiewii Woronow et Zahn 401.
— akjaurense Norrl. 168.
— alatavicum Zahn 52.
— albellipes Schelk. et Zahn 75.
— albidulum Stenstr. 383.
— albipes Dahlst. 347.
— albocinereum Rupr. 424.
— albocostatum Norrl. em. Juxip 140.
— alphostictum Dahlst. 216.
— alpinum L. 157; tab. 20.
— altaicum Naeg. et Peter 598.
— alticaule Litv. et Zahn 536.
— altipes Lindb. f. ex Norrl. 337.
— alupkanum Zahn 488.
— amauranthum Peter 510.
— amaureilema Naeg. et Peter 675.
— amaurobasis Litv. et Zahn 495.
— amblylobum Juxip 236.
— amnoon Naeg. et Peter 466.
— amoeniceps Zahn 648.
— amphileion Pohle et Zahn 125.
— amphitephrodes Sosn. et Zahn 69.
— anfractum Fries 266.
— anocladum Naeg. et Peter 531.

— Hieracium antennarioidiforme Zahn 683.
— apatelioides Zahn 586.
— apatelium Naeg. et Peter 633; tab. 29.
— apatitorum Juxip 288.
— apatorium Naeg. et Peter 519.
— aphanum Juxip 349.
— apiculatum Tausch 161.
— aquilonare (Naeg. et Peter) Zahn 438.
— arctogeton Zahn 126.
— arcuatidens Zahn 263.
— argillaceoides Litv. et Zahn 269.
— armeniacum Naeg. et Peter 465.
— artabirense Zahn 74.
— artvinense Woronow et Zahn 33.
— arvense Naeg. et Peter 492.
— arvicola Naeg. et Peter 517; tab. 35.
— arvorum Naeg. et Peter 458.
— aryslynense Zahn 68.
— asiaticum Naeg. et Peter 419.
— asperrimum Schur 491.
— assimilatum Norrl. 518.
— asterodermum Woronow et Zahn 133.
— astibes Juxip 366.
— atrellum Zahn 189.
— aurantiacum L. 653.
— auratum Fries 84.
— auricula Lam. et DC. ex Naeg. et Peter 670; tab. 32.
— Aurorinii Juxip 379.
— aurosulum Norrl. 646; tab. 40.
— Baenitzii Naeg. et Peter 605.
— bakurianense Fomin et Zahn 53.
— Balansae Boiss. 409.
— barbulatulum Pohle et Zahn 188.
— basifolium (Fries) Almq. 348.
— basileucum Litv. et Zahn 493.
— bauhiniflorum Naeg. et Peter 532.
— beschtaviciforme Juxip 218.
— beschtavicum Litv. et Zahn 47.
— Besserianum Spreng. 461.
— bichloricolor Ganesch. et Zahn 117; tab. 10.
— Biebersteinii Litv. et Zahn 49; tab. 6.
— bimanum Norrl. 174.
— Bobrovii Juxip 289.
— boreum Elfstr. 176.
— Borodinianum Juxip 228; tab. 22.
— brachycephalum Norrl. 620.
— branae Naeg. et Peter 464.
— Brandisianum Zahn. 70.
— brittatense Juxip 146; tab. 19.
— Buhsei Naeg. et Peter 411.
— bupleurifolioides Zahn 143.
— bupleurifolium Tausch 142.
— bupleuroides J. F. Gmel. 395.
— Buschianum Juxip 145; tab. 17.
— caesiiflorioides Juxip 365.
— caesiiflorum Almq. ex Norrl. 376; tab. 33.
— caesiogenum Woloszcz. et Zahn 209.
— caesiomurorum Lindeb. ex Norrl. 344.
— caesium Fries 354.
— caespiticola Norrl. 266.
— callichlorum Litv. et Zahn 44.
— callimorphoides Zahn 630.
— callimorphopsis Zahn 629.
— callimorphum Naeg. et Peter 625.
— calodon Naeg. et Peter 484.
— calodontopsis Litv. et Zahn 485.
— calolepidium Norrl. 657.
— calomastix Naeg. et Peter 516.
— caloprasinum Zahn 73.
— canitiosum Dahlst. 380.
— canum Naeg. et Peter 592.
— carcarophyllum K. Joh. 323; tab. 30.
— cardiobasis Zahn 372.
— cardiophyllum Jord. ex Sudre 324.
— Casparyanum Naeg. et Peter 637.
— caucasicum Naeg. et Peter 403.
— caucasiense Arv.-Touv. 42; tab. 5.
— cauri Juxip 382.
— cercidotelmatodes Juxip 375.
— cereolinum Norrl. 262.
— cernuiforme Naeg. et Peter 646.
— chaetothyrsoides Litv. et Zahn 32.
— chaetothyrsum Litv. et Zahn 30.
— chlorelliceps Norrl. 243.
— chlorellum Sael. et Norrl. 380.
— chlorochromum Sosn. et Zahn 25.
— chloroleucolepium Kozl. et Zahn 54.
— chlorophilum Kozl. et Zahn 51.
— chloroprenanthes Litv. et Zahn 31.
— chlorops Naeg. et Peter 628.
— christoglossum Zahn 539.
— chromolepium Zahn 36.
— chrysophtalmum Norrl. 640.
— cilicicum Naeg. et Peter 684.
— cincinnatum Fries 26; tab. 2.
— cinereostriatum Woronow et Zahn 321.
— colliniforme Naeg. et Peter 599.
— coloratum Elfstr. 171.
— commilitonum Juxip 310.
— comosum Elfstr. 165; tab. 9.
— concinnidens Zahn 58.
— concoloriforme Norrl. 657.
— conferciens Norrl. 579.
— coniciforme Litv. et Zahn 130.
— conicum Arv.-Touv. 128.

— coniops Norrl. 236.
— connatum Norrl. 306.
— constringensiforme Juxip 235.
— contractum Norrl. 554.
— creperiforme Juxip 111.
— crispans Norrl. 371.
— crispum Elfstr. 159; tab. 20.
— crocatum Fries 126; tab. 14.
— cruentiferum Norrl. et Lindb. 107.
— curvescens Norrl. 561.
— curvicollum Norrl. 591.
— curvulatum Zahn 613.
— curvulum Norrl. 518.
— cuspidelliforme Juxip 301.
— cuspidellum Pohle et Zahn 300.
— cymanthodes Kozl. et Zahn 457.
— cymanthum Naeg. et Peter 456.
— cymigerum Reichb. 556; tab. 36.
— cymiramum Schelk. et Zahn 479.
— cymosiforme Naeg. et Peter 512.
— cymosum L. 549.
— cyrtophyllum Norrl. 508.
— czaiense Schischk. et Serg. 98.
— dagoense Juxip 117.
— debilescens Woronow et Zahn 290.
— Dechyi Kozl. et Zahn 104.
— decipiens Tausch 167.
— declivium Juxip 321.
— decurrens Norrl. 181.
— dentatum Hoppe 394.
— denticuliferum Norrl. 555.
— detonsum Norrl. 572.
— diaphanoides Lindeb. 283.
— diaphanoidiceps Woronow et Zahn 54.
— diminuens Norrl. 322.
— discolor Naeg. et Peter 532.
— discoloratum Norrl. 663.
— dissolutum Naeg. et Peter 600.
— distractum Norrl. 330.
— dmitrovense Peter 537.
— dolabratum Norrl. 108.
— Dublizkii B. Fedtsch. et Nevski 602.
— duderhofense Juxip 138.
— durisetum Naeg. et Peter 422.
— echiocephalum Naeg. et Peter 494.
— echiogenes Naeg. et Peter 487.
— echioides Lumn. ex Naeg. et Peter 418; tab. 25.
— Eichvaldii Juxip 378.
— Endaurovae Juxip 218.
— epichlorum Litv. et Zahn 264.
— ericetorum Naeg. et Peter 440.
— ermaniense Juxip 40.
— erythrocarpoides Litv. et Zahn 76.
— erythrocarpum Peter 71; tab. 9.
— erythrochristum Naeg. et Peter 520.
— eusciadium (Naeg. et Peter) Dahlst. 551.
— excubitum Elfstr. 174; tab. 13.
— eximiiforme Dahlst. 194.
— exotericum Jord. ex Boreau 327.
— falcidentatum Juxip 231.
— fallax (Willd.) Naeg. et Peter 423.
— fariniramum Ganesch. et Zahn 264; tab. 25.
— farinodermum Litv. et Zahn 400.
— fastigiatum Naeg. et Peter 453.
— fenno-orbicans Norrl. 309.
— fennoorbicantiforme Juxip 305.
— festinum Jord. ex Boreau 267.
— filiferum Tausch 466.
— finmarkicum Elfstr. 187.
— flagellare (Willd.) Naeg. et Peter 643; tab. 37.
— flagellariforme G. Schneid. 628.
— flexicaule Elfstr. 169.
— floccicomatum Woronow et Zahn 325.
— flocciparum Schelk. et Zahn 106.
— floccipedunculum Naeg. et Peter 437.
— floribundiforme Naeg. et Peter 584.
— floribundoides Zahn 585.
— floribundum Naeg. et Peter 606; tab. 17.
— floridum Naeg. et Peter 632.
— folioliferum Elfstr. 161.
— foliosissimum Woronow et Zahn 34.
— Fominianum Woronow et Zahn 480.
— Freynii Naeg. et Peter 417.
— Frickii Zahn 431.
— frigidellum Pohle et Zahn 319.
— Fritzei F. Schultz 198.
— frondiferum Elfstr. 170.
— frondosum Naeg. et Peter 676.
— fuliginosum Laest. 185; tab. 21.
— fulvescens Naeg. et Peter 618.
— furfuraceoides Zahn 318.
— galbanum Dahlst. ex Anderss. 353; tab. 10.
— Ganeschinii Zahn 238.
— gentile Jord. ex Boreau 333; tab. 31.
— georgicum Fries 60.
— gigantellum Litv. et Zahn 24.
— giganticaule Zahn 569.
— glabriligulatum Norrl. 166.
— glaucescens Bess. 462; tab. 37.
— Glehnii Juxip 306.
— glomeratiforme Zahn 612.
— glomeratum (Fries) Naeg. et Peter 574; tab. 38.
— glomerellum Zahn 70.

Hieracium gnaphalium Naeg. et Peter 634.
— goriense Kozl. et Zahn 95.
— Gorodkowianum Juxip 186; tab. 21.
— gothicifrons Zahn 39.
— grandidens Dahlst. 340.
— granvicum Juxip 305.
— griseum Norrl. 573.
— gudergomiense Juxip 232.
— gudissiense Juxip 225.
— Guentheri Norrl. 270.
— Gustavianum Juxip 255.
— Guttenfeldense Zahn 635.
— gymnogenum Zahn 159.
— gynaeconesaeum Juxip 88.
— haematoglossum Kozl. et Zahn 487.
— Haraldii Norrl. 577.
— heothinum Naeg. et Peter 462.
— heterodontoides Litv. et Zahn 72.
— hispidissimum Rehm. 459.
— Hjeltii Norrl. 322.
— Hohenackeri Naeg. et Peter 403.
— hololeion Maxim. 60; tab. 3.
— homostegium Norrl. 639.
— hopense Juxip 464.
— Hoppeanum Naeg. et Peter 681; tab. 34.
— hylocomum Juxip 328.
— hylogeton Kozl. et Zahn 300.
— hypeuryum Naeg. et Peter 687.
— hypoglaucum Litv. et Zahn 141.
— hypopityforme Juxip 248.
— hypopitys Litv. et Zahn 261.
— hypopogon Litv. et Zahn 50.
— Igoschinae Juxip 289.
— ilyassowoënse Zahn 535.
— imandrense Juxip 139.
— incaniforme Litv. et Zahn 478.
— incanum (M. B.) Naeg. et Peter 398; tab. 34.
— inceptans Norrl. 640.
— inconveniens Juxip 277.
— incurrens Sael. ex Norrl. 242.
— ingricum Naeg. et Peter 454.
— insolens Norrl. 459.
— insolitum Zahn 73.
— intercessum Juxip 369.
— iremelense Juxip 162.
— ischnoadenum Juxip 284.
— isthmicola Norrl. 626.
— jablonicense Woloszcz. 209.
— jailanum Zahn 543.
— juranum Fries 136.
— Kabanovii Juxip 375.
— Kaczurinii Juxip 134.
— kajanense Malmgr. 656.
— karelicum Norrl. 600.
— karelorum Norrl. 141; tab. 15.
— Karjaginii Juxip 227.
— Karpinskyanum Naeg. et Peter 400.
— Kemulariae Juxip 698.
— kiderense Zahn 57.
— Kihlmanii Norrl. 661.
— kirghisorum Juxip 38.
— Knatii Čelak. 113.
— kochtanum Kozl. et Zahn 49.
— Koenigianum Zahn 200.
— Koernickeanum Naeg. et Peter 527.
— Kolenatii Naeg. et Peter 502.
— kolicola Juxip 339; tab. 32.
— konshakovskianum Juxip 379.
— Korshinskyi Zahn 219; tab. 15.
— kosvinskiense Juxip 304.
— kovdaënse Juxip 144.
— Kozlowskyanum Zahn 402.
— Krasanii Woloszcz. 197.
— Kreczetoviczii Juxip 317.
— Krylovii Nevski 130.
— kubanicum Litv. et Zahn 104.
— kubinskense Juxip 279.
— Kulkowianum Zahn 112.
— kumbelicum B. Fedtsch. et Nevski 420.
— Kupfferi Dahlst. 302.
— kuroksarense Juxip 178.
— kusnetzkiense Schischk. et Serg. 129.
— Kuzenevae Juxip 277.
— Lackschewitzii Dahlst. 386.
— ladogense Norrl. 622.
— laevigans Zahn 102.
— laevigatum Willd. 105.
— lailanum Schelk. et Zahn 59; tab. 8.
— lamprocomoides Woronow et Zahn 688.
— lamprocomum Naeg. et Peter 688.
— lamprophtalmum Norrl. 572.
— loncidens Zahn 101.
— lapponicum Fries 109; tab. 2.
— lasiophorum Naeg. et Peter 493.
— lasiothrix Naeg. et Peter 687.
— latens Juxip 256.
— laterale Norrl. 107.
— lateriflorum Norrl. 335.
— latpariense Peter 492.
— Lehbertii Zahn 280.
— lenkoranense Juxip 535.
— lepiduliforme Dahlst. 237.
— lepistoides K. Joh. ex Dahlst. 334.
— leptadenium Dahlst. 552.
— leptocaulon Naeg. et Peter 598.
— leptoclados Naeg. et Peter 520.
— leptogrammoides Juxip 279.

— leptophyes Peter 587.
— leptophyllum Naeg. et Peter 510.
— leptophytomorphum Litv. et Zahn 505.
— leptophytum Naeg. et Peter 533.
— leptoprenanthes Litv. et Zahn 37; tab. 4.
— Lespinassei Kozl. et Zahn 134.
— leucothyrsogenes Kozl. et Zahn 314.
— leucothyrsoides Kozl. et Zahn 223.
— leucothyrsum Litv. et Zahn 106.
— Levieri Peter 650.
— linahamariense Juxip 351.
— linifolium Sael. ex Lindb. 110.
— Lippmae Juxip 386.
— Lipskyanum Juxip 242.
— lissolepium Zahn 103.
— lithuanicum Naeg. et Peter 673.
— litoreum Norrl. 553.
— Litwinowianum Zahn 39.
— ljapinense Juxip 172.
— lomnicense Woloszcz. 197.
— longatum Peter 620.
— longipes C. Koch ex Naeg. et Peter 431.
— longiradiatum Zahn 511.
— longiscapum Boiss. et Kotschy 603.
— longisetum Naeg. et Peter 488.
— longum Naeg. et Peter 502.
— loriense Juxip 145; tab. 18.
— lugdunense Rouy 86.
— lujaurense Norrl. 179.
— luteoglandulosum Sael. ex Norrl. 578.
— lyccense Naeg. et Peter 444.
— lychnaeum Norrl. 655.
— Lydiae Schischk. et Steinb. 514.
— lyratum Norrl. 341.
— macrochaetium Naeg. et Peter 406.
— macrochlorellum Litv. et Zahn 364.
— macrocynum Naeg. et Peter 416.
— macrolepidiforme Zahn 69.
— macrolepioides Zahn 69.
— macrolepis Boiss. 56.
— macrolepium Naeg. et Peter 683.
— macrolygodes Zahn 101.
— macrophyllopodum Zahn 255.
— macroradium Zahn 500.
— maculosum Dahlst. 362.
— magnauricula Naeg. et Peter 674.
— Malmei Dahlst. 389.
— manifestum Juxip 195.
— marginale Naeg. et Peter 468.
— maschukense Litv. et Zahn 415.
— matrense Naeg. et Peter 537.
— maurocybe Juxip 439.
— madianiforme Litv. et Zahn 325.
— medschedsense Zahn 51.
— megalomastix Naeg. et Peter 465.
— Meinshausenianum Juxip 143.
— melachaetum Tausch 455.
— melaneilema Naeg. et Peter 669.
— melanocephalum Tausch 160.
— melanocybe Norrl. 439.
— membranulatum Litv. et Zahn 229.
— Mendelii Naeg. et Peter 678.
— Miansarofii Kozl. et Zahn 62.
— micans Norrl. 573.
— micrastrum Zahn 583.
— microplacerum Norrl. 316.
— microsphaericum Zahn 649.
— microtum Boiss. 35.
— miro-bauhini Zahn 530.
— mirum Naeg. et Peter 495.
— mixopolium Dahlst. 108.
— mnoophorum Naeg. et Peter 514.
— modiciforme Juxip 171.
— mohrungenense Zahn 635.
— mollisetum (Naeg. et Peter) Dahlst. 547.
— monczecola Juxip 175; tab. 19.
— moscovitium Peter 647.
— mukacevense Juxip 281.
— multiceps Naeg. et Peter 484.
— multifrons Brenn. 373.
— multiglandulosum Juxip 144; tab. 16.
— multisetum Naeg. et Peter 687.
— muratovoënse Zahn 612.
— murmanicola Zahn 194.
— nalczikense Juxip 536.
— naniceps Elfstr. 170.
— narymense Schischk. et Serg. 97.
— neglectum Norrl. 568.
— Nenukovii Juxip 252.
— neroikense Juxip 129.
— nesaeum Juxip 280.
— nigrescens Willd. 189.
— nigrisetum Naeg. et Peter 463.
— niphocladum Schelk. et Zahn 52.
— niveolimbatum Juxip 299.
— norrliniiforme Pohle et Zahn 567.
— Obornyanum Naeg. et Peter 523.
— obscuribracteum Naeg. et Peter 453.
— obscuricaule Litv. et Zahn 30.
— obscurum Reichb. 441.
— ochrophyllum Naeg. et Peter 485.
— oioënse Dahlst. 374.
— olympicum Boiss. 76.
— Omangii Elfstr. 115.
— onegense Norrl. 601; tab. 31.
— onosmaceum Zahn 55.
— orbicans Almq. ex Stenstr. 303.
— ornatum Dahlst. 286.

Hieracium orthocladum Zahn 23.
— orthopodum Dahlst. 192.
— osiliae Dahlst. 352.
— Oswaldii Norrl. 127.
— ovaliceps Norrl. 188.
— ovalifrons Woronow et Zahn 315.
— ovatifrons Dahlst. ex Noto 326.
— pachylodes Naeg. et Peter 691.
— Pahnschii Juxip 373.
— panaeoliforme Pohle et Zahn 315.
— Panjutinii Juxip 32; tab. 3.
— pannoniciforme Litv. et Zahn 480.
— pannosum Boiss. 199.
— paragogum Naeg. et Peter 527.
— Pareyssianum Naeg. et Peter 498.
— pasense Juxip 278; tab. 24.
— peczoryense Juxip 428.
— pellucidum Laest. 329.
— pendulum Dahlst. 370.
— perasperum Zahn 482.
— perfoliatum Froel. 146.
— perfugii Juxip 507.
— pergrandidens Zahn 177.
— perileucum Schelk. et Zahn 683.
— permense Zahn 422.
— permicum Zahn 579; tab. 39.
— persimile Dahlst. 302.
— petiolatum Elfstr. 168.
— petrofundii Juxip 263.
— petropavlovskanum Juxip 371.
— Petunnikovii Peter 638.
— philanthrax Stenstr. 390.
— phrygium Zahn 410.
— pilipes Sael. 584.
— pilisquamum Naeg. et Peter 682.
— pilosella L. 692.
— piloselliflorum Naeg. et Peter 648.
— pineum Schischk. et Serg. 427.
— pleuroleucum Dahlst. 326.
— plicatulum Zahn 454.
— pluricaule Schischk. et Serg. 283.
— plurifoliosum Schischk. et Steinb. 98.
— pocuticum Woloszcz. 208.
— podkumokense Juxip 23.
— Pohlei Zahn 217.
— poliophyton Zahn 697.
— poliudovense Juxip 261.
— polymnoon Naeg. et Peter 560.
— polymorphophyllum Elfstr. 186.
— pomoricum Juxip 327.
— Porphyrii Schischk. 97.
— praealtum (Vill.) Naeg. et Peter 441; tab. 36.
— praecurrens Vukot. 211.
— praetermissum Juxip 260.
— praetervisum Juxip 228.
— pratense Tausch 596; tab. 30.
— praticola Surde 83.
— proceriforme Naeg. et Peter 416.
— procerigenum Litv. et Zahn 478.
— procerum (Fries) Naeg. et Peter 409; tab. 35.
— progenitum Norrl. 630.
— prognatum Norrl. 638.
— prolatatum K. Joh. ex Dahlst. 237.
— prolixiforme Norrl. 349; tab. 27.
— prolixum Norrl. 363.
— prolongatum Naeg. et Peter 568.
— proximum Norrl. 308.
— pruiniferum Norrl. 125.
— prussicum Naeg. et Peter 636; tab. 38.
— psammophilum Naeg. et Peter 483.
— pseudauricula Naeg. et Peter 627.
— pseudauriculoides Naeg. et Peter 471.
— pseudobifidum Schur 210.
— pseudobipes Elfstr. 178.
— pseudo-Blyttii Norrl. 664; tab. 39.
— pseudobrachiatum Čelak. 537.
— pseudoconstrictum Zahn 36.
— pseudojuranum Arv.-Touv. 136.
— pseudophyllodes Zahn 173.
— pseudosvaneticum Peter 66.
— pseuduliginosum Zahn 644.
— psilobrachion Woronow et Zahn 534.
— pubens Naeg. et Peter 622.
— pulvinatum Norrl. 666.
— purpureibracteum Zahn 540.
— purpureovittatum Zahn 540.
— purpuristictum Juxip 111.
— puschlachtae Pohle et Zahn 110.
— pycnothyrsum Peter 569.
— pyrsjuense Juxip 172.
— quinquemonticola Juxip 248.
— Raddeanum Zahn 124.
— radiatellum Woronow et Zahn 313.
— ratluense Zahn 72.
— ravusculum Dahlst. 355.
— reducatum Norrl. em. Juxip 137.
— Regelianum Zahn 124.
— regiomontanum Naeg. et Peter 609.
— Rehmannii Naeg. et Peter 566.
— renidescens Norrl. 611.
— retroversilobatum Schelk. et Zahn 319; tab. 29.
— revocans Juxip 337.
— rigidellum Litv. et Zahn 44.
— rigidum Hartm. 111.
— riparium Juxip 369.

— robustum Fries 78.
— rohacsense Kit. 196.
— Rojowskii Rehm. 452.
— Rothianum Zahn 425.
— roxolanicum Rehm. 565.
— rubricymigerum Naeg. et Peter 565.
— rubro-Bauhini Schelk. et Zahn 460.
— rubroonegense Norrl. 658.
— rubropannonicum Litv. et Zahn 486.
— Ruprechtii Boiss. 544.
— rusanum Zahn 570.
— sabiniceps Litv. et Zahn 499.
— sabiniforme Zahn 481.
— sabinopsis Ganesh. et Zahn 421.
— sabulosorum Dahlst. 689.
— Sachokianum Kem.-Nath. 421.
— sagittatum Lindeb. 391.
— samurense Zahn 68.
— sarmentosum Froel. 497; tab. 14.
— sarykamyschense Juxip 225; tab. 18.
— saxifragum Fries 204.
— sbaense Juxip 278; tab. 26.
— scabiosum Sudre 83.
— Schelkownikowii Zahn 404.
— Schellianum Juxip 287.
— schemachense Juxip 467.
— Schmalhausenianum Litv. et Zahn 10.
— Schipczinskii Juxip 255.
— Schischkinii Juxip 230; tab. 23.
— Schliakovii Juxip 313.
— Schultesii (F. Schultz) Naeg. et Peter 678.
— scopulorum Juxip 592.
— scotaiolepis Elfstr. 190.
— scotodes Norrl. 549.
— segevoldense Syr. et Zahn 385.
— semicurvatum Norrl. 190.
— semicymigerum Zahn 513.
— semionegense Norrl. 661.
— semipraecox Zahn 492.
— senescentifrons Elfstr. 178.
— septentrionale Norrl. 444.
— sericicaule Schelk. et Zahn 55.
— serratifolium Jord. ex Boreau 339.
— sershukense Juxip 226.
— signiferum Norrl. 548.
— Silenii Norrl. 270.
— silvicomum Juxip 259.
— silvularum Jord. ex Boreau 341.
— simplicicaule Somm. et. Lev. 57; tab. 7.
— siworkae Juxip 267.
— sobrinatum Litv. 66.
— Soczavae Juxip 180; tab. 11.
— sordidescens Norrl. 258.
— Sosnowskyi Zahn 399.
— spathophyllum Naeg. et Peter 619.
— squarrosulum Norrl. 677.
— stauropolitanum Juxip 457.
— Steinbergianum Juxip 353.
— stellatum Tausch 442.
— stenolepis Lindeb. 377.
— stenomischum Omang 182.
— stenopiforme Pohle et Zahn 181.
— stenozon Zahn 649.
— sterromastix Naeg. et Peter 432.
— streptotrichum Zahn 25.
— strictiramum Naeg. et Peter 483.
— strictissimum Froel. 138.
— stupposipilum Woronow et Zahn 445.
— subambiguum Naeg. et Peter 570.
— subaquilonare Juxip 220.
— subarctoum Norrl. 286.
— subartvinense Juxip 35.
— subasperellum Zahn 247.
— subatriceps Zahn 677.
— subauricula Naeg. et Peter 627.
— subbakurianiense Juxip 65.
— subbauhiniflorum Woronow et Zahn 531.
— subbetuletorum Juxip 290.
— subcompositum Juxip 317.
— subcrassifolium Zahn 307.
— subcymigerum Naeg. et Peter 438.
— suberectum Schischk. et Steinb. 139.
— subfallaciforme Zahn 425.
— subfariniramum Ganesh. et Zahn 268.
— subfiliferum Zahn 466.
— subfloribundum (Naeg. et Peter) Dahlst. 582.
— subgalbanum Juxip 350.
— subhastulatum Zahn 257.
— subhirsutissimum Juxip 89.
— subimandrae Juxip 176.
— subincomptum Zahn 175.
— sublactucaceum Zahn 84.
— sublasiophorum Litv. et Zahn 496.
— sublividum Dahlst. 381.
— submaculosum Dahlst. 361.
— submarginellum Zahn 307; tab. 28.
— submedianum Zahn 282; tab. 27.
— submirum Litv. et Zahn 497.
— subnigrescens Fries 191.
— subnigriceps Zahn 631.
— subobscuriceps Zahn 251.
— subpellucidum Norrl. 285; tab. 16.
— subpollichium (Litv. et Zahn) Juxip 257.
— subrubellum Schelk. et Zahn 539.
— subsimplex Somm. et Lev. 59.
— substoloniferum Naeg. et Peter 467.
— subsvaneticum Litv. et Zahn 67.

Hieracium subumbelliforme Zahn 501.
— subviolascentiforme Pohle et Zahn 229.
— sudavicum Naeg. et Peter 609.
— sudetorum Naeg. et Peter 596.
— suecicum (Fries) Naeg. et Peter 610; tab. 33.
— sulphurelliforme Kozl. et Zahn 41.
— sulphurellum Kozl. et Zahn 41.
— sulphureum Doell 526.
— suomense Norrl. 559.
— svaneticiforme Litv. et Zahn 58.
— Syreistschikovii Zahn 29.
— syrjaenorum Norrl. 547.
— sysolskiense Zahn 581.
— Szovitsii Naeg. et Peter 433.
— tabergense Dahlst. 551.
— taigense Schischk. et Serg. 288; tab. 28.
— tanense Elfstr. 182.
— tanythrix Naeg. et Peter 500.
— tatrense Naeg. et Peter 643.
— teberdaefontis Litv. et Zahn 491.
— teberdense Litv. et Zahn 133.
— tenebricans Norrl. 443.
— tenuiceps Naeg. et Peter 482.
— tenuiglandulosum Norrl. 338.
— tephrantheloides Zahn 634.
— tephranthelum Zahn 566.
— tephrochlorellum Ganesch. et Zahn 426.
— tephrophilum Kozl. et Zahn 223.
— tephropodum Zahn 543.
— Teplouchovii Juxip 239; tab. 24.
— terekianum Litv. et Zahn 33.
— thaumasioides Naeg. et Peter 456.
— thaumasium Naeg. et Peter 457.
— thracicum Naeg. et Peter 498.
— Tilingii Juxip 250.
— tjapomense Norrl. 655.
— torquescens Norrl. 666.
— torticeps Dahlst. 336.
— transbalticum Dahlst. 585.
— transsilvanicum Heuff. 207; tab. 12.
— triangulare Almq. 361.
— tricheilema Naeg. et Peter 674.
— trichobrachium Juxip 114.
— trichocymosum Zahn 548.
— tridentaticeps Zahn 113.
— tridentatum Fr. 102.
— triste Willd. ex Spreng. 9; tab. 1.
— tritum Juxip 226.
— Tschkhubianischwilii Kem.-Nath. 65.
— tubuliflorum Naeg. et Peter 534.
— tunguskanum Ganesch. et Zahn 123; tab. 13.
— turkestanicum Zahn 217.
— tuscheticum Zahn 543.
— tzagwerianum Kozl. et Zahn 48.
— uczanssuense Juxip 224.
— ugandiense Juxip 392.
— umbellaticeps Pohle et Zahn 96.
— umbellatum L. 90.
— umbelliferum Naeg. et Peter 513.
— umbellosum Naeg. et Peter 499.
— umbrosum Jord. 273.
— uralense Elfstr. 173.
— uranopoleos Juxip 318.
— ussense Pohle et Zahn 192.
— Vagae Juxip 370.
— vagum Jord. 85.
— vaidae Juxip 180.
— valmierense Juxip 281.
— variegatisquamum Zahn 75.
— vasconicum Jord. ex Zahn 86.
— Veresczaginii Schischk. et Serg. 127.
— vernicosum Norrl. 665.
— verruculatum Naeg. et Peter 399.
— villosellipes Zahn 62.
— villosum Jacq. 393.
— vindobonae Zahn 431.
— violaceipes Zahn 588.
— violascentiforme Pohle et Zahn 269.
— virelliceps Norrl. 366.
— virenticeps Dahlst. 265.
— virentisquamum Naeg. et Peter. 681.
— virgultorum Jord. 85; tab. 12.
— virosiforme Woronow et Zahn 29.
— virosum Pall. 77.
— vischerae Juxip 135.
— viscidulum Tausch 461.
— vitellicolor Elfstr. 160.
— vitellinum Norrl. 577.
— volhynicum Naeg. et Peter 468.
— voroniense Juxip 193.
— vulgatiforme Dahlst. 231.
— vulgatum (Fries) Almq. 240.
— Wimmeri Uechtr. 360.
— wjasowoënse Zahn 511.
— wolczankense Juxip 260.
— wolgense Zahn 477.
— wologdense Pohle et Zahn 219; tab. 28.
— Woronowianum Zahn 411.
— xanthostigma Norrl. 621.
— Zinserlingianum Juxip 135.
— Zizianum Tausch 508.
Hierochloe R. Br. 2, 59.
— alpina (Liljebl.) Roem. et Schult. 60; tab. 4.
— australis (Schrad.) Roem. et Schult. 60.
— Bungeana Trin. 62.

— glabra Trin. 62.
— odorata (L.) Wahl. 61; tab. 4.
— pauciflora R. Br. 61.
Himantoglossum Spreng. 4, 719.
— carpinum (M. B.) Spreng. 719; tab. 43.
— formosum (Stev.) C. Koch 720; tab. 43.
Hinterhubera Reichb. ex Nym. 8, 549.
Hipericum Neck. 15, 203.
Hippocrepis L. 13, 255.
— biflora Spreng. 257.
— ciliata Willd. 256.
— comosa L. 256.
— unisiliquosa L. 257.
Hippolytia Poljak. 26, 411.
— darvasica (C. Winkl.) Poljak. 415; tab. 18.
— Herderi (Rgl. et Schmalh.) Poljak. 414.
— megacephala (Rupr.) Poljak. 413.
— schugnanica (C. Winkl.) Poljak. 416.
Hippomarathrum Hoffmgg. et Link 16, 252.
— caspicum (DC.) Grossh. 253.
— longilobium (DC.) B. Fedtsch. 254.
— microcarpum (M. B.) B. Fedtsch. 252.
Hippophaestum S. F. Gray 28, 370.
Hippophaë L. 15, 516.
— rhamnoides L. 516.
Hippuris L. 15, 668.
— tetraphylla L. 670.
— vulgaris L. 669.
Hirculus Haw. 9, 159.
Hirschfeldia Moench 8, 470.
— incana (L.) Lagr.-Foss. 471.
Hisutsua DC. 25, 122.
Hohenackeria Fisch. et Mey. 16, 273.
— exscapa (Stev.) K.-Pol. 274; tab. 30.
Holcus L. 2, 237.
— lanatus L. 238; tab. 18.
— mollis L. 239; tab. 18.
— setosus Trin. 238; tab. 18.
Hololachne Ehrenb. 15, 277, 279.
Holopleura Rgl. et Schmalh. 16, 563.
Holopogon Kom. et Nevski 4, 620, 750.
— ussuriensis Kom. et Nevski 620, 751.
Holoschoenus Link 3, 41.
— romanus (L.) Fritsch 42.
— vulgaris Link 41; tab. 3.
Holosteum L. 6, 466.
— glutinosum (M. B.) Fisch. et Mey. 467; tab. 25.
— marginatum C. A. Mey. 468; tab. 25.
— umbellatum L. 467; tab. 25.
Homogyne Cass. 26, 654.
— alpina (L.) Cass. 655.
Honkenya Ehrh. 6, 516.
Hoppea Reichb. 26, 788.
Horaninowia Fisch. et Mey. 6, 272.
— anomala (C. A. Mey.) Moq. 272.
— excellens Iljin 274; tab. 12.
— minor Fisch. et Mey. 273; tab. 10.
— ulicina Fisch. et Mey. 274; tab. 10.
Hordeum L. 2, 722.
— Bogdanii Wilensky 724.
— boreale Scribn. 723.
— brevisubulatum (Trin.) Link 724; tab. 50.
— bulbosum L. 726.
— distichum L. 728.
— hystrix Roth 727.
— leporinum Link 726.
— sibiricum Roshev. 724.
— spontaneum C. Koch 727; tab. 50.
— turkestanicum Nevski 725; tab. 50.
— violaceum Boiss. et Huet 725; tab. 50.
— vulgare L. 728.
Horminum Moench 21, 291.
Hornemannia Willd. 22, 316.
Hosta Tratt. 4, 54.
Hostia Moench 29, 594, 687.
Hoteia Morr. et Decne. 9, 135.
Hottonia L. 18, 252.
— palustris L. 252.
Houttuynia Thunb. 5, 20.
— cordata Thunb. 20.
Hulthemia Dum. 10, 506.
— berberifolia (Pall.) Dum. 506.
— persica Michx. 507.
Hulthemosa Juz. 10, 507, 638.
— guzarica Juz. 507, 639.
— kopetdaghensis (Meffert) Juz. 508.
Humulus L. 5, 382.
— japonicus Sieb. et Zicc. 383; tab. 21.
— lupulus L. 382; tab. 21.
Hutchinsonia R. Br. 8, 548.
— petraea (L.) R. Br. 548.
Hyacintella Schur 4, 407.
— leucophaea (C. Koch) Schur 408; tab. 23.
— Pallasiana (Stev.) Losinsk. 408; tab. 23.
Hyacinthus Tourn. 4, 405.
— Litwinowii Czerniak. 406; tab. 23.
— kopetdaghi Czerniak. 407; tab. 23.
— orientalis L. 407.
— transcaspicus Litv. 406; tab. 23.
Hyalea (DC.) Jaub. et Spach 28, 370, 486.
Hyalea (DC.) Jaub. et Spach 28, 367.
— pulchella (Ledeb.) C. Koch 369.
— tadshicorum (Tzvel.) Soják 368.
Hyalolaena Bge. 16, 563.
— collina Korov. 566.

Hyalolaena depauperata Korov. 565.
— jaxartica Bge. 564; tab. 30.
— paniculata Korov. 565.
Hybidium Fourr. 23, 640.
Hydrangea L. 9, 225.
— paniculata Sieb. 226.
— petiolaris Sieb. et Zucc. 225.
Hydrilla Rich. 1, 294.
— verticillata (L. f.) Rich. 294; tab. 10.
Hydrocharis L. 1, 297.
— asiatica Miq. 298.
— morsus-ranae L. 297; tab. 10.
Hydrocotyle L. 16, 57.
— ramiflora Maxim. 59.
— ranunculoides L. 59.
— vulgaris L. 58.
Hylomecon Maxim. 7. 584.
— vernalis Maxim. 584; tab. 40.
Hymenocrater Fisch. et Mey. 20, 488.
— bituminosus Fisch. et Mey. 490.
— elegans Bge. 489.
Hymenolaena DC. 16, 233.
— alpina Schischk. 235; tab. 18.
— nana Rupr. 234; tab. 18.
— pimpinellifolia Rupr. 233; tab. 15, 18.
Hymenolobus Nutt. 8, 549.
— procumbens (L.) Nutt. ex Torr. et Gr 549.
— puberulus (Rupr.) N. Busch 550.
Hymenolyma Korov. 16, 413.
— bupleuroides (Schrenk) Korov. 414.
— trichophyllum (Schrenk) Korov. 413; tab. 30.
Hymenophyllum Smith 1, 18.
— Wrightii Bosch 18; tab. 1.
Hymenophysa C. A. Mey. 8, 535.
— macrocarpa Franch. 536.
— pubescens C. A. Mey. 536.
Hyoscyamus L. 22, 86.
— albus L. 96.
— bohemicus F. Schmidt 95.
— Camerarii Fisch. et Mey. 90.
— kopetdaghi Pojark. 90.
— niger L. 93.
— pusillus L. 98.
— reticulatus L. 88.
— turcomanicus Pojark. 92.
Hypacanthium Juz. 27, 388.
— echinopifolium (Bornm.) Juz. 360; tab. 29.
Hypecoum L. 7, 576.
— erectum L. 577; tab. 45.
— grandiflorum Benth. 579; tab. 45.
— leptocarpum Hook. f. et Thoms. 578.
— parviflorum Kar. et Kir. 580; tab. 45.
— pendulum L. 581; tab. 45.
— trilobum Trautv. 580; tab. 45.
Hypericon J. F. Gmel. 15, 203.
Hypericum L. 15, 203.
— acutum Moench 241; tab. 11.
— alpestre Stev. 232.
— alpigenum Kit. 255.
— androsaemum L. 214; tab. 9.
— antasiaticum Grossh. 231.
— Ardasenowii R. Keller 254.
— armenum Jaub. et Spach 235.
— ascyron L. 212; tab. 9.
— asperuloides Czern. ex Turcz. 218.
— attenuatum Choisy 236.
— bupleuroides Griseb. 246.
— Buschianum Woronow ex N. Busch 223.
— calycinum L. 211.
— caucasicum (Woronow) Gorschk. 256; tab. 12.
— chrysothyrsum Woronow 232.
— crispum L. 244.
— elegans Steph. ex Willd. 243; tab. 11.
— elongatum Ledeb. 230; tab. 11.
— erectum Thunb. 241.
— fissurale Woronow 236.
— formosissimum Takht. 220.
— Gebleri Ledeb. 213.
— helianthemoides (Spach) Boiss. 229.
— hirsutum L. 227; tab. 10.
— humifusum L. 219; tab. 10.
— inodorum Willd. 217; tab. 9.
— kamtschaticum Ledeb. 237.
— karsianum Woronow 234.
— Komarovii Gorschk. 250.
— lydium Boiss. 231.
— marginatum Woronow 221.
— montanum L. 245.
— mutilum L. 257.
— nachitschevanicum Grossh. 250.
— Nordmannii Boiss. 253.
— nummularioides Trautv. 221.
— olympicum L. 219.
— origanifolium Willd. 247.
— paramushirense Kudo 258.
— perforatum L. 248; tab. 12.
— polygonifolium Rupr. 233.
— ponticum Lipsky 228.
— porphyrandrum Lévl. 258.
— ptarmicifolium Spach 222; tab. 10.
— quadrangulum L. 242; tab. 12.
— sachalinense Lévl. 257.
— scabrum L. 224.
— strictum Maleev 254.
— Theodori Woronow 235.

— venustum Fenzl 245.
— yezoense Maxim. 238.
Hypochaeris L. 29, 199.
— glabra L. 199.
— radiata L. 200; tab. 12.
Hypogomphia Bge. 20, 491.
— elatior (Rgl.) Vass. 492.
— turkestana Bge. 491.
Hypopitys Adans. 18, 20.
— monotropa Crantz 21.
Hipposelinum Britt. et Rose 17, 40.
Hyssopus L. 21, 448.
— ambiguus Iljin 455.
— angustifolius M. B. 451.
— cretaceus Dubjan. 456; tab. 24.
cuspidatus Boriss. 455; tab. 24.
— ferganensis Boriss. 461.
— macranthus Boriss. 457; tab. 25.
— officinalis L. 451.
— seravschanicus (Dub.) Pazij 458; tab. 25.
— tianschanicus Boriss. 452.

Iberidella Boiss. 8, 567.
Iberis L. 8, 550.
— amara L. 552.
— pinnata L. 551.
— saxatilis L. 551.
— taurica DC. 551.
Ibidium Salisb. 4, 637.
Idianthes Desv. 29, 595, 673.
Ikonnikovia Lincz. 18, 378, 745.
— Kaufmanniana (Rgl.) Lincz. 381; tab. 19.
Ilex L. 14, 540.
— aquifolium L. 541; tab. 27.
— colchica Pojark. 542; tab. 27.
— crenata Thunb. 544.
— hyrcana Pojark. 544; tab. 27.
— rugosa Fr. Schmidt 545.
— stenocarpa Pojark. 543.
Iljinia Korov. 6, 309, 877.
— Regelii (Bge.) Korov. 309; tab. 10.
Impatiens L. 14, 624.
— balsamina L. 633.
— brachycentra Kar. et Kir. 632; tab. 33.
— furcillata Hemsl. 627; tab. 33.
— Komarovii Pobed. 531.
— Maackii Hook. 627; tab. 33.
— Nevskii Pobed. 632, 746; tab. 33.
— noli-tangere L. 626.
— parviflora DC. 631; tab. 33.
— Roylei Walp. 633.
— Textori Miq. 628.
Imperata Cyr. 2, 6.
— cylindrica (L.) Beauv. 6; tab. 1.
Imperatoria L. 17, 168, 193.
Imperialis Adans. 4, 302.
Incarvillea Juss. 23, 3.
— Olgae Rgl. 3; tab. 1.
— sinensis Lam. 4.
Indigofera L. 11, 298.
— tinctoria L. 298.
Intybellia Cass. 29, 699, 703.
Intybellia Monn. 29, 595, 661, 673.
Intybus Fries 2, 595, 661.
Inula L. 25, 433.
— acaulis Schott et Kotschy ex Boiss. 471.
— aspera Poir. 452; tab. 27.
— auriculata Boiss. et Bal. 462.
— britannica L. 465.
— caspica Blume 460; tab. 28.
— decurrens M. Pop. 475.
— ensifolia L. 449.
— germanica L. 451.
— glauca C. Winkl. 458; tab. 28.
— grandiflora Willd. 444.
— grandis Schrenk ex Fisch. et Mey. 441.
— Gromczewskii C. Winkl. 463.
— helenium L. 440
— hirta L. 448.
— japonica Thunb. 468.
— Kitamurana Tatewaki ex Honda 456.
— linariifolia Turcz. 468.
— macrolepis Bge. 467.
— magnifica Lipsky 442.
— Mariae Bordz. 457.
— Montbretiana DC. 457.
— multicaulis Fisch. et Mey. 475.
— oculus-Christi L. 461.
— orientalis Lam. 447; tab. 27.
— rhizocephala Schrenk 472.
— Rubtzovii Gorschk. 464.
— sabuletorum Czern. ex Lavr. 453; tab. 27.
— salicina L. 454.
— salsoloides (Turcz.) Ostenf. 476; tab. 29.
— Schmalhausenii C. Winkl. 459.
— Seidlitzii Boiss. 448.
— thapsoides (Willd.) DC. 474.
— vulgaris (Lam.) Trevis. 473; tab. 28.
Ioxylon Raf. 5, 379.
Ipomoea L. 19, 3.
— batatas (L.) Lam. 5.
— hispida (Vahl) Roem. et Schult. 4.
— purpurea (L.) Roth 5.
— sibirica (L.) Pers. 4.
Iris L. 4, 511.
— acutiloba C. A. Mey. 533; tab. 33.
— Alberti Rgl. 553; tab. 34.
— aphyllá L. 551; tab. 34.

Iris baldshuanica B. Fedtsch. 561.
— Bloudowii Ledeb. 550.
— bucharica Foster 570.
— Camillae Grossh. 553.
— caucasica Hoffm. 574.
— coerulea B. Fedtsch. 569.
— darvasica Rgl. 541; tab. 34.
— dichotoma Pall. 530; tab. 32.
— drepanophylla Aitch. et Baker 575.
— elegantissima Sosn. 533; tab. 33.
— ensata Thunb. 518; tab. 31.
— Ewbankiana Foster 537.
— falcifolia Bge. 544.
— flavissima Pall. 545.
— foetidissima L. 523.
— Fominii Woronow 534; tab. 33.
— Fosteriana Aitch. et Baker 560.
— germanica L. 553.
— graminea L. 529; tab. 32.
— Grossheimii Woronow 538.
— halophila Pall. 526.
— Hoogiana Dykes 542; tab. 34.
— humilis M. B. 529; tab. 32.
— hyrcana Woronow 555; tab. 33.
— iberica Hoffm. 532; tab. 33.
— imbricata Lindl. 552; tab. 34.
— Kaempferi Sieb. ex Lemaire 525.
— karategina B. Fedtsch. 542.
— Kolpakowskiana Rgl. 556; tab. 34.
— kopetdagensis Vved. 575.
— Korolkowii Rgl. 541.
— Kuschakewiczii B. Fedtsch. 568.
— laevigata Fisch. 524; tab. 32.
— lazica Albov 514; tab. 32.
— leptorhiza Vved. 576.
— linifolia O. Fedtsch. 564.
— longiscapa Ledeb. 544; tab. 34.
— Ludwigii Maxim. 530.
— lycotis Woronow 538; tab. 33.
— magnifica Vved. 571; tab. 35.
— maracandica Vved. 573.
— maricoides Rgl. 556; tab. 34.
— Medwedewii Fomin 539; tab. 33.
— musulmanica Fomin 527; tab. 32.
— Narbutii O. Fedtsch. 567.
— narynensis O. Fedtsch. 567.
— Nicolaii Vved. 561.
— notha M. B. 528.
— orchioides Carr. 572.
— orientalis Thunb. 520; tab. 31.
— paradoxa Stev. 538; tab. 33.
— parvula Vved. 563.
— Popovii Vved. 562.
— Potaninii Maxim. 549; tab. 34.
— pseudacorus L. 525; tab. 32.
— pseudocaucasica Grossh. 574.
— pumila L. 545; tab. 34.
— reticulata M. B. 555; tab. 33.
— Rosenbachianum Rgl. 562.
— ruthenica Ker.-Gawl. 517.
— scariosa Willd. ex Link 550.
— Schelkownikowii Fomin 537; tab. 33.
— setosa Pall. ex Link 520; tab. 32.
— sibirica L. 519; tab. 31.
— sogdiana Bge. 526.
— songarica Schrenk 516.
— stolonifera Maxim. 540; tab. 35.
— subdecolorata Vved. 568.
— tadshikorum Vved. 563.
— tenuifolia Pall. 515; tab. 32.
— tianschanica (Maxim.) Vved. 515; tab. 31, 32.
— tigridia Bge. 549; tab. 34.
— Timofejewii Woronow 546; tab. 34.
— Tubergeniana Foster 573.
— uniflora Pall. 518; tab. 31.
— variegata L. 552.
— ventricosa Pall. 516.
— vicaria Vved. 569.
— violacea Klatt 528; tab. 32.
— Vvedenskyi Nevski 564.
— warleyensis Foster 570.
— Willmottiana Foster 572.
— Winkleri Rgl. 554.
— Winogradowii Fomin 554; tab. 34.
Isatis L. 8, 203.
— aleppica Scop. 219.
— anceps N. Busch 208, 643.
— apscheronica N. Busch 221.
— araratica Rupr. 210; tab. 11.
— Arnoldiana N. Busch 211.
— Besseri Trautv. 209; tab. 11.
— Boissieriana Reichb. f. 220.
— brachycarpa C. A. Mey. 217; tab. 11.
— Bungeana Seidl. 218.
— canescens DC. 217; tab. 11.
— caucasica (Rupr.) N. Busch 210; tab. 11.
— collina Boiss. 209; tab. 11.
— costata C. A. Mey. 218.
— emarginata Kar. et Kir. 207.
— frutescens Kar. et Kir. 219.
— glauca Auch. 211.
— Grossheimii N. Busch 217; tab. 11.
— hirtocalyx Franch. 219.
— iberica Stev. 221; tab. 11.
— jacutensis N. Busch 214.
— japonica Miq. 213.
— laevigata Trautv. 220.

— lasiocarpa Ledeb. **219.**
— latisiliqua Stev. **208**; **tab. 11.**
— leuconeura Boiss. **211.**
— litoralis Stev. **214**; tab. **11.**
— minima Bge. **222.**
— oblongata DC. **213.**
— ornithorhynchus N. Busch 222; tab. **11.**
— psilocarpa Ledeb. 209; tab. 11.
— reticulata C. A. Mey. 210; tab. 11.
— sabulosa Stev. ex Ledeb. 212; tab. **11.**
— sevangensis N. Busch 212.
— subradiata Rupr. 208.
— taurica M. B. 214; tab. 11.
— tinctoria L. 212; tab. 11.
— trachycarpa Trautv. 220.
Isika Adans. 23, 479.
Iskandera N. Busch 8, 319, 649.
— hissarica N. Busch 320, 649.
Ismelia Cass. 26, 131, 133.
Isnardia L. 15, 567.
Isoetes L. 1, 127.
— asiatica Makino **128**; tab. 6.
— beringensis Kom. **128**; tab. 6.
— echinospora Dur. **128**; tab. 6.
— lacustris L. **127**; tab. 6.
Isopyrum L. 7, 74.
— thalictroides L. **7**, **74**.
Ixeridium (A. Gray) Tzvel. 29, 388.
— chinense (Thunb.) Tzvel. **390.**
— dentatum (Thunb.) Tzvel. **392.**
— gramineum (Fisch.) Tzvel. **391.**
— graminifolium (Ledeb.) Tzvel. **392**; tab. **24.**
— strigosum (Lévl. et Vaniot) Tzvel. **390.**
Ixiolirion Fisch. 4, 489.
— karateginum Lipsky **490.**
— montanum (Labill.) Herb. **491.**
— tataricum (Pall.) Roem. et Schult. **490**; tab. 30.

Jacea auct. 28, 444.
Jacea Juss. 28, 370.
Jalapa Adans. 6, 370.
Jasione L. 24, 444.
— Heldreichii Boiss. et Orph. **448.**
— montana L. **447**; tab. **27.**
Jasminum L. 18, 522.
— fruticans L. **524.**
— nudiflorum Lindl. **523.**
— officinale L. **524.**
— revolutum Sims **525.**
Jaubertia Guillem. 23, 170.
Jeffersonia Bart. 7, 540.
— dubia (Maxim.) Benth. et Hook. 540; **tab. 37.**
Johrenia DC. 17, 49.
— paucijuga (DC.) Bornm. **49**; tab. 5.
Jovibarba Opiz 9, 23.
Jubaea H. B. K. 3, 468.
— spectabilis H. B. K. **468.**
Juglans L. 5, 248.
— fallax Dode **251.**
— manshurica Maxim. **249.**
— regia L. **250.**
Juncastrum Fourr. 3, 543.
Juncellus (Kunth) Clarke 3, 9.
— alopecuroides (Rottb.) Clarke **10.**
— serotinus (Rottb.) Clarke **10**; tab. 1.
Juncinella Fourr. 3, 514.
Juncotypus Dum. 3, 546.
Juncus L. 3, 504.
— acutus L. 543; tab. 29.
— alpigenus C. Koch 542; tab. 30.
— alpinus Vill. 537; tab. 30.
— ambiguus Guss. 518.
— amuricus (Maxim.) V. Krecz. et Gontsch. 516, 623.
— arcticus Willd. 555; tab. **29.**
— arianus V. Krecz. 540.
— atratus Krock. 542; tab. 30.
— atrofuscus Rupr. 529; tab. 28.
— balticus Willd. 555; tab. 29.
— beringensis Buchenau 556; tab. **29.**
— biglumis L. 521; tab. 30.
— brachyspathus Maxim. 553; tab. **29.**
— brachytepalus (Trautv.) V. Krecz. et Gontsch. 547, 630; tab. 29.
— bufonius L. **517.**
— bulbosus L. 537.
— capitatus Wieg. 515.
— castaneus Smith 525; tab. 30.
— compressus Jacq. 527; tab. 28.
— curvatus Buchenau 553.
— decipiens (Buchenau) Nakai 551; tab. 29.
— effusus L. 548; tab. 29.
— epimetrocarpus V. Krecz. et Gontsch. 530; tab. 28 (vide heptopotamicus)
— filiformis L. 552; tab. 29.
— Gerardii Loisel. 528; tab. 28.
— gracillimus (Buchenau) V. Krecz. et Gontsch. 528, 627; tab. 28.
— Haenkei E. Mey. 556; tab. 29.
— heptopotamicus V. Krecz. et Gontsch. 530, 628; tab. 28, 6.
— himalensis Klotzsch 525.
— inflexus L. 546; tab. 29.
— inundatus Drej. 554.

Juncus jaxarticus V. Krecz. et Gontsch. 532, 628; tab. 28.
— Juzepczukii V. Krecz. et Gontsch. 519, 625; tab. 30.
— kamtschatcensis (Buchenau) Kudo 543.
— Kotschyi Boiss. 539.
— lamprocarpus Ehrh. 538; tab. 30.
— Leersii Marss. 551; tab. 29.
— Leschenaultii J. Gay 540; tab. 30.
— leucochlamys Zing. ex V. Krecz. 526, 627; tab. 30.
— littoralis C. A. Mey. 544.
— macrantherus V. Krecz. et Gontsch. 524, 626.
— maritimus Lam. 545.
— minutulus V. Krecz. et Gontsch. 520, 625.
— Muelleri Trautv. 554; tab. 29.
— nastanthus V. Krecz. et Gontsch. 517, 624; tab. 30.
— Nevskii V. Krecz. et Gontsch. 545, 629.
— nipponensis Buchenau 540; tab. 30.
— paniculatus Hoppe ex Mert. et Koch 547; tab. 29.
— prominensis (Buchenau) Miyabe et Kudo 558.
— salsuginosus Turcz. 530; tab. 28.
— Schischkinii Kryl. et Sumn. 522; tab. 30.
— soranthus Schrenk 531; tab. 28.
— sphaerocarpus Nees 516; tab. 30.
— squarrosus L. 532.
— stygius L. 523.
— subulatus Forsk. 557.
— tenageia Ehrh. ex L. f. 515; tab. 30.
— tenuis Willd. 533.
— Thomsonii Buchenau 523; tab. 30.
— triceps Rostk. 526; tab. 30.
— trichodes Steud. 558.
— trifidus L. 533.
— triglumis L. 522; tab. 30.
— Turczaninowii (Buchenau) V. Krecz. 539, 629.
— turkestanicus V. Krecz. et Gontsch. 519, 625.
— tyraicus Pacz. 544.
— virens Buchenau 541; tab. 30.
Jungia Moench 21, 357.
Juniperus L. 1, 174.
— communis L. 179.
— dahurica Pall. 188.
— depressa Stev. 180.
— excelsa M. B. 185.
— foetidissima Willd. 185.
— isophyllos C. Koch 186.
— litoralis Maxim. 183.
— oblonga M. B. 180.
— oxycedrus L. 182.
— polycarpos C. Koch 186.
— pseudosabina Fisch. et Mey. 184.
— rigida Sieb. et Zucc. 182.
— sabina L. 190.
— Sargentii (Henry) Takeda 188.
— schugnanica Kom. 190.
— talassica Lipsky 189.
— seravschanica Kom. 187; tab. 8.
— sibirica Burgsd. 181.
— talassica Lipsky 189.
— turcomanica B. Fedtsch. 187.
— turkestanica Kom. 183.
— virginiana L. 191.
Juno Tratt. 4, 557.
Jurinea Cass. 27, 538.
— Abolinii Iljin 584; tab. 45.
— Abramovii Rgl. et Herd. 652; tab. 50.
— adenocarpa Schrenk 687.
— alata (Desf.) Cass. 673.
— albicaulis Bge. 554.
— algida Iljin 691.
— almaatensis Iljin 690.
— altaica Iljin 559, 717.
— amplexicaulis (S. G. Gmel.) Bobr. 558.
— Androssovii Iljin 590.
— Annae Sosn. 574.
— Antoninae Iljin 656.
— Antonovii C. Winkl. 648.
— apoda Iljin 619; tab. 48.
— arachnoidea Bge. 659; tab. 40, 45.
— armeniaca Sosn. 600; tab. 40.
— asperifolia Iljin 631.
— atripurpurea C. Winkl. ex Iljin 640; tab. 49.
— Auscheriana DC. 644.
— baissunensis Iljin 642.
— baldshuanica C. Winkl. 696.
— bellidioides Boiss. 570; tab. 42.
— bipinnatifida C. Winkl. 628.
— blanda (M. B.) C. A. Mey. 668.
— Bobrovii Iljin 655, 725.
— Botschantzevii Iljin 615, 722.
— bracteata Rgl. et Schmalh. 701.
— bucharica C. Winkl. 611.
— caespitans Iljin 587; tab. 43.
— calcarea Klok. 665.
— Capusii Franch. 588.
— cartaliniana Boiss. 577; tab. 42.
— cephalopoda Iljin 610.
— chaetocarpa Ledeb. 680; tab. 51.
— charcoviensis Klok. 550.
— ciscaucasica (Sosn.) Iljin 561.

— coronopifolia Somm. et Levier 576.
— cretacea Bge. 661.
— creticola Iljin 555, 717.
— cyanoides (L.) Reichb. 546; tab. 40.
— Czilikinoana Iljin 700.
— darvasica Iljin 692.
— Demetrii Iljin 671, 727.
— densisquamea Iljin 693.
— derderioides C. Winkl. 645.
— dshungarica (Rubtz.) Iljin 683.
— Eduardi-Regelii Iljin 587.
— elegans (Stev.) DC. 601.
— elegantissima Iljin 608, 722.
— Eversmannii Bge. 549.
— eximia Tek. 646.
— exuberans (Trautv.) Sosn. 670.
— ferganica (Iljin) Iljin 621; tab. 47.
— Fedtschenkoana Iljin 700; tab. 46.
— filicifolia Boiss. 577.
— foliosa (Iljin) Iljin 689.
— Gorodkovii Iljin 544, 717.
— gracilis Iljin 609.
— granitica Klok. 552.
— Grossheimii Sosn. 675.
— grumosa Iljin 694.
— hamulosa Rubtz. 636.
— helichrysifolia M. Pop. 633; 724.
— Iljinii Grossh. 571.
— impressinervis Iljin 618.
— Kamelinii Iljin 607, 721.
— Kapelkinii O. Fedtsch. 582.
— karabugasica Iljin 647.
— karatavica Iljin 615.
— kasakorum Iljin 662, 725.
— kazachstanica Iljin 562, 718.
— kirghisorum Janisch. 698.
— Knorringiana Iljin 656.
— kokanica Iljin 592.
— Komarovii Iljin 606.
— Krascheninnikovii Iljin 686.
— Kultiassovii Iljin 649.
— kuraminensis Iljin 622; tab. 47.
— lanipes Rupr. 690.
— lasiopoda Trautv. 614.
— laxa Fisch. ex Iljin 556; tab. 41.
— Ledebouri Bge. 663.
— Levieri All. 575.
— Lipskyi Iljin 687.
— lithophila Rubtz. 637.
— longicorollaris (Iljin) Iljin 684; tab. 51.
— Ludmilae Iljin 651.
— Lydiae Iljin 650.
— macranthodia Iljin 591.
— margalensis Iljin 682.
— Mariae Pavl. 702.
— maxima C. Winkl. 627.
— Michelsonii Iljin 666, 726.
— mollissima Klok. 664.
— monticola Iljin 703.
— mugodsharica Iljin 596.
— multiceps Iljin 596; tab. 46.
— multiflora (L.) B. Fedtsch. 579.
— multiloba Iljin 682.
— nivea C. Winkl. 694.
— Olgae Rgl. et Schmalh. 620; tab. 40.
— orientalis Iljin 598.
— Paczoskiana Iljin 564.
— persimilis Iljin 680; tab. 51.
— pineticola Iljin 563, 719.
— Pjataieviae Iljin 629.
— poacea Iljin 695; tab. 40.
— Popovii Iljin 610.
— propinqua Iljin 632; tab. 49.
— psammophila Iljin 612.
— pseudocyanoides Klok. 547.
— pteroclada Iljin 701.
— pulchella (Fisch. et Mey.) DC. 581.
— pumila Albov 569.
— rhizomatoidea Iljin 593; tab. 40.
— robusta Schrenk 605; tab. 44.
— Ruprechtii Boiss. 676.
— salicifolia Grun. 552.
— sangardensis Iljin 641.
— schachimardanica Iljin 585.
— Schischkiniana Iljin 548, 716.
— Semenovii C. Winkl. 635.
— serpenticaulis Iljin 630.
— serratuloides Iljin 638.
— Sintenisii Bornm. 597.
— sordida Stev. 667.
— Sosnovskyi Grossh. 574; tab. 42.
— spectabilis Fisch. et Mey. 674.
— Spiridonovii Iljin 702, 724.
— spissa Iljin 699; tab. 50.
— stenophylla Iljin 586; tab. 43.
— stoechadifolia (M. B.) DC. 580.
— suffruticosa Rgl. 595; tab. 46.
— suidunensis (C. Winkl.) Korsh. 681.
— tadshikistanica Iljin 617, 723.
— Talijevii Klok. 660.
— tanaitica Klok. 566, 719.
— tapetodes Iljin 654; tab. 40, 48.
— tenuiloba Bge. 565.
— thyrsiflora Klok. 567.
— tianschanica Rgl. et Schmalh. 688.
— tortisquamea Iljin 590.
— transhyrcanica Iljin 645.
— transuralensis Iljin 559, 718.

Jurinea Trautvetteriana Rgl. et Schmalh. 608.
— trifurcata Iljin 593.
— venusta Iljin 572, 720.
— Winkleri Iljin 616, 722.
— Woronowii Iljin 570.
— xeranthemoides Iljin 623; tab. 47.
— xerophytica Iljin 549, 716.
— Zakirovii Iljin 634, 723.
Jurinella Jaub. et Spach 27, 704.
— moschus (Habl.) Bobr. 705.
— squarrosa (Fisch. et Mey.) Iljin 707.
— subacaulis (Fisch. et Mey.) Iljin 706.

Kali Adans. 6, 211.
Kalidium Moq. 6, 166.
— caspicum (L.) Ung.-Sternb. 167.
— foliatum (Pall.) Moq. 166; tab. 8.
— Schrenkianum Bge. ex Ung.-Sternb. 168.
Kalimeris Cass. 25, 122.
— incisa (Fisch.) DC. 123.
— integrifolia Turcz. 123.
Kalonymus (G. Beck) Prokh. 14, 566, 744.
— latifolia (Mill.) Prokh. 569; tab. 30.
— leiophloea (Stev.) Prokh. 572; tab. 30.
— macroptera (Rupr.) Prokh. 573.
— Maximowicziana Prokh. 570, 744; tab. 30.
— sachalinensis (Fr. Schmidt) Prokh. 571, 745; tab. 30.
— yesoensis (Koidz.) Prokh. 568.
Kalopanax Miq. 16, 21.
— septemlobum (Thunb.) Koidz. 22.
Karamyschewia Fisch. et Mey. 23, 167.
— hedyotoides Fisch. et Mey. 168; tab. 8.
Karelinia Less. 25, 295.
— caspica (Pall.) Less. 296.
Kaschgaria Poljak. 26, 423.
— brachanthemoides (C. Winkl.) Poljak. 424; tab. 20.
— Komarovii (Krasch.) Poljak. 424.
Kaufmannia Rgl. 18, 203.
— brachyanthera Losinsk. 207; tab. 10.
— Semenovii (Herd.) Rgl. 207: tab. 10.
Kemulariella Tamamsch. 25, 110.
— abchasica (Kem.-Nath.) Tamamsch. 116.
— Albovii Tamamsch. 117.
— caucasica (Willd.) Tamamsch. 112; tab. 6.
— colchica (Albov) Tamamsch. 117.
— rosea (Stev.) Tamamsch. 115.
— Tugana (Albov) Tamamsch. 116; tab. 6.
Kentranthus Neck. 23, 640.
— calcitrapa (L.) Dufr. 641.
— longiflorus Stev. 640; tab. 28.
— ruber (L.) DC. 641.
Kentrophyllum Neck. 28, 581.
Keramocarpus Fenzl 16, 184.
Kerascomion Raf. 16, 376.
Keraselma Neck 15, 309.
Kerria DC. 10, 4.
— japonica (L.) DC. 4.
Kickxia Dum. 22, 176.
— caucasica (Muss.-Puschk.) Kuprian. 178.
— elatine (L.) Dum. 177.
— spuria (L.) Dum. 176.
Kirilowia Bge. 6, 123.
— eriantha Bge. 123; tab. 1, 5.
Klasea Cass. 28, 272.
Knautia L. 24, 12.
— arvensis (L.) Coult. 14.
— dipsacifolia (Host) Gren. et Godr. 13.
— involucrata Somm. et Levier 16; tab. 2.
— montana (M. B.) DC. 16.
— orientalis L. 19.
— tatarica (L.) Litv. 15.
Kochia Roth 6, 127.
— iranica (Hausskn. et Bornm.) Litv. 131; tab. 5.
— Krylovii Litv. 132; tab. 5.
— laniflora (S. G. Gmel.) Borb. 130; tab. 1, 5.
— melanoptera Bge. 133; tab. 5.
— prostrata (L.) Schrad. 128.
— Schrenkiana (Moq.) Iljin 132; tab. 5.
— scoparia (L.) Schrad. 133; tab. 5.
— Sieversiana (Pall.) C. A. Mey. 134; tab. 5.
Koeleria Pers. 2, 320.
— Albovii Domin. 328.
— asiatica Domin. 335; tab. 25.
— askoldensis Roshev. 330; tab. 25.
— atroviolacea Domin 336.
— barbarensis Domin 337.
— Buschiana (Domin) Gontsch. 328.
— caucasica Trin. ex Domin 327; tab. 25.
— Degenii Domin 324; tab. 25.
— Delavignei Czern. 337; tab. 25.
— Fominii Domin 329; tab. 25.
— geniculata Domin 336; tab. 25.
— glauca DC. 324; tab. 25.
— Gorodkowii Roshev. 326.
— gracilis Pers. 330; tab. 25.
— incerta Domin 336.
— Ledebouri Domin 328.
— Litwinowii Domin 325; tab. 25.
— Luerssenii Domin 332; tab. 25.
— monantha Domin 332; tab. 25.
— nitidula Velen. 329; tab. 25.
— poaeformis Domin 329; tab. 25.

— Pohleana (Domin) Gontsch. 323; tab. 25.
— polonica Domin 335; tab. 25.
— sclerophylla P. Smirn. 327; tab. 25.
— seminuda (Trautv.) Domin 331; tab. 25.
— sibirica Domin 331; tab. 25.
— splendens C. Presl 325.
— Thonii Domin 327.
— tokiensis Domin 326.
Koelpinia Pall. 29, 22.
— linearis Pall. 24.
— macrantha C. Winkl. 23.
— tenuissima Pavl. et Lipsch. 26.
— turanica Vass. 25.
Koenigia L. 5, 704.
— islandica L. 704.
Kohlrauschia Kunth 6, 779.
— prolifera (L.) Kunth 779; tab. 55.
Kolomikta Rgl. 15, 186.
Kolpakowskiana Rgl. 4, 489.
Komaroffia Kuntze 7, 71.
Komarovia Korov. 17, 149.
— anisopterum Korov. 149.
Koniga Adans. 8, 364.
— maritima (Desv.) R. Br. 364.
Korolkowia Rgl. 4, 295.
— Sewerzowii Rgl. 295; tab. 18.
Korovinia Nevski et Vved. 17, 211.
— ferganensis Korov. 213.
— microcarpa Korov. 213.
— tenuisecta (Rgl. et Schmalh.) Nevski et Vved. 212; tab. 19, 21.
Korshinskya Lipsky 16, 419.
— bupleuroides Korov. 420.
— Olgae (Rgl. et Schmalh.) Lipsky 420; tab. 15.
Kosopoljanskia Korov. 16, 188.
— turkestanica Korov. 191; tab. 14a, 15.
Kosteletzkia C. Presl 15, 168.
— pentacarpos (L.) Ledeb. 168.
Kozlovia Lipsky 16, 151.
Krascheninnikowia Turcz. 6, 423.
— Davidii Franch. 426; tab. 22.
— japonica Korsh. 425; tab. 22.
— Maximowicziana Franch. et Sav. 429; tab. 22.
— rigida Kom. 425; tab. 22.
— rupestris Turcz. 426; tab. 22.
— silvatica Maxim. 424; tab. 22.
Krasnovia M. Pop. 16, 117, 591.
— longiloba (Kar. et Kir.) M. Pop. 118; tab. 15.
Krylovia Schischk. 25, 129.
— eremophila (Bge.) Schischk. 132; tab. 10.
— limoniifolia (Less.) Schischk. 130; tab. 10.
— Novopokrovskyi (Krasch. et Iljin) Tamamsch. 135.
— Popovii (Botsch.) Tamamsch. 131; tab. 10.
Kudrjaschevia Pojark. 20, 474.
— allotricha Pojark. 475.
— Jacubi (Lipsky) Pojark. 481; tab. 28.
— Korshinskyi (Lipsky) Pojark. 476; tab. 28.
— Nadinae (Lipsky) Pojark. 480.
Kummerovia Schindl. 13, 383.
— stipulacea (Maxim.) Makino 383; tab. 20.
— striata (Thunb.) Schindl. 384; tab. 20.
Kükenthalia Boern. 3, 383.
Kuschakewiczia Rgl. 19, 638, 644.
Kyllinga Rottb. 3, 25.
— brevifolia Rottb. 25.
— kamtschatica Meinsh. 25; tab. 3.

Laburnum Medic. 11, 69.
— anagyroides Medic. 69; tab. 5.
Lachnoloma Bge. 8, 334.
— Lehmannii Bge. 334; tab. 20.
Lachnophyllum Bge. 25, 288.
— gossypinum Bge. 288; tab. 16.
Laciniaria Hill 25, 23.
Lactuca L. 29, 274.
— altaica Fisch. et Mey. 303; tab. 18.
— auriculata DC. 309; tab. 18.
— Chaixii Vill. 289.
— georgica Grossh. 306; tab. 18.
— glauciifolia Boiss. 308; tab. 18.
— indica L. 292; tab. 17.
— mira Pavl. 313.
— Raddeana Maxim. 295; tab. 17.
— rosularis Boiss. 307; tab. 18.
— saligna L. 301; tab. 18.
— sativa L. 299; tab. 18.
— serriola Torner 297; tab. 18.
— sibirica (L.) Benth. ex Maxim. 280, tab. 16.
— spinidens Nevski 312; tab. 18.
— stricta Waldst. et Kit. 288; tab. 17.
— tatarica (L.) C. A. Mey. 282; tab. 16.
— triangulata Maxim. 294; tab. 17.
— undulata Ledeb. 310; tab. 18.
— Wilhelmsiana Fisch. et Mey. 290; tab. 17.
— Winkleri Kirp. 285; tab. 16.
Lactucopsis Sch. Bip. 29, 286.
Ladanum Gilib. 21, 111.
Ladyginia Lipsky 17, 143.
— bucharica Lipsky 144.
Lagedium Soják 29, 279.
Lagenaria Ser. 24, 119.

Lagenaria vulgaris Ser. 119.
Lagochilus Bge. 21, 160.
— acutilobus (Ledeb.) Fisch. et Mey. 164; tab. 9.
— Androssovii Knorr. 169.
— balchanicus Czerniak. 176.
— Bungei Benth. 168.
— cabulicus Benth. 175.
— diacanthophyllus (Pall.) Benth. 170.
— hirtus Fisch. et Mey. 165.
— ilicifolius Bge. 169.
— inebrians Bge. 166.
— kaschgaricus Rupr. 181.
— Knorringianus Pavl. 171.
— kschtutensis Knorr. 166.
— leiacanthus Fisch. et Mey. 171.
— longidentatus Knorr. 175.
— macrodontus Knorr. 179.
— Nevskii Knorr. 172.
— Paulsenii Briq. 178.
— platyacanthus Rupr. 180.
— platycalyx Schrenk ex Fisch. et Mey. 179.
— pubescens Vved. 167, 651.
— pulcher Knorr. 180; tab. 9.
— pungens Schrenk 168.
— seravschanicus Knorr. 177; tab. 9.
— setulosus Vved. 165; 650.
— subhispidus Knorr. 167.
— tianschanicus Pavl. 176.
— usunachmaticus Knorr. 177.
Lagonychium M. B. 11. 13.
— farctum (Banks et Soland.) Bobr. 14: tab. 1.
Lagopsis Bge. 20, 248.
— eriostachya (Benth.) Ik.-Gal. 250.
— flava Kar. et Kir. 249; tab. 16.
— marrubiastrum (Steph.) Ik.-Gal. 248.
— supina (Steph.) Ik.-Gal. 250; tab. 16.
Lagopus Bernh. 11, 232.
Lagoseriopsis Kirp. 29, 373, 726.
— Popovii (Krasch.) Kirp. 374; tab. 20.
Lagoseris auct. 29, 708.
Lagoseris Hoffmgg. et Link 29, 595.
Lagoseris M. B. 29, 703.
Lagoseris M. B. 29, 699.
— aralensis (Bge.) Boiss. 714.
— callicephala Juz. 706; tab. 34.
— glaucescens (C. Koch) Sosn. 710; tab. 34.
— macrantha (Bge.) Iljin 713.
— obovata (Boiss. et Noe) Bornm. 712.
— purpurea (Willd.) Boiss. 705.
— robusta Czer. 704, 759.
— sahendi (Boiss. et Buhse) Czer. 707.
— sancta (L.) K. Maly 708.
Lagotis Gaertn. 22, 500.
— decumbens Rupr. 506; tab. 24.
— glauca Gaertn. 504.
— Ikonnikovii Schischk. 509; tab. 24.
— integrifolia (Willd.) Schischk. 502; tab. 24.
— Korolkowii (Rgl. et Schmalh.) Maxim. 509; tab. 24.
— minor (Willd.) Standl. 505.
— stolonifera (C. Koch) Maxim. 510.
— uralensis Schischk. 503.
Legouzia Durand 24, 427.
— falcata (Ten.) Fritsch 431; tab. 22.
— hybrida (L.) Delarb. 430.
— pentagonia (L.) Thell. 429.
— speculum Veneris (L.) Fisch. ex A. DC. 428.
Lagurostemon Cass. 27, 361, 392.
Lagurus L. 2, 234.
— ovatus L. 234, tab. 17.
Lallemantia Fisch. et Mey. 20, 482.
— baldshuanica Gontsch. 487.
— canescens (L.) Fisch. et Mey. 485.
— iberica (Stev.) Fisch. et Mey. 485.
— peltata (L.) Fisch. et Mey. 484.
— Royleana Benth. 486.
Lamiopsis Opiz 21, 124, 126.
Lamium L. 21, 124.
— album L. 134.
— amplexicaule L. 128.
— barbatum Sieb. et Zucc. 136.
— caucasicum Grossh. 131.
— dumeticola Klok. 135.
— hybridum Vill. 132.
— glaberrimum (C. Koch) Taliev 126.
— maculatum L. 137.
— ordubadicum Grossh. 129.
— Paczoskianum Vorosch. 129.
— purpureum L. 130.
— tomentosum Willd. 133.
Lamprochlaenia Boern. 3, 370.
Lamyra Cass. 28, 219.
— echinocephala (Willd.) Tamamsch. 220; tab. 10.
Lamyropappus Knorr. et Tamamsch. 28, 221.
— schakaptaricus (B. Fedtsch.) Knorr. et Tamamsch. 221.
Lapathum Meisn. 5, 457, 473.
Laportea Gaud. 5, 394.
— bulbifera (Sieb. et Zucc.) Wedd. 395.
Lappa Adans. 27, 93.
Lappula Gilib. 19, 403.
— aktaviensis M. Pop. et Zak. 460, 712.
— anisacantha (Turcz.) Guerke 445; tab. 22.

— balchaschensis M. Pop. ex Pavl. **423**; tab. 20.
— barbata (M. B.) Guerke 461.
— brachycentroides M. Pop. 444.
— ceratophora M. Pop. 417; tab. 20.
— consanguinea (Fisch. et Mey.) Guerke 447; tab. 22.
— coronifera M. Pop. 472, 713.
— cristata (Bge.) B. Fedtsch. 419.
— diploloma (Schrenk) Guerke 420; tab. 20.
— Drobovii M. Pop. ex Pavl. 467; tab. 22, 23.
— echinata Gilib. 449; tab. 22.
— glabrata M. Pop. ex Pavl. 458.
— heteracantha (Ledeb.) Guerke 427; tab. 22.
— intermedia (Ledeb.) M. Pop. 440.
— Korshinskyi M. Pop. 459.
— kulikalonica Zak. 465.
— Lipschitzii M. Pop. 474.
— Lipskyi M. Pop. 427.
— macra M. Pop. 438.
— macrantha (Ledeb.) Guerke 432.
— marginata (M. B.) Guerke 425; tab. 22.
— microcarpa (Ledeb.) Guerke 462; tab. 22.
— occultata M. Pop. 475.
— patula (Lehm.) Aschers. ex Guerke 436; tab. 22.
— petrophila Pavl. 473.
— Popovii Zak. 471.
— rupestris (Schrenk) Guerke 455; tab. 18.
— semialata M. Pop. 468.
— semiglabra (Ledeb.) Guerke 430; tab. 21, 22.
— sericata M. Pop. 457.
— sessiliflora (Boiss.) Guerke **477**.
— sinaica (DC.) Aschers. ex Schweinf. 475.
— spinocarpos (Forsk.) Aschers. ex Kuntze 418; tab. 22.
— stricta (Ledeb.) Guerke 438; tab. 22.
— tadshikorum M. Pop. 452.
— tenuis (Ledeb.) Guerke 442.
— tianschanica M. Pop. et Zak. 451.
— ulacholica M. Pop. 466.
Lappularia Pomel 16, 153, 159.
Lapsana L. 29, 401.
— aipetriensis Vass. 404, 727.
— communis L. 401.
— grandiflora M. B. 403.
— intermedia M. B. 402.
Larbrea St.-Hil. 6, 394.
Larix Mill. 1, 153.
— dahurica Turcz. 156; tab. **7**.
— decidua Mill. 154.
— kamtschatica (Rupr.) Carr. 158.
— olgensis Henry 158.
— sibirica Ledeb. 155; tab. 7, 8.
Laser Borkh. 17, 166.
— trilobum (L.) Borkh. 167.
Laserpitium L. 17, 279.
— affine Ledeb. 282.
— alpinum Waldst. et Kit. 282.
— hispidum M. B. 283; tab. 22.
— latifolium L. 281.
— prutenicum L. 287.
— Stevenii Fisch. et Trautv. 284.
Lasiagrostis Link 2, 71.
— bromoides (L.) Nevski **72**; tab. 6.
— caragana Trin. et Rupr. **72**; tab. 6.
— longearistata (Boiss.) Roshev. et Nevski 73; tab. 6.
— splendens (Trin.) Kunth **72**; tab. 6.
Lasiopogon Cass. 25, 377.
— muscoides (Desf.) DC. 377.
Lasiospermum Fisch. 29, 55.
Lasiospora Cass. 29, 55.
Lasiospora Less. 29, 97.
Lasiotrachyphylla Bernh. 7, 600.
Lathraea L. 22, 803.
— squamaria L. 804.
Lathyrus L. 13, 479.
— alatus (Maxim.) Kom. 514.
— angulatus L. **497**.
— annuus L. 483.
— aphaca L. 480.
— atropatanus (Grossh.) Širj. 512.
— aureus (Stev.) Brandza 518.
— chloranthus Boiss. 486.
— cicera L. 484.
— colchicus Lipsky 484.
— cyaneus (Stev.) C. Koch 512.
— Davidii Hance 506.
— digitatus (M. B.) B. Fedtsch. 511.
— Dominianus Litv. 500.
— Frolovii (Fisch.) Rupr. 514.
— Gmelinii (Fisch.) Fritsch 517.
— hirsutus L. 487.
— humilis Fisch. ex DC. 499.
— inconspicuus L. 498.
— incurvus (Roth) Willd. 504.
— Krylovii Serg. 518.
— laevigatus (Waldst. et Kit.) Fritsch 519.
— latifolius L. 491.
— laxiflorus (Desf.) Kuntze 520.
— Ledebourii Trautv. 510.
— Litvinovii Iljin 495.
— maritimus (L.) Bigel. 507.

Lathyrus miniatus M. B. 494.
— montanus Bernh. 515.
— mulkak Lipsky 491.
— niger (L.) Bernh. 516.
— nissolia L. 481.
— ochrus (L.) DC. 482.
— odoratus L. 488.
— pallescens (M. B.) C. Koch 509.
— palustris L. 501.
— pannonicus (Kramer) Garcke 510.
— pilosus Cham. 502.
— pisiformis L. 505.
— pratensis L. 500.
— quinquenervius (Miq.) Litv. 503.
— roseus Stev. 489.
— rotundifolius Willd. 493.
— sativus L. 485.
— setifolius L. 496.
— silvestris L. 482.
— sphaericus Retz. 497.
— subrotundus Maxim. 516.
— tuberosus L. 490.
— venetus (Mill.) Rouy et Fouc. 513.
— vernus (L.) Bernh. 513.
— Voronovii Bornm. 499.
Laulia Raf. 18, 518.
Launaea auct. 29, 239.
Laurocerasus Roem. 10, 579.
— officinalis Roem. 580; tab. 37.
Laurus L. 7, 572.
— nobilis L. 572; tab. 38.
Lavandula L. 20, 226.
— spica L. 226.
Lavatera L. 15, 76.
— cashemiriana Cambess. 78; tab. 5.
— punctata All. 79; tab. 5.
— thuringiaca L. 76.
— trimestris L. 83.
Laxmannia S. G. Gmel. 23, 192.
Laya Hook. et Arn. 25, 568.
— elegans Torr. et Gray 568.
Lecokia DC. 16, 251.
— cretica (Lam.) DC. 251; tab. 18a.
Ledebouriella Wolff 16, 361.
— multiflora (Ledeb.) Wolff 362; tab. 30.
— seseloides (Hoffm.) Wolff 362; tab. 30.
Ledeburia Link 16, 422.
Ledonia Spach 15, 328.
Ledum L. 18, 27.
— decumbens (Ait.) Small 30.
— hypoleucum Kom. 29.
— palustre L. 28.
Leersia Sw. 2, 48.
— oryzoides (L.) Sw. 48; tab. 3.
Leibnitzia Cass. 28, 592.
— anandria (L.) Turcz. 593.
— Knorringiana (B. Fedtsch.) Pobed. 594; tab. 28.
Lembotropis Griseb. 11, 71.
— nigricans (L.) Griseb. 71; tab. 5.
Lemna L. 3, 492.
— gibba L. 493.
— minor L. 493.
— trisulca L. 493; tab. 26.
Lens Adans. 13, 475.
— culinaris Medic. 478.
— lenticula (Schreb.) Alef. 477.
— nigricans (M. B.) Godr. 476.
— orientalis (Boiss.) Hand-Mazz. 477.
Leobordea Delile 11, 44.
Leonia La Llav. et Lex. 21, 289.
Leontice L. 7, 545.
— Alberti Rgl. 550.
— altaica Pall. 549.
— darwasica Rgl. 550; tab. 37.
— Ewersmannii Bge. 546; tab. 37.
— incerta Pall. 545.
— minor Boiss. 549.
— odessana Fisch. ex Rogow. 550.
— Smirnowii Trautv. 551.
Leontodon L. 29, 204.
— asperrimus (Willd.) Boiss. 216; tab. 13.
— asperus (Waldst. et Kit.) Boiss. 215; tab. 13.
— autumnalis L. 207.
— caucasicus (M. B.) Fisch. 218.
— croceus Haenke 214; tab. 12.
— danubialis Jacq. 211.
— gutzulorum V. Vassil. 210; tab. 10.
— hispidus L. 217; tab. 13.
— keretinus Nyl. 209.
— Kotschyi Boiss. 217; tab. 13.
— pseudotaraxaci Schur 203; tab. 13.
— repens Schur 211; tab. 13.
— rilaensis Hayek 209; tab. 12.
— Schischkinii V. Vassil. 212; tab. 13.
Leontopodium R. Br. 25, 342.
— alpinum Cass. 360.
— brachyactis Gand. 358.
— discolor Beauv. 346.
— kurilense Takeda 348.
— leontopodioides (Willd.) Beauv. 349; tab. 22.
— nanum (Hook. f. et Thoms.) Hand.-Mazz. 359.
— ochroleucum Beauv. 350; tab. 22.
— Palibinianum Beauv. 345.
Leonurus L. 321, 145.

— cardiaca L. 148.
— deminutus V. Krecz. 153.
— heterophyllus Sweet 156.
— glaucescens Bge. 151; tab. 8.
— incanus V. Krecz. et Kuprian. 155.
— macranthus Maxim. 155.
— mongolicus V. Krecz. et Kuprian. 154.
— panzerioides M. Pop. 154, 650.
— quinquelobatus Gilib. 148; tab. 8.
— sibiricus L. 157.
— tataricus L. 153.
— turkestanicus V. Krecz. et Kuprian. 152; tab. 8.
Leopoldia Parl. 4, 409.
— caucasica (Griseb.) Losinsk. 410.
— comosa (L.) Parl. 409; tab. 24.
— longipes (Boiss.) Losinsk. 410; tab. 24.
— tenuiflora (Tausch) Heldr. 410; tab. 24.
Lepargyrea Raf. 15, 518.
Lepechiniella M. Pop. 19, 387.
— alaica M. Pop. 401, 716.
— arassanica (B. Fedtsch.) M. Pop. 392.
— ferganensis M. Pop. 398, 715.
— Korshinskyi M. Pop. 394, 625, 715.
— minuta (Lipsky) M. Pop. 402; tab. 19.
— omphaloides (Schrenk) M. Pop. 397; tab. 18.
— seravschanica (Lipsky) M. Pop. 391; tab. 17.
— transalaica (B. Fedtsch.) M. Pop. 393, 714; tab. 17.
Lepicaune Lapeyr. 29, 595, 627.
Lepicephalus Lag. 24, 25.
Lepidium L. 8, 501.
— amplexicaule Willd. 516.
— apetalum Willd. 508.
— Boissieri N. Busch 505, 651; tab. 28.
— Borsczovii N. Busch 507, 652; tab. 28.
— campestre (L.) R. Br. 506.
— cordatum Willd. ex Stev. 517.
— coronopifolium Fisch. 523.
— crassifolium Waldst. et Kit. 514; tab. 29.
— draba L. 504; tab. 28.
— eremophilum Schrenk 518.
— ferganense Korsh. 519.
— graminifolium L. 524.
— karataviense Rgl. et Schmalh. 519.
— lacerum C. A. Mey. 518.
— latifolium L. 515.
— lyratum L. 517; tab. 29.
— Meyeri Claus 520.
— obtusum Basin. 516.
— perfoliatum L. 512.
— persicum Boiss. 519.
— pinnatifidum Ledeb. 512.
— propinquum Fisch. et Mey. 506.
— pumilum Boiss. et Bal. 514; tab. 29.
— repens (Schrenk) Boiss. 505.
— ruderale L. 511.
— sativum L. 508.
— sibiricum Schweigg. 515.
— soongoricum Schrenk 523.
— Turczaninovii Lipsky 520; tab. 29.
— vesicarium L. 513.
— virginicum L. 512.
— *Lepidocroton* C. Presl 14, 288.
Lepidolopha C. Winkl. 26, 288.
— Fedtschenkoana Knorr. 291.
— filifolia Pavl. 289.
— karatavica Pavl. 291.
— Komarowii C. Winkl. 289.
— mogoltavica Krasch. 290.
— nuratavica Krasch. 290.
Lepidolopsis Poljak. 26, 315.
— turkestanica (Rgl. et Schmalh.) Poljak. 316; tab. 12.
Lepidoseris (Reichb.) Fourr. 29, 595, 695.
Lepigonum Wahl. 6, 556.
Leptaleum DC. 8, 302.
— filifolium (Willd.) DC. 302; tab. 19.
Leptandra Nutt. 22, 494.
Leptanthe Klotzsch 19, 167.
Leptargyreia Schlecht. 15, 518.
Leptasea Haw. 9, 159, 186.
Lepteranthus Neck. 28, 370, 446, 449.
Leptilon Raf. 25, 191.
Leptopyrum Reichb. 7, 72.
— fumarioides (L.) Reichb. 72.
Leptorhabdos Schrenk 22, 526.
— parviflora Benth. 527.
Leptunis Stev. 23, 285.
— trichoides (J. Gay) Schischk. 285; tab. 15.
Lepyrodiclis Fenzl 6, 480.
— holosteoides (C. A. Mey.) Fisch. et Mey. 480; tab. 26.
— stellarioides Schrenk 481; tab. 26.
Lereschia Boiss. 16, 380.
Lerouxia Mérat. 18, 258.
Lespedeza Rich. 13, 376.
— bicolor Turcz. 379; tab. 20.
— cyrtobotrya Miq. 380; tab. 20.
— dahurica (Laxm.) Schindl. 380.
— hedysaroides (Pall.) Kitag. 380.
— tomentosa (Thunb.) Sieb. 382.
Lesquereuxia Boiss. et Reut. 22, 795.
Leucantha S. F. Gray 28, 370, 570.
Leucanthemella Tzvel. 26, 137.
— linearis (Matsum.) Tzvel. 139.

Leucanthemella serotina (L.) Tzvel. 139.
Leucanthemum Mill. 26, 140.
— rotundifolium (Waldst. et Kit.) DC. 142.
— subalpinum (Simonk.) Tzvel. 145.
— vulgare Lam. 143.
Leucoglochin Ehrh. ex Heuff. 3, 301.
Leucojum L. 4, 480.
— aestivum L. 480; tab. 30.
— vernum L. 481.
Leucopoa Griseb. 2, 494.
— albida (Turcz.) V. Krecz. et Bobr. 495; tab. 39
— caucasica (Hack.) V. Krecz. et Bobr. 496.
— karatavica (Bge.) V. Krecz. et Bobr. 496; tab. 39.
— Olgae (Rgl.) V. Krecz. et Bobr. 495.
— sclerophylla (Boiss. et Hohen.) V. Krecz. et Bobr. 497; tab. 39.
Leucorchis E. Mey. 4, 648.
— albidus (L.) E. Mey. 649.
— *Leucostemma* Benth. 6, 421.
Levisticum Hill 17, 40.
— officinale Koch 41.
Liatris Schreb. 25, 23.
— spicata (L.) Willd. 23.
Libanotis L. 16, 471.
— amurensis Schischk. 476.
— buchtormensis (Fisch.) DC. 473.
— calycina Korov. 482.
— condensata (L.) Grantz 480; tab. 30.
— dolichostyla Schrenk 480, 600.
— intermedia Rupr. 474.
— monstrosa (Willd.) DC. 483.
— montana Crantz 477.
— Schrenkiana C. A. Mey. 478, 601.
— seseloides (Fisch. et Mey.) Turcz. 477.
— setifera (Korov.) Schischk. 481.
— sibirica (L.) C. A. Mey. 479.
— transcaucasica Schischk. 475.
Ligularia Cass. 26, 788.
— abakanica Pojark. 810, 887.
— alpigena Pojark. 841.
— altaica DC. 843; tab. 39.
— altissima Pojark. 835.
— arctica Pojark. 817, 891.
— bucovinensis Nakai 818.
— calthifolia Maxim. 800; tab. 37.
— carpatica (Schott, Nym. et Kotschy) Pojark. 848.
— caucasica (M. B.) G. Don 855.
— Correvoniana (Albov) Pojark. 856; tab. 37.
— Fischeri (Ledeb.) Turcz. 802.
— glauca (L.) O. Hoffm. 847; tab. 39.
— heterophylla Rupr. 838.
— Hodgsonii Hook. 798.
— jaluensis Kom. 826.
— karataviensis (Lipsch.) Pojark. 824.
— Kareliniana Stschegl. 821.
— Knorringiana Pojark. 832.
— longipes Pojark. 816, 890.
— Lydiae Minder. 808.
— macrophylla (Ledeb.) DC. 837.
— mongolica (Turcz.) DC. 849; tab. 40.
— narynensis (C. Winkl.) O. et B. Fedtsch. 822; tab. 38.
— Pavlovii (Lipsch.) Cretz. 852; tab. 30.
— renifolia (C. A. Mey.) DC. 853.
— robusta (Ledeb.) DC. 820.
— sachalinensis Nakai 805.
— Schischkinii Rubtz. 825.
— Schmidtii (Maxim.) Makino 850; tab. 40.
— sibirica (L.) Cass. 807.
— sichotensis Pojark. 811, 888.
— songarica (Fisch.) Ling 831.
— splendens (Lévl. et Vaniot) Nakai 803.
— subsagittata Pojark. 812, 889.
— talassica Pojark. 840.
— Thomsonii (Clarke) Pojark. 834.
— thyrsoidea (Ledeb.) DC. 827.
— trichocephala Pojark. 800, 886.
Ligularia Duw. 9, 198.
Ligularia O. Hoffm. 26, 797.
Ligusticum L. 16, 566.
— alatum (M. B.) Spreng. 571.
— arafae Albov 570.
— caucasicum Somm. et Lévier 574.
— discolor Ledeb. 572.
— Fedtschenkoanum Schischk. 576.
— Hultenii Fern. 569.
— mongholicum (Turcz.) Kryl. 573; tab. 34.
— mutellina (L.) Crantz 573.
— physospermifolium Albov 571; tab. 34.
— pumilum Korov. 575; tab. 30.
— purpureopetalum Kom. 569.
— scoticum L. 568.
Ligustridium Spach 18, 518.
Ligustrina Rupr. 18, 516.
— amurensis Rupr. 517; tab. 26.
Ligustrum L. 18, 518.
— japonicum Thunb. 520.
— lucidum Ait. 519.
— Tschonoskii Decne. 520; tab. 26.
— vulgare L. 522.
— yezoense Nakai 521; tab. 26.
Lilac Adans. 18, 502.
Liliacum Renault 18, 502.
Liliarhiza Kellog 4, 318.

Lilium L. 4, 282.
— avenaceum Fisch. 288; tab. 17.
— bulbiferum L. 291.
— callosum Sieb. et Zucc. 291; tab. 17.
— candidum L. 284.
— cernuum Kom. 292.
— cordifolium Thunb. 294.
— dahuricum Ker.-Gawl. 284; tab. 17.
— distichum Nakai 289.
— Hansonii Leichtlin 289.
— Kesselringianum Miscz. 293.
— Ledebouri Boiss. 292; tab. 17.
— martagon L. 288.
— Maximoviczii Rgl. 290.
— monadelphum M. B. 293; tab. 17.
— ponticum C. Koch 293.
— pulchellum Fisch. 287; tab. 17.
— Szovitsianum Fisch. et Avé-Lall. 294.
— tenuifolium Fisch. 291; tab. 17.
— tigrinum Ker.-Gawl. 290.
Limbardia Adans. 25, 433, 476.
Limivasculum Boern. 3, 201, 232.
Limnanthemum S. G. Gmel. 18, 643.
Limnas Trin. 2, 165.
— Stelleri Trin. 165.
— Veresczaginii Kryl. et Schischk. 165; tab. 12.
Limnocrepis Fourr. 29, 595, 608.
Limnopeuce Adans. 15, 668.
Limnorchis Rydb. 4, 664.
— convallariifolius (Fisch. ex Lindl.) Rydb. 665.
— dilatatus (Pursh) Rydb. 665.
— hologlottis (Maxim.) Nevski 666.
Limnoseris Peterm. 29, 595, 630.
Limodorum Rich. 4, 634.
— abortivum (L.) Sw. 635.
Limoniopsis Lincz. 18, 376, 744.
— Overinii (Boiss.) Lincz. 377.
Limonium Mill. 18, 411.
— aperum (L.) Hill ex Kuntze 435.
— Bungei (Claus) Gamajun. 448.
— carnosum (Boiss.) Kuntze 461.
— caspium (Willd.) Gams 449.
— chrysocomum (Kar. et Kir.) Kuntze 431; tab. 21.
— congestum (Ledeb.) Kuntze 423.
— coralloides (Tausch) Lincz. 451; tab. 22.
— dichroanthum (Rupr.) Ik.-Gal. 428.
— drepanostachyum Ik.-Gal. 465; tab. 23.
— Fajzievii Zak. 466, 748.
— ferganense Ik.-Gal. 454.
— Fischeri (Trautv.) Lincz. 427.
— flexuosum (L.) Kuntze 420; tab. 21.
— Gmelinii (Willd.) Kuntze 436.
— Hoeltzeri (Rgl.) Ik.-Gal. 426.
— kaschgaricum (Rupr.) Ik.-Gal. 424.
— latifolium (Smith) Kuntze 442; tab. 22.
— leptolobum (Rgl.) Kuntze 429.
— leptophyllum (Schrenk) Kuntze 459.
— macrorrhizon (Ledeb.) Kuntze 452.
— Meyeri (Boiss.) Kuntze 439.
— Michelsonii Lincz. 430, 747.
— myrianthum (Schrenk) Kuntze 453.
— otolepis (Schrenk) Kuntze 455.
— reniforme (Girard) Lincz. 456.
— Rezniczenkoanum Lincz. 434, 747.
— sareptanum (A. Beck.) Gams 446.
— Semenovii (Herd.) Kuntze 432.
— sinuatum (L.) Mill. 419.
— sogdianum (M. Pop.) Ik.-Gal. 462.
— suffruticosum (L.) Kuntze 458; tab. 22.
— tomentellum (Boiss.) Kuntze 440.
Limosella L. 22, 324.
— aquatica L. 324.
Linaria Mill. 22, 178.
— acutiloba Fisch. ex Reichb. 202.
— albifrons (Sibth. et Smith) Spreng. 225.
— altaica Fisch. 207.
— armeniaca Chav. 218.
— arvensis (L.) Desf. 222.
— Biebersteinii Bess. 197; tab. 7.
— bipartita (Vent.) Willd. 218.
— brachyceras (Bge.) Kuprian. 210.
— Bungei Kuprian. 206; tab. 9.
— buriatica Turcz. 196.
— canadensis (L.) Dum. 218.
— chalepensis (L.) Mill. 217.
— corifolia Desf. 219.
— cretacea Fisch. ex Spreng. 221; tab. 8.
— creticola Kuprian. 221; tab. 8.
— debilis Kuprian. 215.
— dolichocarpa Klok. 208, 818.
— dolichoceras Kuprian. 210.
— dulcis Klok. 209.
— elymaitica (Boiss.) Kuprian. 217.
— euxina Velen. 190.
— genistifolia (L.) Mill. 188; tab. 8.
— grandiflora Desf. 187.
— hepatica Bge. 205.
— incompleta Kuprian. 215.
— japonica Miq. 220.
— kokanica Rgl. 204; tab. 9.
— kopetdaghensis Kuprian. 194; tab. 7.
— kulabensis B. Fedtsch. 205.
— kurdica Boiss. et Hohen. 195.
— lenkoranica Kuprian. 194; tab. 7.
— leptoceras Kuprian. 211.

Linaria lineolata Boiss. 195.
— Loeselii Schweigg. 209.
— macrophylla Kuprian. 222; tab. 8.
— macroura (M. B.) Chav. 216.
— melampyroides Kuprian. 202.
— Meyeri Kuprian. 212.
— micrantha (Cav.) Hoffmg. et Link 224.
— monspessulana (L.) Mill. 219.
— odora (M. B.) Fisch. 208.
— pedicellata Kuprian. 211; tab. 9.
— Pelisseriana (L.) DC. 224.
— pontica Kuprian. 189; tab. 8.
— Popovii Kuprian. 203; tab. 9.
— ramosa (Kar. et Kir.) Kuprian. 207.
— reflexa (L.) Desf. 220.
— ruthenica Blonski 197; tab. 7.
— sabulosa Czern. ex Klok. 190.
— Schelkovnikovii Schischk. ex Grossh. et Schischk. 198.
— schirvanica Fomin 216.
— sessilis Kuprian. 204; tab. 9.
— simplex (Willd.) DC. 223.
— striatella Kuprian. 211; tab. 9.
— syspirensis C. Koch 193; tab. 8.
— transiliensis Kuprian. 206; tab. 9.
— turcomanica Kuprian. 223.
— vulgaris Mill. 201; tab. 7.
— zangezura Grossh. 188.
Lindelofia Lehm. 19, 625.
— angustifolia (Schrenk) Brand 637.
— Capusii (Franch.) M. Pop. 630.
— macrostyla (Bge.) M. Pop. 627; tab. 35.
— Olgae (Rgl. et Smirn.) Brand 629; tab. 35.
— pterocarpa M. Pop. 638.
— stylosa (Kar. et Kir.) Brand 633; tab. 35.
Lindera Adans. 16, 94.
Lindera Aschers. 16, 150.
Lindernia All. 22, 327.
— pyxidaria All. 327.
Linnaea Gronov. ex L. 23, 464.
— borealis L. 465.
Linopsis Reichb. 14, 128.
Linosyris Cass. 25, 173.
— Fominii Kem.-Nath. 176.
— pontica (Lipsky) Novopokr. 177.
— tatarica (Less.) C. A. Mey. 179.
— villosa (L.) DC. 178.
— vulgaris Cass ex Less. 175.
Linum L. 14, 86.
— altaicum Ledeb. 113, 717.
— amurense Alef. 110.
— angustifolium Huds. 95.
— atricalyx Juz. 114; 719.
— austriacum L. 123; tab. 7.
— baicalense Juz. 109, 715; tab. 6.
— basarabicum (Sâvul. et Rayss) Klok. 133.
— bienne Mill. 99.
— boreale Juz. 111, 718; tab. 7.
— brevisepalum Juz. 117, 720.
— catharticum L. 127.
— corymbulosum Rech. 128.
— crepitans Dum. 96.
— Czerniaevii Klok. 135.
— euxinum Juz. 118, 721.
— extraaxillare Kit. 115.
— flavum L. 136.
— gallicum L. 129.
— heterosepalum Rgl. 144; tab. 8.
— hirsutum L. 141; tab. 8.
— humile Mill. 102.
— hypericifolium Salisb. 140.
— jailicola Juz. 93; 714.
— Komarovii Juz. 112, 719; tab. 7.
— lanuginosum Juz. 143, 723; tab. 8.
— linearifolium Jav. 133.
— luteolum M. B. 139.
— macrorhizum Juz. 105, 714; tab. 6.
— Marschallianum Juz. 120, 721; tab. 7.
— mesostylum Juz. 107, 714.
— nervosum Waldst. et Kit. 92.
— Olgae Juz. 145; tab. 8.
— orientale Boiss. 130.
— Pallasianum Schult. 132.
— pallescens Bge. 106; tab. 6.
— perenne L. 116; tab. 7.
— stelleroides Planch. 125.
— subbiflorum Juz. 110, 716; tab. 6.
— tauricum Willd. 131.
— tenuifolium L. 125.
— turcomanicum Juz. 114, 720.
— ucranicum Czern. 134.
— uralense Juz. 135, 722.
— usitatissimum L. 100.
— violascens Bge. 108.
Liparis Rich. 4, 601.
— japonica (Miq.) Maxim. 603.
— Loeselii (L.) Rich. 602; tab. 37.
— Makinoana Schlechter 603; tab. 37.
Lippia L. 19, 697.
— nodiflora (L.) Rich. 697.
Lipskya Nevski 16, 192, 193.
Lipskyella Juz. 27, 357.
— annua (C. Winkl.) Juz. 358.
Liriodendron L. 7, 565.
— tulipiferum L. 565; tab. 39.
Lisaea Boiss. 16, 177.
— armena Schischk. 178; tab. 14.

— heterocarpa (DC.) Boiss. 178; tab. 14.
Listera R. Br. 4, 611.
— brevidens Nevski 613; tab. 38.
— cordata (L.) R. Br. 613; tab. 38.
— major Nakai 614; tab. 38.
— ovata (L.) R. Br. 612.
Lithoon Nevski 12, 98.
Lithosciadium Turcz. 16, 549.
Lithospermum L. 19, 155.
— apulum (L.) Vahl 163.
— arvense L. 165.
— erythrorhizon Sieb. et Zucc. 162.
— officinale L. 162.
— purpureo-coeruleum L. 157.
— tenuiflorum L. f. 164; tab. 9.
— tschimganicum B. Fedtsch. 158; tab. 9.
Littledalea Hemsl. 2, 553.
— alaica (Korsh.) V. Petrov 553; tab. 43.
Littorella Bergius 23, 163.
— uniflora (L.) Aschers. 163.
Litvinovia Woronow 8, 324.
— tenuissima (Pall.) N. Busch 327; tab. 13.
Lloydia Salisb. 4, 365.
— rubroviridis (Boiss. et Kotschy) Baker 366.
— serotina (L.) Reichb. 369.
— triflora (Ledeb.) Baker 366.
Lobelia L. 24, 451.
— Dortmanna L. 451.
— sessilifolia Lamb. 452.
Lobularia Desv. 8, 364.
Lochnera Reichb. 18, 646.
Logfia Cass. 25, 314, 325.
Loiseleuria Desv. 18, 62.
— procumbens (L.) Desv. 62.
Loliolum V. Krecz. et Bobr. 2, 544, 765.
— orientale (Boiss.) V. Krecz. et Bobr. 544; tab. 41.
Lolium L. 2, 545.
— arvense With. 547; tab. 44.
— cuneatum Nevski 547; tab. 44.
— loliaceum (Bory et Chaub.) Hand.-Mazz. 551; tab. 44.
— multiflorum Lam. 551; tab. 44.
— perenne L. 552; tab. 44.
— persicum Boiss. et Hohen. 548; tab. 44.
— remotum Schrank 547; tab. 44.
— rigidum Gaud. 551; tab. 44.
— temulentum L. 546; tab. 44.
Lomatocarum Fisch. et Mey. 16, 385.
Lomatogonium A. Br. 18, 620.
— carinthiacum (Wulfen.) A. Br. 620; tab. 33.
— rotatum (L.) Fries ex Nym. 621; tab. 33.
Lomatopodium Fisch. et Mey. 16, 483, 561.
Londesia Fisch. et Mey. 6, 124.
— eriantha Fisch. et Mey. 124; tab. 5.
Lonicera L. 23, 467.
— Alberti Rgl. 478.
— altaica Pall. 496.
— Altmannii Rgl. et Schmalh. 517.
— asperifolia Hook. f. et Thoms. 522.
— baltica Pojark. 491, 731.
— bracteolaris Boiss. et Buhse 521.
— Buschiorum Pojark. 499, 733.
— caerulea L. 500.
— caprifolium L. 570.
— caucasica Pall. 540.
— Chamissoi Bge. 539.
— chrysantha Turcz. 555.
— cinerea Pojark. 529, 736; tab. 26.
— edulis Turcz. ex Freyn 492.
— etrusca Santi 572.
— floribunda Boiss. et Buhse 551.
— gibbiflora (Rupr.) Dipp. 557.
— Glehnii Fr. Schmidt 532.
— heterotricha Pojark. et Zak. 514.
— hispida Pall. ex Roem. et Schult. 509.
— humilis Kar. et Kir. 525; tab. 26.
— iberica M. B. 503; tab. 25.
— iliensis Pojark. 502, 734.
— japonica Thunb. 568.
— kamtschatica (Sevast.) Pojark. 487, 730.
— Karelinii Bge. ex P. Kir. 530.
— Korolkovii Stapf 548.
— lanata Pojark. 550, 737.
— Maackii Rupr. 565.
— Maximowiszii (Rupr.) Rgl. 535.
— micrantha Trautv. ex Rgl. 547.
— microphylla Willd. ex Roem. et Schult. 482; tab. 24.
— nigra L. 543.
— nummulariifolia Jaub. et Spach 566.
— Olgae Rgl. et Schmalh. 524.
— Pallasii Ledeb. 489.
— pamirica Pojark. 484, 729; tab. 24.
— paradoxa Pojark. 534.
— praeflorens Batal. 515; tab. 26.
— Ruprechtiana Rgl. 563.
— sachalinensis (Fr. Schmidt) Nakai 537.
— Semenovii Rgl. 513.
— simulatrix Pojark. 480, 729; tab. 24.
— stenantha Pojark. 497; tab. 25.
— Steveniana Fisch. ex Pojark. 561, 738.
— subhispida Nakai 511; tab. 26.
— tatarica L. 544.
— tianschanica Pojark. 518.
— Tolmatchevii Pojark. 505, 735; tab. 25.

Lonicera Turczaninowii Pojark. 494, 731.
— xylosteum L. 559.
— zaravschanica (Rehd.) Pojark. 519.
Lophanthus Adans. 20, 275.
— chinensis (Raf.) Benth. 276.
— elegans (Lipsky) Levin 282.
— Krylovii Lipsky 277.
— Lipskyanus Ik.-Gal. et Nevski 280.
— Schrenkii Levin 278.
— Schtschurowskianus (Rgl.) Lipsky 279.
— subnivalis Lipsky 281.
— tschimganicus Lipsky 279.
— urticifolia (Miq.) Makino 529.
Lophioilepis Cass. 28, 52, 69, 89.
Lophochloa Reichb. 2, 337.
— obtusiflora (Boiss.) Gontsch. 338; tab. 25.
— phleoides (Vill.) Reichb. 338; tab. 25.
Lopholoma Cass. 28, 370, 440, 493.
Loranthus L. 5, 407.
— europaeus Jacq. 407; tab. 23.
Loroglossum Rich. 4, 719.
Lotononis DC. 11, 44.
— genistoides (Fenzl) Benth. 44.
Lotophyllum Reichb. 11, 218.
Lotus L. 11, 284.
— aleppicus Boiss. 290.
— angustissimus L. 296; tab. 19.
— caucasicus Kuprian. 291; tab. 19.
— corniculatus L. 291; tab .19.
— frondosus Freyn 295.
— Gebelia Vent. 290; tab. 19.
Louichea L'Hér. 6, 572.
Lowea Lindl. 10, 506.
Loxaniza Raf. 3, 246.
Ludwigia L. 15, 567.
— palustris (L.) Ell. 568; tab. 27.
— prostrata Roxb. 568; tab. 27.
Luerssenia Kuntze 16, 369.
Luffa Adans. 24, 99.
— acutangula (L.) Roxb. 100.
— cylindrica (L.) Roem. 99.
Lunaria L. 8, 335.
— rediviva L. 335.
Lupinaster Adans. 11, 201.
Lupinus L. 11, 48.
— albus L. 50.
— angustifolius L. 49.
— luteus L. 49.
— polyphyllus Lindl. 50.
Luzula DC. 3, 559.
— arcuata Wahl. 569.
— capitata (Miq.) Nakai 572.
— confusa Lindb. 569.
— Forsteri (Smith) DC. 563.
— frigida (Buchenau) Sam. ex Lindm. 573.
— japonica Buchenau 564.
— kamtschadalorum (Sam.) Gorodk. ex Kryl. 568.
— Kjellmanniana Miyabe et Kudo 573.
— macrocarpa (Buchenau) Nakai 564.
— melanocarpa (Michx.) Desv. 567.
— multiflora (Ehrh.) Lej. 572.
— nemorosa E. Mey. 565.
— nivalis Laest. 568.
— oligantha Sam. 574.
— pallescens (Wahl.) Bess. 576.
— parviflora (Ehrh.) Desv. 566.
— pilosa (L.) Willd. 564.
— pseudo-sudetica V. Krecz. 571.
— rufescens Fisch. et E. Mey. 563.
— sibirica V. Krecz. 574, 631.
— silvatica (Huds.) Gaud. 566.
— spicata (L.) DC. 570.
— subpilosa Gilib. 571.
— sudetica (Willd.) DC. 575.
— Wahlenbergii Rupr. 567.
Lychnis L. 6, 692.
— ajanensis Rgl. 693; tab. 42.
— chalcedonica L. 696; tab. 45.
— cognata Maxim. 695; tab. 42.
— fulgens Fisch. 696; tab. 42.
— samojedorum (Sambuk) Gorschk. 694; tab. 42.
— sibirica L. 693; tab. 42.
— vilosula (Trautv.) Gorschk. 694; tab. 42.
— Wilfordii (Rgl.) Maxim. 695.
Lycium L. 22, 77.
— barbarum L. 82.
— dasystemum Pojark. 84.
— flexicaule Pojark. 81.
— kopetdaghi Pojark. 85.
— ruthenicum Murr. 80.
— turcomanicum Turcz. ex Miers 78.
Lycopersicon Mill. 22, 42.
— esculentum Mill. 50.
— Humboldtii (Willd.) Dun. 49.
— peruvianum (L.) Mill. 45.
— pimpinellifolium (Jusl.) Mill. 47.
Lycopersicum Hill 22, 42.
Lycopersicus Neck. 22, 42.
Lycopodium L. 1, 112.
— alpinum L. 122; tab. 6.
— anceps Wallr. 121; tab. 6.
— annotinum L. 117; tab. 6.
— appressum (Desv.) V. Petrov 115; tab. 6.
— chinense Christ. 115; tab. 6.

— clavatum L. 118; tab. 6.
— inundatum L. 116; tab. 6.
— juniperoideum Sw. 116; tab. 6.
— obscurum L. 116; tab. 6.
— pungens La Pyl. 117; tab. 6.
— selago L. 114; tab. 6.
— serratum Thunb. 114; tab. 6.
— sitchense Rupr. 122; tab. 6.
— tristachyum Pursh 121; tab. 6.
Lycopsis L. 19, 311.
— arvensis L. 312.
— orientalis L. 313.
Lycopus L. 21, 591.
— europaeus L. 595.
— exaltatus L. f. 593.
— hirtellus Kom. 596.
— lucidus Turcz. 593.
— Maackianus (Maxim.) Makino 594.
— parviflorus Maxim. 592.
— uniflorus Michx. 592.
Lysichitum Schott 3, 480.
— camtschatcense (L.) Schott 480; tab. 26.
Lysiella Rydb. 4, 663.
— oligantha (Turcz.) Nevski 663; tab. 40.
Lysimachia L. 18, 255.
— barystachys Bge. 263; tab. 14.
— clethroides Duby 266.
— davurica Ledeb. 267.
— dubia Soland. 264.
— Fortunei Maxim. 265.
— japonica Thunb. 259; tab. 14.
— nemorum L. 258; tab. 14.
— nummularia L. 257.
— punctata L. 260.
— verticillaris Spreng. 263.
— vulgaris L. 266.
Lythrum L. 15, 536.
— hyssopifolia L. 545; tab. 26.
— intermedium Ledeb. 551; tab. 26.
— Komarovii Murav. 543; tab. 26.
— linifolium Kar. et Kir. 544; tab. 26.
— nanum Kar. et Kir. 542; tab. 26.
— salicaria L. 548; tab. 26.
— Schelkovnikovii Sosn. 546.
— silenoides Boiss. 547.
— Theodori Sosn. 545.
— thesioides M. B. 544; tab. 26.
— thymifolia L. 547; tab. 26.
— tribracteatum Salzm. 541; tab. 26.
— virgatum L. 551; tab. 26.

Maackia Rupr. et Maxim. 11, 35.
— amurensis Rupr. et Maxim. 35; tab. 3.
Maclura Nutt. 5, 379.
— aurantica Nutt. 379.
Macrocarpium Nakai 17, 317.
Macropodium R. Br. 8, 27.
— nivale (Pall.) R. Br. 28.
— pterospermum Fr. Schmidt 28.
Macroselinum Schur 17, 168, 191.
Macrosema Stev. 12, 479.
Macrosiphon Hochst. 2, 529.
Macrotomia DC. 19, 167.
— densiflora (Ledeb.) Macbr. 171.
— echioides (L.) Boiss. 172.
— euchroma (Royle) Pauls. 169.
— ugamensis M. Pop. 170.
Madia Mol. 25, 567.
— sativa Mol. 567.
Magnolia L. 7, 564.
— grandiflora L. 565; tab. 39.
Mairania Neck. 18, 83.
Majanthemum Wigg. 4, 453.
— bifolium (L.) Fr. Schmidt 453; tab. 27.
— dilatatum (Wood) A. Nels. et Macbr. 454.
Majorana Moench 21, 462.
— hortensis Moench 462.
Malabaila Hoffm. 17, 261.
— dasyantha (C. Koch) Grossh. 265.
— dasycarpa (Rgl. et Schmalh.) Schischk. 262.
— graveolens (M. B.) Hoffm. 265.
— sulcata (C. Koch) Boiss. 261.
Malachium Fries 6, 430.
— aquaticum Fries 430; tab. 22.
Malacocarpus Fisch. et Mey. 14, 150.
— crithmifolius (Retz.) C. A. Mey. 150.
Malacurus Nevski 2, 712.
— lanatus (Korsh.) Nevski 712; tab. 50.
Malaxis Soland. ex Sw. 4, 599.
— paludosa (L.) Sw. 600; tab. 37.
Malcolmia R. Br. 8, 273.
— africana (L.) R. Br. 276; tab. 16.
— bucharica Vass. 276, 647.
— circinata (Bge.) Boiss. 281; tab. 16.
— crenulata (C. A. Mey.) Vass. 284.
— grandiflora (Bge.) Kuntze 280.
— hispida Litv. 275.
— Karelinii Lipsky 281.
— multisiliqua Vass. 282, 647.
— scorpioides (Bge.) Boiss. 282; tab. 16.
— taraxacifolia Balb. 283.
— trichocarpa Boiss. et Buhse 280.
— turkestanica Litv. 275; tab. 16.
Malope L. 15, 25.
— malacoides L. 26; tab. 1.
— trifida Cav. 25.

Malus Mill. 9, 357.
— baccata (L.) Borkh. 369.
— domestica Borkh. 365.
— manshurica (Maxim.) Kom. 371; tab. 22.
— Niedzwetzkyana Dieck 364.
— orientalis Uglitzk. 362, 492; tab. 22.
— Pallasiana Juz. 370; tab. 22.
— praecox (Pall.) Borkh. 360; tab. 22.
— prunifolia (Willd.) Borkh. 366.
— sachalinensis Juz. 372, 493; tab. 22.
— Sieversii (Ledeb.) Roem. 363; tab. 22.
— silvestris Mill. 359.
— turkmenorum Juz. et M. Pop. 364, 493.
Malva L. 15, 31.
— alcea L. 39.
— ambigua Guss. 48.
— armeniaca Iljin 36; tab. 4.
— bucharica Iljin 55; tab. 4.
— crispa L. 73.
— erecta J. et C. Presl 47.
— excisa Reichb. 37.
— Grossheimii Iljin 52.
— leiocarpa Iljin 35· tab. 4.
— mauritiana L. 49.
— meluca Graebn. 74.
— mohileviensis Downar 64; tab. 3.
— moschata L. 40.
— neglecta Wallr. 56.
— nicaeensis All. 53.
— pamiroalaica Iljin 67, 675.
— parviflora L. 63.
— pusilla Smith 60.
— silvestris L. 41.
— verticillata L. 68; tab. 3.
Malvalthaea Iljin 15, 144.
— transcaucasica (Sosn.) Iljin 144; tab. 4.
Malvella Jaub. et Spach 15, 30.
— Scherardiana (L.) Jaub. et Spach 30; tab. 2.
Mandragora L. 22, 75.
— turcomanica Mizgir. 75.
Mannagettaea H. Smith 23, 115.
— ircutensis M. Pop. 115; tab. 6.
Manochlaenia Boern. 3, 356, 419.
Maresia Pom. 8, 58.
— nana (DC.) Batt. 59.
Marianthemum Schrank 24, 133.
Marinella Bub. 22, 533.
Marrubium L. 20, 233.
— alternidens Reichb. f. 236.
— catariifolium Desr. 242.
— goktschaicum N. Pop. 246; tab. 15.
— leonuroides Desr. 242.
— nanum Knorr. 243.
— parviflorum Fisch. et Mey. 240; tab. 15.
— peregrinum L. 245.
— persicum C. A. Mey. 244; tab. 15.
— plumosum C. A. Mey. 241.
— praecox Janka 239.
— propinquum Fisch. et Mey. 244.
— purpureum Bge. 246.
— Turkeviczii Knorr. 247, 516.
— vulgare L. 235.
— Woronowii N. Pop. 239.
Marsilea L. 1, 91.
— aegyptiaca Willd. 91; tab. 3.
— quadrifolia L. 91; tab. 3.
— strigosa Willd. 91.
Martinia Vaniot 25, 122.
Maruta Cass. 26, 9, 63.
Mastrucium Cass. 28, 268.
Matricaria L. 26, 147.
— aurea (L.) Boiss. 151; tab. 7.
— matricarioides (Less.) Porter ex Britt. 150.
— recutita L. 148.
— Tzvelevii Pobed. 149, 871.
Matthiola R. Br. 8, 285.
— albicaulis Boiss. 289.
— bucharica Czerniak. 297, 648.
— chenopodiifolia Fisch. et Mey. 294.
— daghestanica (Conti) N. Busch 291.
— farinosa Bge. 292; tab. 18.
— fragrans Bge. 293.
— incana R. Br. 298.
— integrifolia Kom. 290.
— odoratissima (Pall.) R. Br. 291.
— oxyceras DC. 297.
— robusta Bge. 292.
— runcinata Rgl. 290.
— Stoddartii Bge. 297; tab. 18.
— superba Conti 294.
— tatarica (Pall.) DC. 293.
— *Mattia* Schult. 19, 565, 572.
Mattiastrum Brand 19, 588.
Maukschia Heuff. 3, 163.
Mausolea Bge. 26, 631.
— eriocarpa (Bge.) Poljak. 631.
Mazus Lour. 22, 316.
— japonicus (Thunb.) Kuntze 316.
— stachydifolius (Turcz.) Maxim. 317.
Medicago L. 11, 129.
— agrestis Ten. ex DC. 165; tab. 11.
— arabica All. 166; tab. 11.
— borealis Grossh. 139, 391.
— cancellata M. B. 137; tab. 10.
— coerulea Less. ex Ledeb. 150; tab. 11.
— coronata (L.) Desr. 173.
— daghestanica Rupr. 172; tab. 11.

— denticulata Willd. 168; tab. 11.
— dzawakhetica Bordz. 159; tab. 11.
— falcata L. 140; tab. 10.
— globosa J. et C. Presl 166; tab. 11.
— glutinosa M. B. 156; tab. 10.
— hemicycla Grossh. 154; tab. 10.
— laciniata L. 174; tab. 11.
— lanigera C. Winkl. et B. Fedtsch. 175; tab. 10.
— litoralis Rohde 165; tab. 11.
— lupulina L. 134; tab. 10.
— marina L. 160; tab. 11.
— minima Grufberg 174; tab. 11.
— nigra Willd. 171.
— orbicularis All. 161; tab. 10.
— papillosa Boiss. 158; tab. 11.
— polychroa Grossh. 154; tab. 10.
— praecox DC. 172; tab. 11.
— rigidula Desr. 164; tab. 11.
— romanica Prod. 145.
— rupestris M. B. 136; tab. 10.
— sativa L. 148; tab. 10.
— saxatilis M. B. 138; tab. 11.
— Schischkinii Sumn. 152; tab. 10.
— scutellata All. 161; tab. 10.
— Soleirolii Duby 163.
— tianschanica Vass. 153.
— Trautvetteri Sumn. 152; tab. 10.
— tribuloides Desr. 163; tab. 11.
— virescens Grossh. 157; tab. 10.
Medium Fisch. ex A. DC. 24, 133.
Meehania Britt. 20, 529.
— urticifolia (Miq.) Makino 529.
Megacarpaea DC. 8, 541.
— gigantea Rgl. 545.
— gracilis Lipsky 544.
— megalocarpa (Fisch.) Schischk. 543.
— orbiculata B. Fedtsch. 542.
— schugnanica B. Fedtsch. 544.
Megasperma Lehm. 22, 413.
Megista (Fourr.) Rydb. 22, 63.
Melampyrum L. 22, 533.
— Alboffianum Beauverd 539.
— argyrocomum Fisch. ex Steud. 541.
— arvense L. 540.
— caucasicum Bge. 538; tab. 26.
— chlorostachyum Beauverd 537.
— cristatum L. 536.
— elatius Reut. 542.
— Herbichii Woloszcz. 550.
— laciniatum Koshewn. et Zing. 553.
— nemorosum L. 543.
— polonicum (Beauverd) Soó 545.
— pratense L. 552.
— roseum Maxim. 545; tab. 26.
— saxosum Baumg. 551.
— setaceum (Maxim.) Nakai 546.
— silvaticum L. 549; tab. 26.
Melandrium Roehl. 6, 701.
— adenophorum Schischk. 705, 890.
— affine J. Vahl 719; tab. 44.
— Akinfievii (Schmalh.) Schischk. 709; tab. 45.
— album (Mill.) Garcke 725.
— angustiflorum (Rupr.) Walp. 720; tab. 44.
— apetalum (L.) Fenzl 716; tab. 44.
— apricum (Turcz.) Rohrb. 713.
— astrachanicum Pacz. 726.
— baicalense (Sukacz.) Tolm. 722.
— Balansae Boiss. 724.
— Boissieri Schischk. 725.
— brachypetalum (Hornem.) Fenzl 722; tab. 44.
— erubescens Schischk. 706, 890; tab. 43.
— Fedtschenkoanum (Preobr.) Schischk. 705; tab. 43.
— ferganicum (Preobr.) Schischk. 706; tab. 43.
— firmum (Sieb. et Zucc.) Rohrb. 713.
— gracile Tolm. 721.
— longicarpophorum Kom. 714; tab. 44.
— noctiflorum (L.) Fries 712.
— Olgae Maxim. 712.
— ovalifolium Rgl. et Schmalh. 704; tab. 43.
— quadrilobum (Turcz.) Schischk. 711.
— ruinarum M. Pop. 709.
— sachalinense (Fr. Schmidt) Schischk. 710; tab. 43.
— saxatile (Turcz.) A. Br. 723; tab. 44.
— silvestre (Schk.) Roehl. 724.
— Soczavianum Schischk. 715.
— sordidum (Kar. et Kir.) Rohrb. 715; tab. 44.
— suaveolens (Kar. et Kir.) Schischk. 711; tab. 43.
— tajmyrense Tolm. 721.
— tenellum Tolm. 721; tab. 44.
— triste (Bge.) Fenzl 715; tab. 44.
— viscosum (L.) Čelak. 710.
Melandrum Blytt 6, 701.
Melandryum Reichb. 6, 701.
Melanocarya Turcz. 14, 557.
Melia L. 14, 244.
— azedarach L. 244.
Melica L. 2, 340.
— altissima L. 350.
— atropatana Schischk. 343.
— breviflora Boiss. 344; tab. 26.

Melica canescens (Rgl.) Lavr. 344, 752; tab. 26.
— flavescens (Schur.) Simonk. 346.
— glauca F. Schultz 345.
— Hohenackeri Boiss. 344.
— inaequiglumis Boiss. 343; tab. 26.
— minor Hack. 351; tab. 26.
— nutans L. 351; tab. 26.
— penicillaris Boiss. 342.
— persica Kunth 343.
— picta C. Koch 351; tab. 26.
— secunda Rgl. 352; tab. 26.
— taurica C. Koch 346; tab. 26.
— transsilvanica Schur 345; tab. 26.
— Turczaninowiana Ohwi 349; tab. 26.
— uniflora Retz. 350; tab. 26.
— virgata Turcz. 349.
Melilotus Adans. 11, 176.
— albus Desr. 181; tab. 12.
— altissimus Thuill. 184; tab. 12.
— dentatus (Waldst. et Kit.) Pers. 178; tab. 12.
— hirsutus Lipsky 187.
— indicus (L.) All. 189; tab. 12.
— neapolitanus Ten. 188; tab. 12.
— officinalis (L.) Desr. 180; tab. 12.
— polonicus (L.) Desr. 183; tab. 12.
— suaveolens Ledeb. 180.
— tauricus (M. B.) Ser. 188; tab. 12.
— wolgicus Poir. 182; tab. 12.
Melissa L. 21, 411.
— officinalis L. 411; tab. 21.
Mellissophyllon Adans. 20, 498.
Melittis L. 20, 498.
— melissophyllum L. 499.
Melo Adans. 24, 115.
— dudaim (L.) Sageret ex Roem. 117.
— flexuosus (L.) Sageret ex Roem. 117.
— sativus Sageret ex Roem. 116.
Melongena Mill. 22, 39.
Meniocus Desv. 8, 359.
— linifolius (Steph.) DC. 359.
Menispermum DC. 7, 560.
— dahuricum DC. 563; tab. 38.
Mentha L. 21, 596.
— alaica Boriss. 615.
— aquatica L. 608.
— arvensis L. 604; tab. 32.
— asiatica Boriss. 614.
— crispa L. 609.
— crispata Schrad. 622.
— daghestanica Boriss. 627.
— dahurica Fisch. 610; tab. 32.
— darvasica Boriss. 616.
— haplocalyx Briq. 607.
— interrupta Boriss. 619.
— kopetdaghensis Boriss. 618.
— lapponica Wahl. 606.
— longifolia (L.) Huds. 611; tab. 32.
— micrantha Fisch. 627; tab. 32.
— pamiroalaica Boriss. 617.
— piperita L. 620.
— pulegium L. 624.
— rotundifolia (L.) Huds. 622.
— sachalinensis (Briq.) Kudo 607.
— spicata L. 621.
— vagans Boriss. 614.
Menthella Pérard. 21, 623.
Menyanthes L. 18, 642.
— trifoliata L. 642.
Menziesia J. G. Smith 18, 61.
— pentandra Maxim. 61.
Merckia Fisch. 6, 545.
— physodes (DC.) Fisch. ex Cham. et Schlecht. 545; tab. 31.
Mercurialis L. 14, 295.
— annua L. 298.
— ovata Sternb. et Hoppe 297; tab. 17.
— perennis L. 296.
Merendera Ramond 4, 15.
— candidissima Miscz. ex Grossh. 18.
— Eichleri (Rgl.) Boiss. 18.
— hissarica Rgl. 17.
— Jolantae Czerniak. 19; tab. 2.
— Raddeana Rgl. 16.
— robusta Bge. 18; tab. 2.
— sobolifera C. A. Mey. 16; tab. 2.
— trigyna (Adam) Woronow 17.
Meristotropis Fisch. et Mey. 13, 240.
— erythrocarpa Vass. 241; tab. 13.
— triphylla Fisch. et Mey. 240; tab. 13.
— xanthioides Vass. 241; tab. 13.
Mertensia Roth 19, 238.
— davurica (Sims) G. Don 249.
— dshagastanica Rgl. 257.
— jenissejensis M. Pop. 244.
— kamczatica (Turcz.) DC. 253; tab. 13.
— maritima (L.) S. F. Gray 242.
— Pallasii (Ledeb.) G. Don 244.
— pilosa (Cham.) DC. 251.
— Popovii Rubtz. 247, 706.
— pterocarpa (Turcz.) Tat. et Ohwi 254.
— rivularis (Turcz.) DC. 251; tab. 13.
— serrulata (Turcz.) DC. 255.
— sibirica (L.) G. Don 243.
— stylosa (Fisch.) DC. 248.
— tarbagataica B. Fedtsch. 256.
Mertensianthe M. Pop. 19, 238, 257.

Merwia B. Fedtsch. 17, 62, 74.
Mesocentron Cass. 28, 370, 570.
Mesodiscus Raf. 16, 380.
Mespilus L. 9, 414.
— germanica L. 414; tab. 27.
Metaplexis R. Br. 18, 672.
— japonica (Thunb.) Makino 672; tab. 36.
Metastachys Knorr. 21, 192, 652.
— sagittata (Rgl.) Knorr. 193.
Mibora Adans. 2, 159.
— minima (L.) Desv. 159; tab. 12.
Michauxia L'Hér. 24, 383.
— laevigata Vent. 384; tab. 23.
Micranthemum C. Presl 11, 214.
Macranthes Haw. 9, 145.
Microbiota Kom. 1, 193.
— decussata Kom. 193; tab. 7, 8.
Microcephala Pobed. 26, 152.
— lamellata (Bge.) Pobed. 155.
— subglobosa (Krasch.) Pobed. 156; tab. 5.
— turcomanica (C. Winkl.) Pobed. 155.
Microcerasus Roem. 10, 562.
Microlepia C. Presl 1, 49.
— pilosella S. Moore 50.
— Wilfordii S. Moore 50.
Microlophopsis Czer. 28; 305.
Microlophus Cass. 28, 370, 560.
Micromeles Decne. 9, 406.
— alnifolia (Sieb. et Zucc.) Koehne 406; tab. 27.
Micromeria Benth. 21, 426.
— elegans Boriss. 428.
— elliptica C. Koch 429.
— marifolia (Willd.) Benth. 428.
— serpyllifolia (M. B.) Boiss. 427; tab. 21.
Micropeplis Bge. 6, 352.
Micropetalon Pers. 6, 394.
Micropetalum Tausch 9, 163.
Microphysa Schrenk 23, 286.
— elongata (Schrenk) Pobed. 286; tab. 21.
Micropus L. 25, 300.
— supinus L. 300; tab. 18.
Microstigma Trautv. 8, 298.
— deflexum (Bge.) Juz. 298.
Microstylis Nutt. 4, 600.
— monophyllos (L.) Lindl. 601; tab. 37.
Middendorfia Trautv. 15, 535.
— borysthenica (M. B.) Trautv. 535.
Milium L. 2, 119.
— effusum L. 119; tab. 5.
— Schmidtianum C. Koch 120; tab. 5.
— vernale M. B. 119; tab. 5.
Millefolium C. Koch 26, 75.
Millepora Boiss. 15, 203, 247.
Miltianthus Bge. 14, 158.
Mimulus L. 22, 310.
— guttatus DC. 312.
— moschatus Dougl. 315.
— pilosiusculus H. B. K. 313.
— ringens L. 312.
— stolonifer Novopokr. 314.
— tenellus Bge. 314.
Mindium Juss. 24, 383.
Minuartia L. 6, 482.
— abchasica Schischk. 495.
— adenotricha Schischk. 495; tab. 27.
— aizoides (Boiss.) Bornm. 509.
— Akinfiewii (Schmalh.) Woronow 490.
— arctica (Stev.) Aschers. et Graebn. 515; tab. 28.
— Biebersteinii (Rupr.) Schischk. 500; tab. 27.
— biflora (L.) Schinz et Thell. 516; tab. 28.
— Brotheriana (Trautv.) Woronow 515; tab. 28.
— Buschiana Schischk. 493, 884.
— caucasica (Adam) Mattf. 509; tab. 28.
— dianthifolia (Boiss.) Hand.-Mazz. 499.
— elegans (Cham. et Schlecht.) Schischk. 508.
— glomerata (M. B.) Degen 492; tab. 27.
— gracilipes Kom. 502, 885.
— granuliflora (Fenzl) Grossh. 494.
— Helmii (Fisch.) Schischk. 501; tab. 27.
— hirsuta (M. B.) Hand.-Mazz. 496; tab. 27.
— hybrida (Vill.) Schischk. 488.
— imbricata (M. B.) Woronow 514.
— intermedia (Boiss.) Hand.-Mazz. 491.
— jacutica Schischk. 502.
— Krascheninnikovii Schischk. 493; tab. 27.
— Kryloviana Schischk. 501.
— laricina (L.) Mattf. 513; tab. 28.
— lineata (C. A. Mey.) Bornm. 503.
— Litwinowii Schischk. 504, 885.
— macrocarpa (Pursh) Ostenf. 513; tab. 28.
— Meyeri (Boiss.) Bornm. 489.
— micrantha Schischk. 494, 884.
— oriena (Mattf.) Schischk. 499.
— orthotrichoides Schischk. 506, 886.
— pusilla Schischk. 508, 886.
— Regeliana (Trautv.) Mattf. 488.
— rhodocalyx (Albov) Woronow 514.
— rubella (Wahl.) Graebn. 506; tab. 27.
— sclerantha (Fisch. et Mey.) Thell. 491.
— setacea (Thuill.) Hayek 492; tab. 27.
— stricta (Sw.) Hiern 507; tab. 28.

Minuartia taurica (Stev.) Aschers. et Graebn. 503; tab. 28.
— turcomanica Schischk. 487; tab. 27.
— verna (L.) Hiern 505.
— viscosa (Schreb.) Schinz et Thell. 487.
— Wiesneri (Stapf) Schischk. 490.
— Woronowii Schischk. 495, 884.
Mirabilis L. 6, 370.
— nyctaginea (Michx.) MacMill. 371.
Miscanthus Anderss. 2, 7.
— purpurascens Anderss. 8; tab. 1.
— sacchariflorus (Maxim.) Hack. 7.
— sinensis Anderss. 8.
Miscaria Haw. 9, 174.
Mistyllus C. Presl 11, 216.
Mitchella L. 23, 169.
— undulata Sieb. et Zucc. 169.
Mitella L. 9, 199.
— nuda L. 199.
Mitina Adans. 27, 77.
Modestia Char. et Tamamsch. 28, 223.
— darwasica (C. Winkl.) Char. et Tamamsch. 224.
— mira (Iljin) Char. et Tamamsch. 225; tab. 10.
Moehringia L. 6, 540.
— elongata Schischk. 541, 887.
— lateriflora (L.) Fenzl 541; tab. 31.
— trinervia (L.) Clairv. 540; tab. 31.
— umbrosa (Bge.) Fenzl 542; tab. 31.
Mogoltavia Korov. 17, 214.
— Severzovii (Rgl.) Korov. 214; tab. 1.
Moldavica Adans. 20, 439.
Molinia Schrank 2, 312.
— coerulea (L.) Moench 312; tab. 23.
— litoralis Host 313; tab. 23.
Mollugo L. 6, 374.
— cerviana (L.) Ser. 374; tab. 19.
Moltkea Wettst. 19, 234.
Moltkia Lehm. 19, 234.
— coerulea (Willd.) Lehm. 235.
Molucella L. 21, 181.
— laevis L. 182.
Momordica L. 24, 97.
— balsamita L. 98.
Moneses Salisb. 18, 14.
— uniflora (L.) A. Gray 14.
Monocarpus Adans. 6, 48.
Monochilon Dulac 20, 39.
Monochoria C. Presl 3, 502.
— Korsakowii Rgl. et Maack 502; tab. 26.
— plantaginea (Roxb.) Kunth 503.
Monolepis Schrad. 6, 73.
— asiatica Fisch. et Mey. 73; tab. 2.
— Litwinowii Pauls. 74.
Monotropa L. 18, 20.
Monotropa Nutt. 18, 19.
— uniflora L. 20.
Montia L. 6, 384.
— lamprosperma Cham. 385; tab. 19.
— minor C. C. Gmel. 384; tab. 19.
— rivularis C. C. Gmel. 385.
Moriera Boiss. 8, 557.
Morina L. 24, 2.
— kokanica Rgl. 9; tab. 1.
— Lehmanniana Bge. 7; tab. 1.
— parviflora Kar. et Kir. 4; tab. 1.
— persica L. 8.
Morucodon Salisb. 4, 302.
Morus L. 5, 377.
— alba L. 377; tab. 21.
— nigra L. 377.
Moscharia Forsk. 20, 17.
Mosharia Salisb. 4, 411.
Mosla Buch.-Ham. ex Maxim. 21, 633.
Muehlbergella Feer 24, 442.
Muehlenbergia Schreb. 2, 120.
— Hügelii Trin. 121; tab. 5.
— japonica Steud. 121; tab. 5.
Muffetta Neck. 23, 587.
Mulgedium Cass. 29, 279.
Munbya Boiss. 19, 167, 168.
Muretia Boiss. 16, 415.
— fragrantissima (Lipsky) K.-Pol. 418; tab. 30.
— lutea (M. B.) Boiss. 416.
— oeroilanica Korov. 418, 598.
— transcaspica Korov. 417, 598.
— transitoria Korov. 417.
Murtekias Raf. 14, 391, 404.
Muscari Mill. 4, 412.
— apertum Conr. et Freyn 421.
— coeleste Fomin 422; tab. 24.
— coeruleum Losinsk. 421, 745.
— commutatum Guss. 416.
— dolychanthum Woronow et Tron 417, 744; tab. 24.
— forniculatum Fomin 422; tab. 24.
— leucostomum Woronow 417.
— neglectum Guss. 416; tab. 24.
— pallens M. B. 420; tab. 24.
— paradoxum (Fisch. et Mey.) Baker 421; tab. 24.
— pendulum Trautv. 418; tab. 24.
— polyanthum Boiss. 419; tab. 24.
— racemosum (L.) Mill. 417; tab. 24.
— Steupii Woronow et Tron 418, 745; tab. 24.

— Szowitzianum Baker 420; tab. 24.
— Turkewiczii (Woronow) Losinsk. 420; tab. 24.
— Woronowii Tron et Losinsk. 418, 744; tab. 24.
Muscarimia Kostel. 4, 411.
— muscari (L.) Losinsk. 411; tab. 24.
Myagrum L. 8, 202.
— perfoliatum L. 202; tab. 13.
Mycelis Cass. 29, 335.
— muralis (L.) Dum. 336; tab. 20.
Myconia Neck. ex Sch. Bip. 26, 146.
Mylinum Gaud. 16, 560.
Myobroma Stev. 12, 142.
Myoseris Link 29, 699, 703.
Myosotis L. 19, 352.
— alpestris Schmidt 377.
— amoena (Rupr.) Boiss. 359.
— arvensis Hill 379.
— asiatica Schischk. et Serg. 378.
— baltica Sam. ex Lindem. 367.
— caespitosa Schultz 366.
— collina Hoffm. 382.
— heteropoda Trautv. 379.
— idaea Boiss. et Heldr. 371.
— Krylovii Serg. 374.
— lazica M. Pop. 380; tab. 16.
— litoralis Stev. 368.
— micrantha Pall. ex Lehm. 383.
— palustris Lam. 363.
— propinqua Fisch. et Mey. 360.
— pseudopropinqua M. Pop. 361.
— pseudovariabilis M. Pop. 375, 706; tab. 16.
— refracta Boiss. 384.
— sachalinensis M. Pop. 375.
— sicula Guss. 368.
— silvatica Hoffm. 373.
— sparsiflora Mikan ex Pohl 361.
— suaveolens Waldst. et Kit. 376.
— ucrainica Czern. 372.
— versicolor Smith 384.
Myosoton Moench 6, 430.
Myosurus L. 7, 323.
— minimus L. 323.
Myriactis Less. 25, 52.
— Gmelinii (Fisch. et Mey.) DC. 52; tab. 2.
Myrica L. 5, 243.
— gale L. 243.
— tomentosa (DC.) Aschers. et Graebn. 244; tab. 11.
Myricaria Desv. 15, 321.
— alopecuroides Schrenk 324; tab. 16.
— dahurica (Willd.) Ehrenb. 327.
— elegans Royle 322.
— germanica (L.) Desv. 323.
— longifolia (Willd.) Ehrenb. 326.
— squamosa Desv. 325.
Myrice St.-Lag. 15, 321.
Myriogyne Less. 26, 421.
Myriophyllum L. 15, 662.
— alterniflorum DC. 666.
— isoetophyllum Kom. 664.
— spicatum L. 667; tab. 13.
— ussuriense Maxim. 664.
— verticillatum L. 665; tab. 13.
Myrrhis Mill. 16, 150.
— odorata (L.) Scop. 150.
Myscolus Cass. 29, 11.

Najas L. 1, 269.
— flexilis (Willd.) Rostk. et Schmidt 273; tab. 13.
— foveolata A. Br. ex Magnus 274; tab. 13.
— graminea Delile 275; tab. 13.
— marina L. 270; tab. 13.
— minor All. 273; tab. 13.
— tenuissima A. Br. ex Magnus 274; tab. 13.
Nanophyton Less. 6, 314.
— erinaceum (Pall.) Bge. 314; tab. 14.
Narcissus L. 4, 491.
— caucasicus (Fomin) Gorschk. 492; tab. 30.
— poeticus L. 493.
— pseudo-narcissus L. 492; tab. 30.
— tazetta L. 493.
Nardosmia Cass. 26, 645.
— angulosa Cass. 648.
— Fominii Bordz. 648.
— frigida (L.) Hook. 649.
— glacialis Ledeb. 654.
— Gmelinii (Turcz.) DC. 653.
— japonica Sieb. et Zucc. 652.
— laevigata (Willd.) DC. 650.
— palmata (Ait.) Hook. 651.
— saxatilis Turcz. 647.
Nardurus Reichb. 2, 539.
— Krausei (Rgl.) V. Krecz. et Bobr. 540.
— tenuiflorus (Schrad.) Boiss. 540; tab. 41.
Nardus L. 2, 587.
— stricta L. 587; tab. 44.
Narthecium (Moehr.) Huds. 4, 5.
— caucasicum Miscz. 5.
Narthex Falc. 17, 62, 82.
Nasturtium R. Br. 8, 143.
— officinale (L.) R. Br. 143.

Nathaliella B. Fedtsch. 22, 511.
— alaica B. Fedtsch. 511.
Naumburgia Moench 18, 268.
— thyrsiflora (L.) Reichb. 269.
Nectarobotryum Ledeb. 4, 365.
Nectaroscordum Lindl. 4, 279.
Neesia Spreng. 26, 127.
Negundium Raf. 14, 621.
Negundo Ludwig 14, 621.
Nelumbium Juss. 7, 3.
— nuciferum Gaertn. 3.
Nemauchenes Cass. 92, 595, 697.
Nemostylis Stev. 23, 192.
Nenningia Opiz 24, 133, 166.
Neogaya Meisn. **16, 579.**
Neolindleya Kränzl. 4, 646.
— camtschatica (Cham.) Nevski **646; tab. 40.**
Neopallasia Poljak. 26, 636.
— pectinata (Pall.) Poljak. 537; tab. 31.
Neottia Sw. 4, 617.
— asiatica Ohwi **620; tab. 38.**
— camtschatea (L.) Reichb. f. **619; tab. 38.**
— nidus-avis (L.) Rich. 618; tab. 38.
— papilligera Schlechter 618; tab. 38.
Neottianthe Schlechter 4, 645.
— cucullata (L.) Schlechter 645.
Nepeta L. 20, 286.
— alaghezi Pojark. 366.
— alatavica Lipsky 324.
— amoena Stapf 380; tab. 23.
— badachschanica Kudr. 337.
— betonicifolia C. A. Mey. 361.
— Biebersteiniana (Trautv.) Pojark. 357.
— bracteata Benth. 422; tab. 25.
— brevifolia C. A. Mey. 401.
— bucharica Lipsky 322; tab. 20.
— Buhsei Pojark. 367.
— Buschii Sosn. et Mand. 327.
— cataria L. 349.
— consanguinea Pojark. 347; 519.
— cyanea Stev. 353; tab. 22.
— czegemensis Pojark. 356, 521.
— daenensis Boiss. 425.
— daghestanica Pojark. 399.
— densiflora Kar. et Kir. 314.
— erivanensis Grossh. 420; tab. 24.
— Fedtschenkoi Pojark. 430, 524.
— fissa C. A. Mey. 395.
— formosa Kudr. 344.
— glutinosa Benth. 306; tab. 19.
— Gontscharovii Kudr. 346; tab. 21.
— grandiflora M. B. 359.
— Grossheimii Pojark. 363; tab. 23.
— hajastana Grossh. 373.
— iberica Pojark. 400.
— ispahanica Boiss. 389.
— Knorringiana Pojark. 310, 517; tab. 19.
— kokamirica Rgl. 316.
— kokanica Rgl. 328; tab. 20.
— Komarovii E. Busch 412.
— kopetdaghensis Pojark. 418, 523.
— kubanica Pojark. 355, 520.
— ladanolens Lipsky 338; tab. 20.
— lamiifolia Willd. 405.
— Lipskyi Kudr. 305; tab. 19.
— longibracteata Benth. 303.
— longituba Pojark. 392; tab. 21.
— Mariae Rgl. 317.
— manchuriensis S. Moore 341.
— maussarifii Lipsky 323.
— Meyeri Benth. 384.
— micrantha Bge. 383.
— microcephala Pojark. 432, 525; tab. 25.
— Mussinii Spreng. 369.
— noraschenica Grossh. 378.
— odorifera Lipsky 321.
— Olgae Rgl. 339.
— pallida C. Koch 386.
— pamirensis Franch. 333.
— pannonica L. 406.
— parviflora M. B. 414.
— podostachys Benth. 312.
— pseudofloccosa Pojark. 336.
— pseudokokanica Pojark. 330.
— pulchella Pojark. 318.
— pungens (Bge.) Benth. 429.
— Reichenbachiana Fisch. et Mey. 371; tab. 22.
— saccharata Bge. 388; tab. 23.
— santoana M. Pop. 390; tab. 25.
— satureioides Boiss. 435; tab. 25.
— Schischkinii Pojark. 417, 522.
— schugnanica Lipsky 325.
— sibirica L. 342.
— Sintenisii Bornm. 411; tab. 24.
— somkhetica Kapell. 362.
— Sosnovskyi Asker. 393.
— spathulifera Benth. 427.
— strictifolia Pojark. 365; tab. 22.
— subhastata Rgl. 311; tab. 19.
— sulphurea C. Koch 409.
— supina Stev. 326.
— transcaucasica Grossh. 374.
— transiliensis Pojark. 315.
— Trautvetteri Boiss. et Buhse 396.
— Troitzkii Sosn. 382.
— tytthantha Pojark. 320.
— ucrainica L. 415.

— vakhanica Pojark. 335.
— velutina Pojark. 397; tab. 21.
Nephrophyllidium Gilg 18, 641.
Nerium L. 18, 662.
— oleander L. 662.
Neslia Desv. 8, 602.
— apiculata Fisch. et Mey. 603; tab. 20.
— paniculata (L.) Desv. 602; tab. 20.
Neustruevia Juz. 20, 501, 526.
— karatavica Juz. 502, 527.
Nicandra Adans. 22, 115.
— physaloides (L.) Gaertn. 116.
Nicotiana L. 22, 106.
— rustica L. 108.
— tabacum L. 106.
Niedzwedzkia B. Fedtsch. 23, 5.
— semiretschenskia B. Fedtsch. 5.
Nigella L. 7, 62.
— arvensis L. 68; tab. 6.
— bucharica Schipcz. 71; tab. 6.
— damascena L. 70; tab. 6.
— garidella Spenn. 63; tab. 6.
— glandulifera Freyn et Sint. 70.
— integrifolia Rgl. 71; tab. 6.
— orientalis L. 64; tab. 6.
— oxypetala Boiss. 67; tab. 6.
— persica Boiss. 67; tab. 6.
— sativa L. 69; tab. 6.
— segetalis M. B. 68; tab. 6.
Nigera Bub. 16, 172.
Nikitinia Iljin 27, 712.
— leptoclada (Bornm. et Sint.) Iljin 712.
Nintooa Sweet 23, 568.
Nitraria L. 14, 196.
— Komarovii Iljin et Lava 198.
— Schoberi L. 197.
— sibirica Pall. 197.
Noaea Moq. 6, 266.
— leptoclada (Woronow) Iljin 269.
— minuta Boiss. et Bal. 269.
— mucronata (Forsk.) Aschers. et Schweinf. 266; tab. 13.
Nonea Medic. 19, 317.
— alpestris (Stev.) G. Don 320.
— caspica (Willd.) G. Don 334.
— daghestanica Kusn. 342.
— decurrens (C. A. Mey.) G. Don 341.
— flavescens (G. A. Mey.) Fisch. et Mey. 323.
— intermedia Ledeb. 340.
— karsensis M. Pop. 329.
— lutea (Desr.) Reichb. ex DC. 321.
— macropoda M. Pop. 331; tab. 15.
— melanocarpa Boiss. 332.
— pallens Petrović 336.
— pulla (L.) DC. 337.
— rosea (M. B.) Link 324.
— setosa (Lehm.) Roem. et Schult. 323.
— turcomanica M. Pop. 330.
— ventricosa (Sibth. et Smith) Griseb. 344.
— versicolor (Stev.) Sweet 326.
Nordmannia Ledeb. 19, 291.
Notholaena R. Br. 1, 74.
— Maranthae (L.) R. Br. 74.
Notiosphace Benth. 21, 289.
Notobasis Cass. 28, 215.
— syriaca (L.) Cass. 216.
Novosieversia F. Bolle 10, 244.
— glacialis (Adam) F. Bolle 244; tab. 18.
Nummularia Gilib. 18, 257.
Nummularia Kuntze 18, 268.
Nuphar Smith 7, 12.
— advenum R. Br. 14.
— luteum (L.) Smith 13.
— luteum × N. pumilum 14.
— pumilum (Hoffm.) DC. 13.
Nymphaea L. 7, 6.
— alba L. 7.
— candida J. et C. Presl 8; tab. 1.
— tetragona Georgi 11; tab. 1.
— Wenzelii Maack 12.
Nymphoides Hill 18, 643.
— koreanum Lév. 644.
— peltatum (S. G. Gmel.) Kuntze 643.

Ocimastrum Rupr. 15, 633.
Ocimum L. 21, 641.
— basilicum L. 641.
Octarillum Lour. 15, 518.
Octoceras Bge. 8, 327.
— Lehmannianum Bge. 328; tab. 20.
Odontarrhena C. A. Mey. 8, 344.
Odontites Linn. 22, 649.
— glutinosa (M. B.) Benth. 656; tab. 33.
— litoralis Fries 655; tab. 33.
— salina Kotov 652.
— serotina (Lam.) Dum. 650; tab. 33.
— verna (Bell.) Dum. 652; tab. 33.
Odontolophus Cass. 28, 370, 481.
Oedibasis K.-Pol. 17, 203, 355.
— apiculata (Kar. et Kir.) K.-Pol. 204; tab. 19, 21.
— chaerophylloides (Rgl. et Schmalh.) Korov. 207; tab. 21.
— karatavica Korov. 207, 355.
— platycarpa (Lipsky) K.-Pol. 208; tab. 21.
Oenanthe L. 16, 526.
— abchasica Schischk. 534.

Oenanthe aquatica (L.) Poir. 537.
— banatica Heuff. 535.
— decumbens (Thunb.) K.-Pol. 535.
— Fedtschenkoana K.-Pol. 533.
— fistulosa L. 531.
— heterococca Korov. 533.
— longifoliolata Schischk. 534.
— pimpinelloides L. 530.
— silaifolia M. B. 532.
— Sophiae Schischk. 536.
Oenosciadium Pomel 16, 526.
Oenothera L. 15, 630.
— odorata Jacq. 630.
Ofaiston Rafin. 6, 276.
— monandrum (Pall.) Moq. 276; tab. 14.
Oflifa Cass. 25, 314, 321.
Olea L. 18, 512.
— europaea L. 513.
Olgaea Iljin 28, 43.
— baldschuanica (C. Winkl.) Iljin 50.
— eriocephala (C. Winkl.) Iljin 44.
— lanipes (C. Winkl.) Iljin 47.
— longifolia (C. Winkl.) Iljin 47.
— nidulans (Rupr.) Iljin 48.
— nivea (C. Winkl.) Iljin 45.
— pectinata Iljin 50.
— spinifera Iljin 49.
— Vvedenskyi Iljin 46.
Oligochaeta C. Koch 28, 346.
— divaricata (Fisch. et Mey.) C. Koch 347.
— minima (Boiss.) Briq. 349; tab. 13, 20.
— tomentosa Czer. 348.
Oligosporus Cass. 26, 525.
Olympia Spach 15, 203.
Omalanthus Less. 26, 324.
Omalocline Cass. 29, 595.
Omalotes DC. 26, 324.
Omalotheca Cass. 25, 402.
Omphalodes Moench 19, 607.
— cappadocica (Willd.) DC. 609; tab. 33.
— Kusnetzovii Klok. 613.
— linifolia (L.) Moench 616; tab. 33.
— Lojkae Somm. et Levier 614.
— rupestris Rupr. 615; tab. 32.
— scorpioides (Haenke) Schrank 619; tab. 33.
Omphalothrix Maxim. 22, 640.
— longipes Maxim. 640.
Onagra Adans. 15, 628.
— biennis (L.) Scop. 628.
— muricata (L.) Moench 629.
Onkerma Raf. 3, 305.
Onobrychis Adans. 13, 319.
— altissima Grossh. 346; tab. 16.
— amoena M. Pop. et Vved. 360; tab. 17.
— arenaria (Kit.) DC. 349.
— argyrea Boiss. 355.
— armena Boiss. et Huet. 338.
— atropatana Boiss. 356; tab. 18.
— Biebersteinii Širj. 336; tab. 16.
— Bobrovii Grossh. 363, 554; tab. 17.
— borystenica (Širj.) Klok. 335.
— Buhseana Bge. 356.
— Bungei Boiss. 338.
— cadmea Boiss. 337.
— caput-galli (L.) Lam. 327; tab. 16.
— chorassanica Bge. 360.
— cornuta (L.) Desv. 325.
— Cyri Grossh. 339; tab. 16.
— daghestanica Grossh. 335.
— darwasica Vass. 326.
— Dielsii (Širj.) Vass. 340.
— echidna Lipsky 326; tab. 16.
— ferganica (Širj.) Grossh. 350.
— Gontscharovii Vass. 332.
— gracilis Bess. 332.
— grandis Lipsky 352; tab. 17.
— Grossheimii Kolak. 356.
— hajastana Grossh. 345.
— hamata Vass. 337.
— heterophylla C. A. Mey. 357; tab. 18.
— Hohenackeriana C. A. Mey. 366; tab. 17
— iberica Grossh. 347.
— inermis Stev. 347.
— kachetica Boiss. et Buhse 362.
— Komarovii Grossh. 366; tab. 17.
— Lipskyi Korov. 331.
— major Boiss. et Kotschy 337.
— Majorovii Grossh. 363, 553.
— meschetica Grossh. 364, 554.
— Michauxii DC. 365.
— micrantha Schrenk 328; tab. 16.
— miniata Stev. 340.
— Nemecii Sirj. 348.
— Novopokrovskii Vass. 340.
— oxyodonta Boiss. et Huet 345.
— oxytropoides Bge. 329.
— Pallasii (Willd.) M. B. 359.
— petraea (M. B.) Fisch. 330; tab. 16.
— pulchella Schrenk 328; tab. 16.
— radiata M. B. 364; tab. 17.
— Ruprechtii Grossh. 330.
— schugnanica B. Fedtsch. 352.
— seravschanica B. Fedtsch. 361.
— sibirica Turcz. ex Bess. 349.
— Sintenisii Bornm. 359; tab. 16.
— Sosnovskyi Grossh. 331.
— subacaulis Boiss. 357; tab. 18.

— tanaitica Spreng. 350.
— teheranica Bornm. 358.
— Tournefortii (Willd.) Desv. 365.
— transcaspica V. Nikit. 342.
— transcaucasica Grossh. 345, 555.
— vaginalis C. A. Mey. 363.
— Vassilczenkoi Grossh. 361, 553.
— viciifolia Scop. 341.
Onoclea L. 1, 29.
— sensibilis L. 29.
Ononis L. 11, 94.
— antiquorum L. 100: tab. 7.
— arvensis L. 96; tab. 7.
— intermedia C. A. Mey. ex A. Beck. 101.
— leiosperma Boiss. 101; tab. 7.
— pusilla L. 95; tab. 7.
— repens L. 99; tab. 7.
— spinosa L. 99; tab. 7.
Onopordum L. 28, 228.
— acanthium L. 231; tab. 11.
— armenum Grossh. 234.
— candidum Nabĕl. 238.
— cinereum Grossh. 232.
— Frickii Tamamsch. 236, 605.
— heteracanthum C. A. Mey. 239; tab. 11.
— leptolepis DC. 237.
— Prjachinii Tamamsch. 233, 605.
— seravschanicum Tamamsch. 238.
— tauricum Willd. 233.
Onosma L. 19, 184.
— arenarium Waldst. et Kit. 219.
— armeniacum Klok. 214.
— atrocyaneum Franch. 196; tab. 10.
— azureum Schipcz. 199; tab. 11.
— baldshuanicum Lipsky 209.
— Barsczewskii Lipsky 206.
— brevipilosum Schischk. 200, 704.
— calycinum Stev. 218.
— caucasicum Levin 231; tab. 12.
— dichroanthum Boiss. 210.
— ferganense M. Pop. 200, 703; tab. 10.
— Gmelinii Ledeb. 207.
— gracile Trautv. 225.
— irritans M. Pop. ex Pavl. 208.
— leucocarpum M. Pop. 202.
— Livanovii M. Pop. 226.
— longilobum Bge. 204.
— macrorhizum M. Pop. 227.
— maracandicum Zak. 199.
— microcarpum Stev. ex DC. 215.
— polychromum Klok. 224.
— polyphyllum Ledeb. 194.
— rigidum Ledeb. 229.
— rupestre M. B. 216.
— sericeum Willd. 192.
— setosum Ledeb. 213.
— simplicissimum L. 195.
— stamineum Ledeb. 203.
— tauricum Pall. 230.
— trachycarpum Levin 201; tab. 11.
— tinctorium M. B. 220.
— transrhymnense Klok. 219.
— Visianii Clem. 214.
— zerizaminum Lipsky 205.
Onotrophe Cass. 28, 52, 149.
Ophelia D. Don ex G. Don 18, 625.
— chinensis Bge. ex Griseb. 625.
— diluta (Turcz.) Ledeb. 626.
— tetrapetala (Pall.) Grossh. 627; tab. 33.
— Tscherskyi (Kom.) Grossh. 626.
— Wilfordii Kerner 18, 628.
Ophioglossum L. 1, 92.
— bucharicum B. et O. Fedtsch. 93; tab. 3.
— lusitanicum L. 94.
— thermale Kom. 93; tab. 3.
— vulgatum L. 93.
Ophioscordon Wallr. 4, 142.
Ophrys (L.) Sw. 4, 722.
— apifera Huds. 730; tab. 43.
— caucasica Woronow 727.
— muscifera Huds. 725; tab. 43.
— oestrifera M. B. 729; tab. 43.
— taurica (Agg.) Nevski 728; tab. 43.
— transhyrcana Czerniak. 726.
Oplismenus Beauv. 2, 37.
— Burmannii (Retz.) Beauv. 38.
— undulatifolius (Ard.) Beauv. 37; tab. 2.
Oplopanax Miq. 16, 17.
Opopanax C. Koch 17, 165.
— armeniacum Bordz. 166.
Oporanthus Herb. 4, 485.
Opulus Moench 23, 453.
Orchis L. 4, 677.
— amblyoloba Nevski 707; 753; tab. 42.
— aristata Fisch. ex Lindl. 710.
— baltica Klinge 712.
— caspia Trautv. 688; tab. 41.
— chlorotica Woronow 689; tab. 41.
— coriophora L. 681; tab. 41.
— cruenta Muell. 716.
— euxina Nevski 709; tab. 42.
— Fedtschenkoi Czerniak. 689; tab. 41.
— flavescens C. Koch 701.
— Fuchsii Druce 704; tab. 42.
— iberica M. B. ex Willd. 700; tab. 42.
— latifolia L. 717; tab. 42.
— maculata L. 703; tab. 42.
— majalis Reichb. f. tab. 42.

Orchis mascula L. 683; tab. 41.
— maxima C. Koch 695; tab. 41.
— militaris L. 691; tab. 41.
— morio L. 686.
— pallens L. 684; tab. 41.
— palustris Jacq. 680.
— picta Loisel. 686.
— provincialis Balb. 685; tab. 41.
— punctulata Stev. ex Lindl. 696; tab. 41.
— purpurea Huds. 694; tab. 41.
— romana Seb. et Maur. 702.
— salina Turcz. ex Lindl. 713.
— sambucina L. 702.
— sanasunitensis Fleischm. 715.
— Schelkovnikovii Woronow 697; tab. 41.
— simia Lam. 693; tab. 41.
— Traunsteineri Saut. ex Reichb. f. 711.
— tridentata Scop. 691.
— triphylla C. Koch 708; tab. 42.
— umbrosa Kar. et Kir. 714; tab. 42.
— ustulata L. 690; tab. 41.
— viridifusca Albov 685.
Oreophila D. Don 29, 202.
Oreorchis Lindl. 4, 609.
— patens Lindl. 609.
Oreoselinum Adans. 17, 168, 188.
Origanum L. 21, 463.
— kopetdaghensis Boriss. 469.
— tyttanthum Gontsch. 466; tab. 26.
— vulgare L. 464.
Oritrophium Torr. et Gray 25, 104.
Orlaya Hoffm. 16, 181.
— grandiflora (L.) Hoffm. 182; tab. 14.
— platycarpos (L.) Koch 183; tab. 14.
Ormopterum Schischk. 16, 363, 597.
— turcomanicum (Korov.) Schischk. 363; tab. 30.
Ormosciadium Boiss. 17, 275.
— pulchrum Schischk. 276; tab. 21.
Ornithogalum L. 4, 379.
— arcuatum Stev. 391; tab. 22.
— arianum Lipsky 390; tab. 22.
— Balansae Boiss. 385; tab. 22.
— Boucheanum (Kunth) Aschers. 393; tab. 22.
— brachystachys C. Koch 392.
— fimbriatum Willd. 383.
— Fischerianum Krasch. 392.
— flavescens Lam. 390; tab. 22.
— graciliflorum C. Koch 389.
— hyrcanum Grossh. 389.
— magnum Krasch. et Schischk. 391, 743.
— nutans L. 393.
— platyphyllum Boiss. 389; tab. 22.
— pyrenaicum L. 391; tab. 22.
— refractum Kit. 383.
— Schelkovnikovii Grossh. 390.
— Schischkinii Krasch. 386, 743.
— Schmalhausenii Albov 385.
— sigmoideum Freyn et Sint. 386; tab. 22.
— Sintenisii Freyn 384; tab. 22.
— tenuifolium Guss. 383; tab. 22.
— transcaucasicum Miscz. 384.
— umbellatum L. 382.
— Woronowii Krasch. 384, 743; tab. 22.
Ornithopodium Scop. 13, 247.
Ornithopus L. 13, 245.
— compressus L. 246.
— perpusillus L. 246.
— sativus Broth. 246.
Orobanche L. 23, 30.
— aegyptiaca Pers. 60; tab. 4.
— alba Steph. 93; tab. 5.
— alsatica Kirschleg. 110.
— amoena C. A. Mey. 74; tab. 5.
— amurensis (G. Beck) Kom. 74; tab. 5.
— Androssovii Novopokr. 53.
— arenaria Borkh. 50.
— ariana Gontsch. 75.
— armena Tzvel. 105; 686.
— badchysensis Novopokr. et V. Nikit. 82.
— Borissoviae Novopokr. 55.
— brachypoda Novopokr. 64.
— brassicae Novopokr. 62.
— brevidens Novopokr. 81.
— Bungeana G. Beck 48.
— caesia Reichb. 23, 46.
— camptolepis Boiss. et Reut. 83; tab. 5.
— cernua Loefl. 79.
— cilicica G. Beck 57.
— Clarkei Hook. f. 76.
— coelestis Boiss. 54; tab. 4.
— coerulescens Steph. 71.
— colorata C.Koch 113; tab. 6.
— connata C. Koch 104.
— crenata Forsk. 98.
— cumana Wallr. 79; tab. 5.
— dalmatica (G. Beck) Tzvel. 66.
— flava Mart. 108.
— gamosepala Reut. 87; tab. 5.
— gigantea (G. Beck) Gontsch. 84.
— glabricaulis Tzvel. 86, 685.
— glaucantha Trautv. 72.
— gracilis Smith 112; tab. 6.
— Grigorjevii Novopokr. 95.
— Grossheimii Novopokr. 106.
— Hansii Kerner 76.

— hederae Duby 104; tab. 5.
— hirtiflora (Reut.) Tzvel. 59.
— Hohenackeri (Reut.) Tzvel. 63.
— hymenocalyx Reut. 92.
— iberica (G. Beck) Tzvel. 49.
— inulae Novopokr. et Abramov 108; tab. 5.
— karatavica Pavl. 54.
— Kelleri Novopokr. 56; tab. 4.
— Korshinskyi Novopokr. 72.
— Kotschyi Reut. 83.
— Krylowii G. Beck 109.
— kurdica Boiss. et Hausskn. 105.
— Linczevskyi Novopokr. 89.
— lutea Baumg. 88; tab. 5.
— major L. 107.
— minor Sutt. 102.
— Mutelii F. Schultz 61.
— nana Noë 69.
— orientalis G. Beck 65.
— Owerinii G. Beck 97.
— oxyloba (Reut.) G. Beck 64.
— pallidiflora Wimm. et Grab. 94.
— picridis F. Schultz 101.
— pulchella (C. A. Mey.) Novopokr. 23, 47; tab. 4.
— purpurea Jacq. 23, 45.
— pycnostachya Hance 73.
— Raddeana G. Beck 96; tab. 5.
— ramosa L. 70; tab. 4.
— reticulata Wallr. 95.
— rosea Tzvel. 111, 687.
— Schelkovnikovii Tzvel. 90, 685.
— septemloba (G. Beck) Tzvel. 58; tab. 4.
— serratocalyx G. Beck 58.
— Sintenisii G. Beck 77.
— sogdiana Novopokr. 66; tab. 4.
— solenanthi Novopokr. et Pissjauk. 78.
— Solmsii Clarke 81.
— sordida C. A. Mey. 85; tab. 5.
— sulphurea Gontsch. 78.
— teucrii Holand. 89.
— transcaucasica Tzvel. 97, 686.
— uralensis G. Beck 46.
— versicolor F. Schultz 103.
— vitellina Novopokr. 86.
— vulgaris Poir. 91.
Orobus L. 13, 507.
Orostachys (DC.) Fisch. 9, 108.
— cartilaginea Boriss. 112, 482.
— fimbriata (Turcz.) Berger 113; tab. 6.
— malacophylla (Pall.) Fisch. 110; tab. 6.
— spinosa (L.) C. A. Mey. 110.
— thyrsiflora Fisch. 112; tab. 6.
Orphanidesia Boiss. et Bal. 18, 77.
Orthantha (Benth.) Kerner 22, 647.
— Aucheri (Boiss.) Wettst. 648.
— lutea (L.) Kerner ex Wettst. 647.
Orthocentron Cass. 28, 52, 150.
Orthodon Benth. 21, 633.
— grosseserratum (Maxim.) Kudo 633.
Orthurus Juz. 10, 262, 616.
— heterocarpus Boiss. 262.
— kokanicus (Rgl. et Schmalh.) Juz. 263; tab. 19.
Orythia Baker 4, 362.
Oryza L. 2, 47.
— sativa L. 47; tab. 3.
Oryzopsis Michx. 2, 112.
— coerulescens (Desf.) Hack. 114; tab. 5.
— ferganensis Litv. 116, 742; tab. 5.
— holciformis (M. B.) Hack. 114; tab. 5.
— kokanica (Rgl.) Roshev. 115, 742; tab. 5.
— kopetdagensis Roshev. 117, 743; tab. 5.
— latifolia Roshev. 116, 743; tab. 5.
— molinioides (Boiss.)Hack. ex Pauls. 117; tab. 5.
— purpurascens Hack. ex Pauls. 117; tab. 5.
— songorica (Trin. et Rupr.) B. Fedtsch. 115; tab. 5.
— virescens (Trin.) G. Beck 113; tab. 5.
Osmorhiza Raf. 16, 148.
— aristata (Thunb.) Makino et Vabe 149.
Osmathamnus DC. 18, 40.
Osmunda L. 1, 87.
— cinnamomea L. 88.
— Claytoniana L. 88; tab. 3.
— regalis L. 88.
Ossea Lunell 17, 331.
Ostericium Hoffm. ex Endl. 17, 10.
Ostericum Hoffm. 17, 10.
— palustre Bess. 10.
Ostrowskia Rgl. 24, 382.
— magnifica Rgl. 382; tab. 23.
Ostrutium Link 17, 168.
Ostrya Scop. 5, 253.
— carpinifolia Scop. 254; tab. 12.
Otanthus Hoffmgg. et Link 26, 127.
— maritimus (L.) Hoffmgg. et Link 128.
Otites Adans. 6, 685.
Otostegia Benth. 21, 182.
— bucharica B. Fedtsch. 186.
— Fedtschenkoana Kudr. 186.
— Olgae (Rgl.) Korsh. 184.
— sogdiana Kudr. 185.
Ottelia Pers. 1, 297.
— alismoides (L.) Pers. 297; tab. 10.
Ovilla Adans. 24, 444.

Oxalis L. 14, 76.
— acetosella L. 79.
— corniculata L. 77.
— obtriangulata Maxim. 81.
— pes-caprae L. 82.
— stricta L. 78.
— violacea L. 81.
Oxycoccus Adans. 18, 103.
— quadripetalus Gilib. 103.
— microcarpus Turcz. 104.
Oxygraphis Bge. 7, 325.
— Chamissonis (Schlecht.) Freyn 328; tab. 34.
— glacialis (Fisch.) Bge. 326; tab. 34.
— vulgaris Freyn 327; tab. 34.
Oxylapathon St. Lager 5, 444.
Oxynepeta Bge. 20, 413.
Oxyria Hill 5, 444.
— digyna (L.) Hill 444.
Oxytropis DC. 13, 1.
— aciphylla Ledeb. 225; tab. 9.
— Adamsiana (Trautv.) Vass. 69.
— adscendens Gontsch. 154.
— ajanensis (Rgl. et Til.) Bge. 76.
— alajica Drob. 47, 545.
— albana Stev. 35.
— albiflora Bge. 81.
— albovillosa B. Fedtsch. 159.
— alpestris Schischk. 70.
— alpicola Turcz. 92.
— alpina Bge. 86.
— altaica (Pall.) Pers. 88.
— ambigua (Pall.) DC. 85; tab. 2.
— ammophila Turcz. 102.
— ampullata (Pall.) Pers. 181.
— anadyrensis Vass. 212.
— approximata Less. 65.
— arassanica Gontsch. 162, 551; tab. 7.
— arctica R. Br. 87; tab. 1.
— argentata (Pall.) Pers. 72.
— aspera Gontsch. 126, 548; tab. 6.
— assiensis Vass. 179.
— astragaloides Boriss. 131.
— atbaschi Saposhn. 24.
— aulieatensis Vved. 111, 546.
— aurea Vass. 110.
— austrosachalinensis Vass. 75.
— avis Saposhn. 26.
— baicalia (Pall.) Pers. 194.
— baissunensis Vass. 112; tab. 4.
— baldshuanica V. Fedtsch. 40.
— bella B. Fedtsch. 227.
— Bellii (Britt.) Palib. 196.
— biloba Saposhn. 141.
— Bobrovii B. Fedtsch. 152, 550; tab. 5.
— boguschi B. Fedtsch. 10.
— brachycarpa Vass. 139.
— bracteata Basil. 180.
— bracteolata Vass. 67.
— brevicaulis Ledeb. 123.
— caespitosa (Pall.) Pers. 185.
— caespitosula Gontsch. 160, 551; tab. 7.
— campanulata Vass. 100; tab. 3.
— campestris (L.) DC. 60; tab. 2.
— cana Bge. 45.
— candicans (Pall.) DC. 100.
— canopatula Vass. 124; tab. 6.
— Capusii Franch. 115.
— carpathica Uechtr. 16.
— caucasica Rgl. 35.
— caulescens Gontsch. 153, 550.
— chantengriensis Vass. 29.
— chesneyoides Gontsch. 151.
— chiliophylla Royle 218.
— chionobia Bge. 204; tab. 11.
— chionophylla Schrenk 92.
— chorgossica Vass. 47.
— cobresietorum Vass. 161.
— coerulea (Pall.) DC. 22.
— columbina Vass. 182; tab. 8.
— coluteoides Vass. 165; tab. 4.
— confusa Bge. 99.
— crassiuscula Boriss. 163.
— cretacea Basil. 45, 544.
— cuspidata Bge. 126.
— cyanea M. B. 31.
— czapandaghi B. Fedtsch. 27, 543.
— dasypoda Rupr. ex Boiss. 36.
— deflexa (Pall.) DC. 43.
— dichroantha Schrenk 148.
— didymophysa Bge. 159.
— Dorogostajskyi Kuzen. 63.
— dubia Turcz. 194.
— echidna Vved. 224.
— erecta Kom. 78.
— eriocarpa Bge. 186.
— Fedtschenkoana Vass. 171.
— ferganensis Vass. 113.
— Fetissovii Bge. 203; tab. 10.
— filiformis DC. 24.
— floribunda (Pall.) DC. 142.
— Fominii Grossh. 102; tab. 12.
— frigida Kar. et Kir. 94.
— fruticulosa Bge. 135.
— Gebleri Fisch. 81.
— glabra (Lam.) DC. 41.
— glandulosa Turcz. 217.
— glareosa Vass. 49.

— globiflora Bge. 17.
— Gmelinii Fisch. 64.
— Gorbunovii Boriss. 49.
— **grandiflora (Pall.) DC. 93; tab. 2.**
— guntensis B. Fedtsch. 13.
— gymnogyne Bge. 171.
— heteropoda Bge. 46.
— heterotricha Turcz. 196; tab. 11.
— Hippolyti Boriss. 66.
— hirsuta Bge. 146.
— hirsutiuscula Freyn 48.
— humifusa Kar. et Kir. 29.
— hystrix Schrenk 222.
— immersa (Baker) Bge. ex B. Fedtsch. 25.
— inaria (Pall.) DC. 201.
— includens Basil. 229.
— integripetala Bge. 125.
— intermedia Bge. 187.
— iskanderica B. Fedtsch. 151.
— jucunda Vved. 149, 549.
— kamtschatica Hult. 83.
— karataviensis Pavl. 175.
— Karjaginii Grossh. 36.
— kasbeki Bge. 31.
— katangensis Basil. 63.
— kawasimensis Sugaw. 91.
— ketmenica Saposhn. 82.
— Komarovii Vass. 228; tab. 12.
— kopetdagensis Gontsch. 40, 544.
— Krylovii Schipcz. 21; tab. 1.
— kubanensis Leskov 61.
— Kusnetzovii Kryl. 86.
— Ladyginii Kryl. 22.
— lanata (Pall.) DC. 197.
— lapponica (Wahl.) J. Gay 17.
— lasiocarpa Gontsch. 150.
— lazica Boiss. 71.
— Lehmannii Bge. 8.
— leptophylla (Pall.) DC. 184.
— leptophysa Bge. 172.
— leucantha (Pall.) Bge. 211.
— leucicyanea Bge. 147.
— leucotricha Turcz. 183.
— Linczevskii Gontsch. 128, 549; tab. 6.
— Lipskyi Gontsch. 121, 547.
— lithophila Vass. 119; tab. 5.
— litoralis Kom. 77.
— Litvinovii B. Fedtsch. 145.
— longibracteata Kar. et Kir. 76.
— longipes Fisch. 167.
— longirostra DC. 93.
— lupinoides Grossh. 556.
— macrobotrys Bge. 143.
— macrocarpa Kar. et Kir. 108.
— macrodonta Gontsch. 110, 545.
— macrosema Bge. 79.
— maidantalensis B. Fedtsch. 162, 552.
— malocophylla Bge. 180.
— mandshurica Bge. 23.
— marina Vass. 70.
— Martjanovii Kryl. 80.
— masarensis Vass. 97.
— **Maydelliana Trautv. 64.**
— megalorrhyncha Nevski 130.
— Meinshausenii C. A. Mey. 103.
— melaleuca Bge. 79.
— melanotricha Bge. 30.
— merkensis Bge. 27; tab. 5.
— Mertensiana Turcz. 166.
— Meyeri Bge. 90.
— Michelsonii B. Fedtsch. 9.
— microcarpa Gontsch. 8, 542.
— **microphylla (Pall.) DC. 216.**
— **microsphaera Bge. 158; tab. 7.**
— Middendorfii Trautv. 207.
— mixotriche Bge. 183.
— mugodsharica Bge. 114.
— mumynabadensis B. Fedtsch. 120.
— muricata (Pall.) DC. 215.
— myriophylla (Pall.) DC. 197.
— Niedzweckiana M. Pop. 125, 548.
— nigrescens (Pall.) Fisch. 164.
— nitens Turcz. 177.
— nivea Bge. 90.
— nuda Basil. 187.
— nutans Bge. 37.
— ochotensis Bge. 91.
— ochroleuca Bge. 44.
— oligantha Bge. 205.
— ornata Vass. 119.
— Owerinii Bge. 89.
— oxyphylla (Pall.) DC. 195.
— pagobia Bge. 28.
— Pallasii Pers. 105.
— pauciflora Bge. 15.
— pellita Bge. 205.
— penduliflora Gontsch. 38, 543.
— physocarpa Ledeb. 215.
— piceetorum Vass. 88.
— pilosa (L.) DC. 104.
— pilosissima Vved. 116, 547.
— platonychia Bge. 50; tab. 8.
— platysema Schrenk 15.
— podoloba Kar. et Kir. 144.
— polyphylla Ledeb. 221.
— Poncinsii Franch. 178.
— prostrata (Pall.) DC. 198.
— proxima Boriss. 136.

Oxytropis pseudofrigida Saposhn. 179.
— pseudoleptophysa Boriss. 175.
— puberula Boriss. 43; tab. 1.
— pulvinata Saposhn. 14.
— pulvinoides Vass. 147.
— pumila Fisch. ex DC. 199.
— pumilio (Pall.) Ledeb. 165.
— recognita Bge. 62.
— retusa Matsum. 85.
— revoluta Ledeb. 166.
— rhynchophysa Schrenk 202; tab. 8, 10.
— riparia Litv. 42.
— robusta M. Pop. 109, 545.
— rosea Bge. 120.
— roseiformis B. Fedtsch. 121.
— roseililacina Vass. 127.
— rostrata Vass. 46.
— rubriargillosa Vass. 113.
— rubricaudex Hult. 84.
— Rübsaamenii B. Fedtsch. 26.
— rukutamensis Sugaw. 75.
— rupifraga Bge. 25.
— ruthenica Vass. 78.
— sachalinensis Miyabe et Tatewaki 83.
— salina Vass. 50.
— samurensis Bge. 32.
— Saposhnikovii Kryl. 21.
— sarkandensis Vass. 140, 549.
— saurica Saposhn. 206.
— savellanica Bge. ex Boiss. 14.
— scabrida Gontsch. 129.
— Schischkinii Vass. 66.
— Schmidtii Meinsh. 208.
— Schrenkii Trautv. 141.
— selengensis Bge. 201.
— Semenovii Bge. 122.
— seravschanica Gontsch. 169, 552.
— sericea (Lam.) Simonk. 71.
— setosa (Pall.) DC. 188.
— Severzovii Bge. 157.
— silvatica (Pall.) DC. 101.
— songorica (Pall.) DC. 98; tab. 3.
— sordida (Willd.) Pers. 59; tab. 2.
— spicata (Pall.) O. et B. Fedtsch. 99.
— spinifer Vass. 222.
— squamulosa DC. 213.
— stenophylla Bge. 186.
— strobilacea Bge. 68.
— Stukovii Palib. 199.
— suavis Boriss. 152.
— subcapitata Gontsch. 123, 547.
— submutica Bge. 169.
— subverticillaris Ledeb. 200.
— sulphurea (Fisch.) Ledeb. 61.
— susamyrensis B. Fedtsch. 149.
— tachtensis Franch. 115.
— talassica Gontsch. 6.
— talgarica M. Pop. 112, 546.
— tenuirostris Boriss. 129.
— tenuissima Vass. 144.
— terekensis B. Fedtsch. 203; tab. 11
— teres (Lam.) DC. 143.
— tianschanica Bge. 134; tab. 5.
— Tilingii Bge. 97.
— todomoshiriensis Miyabe et Miyake 69
— tomentosa Gontsch. 130.
— tragacanthoides Fisch. 223.
— trajectorum B. Fedtsch. 161.
— transalaica Vass. 182.
— transversa Vass. 139.
— Trautvetteri Meinsh. 211.
— trichocalycina Bge. 226.
— trichophysa Bge. 214.
— trichosphaera Freyn 226.
— triphylla (Pall.) Pers. 191.
— tschimganica Gontsch. 7.
— tschujae Bge. 82.
— tyttantha Gontsch. 9.
— ugamica Gontsch. 170.
— uralensis (L.) DC. 68.
— vermicularis Freyn 30.
Ozodia Wight et Arn. 16, 541.
Ozophyllum Dum. ex Kuntze 3, 534.

Pachylepis Less. 29, 595.
Pachyloma Spach 7, 351, 475.
Pachyphragma (DC.) N. Busch 8, 575.
— macrophyllum (Hoffm.) N. Busch 575; tab. 32.
Pachypleurum Ledeb. 16, 579.
— alpinum Ledeb. 579; tab. 37.
— gayoides (Rgl. et Schmalh.) Schischk. 582; tab. 34, 37.
— mucronatum (Schrenk) Schischk. 581; tab. 34, 37.
Pachypteris Kar. et Kir. 8, 223.
Pachypterygium Bge. 8, 223.
— brevipes Bge. 224; tab. 12.
— densiflorum Bge. 227; tab. 12.
— multicaule (Kar. et Kir.) Bge. 227; tab. 12.
Pachysandra Michx. 14, 503.
— terminalis Sieb. et Zucc. 504; tab. 16.
Padus Mill. 10, 575.
— asiatica Kom. 578.
— Maackii (Rupr.) Kom. 579; tab. 37.
— racemosa (Lam.) Gilib. 576.
— ssiori (Fr. Schmidt) Schneid. 576.

-- virginiana (L.) Mill. **575.**
Paederotella (Wulff) Kem.-Nath. 22, 329, 492.
Paeonia L. 7, 24.
— abchasica Miscz. 33.
— albiflora Pall. 25.
— anomala L. **33, tab. 3.**
— Biebersteiniana Rupr. **35.**
— caucasica Schipcz. **28; tab. 3.**
— hybrida Pall. 34.
— japonica Miyabe et Takeda 27.
— macrophylla Lomak. 30.
— Mlokosewitschii Lomak. 26.
— obovata **Maxim. 27.**
— oreogeton S. Moore **26.**
— tenuifolia L. 34.
— tomentosa (Lomak.) N. Busch 30; tab. 3.
-— triternata Pall. 28.
— vernalis Mandl 27.
— Wittmanniana Hartwiss et **Lindl. 29;** tab. 3.
Paleya Cass. 29, 595.
Palimbia Bess. 17, 45.
— rediviva (Pall.) Thell. 45; tab. 6.
Palimbia DC. 17, 168.
Paliurus Mill. 14, 635.
— spina-Christi Mill. 636; tab. 34.
Pallasia L. f. 5, 530.
Pallenis Cass. 25, 508.
— spinosa (L.) Cass. 508.
Panax L. 16, 34.
— schin-seng Nees 34; tab. 2.
Pancratium L. 4, 494.
— maritimum L. 494; tab. 30.
Panderia Fisch. et Mey. 6, 119.
— pilosa Fisch. et Mey. 120.
— turkestanica Iljin 120; tab. 5.
Panicularia Fabr. 2, 449.
Panicum L. 2, 35.
— acroanthum Steud. 36.
— capillare L. 36.
— miliaceum L. 36; tab. 2.
— miliare Lam. 37.
Panzeria Moench 21, 157.
— argyracea Kuprian. 158.
— canescens Bge. 159.
— lanata (L.) Bge. 159.
Papaver L. 7, 598.
— ajanense M. Pop. 607, 748.
— alboroseum Hult. 608.
— ambiguum M. Pop. 638, 749; tab. 41.
— angustifolium Tolm. 614.
— anomalum Fedde 604.
— arenarium M. B. 642; tab. 42.
— argemone L. 630.
— Belangeri Boiss. 633; tab. 42.
— bipinnatum C. A. Mey. 643; tab. 42.
— bracteatum Lindl. 617; tab. 41.
— canescens Tolm. 610.
— **caucasicum** M. B. 624.
— **chelidoniifolium** Boiss. et Buhse 642; tab. 41.
— commutatum Fisch. et Mey. 637; tab. 41.
— croceum Ledeb. 607.
— **dubium L. 639.**
— **hybridum** L. 629; tab. 42.
— **intermedium** DC. 618.
— **involucratum** M. Pop. 609, 748; tab. 41.
— **lacerum M. Pop. 641, 749.**
— laevigatum M. B. 640.
— lapponicum (Tolm.) Nordh. 615.
— lasiothrix Fedde 619.
— Ledebourianum Lundstr. 607.
— leiocarpum Turcz. 604.
— Lisae N. Busch 603.
— Litwinowii Fedde ex Bornm. 640; tab. 42.
— macrostomum Boiss. et Huet 634.
— microcarpum DC. 608.
— monanthum Trautv. 622.
— nivale Tolm. 615.
— nudicaule L. 606.
— ocellatum Woronow 628; tab. 42.
— oreophilum Rupr. 622.
-— orientale L. 619.
— paucifoliatum (Trautv.) Fedde 620.
— pavoninum Schrenk 627; tab. 42.
— persicum Lindl. 625.
— pseudocanescens M. Pop. 610, **749.**
— pseudo-Stubendorfii M. Pop. 605, 748.
— pulvinatum Tolm. 614.
— radicatum Rottb. 613.
— rhoeas L. 636.
— Roopianum (Bordz.) Sosn. 626.
— rubro-aurantiacum Fisch. ex Steud. 606.
— setigerum DC. 645.
— somniferum L. 644.
— strigosum (Boenn.) Schur 637.
— Stubendorfii Tolm. 605.
— talyschense Grossh. 645.
— tenellum Tolm. 606.
— tianschanicum M. Pop. 609, 748.
Pappagrostis Roshev. 2, 230, 749.
— pappophorea (Hack.) Roshev. 231; tab. 12.
Pappophorum Schreb. 2, 296.
-- boreale Griseb. 297; tab. 23.
— persicum (Boiss.) Steud. **297**; tab. 23.
Paracaryum Boiss. 19, 588.

Paracaryum Bungei (Boiss.) Brand 605; tab. 29.
— gracile Czerniak. 600; tab. 31.
— himalayense (Klotsch) Clarke 597; tab. 30.
— incanum (Ledeb.) Boiss. 593; tab. 31.
— intermedium (Fresen.) Lipsky 602; tab. 30.
— karataviense Pavl. ex M. Pop. 597, 717.
— laxiflorum Trautv. 594; tab. 29.
— strictum (C. Koch) Boiss. 601; tab. 31.
— turcomanicum Bornm. et Sint. 599; tab. 30.
Paracaryum Brand 19, 588.
Paracynoglossum M. Pop. 19, 673, 717.
— asperrimum (Nakai) M. Pop. 679.
— denticulatum (DC.) M. Pop. 674.
— imeretinum (Kusn.) M. Pop. 676; tab. 39.
Parageum Nakai et Hara 10, 247.
— calthifolium (Menzies) Nakai et Hara 248; tab. 18.
Paraixeris Nakai 29, 396.
— denticulata (Houtt.) Nakai 397; tab. 24.
— pinnatipartita (Makino) Tzvel. 398.
— saxatilis (A. Baran.) Tzvel. 400.
— serotina (Maxim.) Tzvel. 399.
Paramicrorhynchus Kirp. 29, 236, 725.
— procumbens (Roxb.) Kirp. 237; tab. 15.
Paraquilegia Drum. et Hutch. 7, 75.
— anemonoides (Kar. et Kir.) Schipcz. 78.
— caespitosa (Boiss. et Hoh.) Drum. et Hutch. 77.
— grandiflora (Fisch.) Drum. et Hutch. 76.
— microphylla (Royle) Drum. et Hutch. 77.
Parasyringa W. W. Smith 18, 518.
Parentucellia Viv. 22, 641.
— flaviflora (Boiss.) Nevski 645; tab. 32.
— latifolia (L.) Caruel 642; tab. 32.
— viscosa (L.) Caruel 646; tab. 32.
Parietaria L. 5, 398.
— alsinifolia Delile 402.
— chersonensis (O. Lang et Szov.) Doerfl. 401; tab. 22.
— erecta Mert. et Koch 400; tab. 22.
— judaica Strand. 400.
— micrantha Ledeb. 402; tab. 22.
— officinalis L. 399; tab. 22.
Paris L. 4, 469.
— hexaphylla Cham. 470.
— incompleta M. B. 470; tab. 29.
— manshurica Kom. 473, 748.
— quadrifolia L. 469; tab. 29.
Parnassia L. 9, 215.
— bifolia Nekras. 218.
— Kotzebuei Cham. et Schlecht. 218.
— Laxmannii Pall. ex Schult. 219.
— palustris L. 216.
Paronychia Adans. 6, 565.
— cephalotes (M. B.) Bess. 565; tab. 34.
— kurdica Boiss. 566; tab. 34.
Parrotia C. A. Mey. 9, 272.
— persica (DC.) C. A. Mey. 272; tab. 16.
Parrya R. Br. 8, 156.
— ajanensis N. Busch 261, 646.
— albida M. Pop. ex P. Baran. 263.
— arctica R. Br. 261.
— asperrima M. Pop. 259.
— Beketovii Krassn. 262.
— eriocalyx Rgl. et Schmalh. 272.
— excapa C. A. Mey. 271.
— fruticulosa Rgl. et Schmalh. 263.
— Golenkinii Lipsch. 266.
— karatavica Lipsch. 264.
— microcarpa Ledeb. 270.
— nudicaulis (L.) Rgl. 260.
— pinnatifida Kar. et Kir. 262.
— pulvinata M. Pop. 270, 646; tab. 15.
— runcinata (Rgl. et Schmalh.) N. Busch 264; 645.
— schugnana Lipsch. 260.
— siliquosa Krassn. 269.
— stenocarpa Kar. et Kir. 265.
— stenophylla M. Pop. 266, 646.
— subsiliquosa M. Pop. 269, 645.
— surculosa N. Busch 270, 647; tab. 15.
— turkestanica (Korsh.) N. Busch 259, 644.
Parthenium L. 25, 530.
— argentatum A. Gray 530.
Parthenocissus Planch. 14, 707.
— inserta (Kerner) Fritsch 709; tab. 39.
— quinquefolia (L.) Planch. 709; tab. 39.
— tricuspidata (Sieb. et Zucc.) Planch. 708.
Paspalum L. 2, 25.
— digitaria Poir. 25; tab. 2.
— dilatatum Poir. 26.
— Thunbergii Kunth ex Steud. 26.
Pastinaca L. 17, 215.
— armena Fisch. et Mey. 221; tab. 20.
— aurantiaca (Albov) Kolak. 221.
— pimpinellifolia M. B. 218.
— sativa L. 217.
— silvestris Gars. 216.
— umbrosa Stev. ex DC. 217; tab. 20.
Pastinacopsis Golosk. 17, 273.
— glacialis Golosk. 274; tab. 23.
Patrinia Juss. 23, 587.
— gibbosa Maxim. 593.
— intermedia (Hornem.) Roem. et Schult. 590; tab. 28.
— rupestris (Pall.) Dufr. 590; tab. 28.

— scabiosifolia Fisch. 588.
— sibirica (L.) Juss. 589; tab. 28.
Pecten Lam. 16, 139.
Pectinaria Bernh. 16, 139.
Pedicularis L. 22, 687.
— abrotanifolia M. B. ex Stev. 730.
— achilleifolia Steph. 761.
— acmodonta Boiss. 764.
— Adamsii Hult. 782.
— adunca M. B. ex Stev. **774**.
— alatauica Stadlm. ex Vved. **755**, 813; tab. 37.
— Alberti Rgl. 786; tab. 39.
— altaica Steph. ex Stev. **771**.
— amoena Adam ex Stev. 709; tab. 35.
— amoeniflora Vved. **721**, 810.
— arguteserrata Vved. 706; 809.
— atripurpurea Nordm. 789; tab. 38.
— balkharica E. Busch **791**.
— brachystachys Bge. 746.
— capitata Adam 792.
— caucasica M. B. **712**.
— Chamissonis Stev. 704; tab. 39.
— cheilanthifolia Schrenk 713.
— chroorrhyncha Vved. 767, 815.
— compacta Steph. **748**.
— condensata M. B. 788.
— crassirostris Bge. 705.
— daghestanica Bonati 765.
— dasyantha Hadac 782.
— dasystachys Schrenk **749**.
— dolichorrhiza Schrenk 757.
— dubia B. Fedtsch. 763.
— elata Willd. 744.
— eriophora Turcz. 708; tab. 35.
— exaltata Bess. ex Bge. 787.
— fissa Turcz. 758.
— flava Pall. 760.
— grandiflora Fisch. 794.
— grandis M. Pop. 756, 814.
— Hacquetii Graf 788.
— hirsuta L. 784; tab. 35.
— hyperborea Vved. **777**, 817.
— incarnata L. **747**.
— inconspicua Vved. **724**, 811.
— interrupta Steph. ex Willd. **715**.
— karatavica Pavl. **727**.
— Karoi Freyn 776.
— Kaufmannii Pinzger 764.
— Korolkovii Rgl. 707; tab. 35.
— Krylovii Bonati 762.
— Kuznetzovii Kom. 732; tab. 37.
— labradorica Wirsing. 738; tab. 38.
— Langsdorfii Fisch. ex Stev. 783.
— lapponica L. **735**.
— lasiostachys Bge. **759**.
— longiflora J. Rudolph 699.
— Ludwigii Rgl. 729.
— macrochila Vved. **705**.
— mandshurica Maxim. **756**.
— Mariae Rgl. 772.
— Maximowiczii Krassn. **728**.
— myriophylla Pall. 729.
— nasuta M. B. ex Stev. **741**.
— Nordmanniana Bge. **745**; tab. 38.
— Oederi Vahl 785.
— Olgae Rgl. **720**.
— Pallasii Vved. 781, 817; tab. 35.
— palustris L. **775**.
— Panjutinii E. Busch 790.
— peduncularis M. Pop. 700; tab. 36.
— Pennellii Hult. **777**
— physocalyx Bge. 750.
— platyrrhyncha Schrenk 716.
— pontica Boiss. 712.
— Popovii Vved. 726, 812.
— proboscidea Stev. **745**; tab. 36.
— pubiflora Vved. **754**, 812; tab. 37.
— pulchra Pauls. 722.
— pycnantha Boiss. 719.
— resupinata L. 737.
— rhinanthoides Schrenk 700.
— rubens Steph. ex Willd. 760; tab. 39.
— sceptrum-carolinum L. 793.
— schistostegia Vved. 770.
— schugnana B. Fedtsch. 772.
— Semenovii Rgl. 725.
— sibirica Vved. 767, 816.
— Sibthorpii Boiss. 766.
— songarica Schrenk 753.
— spicata Pall. 731; tab. 37.
— striata Pall. **743**.
— subrostrata C. A Mey. **711**.
— sudetica Willd. 739.
— sylvatica L. 773.
— talassica Vved. **762**, 814.
— tianschanica Rupr. 703.
— tristis L. 736.
— uliginosa Bge. 742.
— uralensis Vved. 768, 816.
— venusta Schang. ex Bge. **769**.
— Verae Vved. 723, 810; tab. 36.
— verticillata L. **714**.
— villosa Ledeb. **740**.
— violascens Schrenk **709**.
— Vlassoviana Stev. **776**.
— Waldheimii Bonati **727**.
— Wilhelmsiana Fisch. ex M. B. **791**.

Pedicularis Willdenovii Vved. 778.
— yezoënsis Maxim. 737.
— zeravschanica Rgl. 723.
Pedina Stev. 16, 443.
Peganum L. 14, 147.
— harmala L. 148.
— **nigellastrum** Bge. 149.
Pelargonium L'Hér. 14, 72.
— Endlicherianum Fenzl 73.
Peltaria Jacq. 8, 594.
— Aucheri Boiss. 595.
— turkmena Lipsky 595.
— Voronovii N. Busch 596.
Peltariopsis N. Busch 8, 571.
— Grossheimii N. Busch 571.
Pennisetum Rich. 2, 44.
— flaccidum Griseb. 44; tab. 2.
— orientale Rich. 45; tab. 2.
Pentagonia Heist. ex Fabr. 22, 115.
Pentanema Cass. 25, 477.
— Albertoregelia (C. Winkl.) Gorschk. 479; tab. 29.
— divaricatum Cass. 480.
— glanduligerum (Krasch.) Gorschk. 485; tab. 29.
— parietarioides (Nevski) Gorschk. 484.
— propinquum (Nevski) Gorschk. 484.
— rupicola (Krasch.) Gorschk. 483.
Pentastemon L'Hér. 22, 308.
— frutescens Lamb. 309.
Pentataenium Tamamsch. 17, 260.
Penthorum Gronov. 9, 132.
— chinense Pursh 132; tab. 7.
— humile Rgl. et Maack ex Rgl. 133.
Peplis L. 15, 532.
— alternifolia M. B. 533.
— hyrcanica Sosn. 534.
— portula L. 534.
Peracarpa Hook. f. et Thoms. 24, 380.
— circaeoides (Fr. Schmidt) Feer 380; tab. 23.
Peramium Salisb. 4, 639.
Pereuphora Hoffmgg. 28, 268.
Periclymenum Mill. 23, 569.
Perilla L. 23, 630.
— nankinensis (Lour.) Decne. 632.
— ocymoides L. 631.
Periploca L. 18, 665.
— graeca L. 665.
— sepium Bge. 666.
Perovskia Kar. 21, 374.
— abrotanoides Kar. 375.
— angustifolia Kudr. 379.
— Kudrjaschevii Gorschk. et Pjat. 376.
— Linczevskii Kudr. 380.
— scrophulariifolia Bge. 377.
— virgata Kudr. 379.
Perplexia Iljin 27, 708, 727.
— microcephala (Boiss.) Iljin 709, 727.
Persica Mill. 10, 600.
— ferganensis (Kostina et Rjab.) Kov. et Kostina 603.
— vulgaris Mill. 601.
Persicaria Tourn. 5, 643.
Perularia Lindl. 4, 650.
— fuscescens (L.) Lindl. 651; tab. 40.
— ussuriensis (Maxim.) Schlechter 652; tab. 40.
Pervinca Adans. 18, 646.
Petasites Mill. 26, 642.
— albus (L.) Gaertn. 644.
— hybridus (L.) Gaertn. 643.
— spurius (Retz.) Reichb. 643.
Petillium L. 4, 317.
Petrocoma Rupr. 6, 726.
— Hoefftiana (Fisch.) Rupr. 729; tab. 45.
Petroselinum Hoffm. 16, 373.
— crispum (Mill.) Nym. 374.
Petroselinum Reichb. 17, 168, 197.
Petrosimonia Bge. 6, 316.
— brachiata (Pall.) Bge. 325.
— brachyphylla (Bge.) Iljin 323.
— crassifolia (Pall.) Bge. 319.
— glauca (Pall.) Bge. 324.
— glaucescens (Bge.) Iljin 323.
— hirsutissima (Bge.) Iljin 323; tab. 16.
— Litwinowii Korsh. 319.
— monandra (Pall.) Bge. 317; tab. 16.
— sibirica (Pall.) Bge. 326.
— squarrosa (Schrenk) Bge. 325.
— triandra (Pall.) Simonk. 318; tab. 16.
Peucedanum L. 17, 168.
— Adae Woronow 199.
— baicalense (Redow.) C. Koch 181.
— borysthenicum Klok. 197, 354.
— calcareum Albov 174.
— caucasicum (M. B.) C. Koch 187.
— cervaria (L.) Cuss. 189.
— cervariifolium C. A. Mey. 190.
— eryngiifolium Kom. 183.
— elegans Kom. 183.
— falcaria Turcz. 184.
— hissaricum Korov. 178.
— hystrix Bge. 186; tab. 8.
— latifolium (M. B.) DC. 191.
— Lubimenkoanum Kotov 198.
— luxurians Tamamsch. 176, 353.
— m :rophyllum Schischk. 192.

— mogoltavicum Korov. 178.
— Morissonii Bess. 175.
— oreoselinum (L.) Moench 188.
— ostruthium (L.) C. Koch 193.
— palimbioides Boiss. 196.
— palustre (L.) Moench 201.
— paucifolium Ledeb. 199.
— pauciradiatum Tamamsch. 200.
— podolicum (Bess.) Eichw. 195.
— polyanthum Korov. 202.
— pschavicum Boiss. 196.
— puberulum Turcz. 186.
— Renardii Rgl. et Schmalh. 177; tab. 8.
— ruthenicum M. B. 173.
— salinum Pall. 184.
— Schottii Bess. 194.
— Sintenisii Wolff 191.
— songoricum Schischk. 176, 354.
— tauricum M. B. 174.
— terebinthaceum Fisch. 182.
— transiliense Herd. 202; tab. 18.
— turcomanicum Schischk. 177; tab. 18.
— vaginatum Ledeb. 185.
— Zedelmeyerianum Manden. 187.
Pfundia Opiz 7, 351, 475.
Phaca L. 12, 11.
Phacelia Juss. 19, 95.
— tanacetifolia Benth. 96.
Phacellanthus Sieb. et Zucc. 23, 116.
— tubiflorus Sieb. et Zucc. 116; tab. 6.
Phaecasium Cass. 29, 595, 673.
Phaenixopus Cass. 29, 317.
Phaeopappus (DC.) Boiss. 28, 359.
Phagnalon Cass. 25, 379.
— Androssovii B. Fedtsch. 380.
— darvasicum Krasch. 379.
Phalacrachena Iljin 28, 579.
— calva (Ledeb.) Iljin 581; tab. 27.
— inuloides (Fisch. ex Schmalh.) Iljin 580.
Phalacroderis DC. 29, 595.
Phalacroloma Cass. 25, 191.
Phalaris L. 2, 51.
— brachystachys Link 54; tab. 4.
— bulbosa L. 52; tab. 4.
— canariensis L. 53; tab. 4.
— minor Retz. 53; tab. 4.
— paradoxa L. 54; tab. 4.
— truncata Guss. 51; tab. 4.
Phalolepis Cass. 28, 340, 370, 538.
Pharbitis Choisy 19, 3.
Phaseolus L. 13, 354.
— acutifolius A. Gray 535.
— angularis (Willd.) W. Wight 537.
— aureus Roxb. 537.
— coccineus L. 536.
— vulgaris L. 535.
Phellandrium L. 16, 526, 537.
Phellodendron Rupr. 14, 232.
— amurense Rupr. 232; tab. 13.
Phellopterus Benth. 17, 42.
Phelypaea auct. 23, 30.
Phelypaea Desf. 23, 28.
— coccinea (M. B.) Poir. 28.
— Helenae Popl. 29.
— Tournefortii Desf. 30; tab. 3.
Philadelphus L. 9, 220.
— caucasicus Koehne 220.
— latifolius Schrad. ex DC. 223.
— Schrenkii Rupr. et Maxim. 222.
— tenuifolius Rupr. et Maxim. 221; tab. 13.
Phillyrea L. 18, 510.
— angustifolia L. 512.
— latifolia L. 511.
— Medwedewii Sred. 511.
Phippsia R. Br. 2, 446.
— algida (Soland.) R. Br. 447; tab. 33.
— concinna (Fries) Lindeb. 447; tab. 33.
Phleboanthe Tausch 20, 17.
Phleum L. 2, 127.
— alpinum L. 135; tab. 9.
— arenarium L. 129; tab. 9.
— Boissieri Bornm. 129; tab. 9.
— echinatum Host 132; tab. 9.
— graecum Boiss. et Heldr. 130; tab. 9.
— Michelii All. 131; tab. 9.
— montanum C. Koch 131; tab. 9.
— paniculatum Huds. 130; tab. 9.
— phleoides (L.) Simonk. 131; tab. 9.
— pratense L. 132; tab. 9.
— tenue Schrad. 129; tab. 9.
Phlojodicarpus Turcz. 17, 49.
— sibiricus (Steph.) K.-Pol. 50; tab. 5, tab. 7.
— villosus Turcz. 51; tab. 7.
Phlomidopsis Link 21, 85.
Phlomidoschema (Benth.) Vved. 21, 242.
— parviflorum (Benth.) Vved. 243.
Phlomis L. 21, 57.
— agraria Bge. 103.
— alaica Knorr. 106; tab. 6.
— alpina Pall. 92.
— angrenica Knorr. 73.
— betonicifolia Rgl. 79.
— brachystegia Bge. 86; tab. 5.
— bucharica Rgl. 65.
— cancellata Bge. 83; tab. 4.
— canescens Rgl. 92; tab. 6.
— caucasica Rech. f. 65.

Phlomis cyclodon Knorr. 80.
— Drobovii M. Pop. 78; 646.
— dszumrutensis Afan. 95.
— ferganensis M. Pop. 90.
— fruticetorum Gontsch. 74.
— fruticosa L. 64.
— hybrida Zelen. 104.
— hypanica Schost. 102.
— hypoleuca Vved. 78, 646.
— Knorringiana M. Pop. 107, 648.
— kopetdagensis Knorr. 82.
— lenkoranica Knorr. 83.
— linearifolia Zak. 75.
— maeotica Schost. 101.
— majkopensis (Novopokr.) Grossh. 85.
— Maximoviczii Rgl. 105.
— nubilans Zak. 71.
— Olgae Rgl. 67; tab. 4.
— oreophila Kar. et Kir. 95; tab. 6.
— Ostrovskiana Rgl. 89; tab. 5.
— pratensis Kar. et Kir. 94.
— pseudopungens Knorr. 81, 647.
— puberula Kryl. et Serg. 101.
— pungens Willd. 80.
— Regelii M. Pop. 77.
— salicifolia Rgl. 76; tab. 4.
— scythica Klok. 104.
— Sewerzovii Rgl. 75.
— spinidens Nevski 68.
— taurica Hartwiss. 84.
— tenuis Knorr. 72.
— thapsoides Bge. 67.
— tomentosa Rgl. 66.
— tschimganica Vved. 90, 647.
— tuberosa L. 99.
— tytthaster Vved. 93, 648.
— urodonta M. Pop. 96.
— Vavilovii M. Pop. 86.
— Zenaidae Knorr. 73.
Phlomoides Moench 21, 85.
Phlox L. 19, 93.
— sibirica L. 93.
Phlyarodoxa S. Moore 18, 518.
Phoenix L. 3, 466.
— canariensis Hort. 467; tab. 25.
— silvestris Roxb. 467; tab. 25.
Phoenixopus Reichb. 29, 317.
Phoenopus DC. ex Nym. 29, 317.
Pholiurus Trin. 2, 588.
— glabriglumis Nevski 589.
— incurvus (L.) Schinz et Thell. 588; tab. 4.
— pannonicus (Host) Trin. 388; tab. 44.
— persicus (Boiss.) A. Camus 589; tab. 44.
Phragmites Adans. 2, 303.
— communis Trin. 304; tab. 23.
— isiaca (Delile) Kunth 305.
— serotina Kom. 305, 751.
Phrygia S. F. Gray 28, 370, 440, 449.
Phryma L. 23, 132.
— leptostachya L. 132.
Phryne Bub. 8, 55.
— Huetii (Boiss.) O. E. Schulz 56.
Phteirospermum Bge. 22, 555.
— chinense Bge. 555.
Phtheirotheca Maxim. ex Rgl. 7, 552.
Phuopsis (Griseb.) Hook. f. 23, 192.
— stylosa (Trin.) Hook. f. 192; tab. 10.
Phyllanthus L. 14, 287.
— ussuriensis Rupr. et Maxim. 287; tab. 15.
Phyllitis Ludwig 1, 60.
— scolopendrium (L.) Newm. 60.
Phyllocara Gusul. 19, 314.
— Aucheri (DC.) Gusul. 315.
Phyllodoce Link 11, 1.
Phyllodoce Salisb. 18, 63.
— aleutica (Spreng.) Heller 64.
— coerulea (L.) Bab. 64.
Phyllodolon Salisb. 4, 195.
Phylloschoenus Fourr. 3, 534.
Phyllospadix Hook. 1, 268.
— Scouleri Hook. 268.
Phyllostachys Sieb. et Zucc. 2, 732.
— aurea Carr. 736.
— bambusoides Sieb. et Zucc. 733.
— Marliacea Mitf. 734.
— mitis A. et C. Rivière 735.
— nigra Munro 735.
— puberula (Miq.) Makino 734.
— pubescens Houz. 735.
— sulphurea A. et C. Rivière 736.
Phymatis E. Meyer 16, 378.
Physaliastrum Makino 22, 60.
— echinatum (Yatabe) Makino 61; tab. 3.
Physalidium Fenzl 8, 200.
— graellsiifolium Lipsky 201.
Physalis L. 22, 62.
— alkekengi L. 64.
— glabripes Pojark. 65.
— ixocarpa Brot. et Hornem. 68.
— peruviana L. 70.
— praetermissa Pojark. 67; tab. 3.
— pubescens L. 69.
Physalodes Boehm. 22, 115.
Physcium Lour. 1, 295.
Physiglochis Neck. 3, 163.
Physocarpon Neck. 6, 701.
Physocarpus Maxim. 9, 282.

— amurensis Maxim. 282; tab. 17.
— opulifolia (L.) Maxim. 283.
— ribesifolia Kom. 283.
Physocaulis (DC.) Tausch 16, 93.
— nodosus (L.) Tausch 93.
Physochlaena Miers 22, 103.
Physochlaina G. Don 22, 103.
— orientalis (M. B.) G. Don 104.
— physaloides (L.) G. Don 104; tab. 4.
— Semenowii Rgl. 105.
Physolophium Turcz. 17, 32.
Physophora Link 16, 223.
Physoptychis Boiss. 8, 363.
— gnaphalodes (DC.) Boiss. 363.
Physopyrum M. Pop. 5, 501.
Physospermum Cuss. 16, 223.
Phyteuma L. 24, 388.
— nigrum Fr. Schmidt 392.
— orbiculare L. 393.
- spicatum L. 390; tab. 24.
— tetramerum Schur 394.
- Vagneri Kerner 391.
Phytolacca L. 6, 373.
— americana L. 373; tab. 19.
Picea Dietr. 1, 142.
— canadensis (Mill.) Britt. 149.
— Engelmannii Engelm. 149.
— excelsa Link 144.
— fennica Rgl. 145.
— Glehnii Mast. 148.
— jezoensis (Sieb. et Zucc.) Carr. 150; tab. 7, 8.
— kamtchatkensis Lacas. 153.
— koraiensis Nakai 146; tab. 7.
— obovata Ledeb. 145.
— orientalis (L.) Link. 147; tab. 7.
— pungens Engelm. 150.
— Schrenkiana Fisch. et Mey. 147; tab. 7, tab. 8.
— tianschanica Rupr. 147.
Picnomon Cass. 28, 218.
— acarna (L.) Cass. 218.
Picraena Stev. 12, 549.
Picridium Desf. 29, 260.
Picris L. 29, 219.
— canescens (Stev.) V. Vassil. 227.
— hieracioides L. 221; tab. 14.
— japonica Thunb. 221; tab. 14.
— kamtschatica Ledeb. 222; tab. 14.
— pauciflora Willd. 224; tab. 14.
— rigida Ledeb. 222; tab. 14.
— similis V. Vassil. 223; tab. 14.
— strigosa M. B. 226; tab. 14.
Pilea Lindl. 5, 396.
— peploides Hook. et Arn. 396.
— viridissima Makino 397.
Pilopleura Schischk. 17, 293, 358.
— Kozo-Poljanskii Schischk. 293.
Pilosanthus Neck. 25, 23.
Pilostemon Iljin 27, 709.
— filifolia (C. Winkl.) Iljin 711.
— karategini (Lipsky) Iljin 710.
Pilularia L. 1, 90.
— globulifera L. 90; tab. 4.
Pimpinella L. 16, 422.
— affinis Ledeb. 441.
— anthriscoides Boiss. 432.
— armena Schischk. 444.
— aromatica M. B. 440; tab. 28.
— confusa Woronow 438.
— daghestanica Schischk. 437.
— dissecta Retz. 428.
— Grossheimii Schischk. 438.
— Idae Takht. 437.
— Korshinskyi Schischk. 444.
— lithophila Schischk. 439; tab. 27.
— Litvinovii Schischk. 440.
— major (L.) Huds. 431.
— nudicaulis Trautv. 430.
— peucedanifolia Fisch. 432.
— puberula (DC.) Boiss. 442; tab. 27.
— ramosa Schischk. 443.
— rhodantha Boiss. 430.
— saxifraga L. 427.
— taurica (Ledeb.) Steud. 441.
— Thellungiana Wolff 429.
— titanophila Woronow 436.
— tomiophylla (Woronow) Stank. 436.
— turcomanica Schischk. 439.
Pinardia auct. 26, 131.
Pinguicula L. 23, 118.
— alpina L. 119.
— bicolor Woloszcz. 120.
— glandulosa Trautv. et Mey. 122.
— spathulata Ledeb. 122.
— variegata Turcz. 119.
— villosa L. 120.
— vulgaris L. 121.
Pinus L. 1, 159.
— armena C. Koch 171.
— eldarica Medw. 172.
— funebris Kom. 167.
— hamata Sosn. 170; tab. 7.
— Kochiana Klotzsch 170.
— koraiensis Sieb. et Zucc. 162; tab. 7, 8.
— Pallasiana Lamb. 166.
— pinea L. 165.
— pithyusa Strangw. 171; tab. 7.

Pinus pumila (Pall.) Rgl. 164; tab. 7.
— sibirica (Rupr.) Mayr 163; tab. 7.
— silvestris L. 167; tab. 7, 8.
— Stankeviczii (Sukacz.) Fomin 172.
— strobus L. 165.
Piptanthus D. Don 11, 36.
— nanus M. Pop. 36.
Piptoceras Cass. 28, 370, 560.
Piptoclaina G. Don 19, 153.
Piptoptera Bge. 6, 339.
— turkestana Bge. 340; tab. 14.
Pistacia L. 14, 519.
— mutica Fisch. et Mey. 524; tab. 28.
— vera L. 520; tab. 28.
Pisum L. 13, 520.
— arvense L. 524.
— Aucheri Jaub. et Spach 523.
— elatius M. B. 524.
— formosum (Stev.) Boiss. 522.
— humile Boiss. et Noe 523.
— sativum L. 525.
Pittosporum Dryand ex Gaertn. 9, 270.
— tobira Dryand 271.
Plagiobasis auct. 28, 250.
Plagiobasis Schrenk 28, 255.
— centauroides Schrenk 256; tab. 13.
Plagiorhegma Maxim. 7, 540.
Plagiospermum Oliv. 10, 580.
Plantago L. 23, 134.
— arachnoidea Schrenk 146; tab. 7.
— atrata Hoppe 152.
— camtschatica Link 146.
— canescens Adam 152.
— coronopus L. 142.
— Cornuti Gouan 141.
— depressa Willd. 145.
— gentianoides Sibth. et Smith 155.
— indica L. 161.
— Komarovii Pavl. 147.
— lachnantha Bge. 160.
— lagocephala Bge. 160; tab. 7.
— lanceolata L. 156.
— Loeflingii L. 159.
— macrocarpa Cham. et Schlecht. 142.
— major L. 138.
— maritima L. 143.
— maxima Juss. ex Jacq. 151.
— media L. 148; tab. 7.
— minuta Pall. 158.
— notata Lag. 159.
— ovata Forsk. 157.
— polysperma Kar. et Kir. 140.
— psyllium L. 162.
— Schwarzenbergiana Schur 145.
— stepposa Kupr. 148; tab. 8.
— tenuiflora Waldst. et Kit. 139.
Platanthera Rich. 4, 652.
— bifolia (L.) Rich. 656; tab. 40.
— chlorantha Cust. ex Reichb. 658; tab. 40.
— cornu bovis Nevski 662, 752; tab. 40.
— extremiorientalis Nevski 657, 753.
— Freynii Kränzl. 659; tab. 40.
— ophrydioides Fr. Schmidt 661; tab. 40.
— sachalinensis Fr. Schmidt 660; tab. 40.
— tipuloides (L. f.) Lindl. 660.
Platanus L. 9, 274.
— cuneata Willd. 275.
— digitata Gord. 276; tab. 16.
— occidentalis L. 279.
— orientalior Dode 276; tab. 16.
— orientalis L. 275.
Platychaeta Boiss. 25, 487.
Platycodon A. DC. 24, 438.
— grandiflorus (Jacq.) A. DC. 438; tab. 26.
Platylophus Cass. 28, 370, 440, 449.
Platyraphium Cass. 28, 219.
Platyspermum Hoffm. 17, 288.
Platytaenia Nevski et Vved. 17, 268.
— bucharica (B. Fedtsch.) Schischk. 271; 356.
— depauperata Schischk. 270, 356.
— heterodonta Korov. 271.
— Komarovii (Manden.) Schischk. 272; 357.
— pamirica (Lipsky) Nevski et Vved. 270; tab. 23.
— pimpinelloides Nevski 269.
— Rubtzovii Schischk. 273, 357; tab. 23.
Plectranthus L'Hér. 21, 638.
— excisus Maxim. 639.
— glaucocalyx Maxim. 639.
— serra Maxim. 640.
Pleuropogon R. Br. 2, 353.
— Sabinii R. Br. 353; tab. 27.
Pleuropteropyrum H. Gross 5, 661.
Pleurosoriopsis Fomin 1, 70.
— Makinoi (Christ.) Fomin 70; tab. 3.
Pleurospermum Hoffm. 16, 229.
— austriacum (L.) Hoffm. 229.
— uralense Hoffm. 231.
— camtschaticum Hoffm. 232.
Plumaria Opiz 6, 848.
Plumbagella Spach 18, 297.
— micrantha (Ledeb.) Spach 297; tab. 16.
Plumbago L. 18, 295.
— europaea L. 296; tab. 16.
Pneumaria Hill 19, 238.
Poa L. 2, 366.
— abbreviata R. Br. 412; tab. 31.

— acuminata Ovcz. 416.
— Alberti Rgl. 412; tab. 31.
— alpigena (Fries) Lindm. 390; tab. 29.
— alpina L. 411; tab. 31.
— altaica Trin. 399; tab. 30.
— angustifolia L. 388.
— angustiglumis Roshev. 395; tab. 29.
— annua L. 379; tab. 28.
— araratica Trautv. 402; tab. 30.
— arctica R. Br. 410; tab. 30.
— argunensis Roshev. 404; tab. 30.
— attenuata Trin. 403; tab. 30.
— bactriana Roshev. 376; tab. 28.
— bakuensis Litw. 383, 753.
— bedeliensis Litw. 415, 755; tab. 31.
— botryoides Trin. 403; tab. 30.
— bracteosa Kom. 422; tab. 31.
— bucharica Roshev. 384; tab. 28.
— bulbosa L. 376; tab. 28.
— calliopsis Litv. 414, 755; tab. 31.
— caucasica Trin. 412; tab. 31.
— Chaixii Vill. 385; tab. 28.
— compressa L. 408; tab. 30.
— dahurica Trin. 404.
— densa Troitzky 378; tab. 28.
— densissima Roshev. 418; tab. 31.
— disiecta Ovcz. 415.
— dschungarica Roshev. 415; tab. 31.
— dshilgensis Roshev. 377; tab. 28.
— eminens C. Presl 425; tab. 32.
— Fedtschenkoi Roshev. 421; tab. 31.
— Ganeschinii Roshev. 398; tab. 29.
— glabriflora Roshev. 376; tab. 28.
— glauca Vahl 398; tab. 29.
— glauciculmis Ovcz. 402.
— Gorbunovii Ovcz. 417.
— hissarica Roshev. 416; tab. 31.
— hybrida Gaud. 386; tab. 28.
— iberica Fisch. et Mey. 383; tab. 28.
— insignis Litv. 384, 753.
— ircutica Roshev. 387; tab. 28.
— irrigata Lindm. 390; tab. 29.
— karateginensis Roshev. 416.
— Komarovii Roshev. 422; tab. 31.
— lanatiflora Roshev. 409; tab. 30.
— laudanensis Roshev. 418; tab. 31.
— Lipskyi Roshev. 421; tab. 31.
— Litwinowiana Ovcz. 417.
— longifolia Trin. 384.
— macrocalyx Trautv. et Mey. 424; tab. 32.
— malacantha Kom. 422; tab. 31.
— marginata Ovcz. 417.
— Mariae Reverd. 396.
— masenderana Freyn et Sint. 400; tab. 30.
— Maydelii Roshev. 396; tab. 29.
— Meyeri Trin. 383, 753.
— naltchikensis Roshev. 413; tab. 31.
— nemoralis L. 400; tab. 300.
— Nevskii Roshev. 377; tab. 28.
— nivicola Kom. 379; tab. 28.
— ochotensis Trin. 395; tab. 29.
— palustris L. 397; tab. 29.
— pamirica (Litv.) Roshev. 414, 755; tab. 31.
— penicillata Kom. 396; tab. 29.
— petraea Trin. 424.
— petschorica Roshev. 410; tab. 30.
— pinegensis Roshev. 393; tab. 29.
— platyantha Kom. 423; tab. 31.
— podolica Blocki 399; tab. 29.
— pratensis L. 388; tab. 29.
— pruinosa Korotki 394.
— pseudoabbreviata Roshev. 413; tab. 31.
— radula Franch. et Savat. 397; tab. 29.
— relaxa Ovcz. 402.
— remota Forsel. 385; tab. 28.
— Reverdattoi Roshev. 407; tab. 30.
— rhomboidea Roshev. 409, 754; tab. 30.
— sabulosa Turcz. 394.
— sajanensis Roshev. 393.
— setacea Hoffm. 389; tab. 29.
— sibirica Roshev. 380; tab. 28.
— silvicola Guss. 387; tab. 28.
— sinaica Steud. 378; tab. 28.
— Smirnovii Roshev. 424; tab. 32.
— Soczawai Roshev. 410.
— sphondylodes Trin. ex Bge. 407; tab. 30.
— stepposa (Kryl.) Roshev. 401, 754; tab. 30.
— sterilis M. B. 408; tab. 30.
— strigosa Hoffm. 389; tab. 29.
— subfastigata Trin. 425; tab. 32.
— subglabriflora Roshev. 394.
— supina Schrad. 379.
— taimyrensis Roshev. 409.
— Tanfiljewii Roshev. 413; tab. 31.
— tibetica Munro ex Duthie 425; tab. 32.
— Tolmatchewii Roshev. 411; tab. 30.
— transbaicalica Roshev. 404; tab. 30.
— tremuloides Litv. 421, 756.
— tristis Trin. ex Bge. 400; tab. 30.
— trivialiformis Kom. 423.
— trivialis L. 386; tab. 28.
— turfosa Litv. 389.
— udensis Trautv. 414; tab. 31.
— urjanchaica Roshev. 395; tab. 29.
— ussuriensis Roshev. 394, 754; tab. 29.
— versicolor Bess. 399; tab. 30.

Poa Woronowii Roshev. 387; tab. 28.
— Zaprjagajevii Ovcz. 377; tab. 28.
Poacynum Baill. 18, 660.
— Hendersonii (Hook. f.) Woodson 661; tab. 35.
— pictum (Schrenk) Baill. 660; tab. 35.
Pocockia Ser. 11, 120.
Podagraria Hill 16, 451.
Podanthum Boiss. 24, 396.
Podospermum DC. 29, 37.
Pogonia Juss. 4, 610.
— japonica Reichb. f. 611; tab. 38.
Pogonostylis Bertol. 3, 95.
Polakia Stapf 21, 272.
Polemonium L. 19, 78.
— acutiflorum Willd. ex Roem. et Schult. 81; tab. 5.
— boreale Adam 88.
— caucasicum N. Busch 82.
— coeruleum L. 83.
— humile Willd. ex Roem. et Schult. 87.
— hyperboreum Tolm. 91; tab. 5.
— liniflorum V. Vassil. 84.
— majus Tolm. 91.
— pacificum V. Vassil. 81; tab. 5.
— parviflorum Tolm. 92; tab. 5.
— pseudopulchellum V. Vassil. 86; tab. 5.
— pulchellum Bge. 85; tab. 5.
— sibiricum D. Don 83.
— villosum J. Rudolph ex Georgi 80.
Polgidon Raf. 16, 94.
Pollichia Medic. 19, 690.
Pollichia Schrank 21, 124, 126.
Pollinia Trin. 2, 10.
— imberbis Nees ex Steud. 10.
— nuda Trin. 11; tab. 1.
Polycarpon Loefl. 6, 564.
— tetraphyllum L. 564; tab. 34.
Polycnemum L. 6, 29.
— arvense L. 29.
— Heuffelii O. Lang 30; tab. 2.
— majus A. Br. 31.
— perenne Litv. 31.
— verrucosum O. Lang 30.
Polycyrtus Schlecht. 17, 62.
Polygala L. 14, 246.
— alpicola Rupr. 258.
— amarella Crantz 257.
— amoenissima Tamamsch. 262.
— anatolica Boiss. et Heldr. 265.
— andrachnoides Willd. 255.
— austriaca Crantz 256.
— caucasica Rupr. 259.
— colchica Tamamsch. 263, 733.
— comosa Schkuhr 261.
— cretacea Kotov 262.
— Hohenackeriana Fisch. et Mey. 253.
— hybrida DC. 260.
— japonica Houtt. 252.
— Kemulariae Tamamsch. 265, 733.
— leucothyrsa Woronow 264.
— major Jacq. 262.
— Mariamae Tamamsch. 263.
— moldavica Kotov 261.
— papilionacea Boiss. 253.
— pruinosa Boiss. 256.
— pseudohospita Tamamsch. 255.
— sibirica L. 250.
— Stocksiana Boiss. 254.
— suanica Tamamsch. 258.
— subamara Fritsch 257.
— supina Schreb. 254.
— tenuifolia Willd. 251.
— transcaucasica Tamamsch. 265.
— Urartu Tamamsch. 264.
— vulgaris L. 259.
Polygonatum Adans. 4, 456.
— acuminatifolium Kom. 465.
— Desoulavyi Kom. 463.
— glaberrimum C. Koch 465.
— humile Fisch. ex Maxim. 463; tab. 28.
— involucratum Maxim. 460; tab. 28.
— latifolium Desf. 464; tab. 28.
— Maximoviczii Fr. Schmidt 464.
— multiflorum (L.) All. 466; tab. 28.
— obtusifolium Miscz. 466.
— officinale All. 463; tab. 28.
— ovatum Micz. 465.
— polyanthemum (M. B.) Dietr. 466.
— roseum (Ledeb.) Kunth 458.
— Sewerzovii Rgl. 460; tab. 28.
— sibiricum Redoute 459; tab. 28.
— stenophyllum Maxim. 458; tab. 28.
— verticillatum (L.) All. 459.
Polygonum L. 5, 594.
— abbreviatum Kom. 679, 724; tab. 47.
— acerosum Ledeb. ex Meisn. 625; tab. 42.
— acetosellum Klok. 622.
— acetosum M. B. 622.
— ajanense (Nakai) Grig. 666; tab. 46.
— alatum A. Hamilt. ex D. Don 640.
— alopecuroides Turcz. 680; tab. 47.
— alpestre C. A. Mey. 609; tab. 39.
— alpinum All. 663.
— ammanioides Jaub. et Spach 608; tab. 39.
— amphibium L. 645; tab. 45.
— angustifolium Pall. 665; tab. 46.
— araraticum Kom. 621, 720.

— arenarium Waldst. et Kit. 637; tab. 44.
— argyrocoleum Steud. 634; tab. 44.
— arianum Grig. 602; tab. 38.
— Aschersonianum H. Gross 629; tab. 42.
— attenuatum V. Petrov 684, 725.
— aviculare L. 614; tab. 41.
— baldshuanicum Rgl. 699; tab. 49.
— belophyllum Litv. 690; tab. 48.
— biaristatum Aitch. et Hemsl. 607; tab. 39.
— bistorta L. 681; tab. 47.
— Bornmülleri Litv. 639.
— bucharicum Grig. 664; tab. 46.
— Bungeanum Turcz. 647.
— buxifolium Nutt. ex Bong. 613; tab. 40.
— calcatum Lindm. 618; tab. 41.
— carneum C. Koch 682.
— caspicum Kom. 623, 721.
— convolvulus L. 694.
— coriarium Grig. 663.
— corrigioloides Jaub. et Spach 624; tab. 42.
— cretaceum Kom. 631, 721.
— cuspidatum Sieb. et Zucc. 701.
— deciduum Boiss. et Noë 637.
— dentato-alatum Fr. Schmidt 696; tab. 49.
— dissitiflorum Hemsl. 693.
— divaricatum L. 664.
— dumetorum L. 695; tab. 49.
— ellipticum Willd. ex Spreng. 677; tab. 47.
— equisetiforme Sibth. et Smith 610; tab. 40.
— excurrens Steward 657.
— fibrilliferum Kom. 607, 718; tab. 39.
— floribundum Schlecht. ex Meisn. 624.
— foliosum Lindb. f. 655; tab. 45.
— fusco-ochreatum Kom. 613, 719; tab. 40.
— gracilius Ledeb. 630; tab. 43.
— graminifolium Wierzb. 621.
— heterophyllum Lindm. 614; tab. 41.
— hissaricum M. Pop. 669.
— humifusum Pall. ex Ledeb. 618; tab. 41.
— hydropiper L. 656; tab. 45.
— imeretinum Kom. 652, 722.
— inflexum Kom. 632, 721; tab. 43.
— intercedens V. Petrov 678, 723; tab. 45.
— intricatum Kom. 657, 723; tab. 45.
— Janatae Klok. 638.
— junceum Ledeb. 634.
— Kitaibelianum Sadl. 629.
— Laxmannii Lepech. 665.
— lencoranicum Kom. 613, 720.
— limosum Kom. 670; tab. 46.
— linicola Sutul. 650; tab. 45.
— litorale Meisn. 612; tab. 40.
— luxurianus Grig. 664, 723; tab. 46.
— luzuloides Jaub. et Spach 603.
— Maackianum Rgl. 692; tab. 48.
— manshuriense V. Petrov 680.
— maritimum L. 610; tab. 40.
— minus Huds. 652; tab. 45.
— mite Schrank 656.
— molliiforme Boiss. 639.
— multiflorum Thunb. 699.
— myrtillifolium Kom. 608, 718; tab. 39.
— neglectum Bess. 618; tab. 41.
— nitens (Fisch. et Mey.) V. Petrov 683, 724.
— nodosum Pers. 649; tab. 45.
— novo-ascanicum Klok. 631.
— ochotense V. Petrov 687, 726; tab. 47.
— ochreatum L. 666.
— orientale L. 648.
— oxanum Kom. 633, 722; tab. 43.
— oxyspermum Mey. et Bge. 611.
— pacificum V. Petrov 682; tab. 47.
— paludosum Kom. 689, 726; tab. 48.
— pamiricum Korsh. 671; tab. 46.
— pamiroalaicum Kom. 609, 719; tab. 39.
— paronychioides C. A. Mey. ex Hohen 604; tab. 38.
— patulum M. B. 629; tab. 44.
— pauciflorum Maxim. 694.
— perfoliatum L. 688; tab. 49.
— persicaria L. 651; tab. 45.
— polycnemoides Jaub. et Spach 624; tab. 42.
— posumbu A. Hamilt. 658.
— propinquum Ledeb. 621; tab. 41.
— pseudoarenarium Klok. 638; tab. 44.
— pulvinatum Kom. 607, 717.
— Rayi Bab. 611; tab. 40.
— Regelianum Kom. 683.
— relictum Kom. 670.
— Roberti Loisel. 612.
— runcinatum A. Hamilt. 640.
— rupestre Kar. et Kir. 609; tab. 39.
— sachalinense Fr. Schmidt 701.
— salsugineum M. B. 625; tab. 42.
— samarense H. Gross 626.
— scabrum Moench 648.
— schugnanicum Kom. 684, 725.
— senticosum Meisn. 689.
— sericeum Pall. 669.
— setosum Jacq. 603; tab. 38.
— sibiricum Laxm. 671.
— Sieboldii Meisn. 689; tab. 48.
— songoricum Schrenk 669; tab. 46.
— strigosum R. Bg. 691; tab. 48.
— subauriculatum V. Petrov 678.
— Thunbergii Sieb. et Zucc. 692; tab. 48.

Polygonum thymifolium Jaub. et Spach 603; tab. 38.
— tiflisiense Kom. 633, 720.
— tinctorium Ait. 646.
— tripterocarpum A. Gray 672; tab. 46.
— ussuriense (Rgl.) Nakai ex Mori 690; tab. 48.
— venosum Steward 623.
— viscoferum Makino 661.
— viscosum A. Hamilt. ex D. Don 647.
— viviparum L. 674; tab. 47.
— Weyrichii Fr. Schmidt 700; tab. 49.
— zeaense Kom. 679, 724.
Polylophium Boiss. 17, 278.
— Panjutinii Manden. et Schischk. 279.
Polypodium L. 1, 84.
— lineare Thunb. 86.
— serratum (Willd.) Futo 85.
— virginianum L. 85; tab. 3.
— vulgare L. 85.
Polypogon Desf. 2, 163.
— demissus Steud. 164; tab. 12.
— maritimus Willd. 165; tab. 12.
— monspeliensis (L.) Desf. 164; tab. 12.
Polyspermum N. Busch 8, 549.
Polystichum Roth 1, 45.
— angulare (C. Presl) Fomin 47.
— Braunii Fée 48.
— craspedosorum (Maxim.) Diels 47; tab. 1.
— lobatum (Sw.) C. Presl 48.
— lonchitis (L.) Roth 46.
— microchlamys Kodama 49.
— tripteron (Kuntze) C. Presl 46.
— Woronowii Fomin 47.
Polytaxis Bge. 27, 535.
— Lehmannii Bge. 536.
— Winkleri Iljin 537.
Poncirus Raf. 14, 234.
— trifoliata (L.) Raf. 234; tab. 14.
Pongatium Juss. 24, 449.
Pongelion Adans. 14, 241.
Popoviocodonia Fed. 24, 372, 470.
— stenocarpa (Trautv. et Mey.) Fed. 374.
— Uyemurae (Kudo) Fed. 373; tab. 22.
Populus L. 5, 216.
— alba L. 225.
— amurensis Kom. 240; tab. 10.
— ariana Dode 222.
— baikalensis Kom. 240.
— balsamifera L. 241.
— berolinensis Dipp. 234.
— Bolleana Lauche 224.
— candicans Ait. 242.
— canescens Smith 226.
— cataracti Kom. 235.
— Davidiana Dode 227.
— deltoides Marsh. 234.
— densa Kom. 237.
— diversifolia Schrenk 221.
— euphratica Oliv. 222; tab. 10.
— kanjilaliana Dode 233.
— koreana Rehd. 239.
— Krauseana Dode 234.
— laurifolia Ledeb. 236; tab. 10.
— Litwinowiana Dode 222.
— Maximowiczii A. Henry 238.
— moskoviensis Schroed. 241.
— nigra L. 228; tab. 10.
— nivea Willd. 224.
— pamirica Kom. 236.
— pilosa Rehd. 240.
— pruinosa Schrenk 223; tab. 10.
— pyramidalis Rozier 230.
— Sieboldii Miq. 228.
— Simonii Carr. 241.
— suaveolens Fisch. 238; tab. 10.
— tadshikistanica Kom. 235.
— talassica Kom. 237.
— tremula L. 226; tab. 10.
— usbekistanica Kom. 234.
— ussuriensis Kom. 239; tab. 10.
Porphyroscias Miq. 17, 11, 26.
Portulaca L. 6, 386.
— oleracea L. 386.
Potamogeton L. 1, 230.
— acutifolius Link 242; tab. 12.
— alpinus Balb. 252; tab. 12.
— amblyophyllus C. A. Mey. 237.
— asiaticus A. Benn. 251.
— crispus L. 240; tab. 12.
— cristatus Rgl. et Maack 248; tab. 12.
— digynus Wall. 253.
— distinctus A. Benn. 259.
— filiformis Pers. 236; tab. 12.
— Franchetii Baagoe ex A. Benn. 253.
— Friesii Rupr. 245; tab. 12.
— Fryeri A. Benn. 254.
— Henningii A. Benn. 241.
— heterophyllus Schreb. 256; tab. 12.
— javanicus Hassk. 251; tab. 12.
— limosellifolius Maxim. ex Korsh. 246; tab. 12.
— lucens L. 257; tab. 12.
— Maackianus A. Benn. 240; tab. 12.
— malainus Miq. 258; tab. 12.
— manschuriensis A. Benn. 243.
— miduhikimo Makino 252.
— natans L. 255; tab. 12.

— nodosus Poir. 254; tab. 12.
— obtusifolius Mert. et Koch 246; tab. 12.
— oxyphyllus Miq. 243; tab. 12.
— pamiricus Baagoe 237.
— panormitanus Biv.-Bern. 246; tab. 12.
— pectinatus L. 239; tab. 12.
— perfoliatus L. 260; tab. 12.
— proelongus Wulf. 259; tab. 12.
— pusillus L. 247; tab. 12.
— rutilus Wolfg. 245; tab. 12.
— sibiricus A. Benn. 242.
— subretusus Hagstr. 238.
— subsibiricus Hagstr. 243.
— tenuifolius Raf. 253; tab. 12.
— trichoides Cham. et Schlecht. 244.
— vaginatus Turcz. 238.
— zosterifolius Schum. 241; tab. 12.
Potentilla L. 10, 78.
— acaulis L. 210; tab. 15.
— adenophylla Boiss. et Hohen. 208, tab. 15.
— adscharica Somm. et Levier 188.
— aemulans Juz. 214, 614.
— agrimonioides M. B. 132.
— alba L. 91; tab. 8.
— Alexeenkoi Lipsky 122.
— amurensis Maxim. 166.
— anadyrensis Juz. 191, 612.
— anserina L. 221; tab. 16.
— approximata Bge. 125.
— arenaria Brokh. 208.
— arenosa (Turcz.) Juz. 137; tab. 10.
— argentea L. 145; tab. 10.
— argenteiformis Kauffm. 151; tab. 11.
— arnavatensis Th. Wolf 174.
— asiatica Juz. 182.
— asperrima Turcz. 170.
— astracanica Jacq. 163; tab. 12.
— astragalifolia Bge. 159; tab. 11.
— baltistana Th. Wolf 124.
— biflora Willd. 84; tab. 7.
— bifurca L. 81; tab. 7.
— brachypetala Fisch. 90; tab. 8.
— Bungei Boiss. 174.
— Callieri Juz. 164; tab. 12.
— Camillae Kolak. 92.
— canescens Bess. 147; tab. 10.
— caucasica Juz. 185, 611.
— centugrana Maxim. 168; tab. 13.
— chinensis Ser. 127.
— chrysantha Trev. 181; tab. 14.
— conferta Bge. 131; tab. 9.
— Crantzii (Crantz) Beck. 202; tab. 15.
— cryptophila Bornm. 179.
— cryptotaeniae Maxim. 168; tab. 13.
— dealbata Bge. 142.
— depressa Willd. ex Schlecht. 208, tab. 15.
— desertorum Bge. 173.
— discolor Bge. 123.
— divina Albov 89; tab. 8.
— Egedii Wormsk. 223.
— elatior Willd. ex Schlecht. 86.
— elegans Cham. et Schlecht. 198.
— emarginata Pursh 198.
— erecta (L.) Hampe 218; tab. 16.
— eremica Th. Wolf 148.
— Eversmanniana Fisch. ex Claus 119.
— evestita Th. Wolf 140.
— Fedtschenkoana Siegfr. ex Th. Wolf 126.
— filipendula Willd. ex Schlecht. 153; tab. 11.
— flabellata Rgl. et Schmalh. 176.
— flagellaris Willd. ex Schlecht. 220.
— foliosa Somm. et Levier 94.
— fragarioides L. 211; tab. 16.
— fragiformis Willd. 193; tab. 14.
— Freyniana Bornm. 217; tab. 16.
— gelida C. A. Mey. 199; tab. 15.
— geoides M. B. 97.
— ghalghana Juz. 89.
— glaucescens Willd. ex Schlecht. 209; tab. 15.
— Goldbachii Rupr. 182; tab. 14.
— Heidenreichii Zimm. 170.
— heptaphylla L. 204; tab. 15.
— hololeuca Boiss. 126.
— holopetala Turcz. 181.
— humifusa Willd. ex Schlecht. 207; tab. 15.
— imbricata Kar. et Kir. 84; tab. 7.
— impolita Wahl. 146.
— inquinans Turcz. 95; tab. 9.
— intermedia L. 169; tab. 13.
— jacutica Juz. 141, 611.
— jailae Juz. 96, 609.
— Komaroviana Th. Wolf 177.
— Kryloviana Th. Wolf 155.
— kulabensis Th. Wolf 98.
— Kuznetzowii (Gowor.) Juz. 137.
— lapponica (Nyl.) Juz. 117.
— leucophylla Pall. 134, tab. 10.
— leucopolitana P. J. Muell. ex F. Schultz 151.
— Lipskyana Th. Wolf 189.
— Lomakinii Grossh. 133.
— longipes Ledeb. 187; tab. 14.
— macrantha Ledeb. 141.

Potentilla malacotricha Juz. 121, 610.
— megalantha Takeda 194.
— Meyeri Boiss. 147.
— micrantha Ramon 92; tab. 8.
— mollissima Lehm. 149.
— Moorcroftii Wall. 83.
— multifida L. 114; tab. 9.
— nervosa Juz. 135, 610; tab. 10.
— nivea L. 135.
— Nordmanniana Ledeb. 180.
— norvegica L. 167; tab. 13.
— nudicaulis Willd. ex Schlecht. 153.
— nurensis Boiss. et Hausskn. 162.
— orbiculata Th. Wolf 186.
— orientalis Juz. 82.
— pacifica Howell 222; tab. 16.
— pamirica Th. Wolf 177.
— pamiroalaica Juz. 121, 609; tab. 9.
— pimpinelloides L. 156.
— polyschista Boiss. 123.
— porphyrantha Juz. 178.
— procumbens L. 224; tab. 17.
— pulchella R. Br. 113.
— Raddeana Juz. 175.
— recta L. 160; tab. 12.
— Regeliana Th. Wolf 173.
— reptans L. 219; tab. 16.
— rigidula Th. Wolf 159.
— rupestris L. 94.
— Ruprechtii Boiss 178.
— sanguisorba Willd. ex Schlecht. 155.
— Schrenkiana Rgl. 188.
— Schurii Fuss ex Zimm. 203.
— Seidlitziana Bien. 201.
— semiglabrata Juz. 83.
— semilaciniosa Borb. 162; tab. 12.
— sericata Th. Wolf 190.
— sericea L. 120; tab. 9.
— Sommieri Siegfr. et Keller 150.
— soongorica Bge. 118.
— sphenophylla Th. Wolf 186.
— Sprengeliana Lehm. 212.
— stipularis L. 192; tab. 14.
— stolonifera Lehm. 215.
— strigosa Pall. ex Pursh 128; tab. 9.
— subpalmata Ledeb. 201.
— supina L. 165; tab. 13.
— svanetica Siegfr. et Keller 150.
— Szovitsii Th. Wolf 185.
— tanacetifolia Willd. ex Schlecht. 152.
— taurica Willd. 164; tab. 12.
— tephroleuca Th. Wolf 200.
— thyrsiflora Zimm. 150.
— tianschanica Th. Wolf 98.
— Tollii Trautv. 191.
— transcaspica Th. Wolf 162.
— Tranzschelii Juz. 213, 612.
— umbrosa Stev. 179; tab. 14.
— uniflora Ledeb. 138.
— ussuriensis Juz. 213, 613.
— Vahliana Lehm. 139.
— verticillaris Steph. ex Willd. 117; tab. 9.
— villosa Pall. 139.
— viscosa J. Don. 154; tab. 11.
— Wibeliana Th. Wolf 149.
Poterium L. 10, 429.
— lasiocarpum Boiss. et Hausskn. ex Boiss. 431.
— polygamum Waldst. et Kit. 430.
— sanguisorba L. 429; tab. 26.
Pragmotessara Pierre 14, 553.
Prangos Lindl. 16, 263.
— acaulis (DC.) Bornm. 265.
— arcis-romanae Boiss. et Huet 267.
— bucharica B. Fedtsch. 268.
— cylindrocarpa Korov. 271.
— Fedtschenkoi (Rgl. et Schmalh.) Korov. 272.
— ferulacea (L.) Lindl. 265.
— isphairamica B. Fedtsch. 267, 594.
— latiloba Korov. 275.
— Lipskyi Korov. 269.
— lophotera Boiss. 269.
— pabularia Lindl. 270.
— seravschanica (Rgl. et Schmalh.) Korov. 271.
— tschimganica B. Fedtsch. 266.
— uloptera DC. 268.
Prenanthes L. 29, 262.
— abietina (Boiss.) Kirp. 270; tab. 23.
— angustifolia Boulos 273; tab. 23.
— cacaliifolia (M. B.) Beauverd 266; tab. 23.
— Maximowiczii Kirp. 269; tab. 23.
— pontica (Boiss.) Leskov 268; tab. 23.
— purpurea L. 264; tab. 23.
— Tatarinowii Maxim. 265; tab. 23.
Primula Kuntze 18, 208.
Primula L. 18, 111.
— abchasica Sosn. 141.
— algida Adam 158; tab. 6.
— amoena M. B. 142; tab. 5.
— arctica Koidz. 199.
— auriculata Lam. 183.
— baldschuanica B. Fedtsch. 160; tab. 6.
— Bayerni Rupr. 190.

— borealis Duby **171**; tab. **7**.
— carpathica (Griseb. et Schenk) Fuss 151.
— cordifolia Rupr. 152.
— cortusoides L. 130; tab. 4.
— cuneifolia Ledeb. 188.
— darialica Rupr. 157; tab. 6.
— drosocalyx Poljak. et Lincz. 125.
— elatior (L.) Hill. 150.
— Eugeniae Fed. 124; tab. 4.
— eximia Greene 200.
— farinifolia Rupr. 157.
— farinosa L. 161; tab. 6.
— Fedtschenkoi Rgl. 189; tab. 8.
— finmarchica Jacq. 175.
— fistulosa Turkev. 171; tab. 7.
— flexuosa Turkev. 187; tab. 8.
— gigantea Jacq. 164.
— Halleri J. F. Gmel. 181; tab. 7.
— heterochroma Stapf 141.
— Iljinskii Fed. 176, 726; tab. 7.
— Juliae Kusn. 132; tab. 4.
— Kaufmanniana Rgl. 126; tab. 4.
— Knorringiana Fed. 177, 726; tab. 7.
— Komarovii Losinsk. 437.
— Kusnetzovii Fed. 144, 723.
— lactiflora Turkev. 125.
— longipes Freyn et Sint. 191.
— longiscapa L. 168; tab. 6.
— luteola Rupr. 184.
— macrocalyx Bge. 146; tab. 5.
— Matsumurae Petitm. 173; tab. 7.
— megaseifolia Boiss. et Bal. ex Boiss. 133; tab. 5.
— Meyeri Rupr. 143.
— minima L. 201; tab. 9.
— Minkwitziae W. W. Smith 129.
— Moorkroftiana Wall. 197; tab. 9.
— nivalis Pall. 192; tab. 9.
— Olgae Rgl. 185; tab. 8.
— ossetica Kusn. 156.
— Pallasii Lehm. 149; tab. 5.
— pamirica Fed. 178, 724; tab. 7.
— patens Turcz. 131; tab. 4.
— pinnata M. Pop. et Fed. 167, 725.
— poloninensis (Domin) Fed. 150, 724.
— pseudoelatior Kusn. 155.
— pulverea Fed. 198, 727.
— renifolia Volgun. 134; tab. 5.
— Ruprechtii Kusn. 148.
— sachalinensis Nakai 162; tab. 6.
— sibirica Jacq. 174; tab. 7.
— Sibthorpii Hoffmgg. 137; tab. 5.
— stricta Hornem. 169.
— Tournefortii Rupr. 186.
— tschuktschorum Kjellm. 198; tab. 9.
— turkestanica (Rgl.) E. A. White 196.
— veris L. 145.
— vulgaris Huds. 135.
— Warshenewskiana B. Fedtsch. 182; tab. 8.
— Woronowii Losinsk. 138.
— xanthobasis Fed. 195, 727; tab. 9.
Prinsepia Royle 10, 580.
— sinensis (Oliv.) Kom. 583; tab. 37.
Prionitis Adans. 16, 382.
Prismatanthus Hook. et Arn. 22, 795.
Prismatocarpus L'Hér. 24, 427.
Proboscidea Moench 23, 18.
— louisiana (Mill.) Woot. et Standl. 18.
Proselias Stev. 12, 594.
Proteinia Reichb. 6, 861.
Proteocarpus Boern. 3, 422.
Prunella L. 20, 494.
— grandiflora (L.) Jacq. 496.
— laciniata L. 497.
— vulgaris L. 495.
Prunus L. 10, 510.
— caspica Koval. et Ekim. tab. 32.
— divaricata Ledeb. 516; tab. 32.
— domestica L. 515.
— ferganica Lincz. 519.
— salicina Lindl. 520.
— Simonii Carr. 520.
— spinosa L. 511; tab. 32.
— ussuriensis Koval. et Kost. 521.
Psammogeton Edgew. 16, 164.
— Borsczovii (Rgl. et Schmalh.) Lipsky 166; tab. 12.
— canescens (DC.) Vatke 166.
— setifolium Boiss. 165.
Psammoseris Boiss. 29, 595.
Psathyrostachys Nevski 2, 712.
— daghestanica (Alex.) Nevski 715.
— fragilis (Boiss.) Nevski 716; tab. 50.
— juncea (Fisch.) Nevski 714; tab. 50.
— Kronenburgii (Hack.) Nevski 713.
— lanuginosa (Trin.) Nevski 714; tab. 50.
— rupestris (Alex.) Nevski 715.
Psedera Neck. 14, 707.
Psephellus Cass. 28, 370, 420.
Pseuderemostachys M. Pop. 20, 500.
— Sewertzowii (Herd.) M. Pop. 500.
Pseudo-acacia Medic. 11, 305.
Pseudobetckea (Hoeck) Lincz. 23, 681.
— caucasica (Boiss.) Lincz. 681; tab. 34.
Pseudocamelina N. Busch 8, 570.
— Szovitsii (Boiss.) N. Busch 570.

Pseudo-Chamomilla DC. 26, 181.
Pseudodictamnus Adans. 21, 187.
Pseudodiphryllum Nevski 4, 649, 752.
— Chorisianum (Cham.) Nevski 650; tab. 40.
Pseudohandelia Tzvel. 26, 361, 878.
— umbellifera (Boiss.) Tzvel. 363; tab. 15.
Pseudolinosyris Novopokr. 25, 180.
— Grimmii (Rgl. et Schmalh.) Novopokr. 181; tab. 11.
— microcephala (Novopokr.) Tamamsch. 182.
— Sintenisii (Bornm.) Tamamsch. 183.
Pseudolysimachion Opiz 22, 329, 367.
Pseudosedum (Boiss.) Berger 9, 99.
— bucharicum Boriss. 105.
— campanuliforum Boriss. 107.
— condensatum Boriss. 102.
— Fedtschenkoanum Boriss. 107.
— ferganense Boriss. 105; tab. 6.
— karatavicum Boriss. 108.
— Lievenii (Ledeb.) Berger 101.
— longidentatum Boriss. 102.
— multicaule (Boiss. et Buhse) Boriss. 106; tab. 6.
Pseudosempervivum (Boiss.) Grossh. 8, 571.
— karsianum (N. Busch) Grossh. 572.
Pseudospermum S. F. Gray 16, 223.
Pseudostellaria Pax 6, 423.
Pseudovesicaria (Boiss.) Rupr. 8, 170.
— digitata (C. A. Mey.) Rupr. 170.
Pseudovyenomus Nakai 14, 558.
Psilochaenia Nutt. 29, 595.
Psilonema C. A. Mey. 8, 358.
Psilostemon DC. 19, 291.
Psilurus Trin. 2, 552.
— aristatus (L.) Duv.-Youv. 553; tab. 44.
Psolanum Neck. 22, 42.
Psoralea L. 11, 300.
— acaulis Stev. 301; tab. 21.
— bituminosa L. 301; tab. 21.
— drupacea Bge. 300; tab. 21.
Psychrogeton Boiss. 25, 191.
Psylliostachys (Jaub. et Spach) Nevski 18, 467.
— anceps (Rgl.) Roshk. 473; tab. 23.
— Androssovii Roshk. 474, 749.
— leptostachya (Boiss.) Roshk. 468; tab. 23.
— myosuroides (Rgl.) Roshk. 470.
— spicata (Willd.) Nevski 471; tab. 23.
— Suvorovii (Rgl.) Roshk. 469; tab. 23.
Psyllium Juss. 23, 161.
Psyllophora Ehrh. 3, 198.
Ptarmica DC. 26, 104.
Ptelea L. 14, 233.
— trifoliata L. 14, 233; tab. 13.
Pteranthus Forsk. 6, 572.
— dichotomus Forsk. 572.
Pteridium Gled. 1, 83.
— aquilinum (L.) Kuhn 83.
Pteris L. 1, 80.
— cretica L. 83.
Pterocarya Kunth 5, 248.
— fraxinifolia (Lam.) Spach 248; tab. 11.
Pterocephalus Vaill. 24, 49.
— afghanicus (Aitch. et Hemsl.) Boiss. 49.
— khorossanicus Czerniak. 50.
— fruticulosus Korov. 50.
— plumosus (L.) Coult. 51.
Pterochaeta Boiss. 25, 487.
Pterococcus Pall. 5, 530.
Pterolobium Andrz. 8, 575.
Pteroneurum DC. 8, 169.
Pterotheca Cass. 29, 699, 708.
Pterygocalyx Maxim. 18, 536.
Ptilagrostis Griseb. 2, 74.
— concinna (Hook.)Roshev. 75.
— mongolica (Turcz.) Griseb. 75; tab. 6.
— purpurea (Griseb.) Roshev. 76; tab. 6.
— subsessiliflora (Rupr.) Roshev. 74; tab. 6.
Ptilotrichum C. A. Mey. 8, 360.
— canescens C. A. Mey. 360.
— elongatum C. A. Mey. 361.
Puccinellia Parl. 2, 460.
Pueraria DC. 13, 531.
— hirsuta (Thunb.) Schneid. 531.
Pulegium Mill. 21, 623.
Pulicaria Gaertn. 25, 487.
— dysenterica (L.) Gaertn. 490; tab. 30.
— gnaphalodes (Vent.) Boiss. 496; tab. 31.
— prostrata (Gilib.) Aschers. 489; tab. 30.
— salviifolia Bge. 495; tab. 31.
— uliginosa Stev. 491; tab. 30.
Pulmonaria L. 19, 344.
— angustifolia L. 350.
— Filarszkyana Jáv. 349.
— mollissima A. Kerner 347.
— officinalis L. 345.
— rubra Schott 348.
Pulsatilla Adans. 7, 285.
— ajanensis Rgl. 297; tab. 18.
— albana (Stev.) Bercht. et Presl 304.
— ambigua (Turcz.) Juz. 307; tab. 19.
— andina (Rupr.) Woronow 305.

— armena (Boiss.) Rupr. 306.
— aurea (N. Busch) Juz. 287.
— Bungeana C. A. Mey. 298; tab. 19.
— campanella Fisch. ex Rgl. 306, tab. 19.
— cernua (Thunb.) Bercht. 303; tab. 19.
— chinensis (Bge.) Rgl. 290; tab. 18.
— dahurica (Fisch.) Spreng. 303; tab. 19.
— flavescens (Zucc.) Juz. 296.
— georgica Rupr. 305.
— grandis Wend. 293; tab. 18.
— Halleri (All.) Willd. 290; tab. 18.
— Kostyczewii (Korsh.) Juz. 288; tab. 18.
— montana (Hoppe) Reichb. 302; tab. 19.
— multifida (Pritz.) Juz. 296; tab. 18.
— nigricans Störck 302.
— patens (L.) Mill. 295; tab. 18.
— pratensis (L.) Mill. 301; tab. 19.
— Sukaczewii Juz. 301, 741.
— tenuiloba (Hayek) Juz. 298.
— Turczaninovii Kryl. et Serg. 294.
— vernalis (L.) Mill. 289; tab. 18.
— violacea Rupr. 305; tab. 19.
Punica L. 15, 553.
— granatum L. 553.
Puschkinia Adam 4, 394.
— hyacinthoides Baker 394.
— scilloides Adam 394.
Pycnostelma Bge. 18, 671.
— paniculata (Bge.) Schum. 671; tab. 36.
Pycreus Beauv. 3, 4.
— colchicus C. Koch 8; tab. 1.
— eragrostis (Vahl) Palla 7.
— flavescens (L.) Beauv. ex Reichb. 5.
— globosus (All.) Reichb. 6.
— limosus (Maxim.) Schischk. 8.
— nilagiricus (Hochst.) Schischk. 6; tab. 1.
— pratorum (Korotki) Schischk. 8.
— Rehmannii (Boiss.) Palla ex Grossh. 7.
— setiformis (Korsh.) Schischk. 5
— tremulus (Poir.) Clarke 9.
Pyracantha Roem. 9, 413.
— coccinea Roem. 414.
Pyramistigma Elk. 7, 623.
Pyrethrum Zinn. 26, 184.
— abrotanifolium Bge. ex Ledeb. 250.
— alatavicum (Herd.) O. et B. Fedtsch. 249.
— alpinum (L.) Schrank 228.
— arassanicum (C. Winkl.) O. et B. Fedtsch. 259.
— aromaticum (Rupr. ex Boiss.) Tzvel. 222.
— Aucherianum DC. 239.
— balsamita (L.) Willd. 196.
— balsamitoides (Nábělek) Tzvel. 198.
— chamaemelifolium (Somm. et Levier) Sosn. 219.
— cinerariifolium Trev. 213.
— Clusii Fisch. ex Reichb. 234.
— coccineum (Willd.) Vorosch. 218.
— corymbiforme Tzvel. 234, 873.
— corymbosum (L.) Willd. 232.
— daghestanicum (Rupr. ex Boiss.) Rupr. ex Fler. 222.
— djilgense (Franch.) Tzvel. 260.
— fruticulosum Biehl. 209.
— Galae M. Pop. 242.
— glanduliferum Somm. et Levier 205.
— Grossheimii Sosn. 206.
— Heldreichianum Fenzl ex Tchih. 231.
— hissaricum Krasch. 241.
— Karelinii Krasch. 254.
— Kelleri (Kryl. et Plotn.) Krasch. 212.
— Komarovii Sosn. 224.
— Kotschyi Boiss. 229.
— Krylovianum Krasch. 250.
— kubense Grossh. 209.
— lanuginosum (Sch. Bip. et Herd.) Tzvel. 252.
— leontopodium (C. Winkl.) Tzvel. 260; tab. 9.
— leptophyllum Stev. 211; tab. 8.
— macrophyllum (Waldst. et Kit.) Willd. 194.
— majus (Desf.) Tzvel. 198.
— Marionii Albov 216.
— Mikeschinii Tzvel. 243, 873.
— neglectum Tzvel. 257, 874.
— ordubadense Manden. 230.
— oxylepis (Bordz.) Tzvel. 238.
— parthenifolium Willd. 203.
— parthenium (L.) Smith 204.
— peucedanifolium (Sosn.) Manden. 202.
— poteriifolium Ledeb. 235.
— pulchellum Turcz. 251.
— pulchrum Ledeb. 253.
— punctatum (Desr.) Bordz. ex Sosn. 215.
— pyrethroides (Kar. et Kir.) B. Fedtsch. ex Krasch. 256.
— roseum (Adam) M. B. 217.
— Semenovii (Herd.) C. Winkl. ex O. et B. Fedtsch. 243; tab. 10.
— sericeum (Adam) M. B. 237.
— sevanense Sosn. 207.
— silaifolium Stev. 220.
— songaricum Tzvel. 255, 874; tab. 8.
— sorbifolium Boiss. 195.
— tianschanicum Krasch. 247; tab. 9.
— transiliense (Herd.) Rgl. et Schmalh. 258.

Pyrethrum tricholobum Sosn. ex Manden. 223.
— urgutense M. Pop. 874.
Pyrola L. 18, 8.
— grandiflora Rad. 12.
— incarnata Fisch. ex DC. 11.
— media Sw. 12.
— minor L. 13.
— renifolia Maxim. 9.
— rotundifolia L. 10.
— virescens Schweigg. 9.
Pyrus L. 9, 336.
— asiae-mediae M. Pop. 342.
— Balansae Decne. 339.
— Boissieriana Buhse 341; tab. 20.
— bucharica Litv. 356; tab. 21.
— communis L. 338.
— elaeagrifolia Pall. 346; tab. 20.
— Grossheimii Fed. 345.
— Korshinskyi Litv. 352; tab. 21.
— oxyprion Woronow 351; tab. 21.
— Raddeana Woronow 351.
— Regelii Rehd. 355; tab. 21.
— salicifolia Pall. 348; tab. 20.
— Sosnovskii Fed. 350.
— syriaca Boiss. 350; tab. 21.
— Takhtadzhianii Fed. 349.
— taochia Woronow 347; tab. 20.
— turcomanica Maleev 340.
— ussuriensis Maxim. 341; tab. 20.
— zangezura Maleev 352.

Quercus L. 5, 322.
— araxina (Trautv.) Grossh. 333; tab. 17.
— castaneifolia C. A. Mey. 329; tab. 17.
— dentata Thunb. 352.
— erucifolia Stev. 342; tab. 18.
— Hartwissiana Stev. 337; tab. 17.
— hypochrysa Stev. 347.
— iberica Stev. ex M. B. 343; tab. 18.
— ilex L. 332.
— imeretina Stev. ex Woronow 341, 713; tab. 17.
— Kozlowskyi Woronow ex Grossh. 350.
— longifolia C. Koch 347; tab. 18.
— longipes Stev. 342; tab. 18.
— macranthera Fisch. et Mey. 330; tab. 17.
— mongolica Fisch. ex Turcz. 351.
— occidentalis J. Gay 328.
— pedunculiflora C. Koch 341; tab. 18.
— petraea Liebl. 348.
— pontica C. Koch 332; tab. 17.
— pubescens Willd. 349.
— robur L. 339.
— rubra L. 353.
— suber L. 327.
— Woronowii Maleev 334, 714; tab. 18.
Querezia L. 6, 481.
Queria L. 6, 481.
Queria Loefl. 6, 481.
— hispanica L. 481.
Quinaria Raf. 14, 707.
Quinquelocularia C. Koch 24, 133, 162.

Raclathris Raf. 19, 548.
Radiola Roth 14, 85.
— linoides Roth 85.
Ramischia Opiz 18, 15.
— obtusata (Turcz.) Freyn 16.
— secunda (L.) Garcke 16.
Ranunculus L. 7, 351.
— abchasicus Freyn 444.
— acer L. 461.
— acutidentatus Rupr. 432.
— acutilobus Ledeb. 431; tab. 27.
— affinis R. Br. 388.
— afghanicus Aitch. et Hemsl. 493; tab. 31.
— Ageri Bertol. 488.
— alajensis Ostenf. 490.
— Alberti Rgl. et Schmalh. 401.
— Allemannii Br.-Bl. 381.
— alpigenus Kom. 504; tab. 31.
— altaicus Laxm. 400.
— ampelophyllus Somm. et Lev. 430; tab. 27.
— amurensis Kom. 363.
— anadyriensis Ovcz. 386.
— anemonifolius DC. 455.
— arachnoideus C. A. Mey. 405.
— arvensis L. 478.
— aureopetalus Kom. 491.
— auricomus L. 377.
— baidarae Rupr. 435.
— baldshuanicus Rgl. ex Kom. 472.
— balkharicus N. Busch 451.
— borealis Trautv. 463.
— brachylobus Boiss. et Hohen. 440; tab. 28.
— Brotherusii Freyn 392; tab. 24.
— brutius Ten. 418.
— Buhsei Boiss. 424; tab. 26.
— bulbosus L. 449.
— Buschii Ovcz. 443, 744; tab. 28.
— cassubicus L. 379; tab. 24.
— caucasicus M. B. 423; tab. 26.
— chinensis Bge. 409.
— chius DC. 476; tab. 30.
— cicutarius Schlecht. 487.

— constantinopolitanus Urv. 452.
— crassifolius (Rupr.) Grossh. 436; tab. 28.
— czimganicus Ovcz. 493, 747.
— dilatatus Ovcz. 497, 747.
— dissectus M. B. 459.
— dolosus Fisch. et Mey. 409; tab. 24.
— dzhavakheticus Ovcz. 438, 744.
— elbrusensis Boiss. 504; tab. 31.
— elegans C. Koch 456.
— Eschscholtzii Schlecht. 398.
— flammula L. 363.
— fraternus Schrenk 402.
— gelidus Kar. et Kir. 406.
— glabriusculus Rupr. 462.
— Gmelinii DC. 354; tab. 23.
— grandifolius C. A. Mey. 468.
— gymnadenus Somm. et Levier 443.
— Helenae Albov 410.
— Huetii Boiss. 459.
— hyperboreus Rottb. 358; tab. 23.
— illyricus L. 489.
— jacuticus Ovcz. 470, 746; tab. 27.
— japonicus Thunb. 470.
— Komarovii Freyn 497; tab. 33.
— kopetdaghensis Litv. 421.
— Kotschyi Boiss. 458.
— Krasnovii Ovcz. 376, 742.
— Krylovii Ovcz. 385, 741; tab. 24.
— laetus Wall. 472.
— lanuginosiformis Selin 464.
— lanuginosus L. 467.
— lapponicus L. 360; tab. 23.
— lasiocarpus C. A. Mey. 392.
— lateriflorus DC. 474.
— leptorrhynchus Aitch. et Hemsl. 503; tab. 32.
— linearilobus Bge. 499; tab. 32.
— lingua L. 362.
— Lojkae Somm. et Levier 445.
— lomatocarpus Fisch. et Mey. 478; tab. 30.
— longicaulis C. A. Mey. 375.
— megacarpus Koch 380.
— Meinshausenii Schrenk 505.
— merovensis Grossh. 429.
— Meyerianus Rupr. 447; tab. 29.
— mindshelkensis B. Fedtsch. 492; tab. 31.
— monophyllus Ovcz. 382; tab. 24.
— muricatus L. 476; tab. 30.
— napellifolius DC. 428.
— natans C. A. Mey. 357; tab. 23.
— neapolitanus Ten. 451.
— nemorosus DC. 448; tab. 29.
— nivalis L. 397.
— obesus Trautv. 457.
— ophioglossifolius Vill. 368.
— oreophilus M. B. 437.
— orientalis L. 484.
— osseticus Ovcz. 430, 743.
— oxyspermus M. B. 483.
— Pallasii Schlecht. 360; tab. 23.
— pamiri Korsh. 394.
— paucidentatus Schrenk 490.
— pauperculus Ovcz. 398, 742.
— pedatifidus Smith 398.
— pedatus Waldst. et Kit. 494.
— pinnatisectus M. Pop. 498; tab. 32.
— platyspermus Fisch. 505; tab. 33.
— polyanthemus L. 446; tab. 29.
— polyphyllus Kit. 359.
— polyrhizus Steph. 390.
— Popovii Ovcz. 393, 741; tab. 24.
— propinquus C. A. Mey. 461.
— pseudobulbosus Schur 450.
— pseudohirculus Schrenk 375.
— pulchellus C. A. Mey. 374.
— pulsatillifolius Litv. 488.
— pygmaeus Wahl. 404.
— Raddeanus Rgl. 420.
— radicans C. A. Mey. 356; tab. 23.
— recurvatus Poir. 422.
— Regelianus Ovcz. 506; tab. 33.
— repens L. 423.
— reptabundus Rupr. 355; tab. 23.
— reptans L. 364.
— rigescens Turcz. 389.
— rubrocalyx Rgl. ex Kom. 404; tab. 25.
— rufosepalus Franch. 407; tab. 25.
— samojedorum Rupr. 358; tab. 23.
— sardous Crantz 450.
— Sartorianus Boiss. et Heldr. 437.
— sceleratus L. 408.
— Schaftoanus (Aitch. et Hemsl.) Boiss. 473.
— Severzovii Rgl. 500.
— Smirnovii Ovcz. 467, 745.
— Sommieri Albov 419.
— songoricus Schrenk 402; tab. 25.
— stenopetalus Ovcz. 491, 746; tab. 31.
— Stevenii Andrz. 471.
— strigillosus Boiss. et Huet 367.
— subcorymbosus Kom. 469; tab. 27.
— submarginatus Ovcz. 439, 745; tab. 29.
— subtilis Trautv. 386.
— sulphureus Soland. 399.
— suukensis N. Busch 387.
— tenuilobus Rgl. ex Kom. 500.
— trachycarpus Fisch. et Mey. 477; tab. 30.
— Trautvetterianus Rgl. 403; tab. 25.

Ranunculus trisectilis Ovcz. 427, 743; tab. 26.
— turkestanicus Franch. 494; tab. 32.
— ussuriensis Kom. 391.
— villosus DC. 455.
— Walteri Rgl. ex Freyn 498.
Raphanus L. 8, 492.
— maritimus Smith 493.
— raphanistroides (Makino) Sinsk. 494.
— raphanistrum L. 494.
— rostratus DC. 494.
— sativus L. 493.
Rapinia Lour. 24. 449.
Rapistrum Desv. 8, 471.
— perenne L. 472.
— rugosum (L.) All. 472.
Rapunculus Fourr. 24, 133, 301.
Razoumowskia Hoffm. 5, 407.
— oxycedri (DC.) F. Schultz 408; tab. 23.
Reaumuria L. 15, 277.
— badhysi Korov. 290.
— cistoides Adam 282; tab. 14.
— fruticosa Bge. ex Boiss. 288.
— kaschgarica Rupr. 287.
— Kuznetzovii Sosn. et Manden. 284.
— oxiana (Ledeb.) Boiss. 289.
— persica Boiss. 288.
— reflexa Lipsky 281; tab. 14.
— sogdiana Kom. 281; tab. 14.
— soongorica (Pall.) Maxim. 279; tab. 14.
— turkestanica Gorschk. 282; tab. 14.
— Zakirovii Gorschk. 287, 679.
Redovskia Cham. et Schlecht. 8, 92.
— sophiifolia Cham. et Schlecht. 92.
Reichardia Roth 29, 260.
— dichotoma (Vahl) Freyn 261; tab. 19.
Renarda Rgl. 16, 233.
Reseda L. 8, 607.
— Aucheri Boiss. 612.
— brevipedunculata N. Busch 612, 654; tab. 34.
— bucharica Litv. 612; tab. 34.
— dshebeli Czerniak. 615, 653; tab. 34.
— globulosa Fisch. et Mey. 611.
— hemithamnodes Czerniak. 616, 653; t. 34.
— inodora Reichb. 609.
— lutea L. 610.
— luteola L. 608.
— microcarpa Muell. Arg. 615; tab. 34.
— odorata L. 610.
Restella Pobed. 15, 514.
— Alberti (Rgl.) Pobed. 515.
Reutera Boiss. 16, 448.
— aurea (DC.) Boiss. 449; tab. 29.
— Bobrovii Woronow 449, 599; tab. 29.
Reynoutria Houtt. 5, 696.
Rhabdotheca Cass. 29, 239.
— Korovinii (M. Pop.) Kirp. 240; tab. 15.
Rhagadiolus Scop. 29, 233.
— edulis Gaertn. 234.
— hebelaenus (DC.) Vass. 234.
Rhamnoides Moench 15, 516.
Rhamnus L. 14, 645.
— alaternus L. 653.
— baldschuanica Grub. 672; tab. 37.
— cathartica L. 660; tab. 37.
— coriacea (Rgl.) Kom. 671.
— dahurica Pall. 656; tab. 36.
— depressa Grub. 655; tab. 35.
— diamantiaca Nakai 662; tab. 36.
— dolichophylla Gontsch. 664; tab. 37.
— erythroxylon Pall. 666; tab. 36.
— imeretina Booth 654; tab. 35.
— microcarpa Boiss. 655; tab. 35.
— minuta Grub. 673.
— Pallasii Fisch. et Mey. 669; tab. 37.
— parvifolia Bge. 663; tab. 36.
— Sintenisii Rech. f. 671; tab. 37.
— songorica Gontsch. 673.
— spathulifolia Fisch. et Mey. 670.
— tinctoria Waldst. et Kit. 665.
— ussuriensis Ja. Vassil. 659; tab. 36.
Rhamphicarpa Benth. 22, 529.
— Medwedewii Albov 529.
Rhaphidophyton Iljin 6, 271, 877.
— Regelii (Bge.) Iljin 271; tab. 13.
Rhaponticum Adans. 28, 308.
— aulieatense Iljin 320.
— carthamoides (Willd.) Iljin 311.
— integrifolium C. Winkl. 315.
— karatavicum Rgl. et Schmalh. 321.
— lyratum C. Winkl. ex Iljin 314.
— namanganicum Iljin 321.
— nanum Lipsky 322.
— nitidum Fisch. 319.
— pulchrum Fisch. et Mey. 316.
— Satzyperovii Sosk. 317.
— serratuloides (Georgi) Bobr. 313.
— uniflorum (L.) DC. 318.
Rhaptocalymma Boern. 3, 187.
Rheum L. 5, 482.
— altaicum Losinsk. 485; tab. 28.
— compactum L. 485; tab. 28.
— cordatum Losinsk. 498; tab. 29.
— darvasicum Titov 491; tab. 29.
— Fedtschenkoi Maxim. ex Rgl. 494; tab. 29.
— hissaricum Losinsk. 498.
— Korshinskyi Titov 491; tab. 29.

— lobatum Litv. ex Losinsk. 500.
— lucidum Losinsk. 491; tab. 29.
— macrocarpum Losinsk. 499; tab. 29.
— Maximowiczii Losinsk. 497; tab. 29.
— nanum Siew. 493; tab. 29.
— orientale Losinsk. 489; tab. 28.
— palmatum L. 500.
— plicatum Losinsk. 499.
— reticulatum Losinsk. 492; tab. 29.
— rhaponticum L. 501.
— rhizostachyum Schrenk 492.
— ribes L. 497; tab. 29.
— rupestre Litv. 490; tab. 28.
— tataricum L. f. 489; tab. 28.
— turkestanicum Janisch. 490; tab. 28.
— undulatum L. 484; tab. 28.
— Wittrockii Lundstr. 486; tab. 28.
Rhinactina Less. 25, 129.
Rhinactinidia Novopokr. 25, 129.
Rhinanthus L. 22, 686.
Rhinanthus L. 22, 659.
— aestivalis (Zing.) Schischk. et Serg. 665.
— alpinus Baumg. 678.
— angustifolius J. G. Gmel. 675.
— apterus (Fries) Ostenf. 673.
— borealis (Sterneck) Druce 679.
— colchicus Vass. 681.
— cretaceus Vass. 666.
— ferganensis Vass. 672; tab. 34.
— groenlandicus (Ostenf.) Chabert 678.
— major L. 680; tab. 34.
— mediterraneus (Sterneck) Adamov. 682.
— minor L. 676; tab. 34.
— montanus Saut. 664.
— nigricans Meinsh. 675.
— osilensis (Ronn. et Saars.) Vass. 684.
— patulus (Sterneck) Thell. et Schinz 680.
— pectinatus (Behrend.) Vass. 668; tab. 34.
— ponticus (Sterneck) Vass. 667.
— rumelicus Velen. 683.
— rusticulus (Chabert) Druce 677.
— sachalinensis Vass. 674.
— Schischkinii Vass. 683; tab. 34.
— songaricus (Sterneck) B. Fedtsch. 671.
— subulatus (Sterneck) Soo 668.
— vernalis (Zing.) Schischk. et Serg. 666.
Rhinopetalum Fisch. 4, 296.
— arianum Losinsk. et Vved. 298, 738; tab. 18.
— bucharicum (Rgl.) Losinsk. 298.
— gibbosum (Boiss.) Losinsk. et Vved. 297.
— Karelinii Fisch. 297.
— stenantherum Rgl. 301; tab. 18.
Rhizocephalus Boiss. 2, 126.
— turkestanicus (Litv.) Roshev. 127, 743.
Rhodiola L. 9, 24.
— algida (Ledeb.) Fisch. et Mey. 37.
— arctica Boriss. 30, 471; tab. 3.
— atropurpurea (Turcz.) Trautv. et Mey. 33.
— borealis Boriss. 32, 474.
— coccinea (Royle) Boriss. 41.
— gelida Schrenk 41.
— heterodonta (Hook. et Thoms.) Boriss. 32; tab. 3.
— iremelica Boriss. 31, 472.
— kaschgarica Boriss. 39, 476; tab. 3.
— Kirilowii Rgl. ex Maxim. 34.
— Komarovii Boriss. 38, 476.
— linearifolia Boriss. 35, 474.
— Litvinovii Boriss. 45, 478; tab. 3.
— pamiroalaica Boriss. 40, 477; tab. 3.
— pinnatifida Boriss. 36, 475.
— quadrifida (Pall.) Fisch. et Mey. 39.
— recticaulis Boriss. 42, 478.
— rosea L. 29.
— sachalinensis Boriss. 31, 473; tab. 3.
— Semenovii (Rgl. et Herd.) Boriss. 28.
— Stephanii (Cham.) Trautv. et Mey. 36.
— viridula Boriss. 510.
Rhododendron L. 18, 31.
— Adamsii Rend. 43.
— aureum Georgi 38; tab. 1.
— caucasicum Pall. 39.
— dahuricum L. 48.
— hypopitys Pojark. 37, 722; tab. 1.
— kamtschaticum Pall. 58.
— Kotschyi Simk. 44.
— lapponicum (L.) Wahl. 46.
— Ledebouri Pojark. 52, 722; tab. 2.
— mucronulatum Turcz. 47; tab. 2.
— luteum Sweet 57.
— parvifolium Adam 45.
— ponticum L. 34.
— Redowskianum Maxim. 60.
— Schlippenbachii Maxim. 56.
— sichotense Pojark. 50, 722; tab. 2.
— Smirnovii Trautv. 36.
— Tschonoskii Maxim. 55.
— Ungernii Trautv. 35.
Rhus L. 14, 527.
— coriaria L. 528; tab. 29.
— hirta (L.) Sudw. 530.
— javanica L. 531.
Rhynchocorys Griseb. 22, 686.
— elephas (L.) Griseb. 687.
— orientalis (L.) Benth. 686.
Rhynchopappus Dulac 29, 595, 687, 695.
Rhynchopera Boern. 3, 157.

Rhynchospora Vahl 3, 102.
— alba (L.) Vahl 103; tab. 5.
— caucasica Palla 104; tab. 5.
— Fauriei Franch. 104; tab. 5.
— fusca (L.) Ait. 103; tab. 5.
Rhynchostylis Tausch 16, 94.
Rhytispermum Link 19, 163.
Ribes L. 9, 226.
— alpinum L. 260; tab. 15.
— altissimum Turcz. 243; tab. 14.
— atropurpureum C. A. Mey. 244.
— aureum Pursh. 266.
— Biebersteinii Berlin 245.
— diacantha Pall. 264.
— dikuscha Fisch. 254; tab. 14.
— fragrans Pall. 256.
— graveolens Bge. 256.
— heterotrichum C. A. Mey. 258.
— hispidulum Pojark. 238.
— horridum Rupr. 247; tab. 15.
— Janczewskii Pojark. 252.
— kolymense Kom. ex Pojark. 251.
— Komarovii Pojark. 263.
— latifolium Jancz. 242.
— lucidum Kit. 264; tab. 15.
— malvifolium Pojark. 246.
— manschuricum (Maxim.) Kom. 234; tab. 14.
— Maximoviczianum Kom. 260.
— melananthum Boiss. et Hohen. ex Kotschy 258.
— Meyeri Maxim. 241; tab. 13.
— nigrum L. 252; tab. 14.
— orientale Desf. 257.
— Palczewskii Pojark. 237.
— pallidiflorum Pojark. 242; tab. 14.
— pauciflorum Turcz. 248.
— procumbens Pall. 255; tab. 14.
— pubescens Hedl. 239; tab. 13.
— pulchellum Turcz. 266.
— rubrum L. 237; tab. 13.
— sachalinense Nakai 246; tab. 15.
— sanguineum Pursh 266.
— saxatile Pall. 265; tab. 15.
— scandicum Hedl. 240.
— triste Pall. 232; tab. 13.
— turbinatum Pojark. 252.
— ussuriense Jancz. 248.
— villosum Wall. 259; tab. 15.
— vulgare Lam. 233; tab. 13.
Richteria Kar. et Kir. 26, 184, 248.
Ricinoides Moench 14, 288.
Ricinus L. 14, 300.
— communis L. 300.
Rindera Pall. 19, 565.
— austroechinata M. Pop. 582.
— baldshuanica Kusn. 581.
— coechinata M. Pop. 583.
— cyclodonta Bge. 580.
— echinata Rgl. 581.
— ferganica M. Pop. 571; tab. 28.
— holochiton M. Pop. 585, 716.
— lanata (Lam.) Bge. 573.
— oblongifolia M. Pop. 584.
— ochroleuca Kar. et Kir. 577; tab. 28.
— oschensis M. Pop. 572; tab. 28.
— Korshinskyi (Lipsky) Brand 574.
— tetraspis Pall. 579.
— tianschanica M. Pop. 569.
— tschotkalensis M. Pop. 570.
— turkestanica Kusn. 583.
— umbellata (Waldst. et Kit.) Bge. 572.
Ritro Boiss. 27, 40.
Robinia L. 11, 305.
— hispida L. 306.
— pseudacacia L. 306.
— viscosa Vent. 309.
Rochelia Reichb. 19, 548.
— Bungei Trautv. 554.
— campanulata M. Pop. et Zak. 559.
— cardiosepala Bge. 563; tab. 27.
— claviculata M. Pop. et Zak. 560.
— disperma (L. f.) C. Koch 551.
— jackabaghi Pavl. 560.
— karsensis M. Pop. 557.
— leiocarpa Ledeb. 557; tab. 27.
— macrocalyx Bge. 562; tab. 27.
— peduncularis Boiss. 561; tab. 27.
— persica Bge. ex Boiss. 553.
— retorta (Pall.) Lipsky 552.
Rodigia Spreng. 29, 595.
Roegneria C. Koch 2, 599.
— Abolinii (Drob.) Nevski 611.
— amurensis (Drob.) Nevski 606.
— angustiglumis Nevski 618.
— borealis (Turcz.) Nevski 624.
— brachypodioides Nevski 617.
— Buschiana (Roshev.) Nevski 620.
— canina (L.) Nevski 617.
— caucasica C. Koch 604.
— ciliaris (Trin.) Nevski 607; tab. 46.
— confusa (Roshev.) Nevski 605.
— curvata Nevski 608; tab. 46.
— czimganica (Drob.) Nevski 604; tab. 46.
— Drobovii Nevski 613; tab. 45.
— fibrosa (Schrenk) Nevski 625.
— himalayana Nevski 606; tab. 46.

— interrupta Nevski 611.
— jacutensis (Drob.) Nevski 625.
— kamczadalorum Nevski 619.
— Komarovii Nevski 615; tab. 45.
— leptoura Nevski 623.
— macrochaeta Nevski 612.
— macroura (Turcz.) Nevski 627; tab. 45.
— oschensis (Roshev.) Nevski 619.
— panormitana (Betrol.) Nevski 612.
— pendulina Nevski 616; tab. 45.
— pubescens (Trin.) Nevski 626.
— sajanensis Nevski 624.
— scandica Nevski 624; tab. 45.
— Schrenkiana (Fisch. et Mey.) Nevski 605.
— schugnanica Nevski 604; tab. 46.
— sclerophylla Nevski 614.
— taigae Nevski 616.
— tianschanica (Drob.) Nevski 615.
— transbaicalensis Nevski 619.
— troctolepis Nevski 623.
— Turczaninovii (Drob.) Nevski 607.
— turuchanensis (Reverd.) Nevski 626.
— ugamica (Drob.) Nevski 613.
— uralensis Nevski 614; tab. 45.
— viridiglumis Nevski 616.
Roemeria Medic. 7, 596.
— hybrida (L.) DC. 596; tab. 40.
— refracta (Stev.) DC. 597; tab. 40.
Rokejeka Forsk. 6, 738.
Roripa Scop. 8, 135.
— amphibia (L.) Bess. 138.
— anceps (Wahl.) Grossh. 137.
— austriaca (Crantz) Bess. 139.
— brachycarpa (C. A. Mey.) Woronow 138.
— globosa (Turcz.) Vass. 140.
— hispida (DC.) Britt. 139.
— microsperma (DC.) Vass. 140.
— palustris (Leyss.) Bess. 137.
— silvestris (L.) Bess. 136.
Rosa L. 10, 431.
— acicularis Lindl. 449; tab. 28.
— Afzeliana Fries 501.
— agrestis Savi 497.
— alaica Juz. 465; 638.
— alba L. 485.
— Alberti Rgl. 452; tab. 29.
— Alexeenkoi Crép. 504, 638.
— alticola Boulenger 500.
— amblyotis C. A. Mey. 460.
— arvensis Huds. 443.
— Banksiae R. Br. 444.
— Beggeriana Schrenk 462; tab. 29.
— bellicosa Nevski 467.
— bifera Poir. 484.
— Boissieri Crép. 490; tab. 30.
— Bungeana Boiss. et Buhse 482; tab. 27.
— canina L. 502; tab. 31.
— caryophyllacea Boiss. 493.
— centifolia L. 484.
— chinensis Jacq. 444.
— cinnamomea L. 454; tab. 28.
— coriifolia Fries 502, tab. 31.
— corymbifera Borkh. 503.
— damascena Mill. 484.
— davurica Pall. 459; tab. 28.
— Ecae Aitch. 475.
— eglanteria L. 492; tab. 30.
— elasmacantha Trautv. 473.
— elymaitica Boiss. et Hausskn. 491.
— Fedtschenkoana Rgl. 465; tab. 29.
— foetida Herrm. 477.
— gallica L. 483.
— glabrifolia C. A. Mey. ex Rupr. 456.
— glauca Pourr. 487.
— glutinosa Smith 499; tab. 30.
— gorinkensis Bess. 456.
— haemisphaerica Herrm. 481.
— hirtissima Lonacz. 490.
— hissarica Slob. 468; tab. 29.
— horrida Fisch. 498; tab. 30.
— iberica Stev. 494; tab. 30.
— jacutica Juz. 460; 637.
— Jundzillii Bess 487.
— kamtschatica Vent. 448.
— kaschgarica Rupr. 462.
— Klukii Bess. 504.
— kokanica Rgl. 476.
— Korshinskiana Bouleng. 469.
— kuhitangi Nevski 466.
— lacerans Boiss. et Buhse 463.
— laxa Retz. 461; tab. 28.
— leucantha M. B. 505.
— maracandica Bge. 465.
— Marretii Lévl. 460.
— Maximowicziana Rgl. 442; tab. 27.
— micrantha Smith 497.
— mollis Smith 489.
— multiflora Thunb. 443.
— myriacantha DC. ex Lam. 474.
— nanothamnus Bouleng. 467.
— oxyacantha M. B. 450.
— oxyodon Boiss. 453.
— piptocalyx Juz. 463, 637.
— platyacantha Schrenk 474; tab. 27.
— pomifera Herrm. 489; tab. 31.
— Roopae Lonacz. 454.

Rosa rugosa Thunb. 447; tab. 28.
— Schischkinii Juz. 451, 638.
— Schrenkiana Crép. 452.
— spinosissima L. 470; tab. 27.
— svanetica Crép. 491.
— tomentosa Smith 489; tab. 31.
— turbinata Aitch. 485.
— turkestanica Rgl. 477.
— tuschetica Boiss. 499.
— ussuriensis Juz. 451, 637.
— Webbiana Wall. ex Royle 464.
— Woronowii Lonacz. 505.
Roscyna Spach 15, 203, 212.
Rosenbachia Rgl. 20, 17.
Rosmarinus L. 20, 70.
— officinalis L. 71.
Rosularia (DC.) Stapf 9, 114.
— alpestris (Kar. et Kir.) Boriss. 129.
— elymaitica (Boiss. et Hausskn.) Berger 121.
— glabra (Rgl. et C. Winkl.) Berger 124.
— hissarica Boriss. 122, 483; tab. 7.
— kokanica (Rgl. et Schmalh.) Boriss. 128.
— Lipskyi Boriss. 121, 483.
— lutea Boriss. 128, 484.
— paniculata (Rgl. et Schmalh.) Berger 123.
— persica (Boiss.) Berger 120.
— pilosa (M. B.) Boriss. 117.
— platyphylla (Schrenk) Berger 131; tab. 7.
— radiciflora (Steud.) Boiss. 120.
— Schischkinii Boriss. 130, 484; tab. 7.
— sempervivoides (Fisch.) Boriss. 118.
— sempervivum (M. B.) Berger 119; tab. 7.
— subspicata (Freyn et Sint.) Boriss. 123.
— tadzhikistana Boriss. 130, 485.
— turkestanica (Rgl. et C. Winkl.) Berger 127.
Rotala L. 15, 526.
— densiflora (Roth) Koehne 526.
— indica (Willd.) Koehne 527.
Roucela Dum. 24, 133, 189.
Rubacer Rydb. 10, 5.
— odoratum (L.) Rydb. 5.
Rubia L. 23, 382.
— chinensis Rgl. et Maack 391.
— chitralensis Ehrend. 400.
- cordifolia L. 387.
— cretacea Pojark. 396, 721; tab. 22.
— deserticola Pojark. 399, 722.
— dolichophylla Schrenk 404.
— florida Boiss. 415.
— iberica (Fisch. ex DC.) C. Koch 394; tab. 22.
— jesoensis (Miq.) Miyabe et Miyake 407.
— Komarovii Pojark. 403, 724; tab. 23.
— Krascheninnikovii Pojark. 406, 724; tab. 22.
— laxiflora Gontsch. 412.
— Rechingeri Ehrend. 416.
— Regelii Pojark. 402, 723.
— Rezniczenkoana Litv. 397; tab. 23.
— rigidifolia Pojark. 414, 726; tab. 23.
— schugnanica B. Fedtsch. 413, 725.
— tatarica (Trev.) Fr. Schmidt 408.
— tibetica Hook. f. 401.
— tinctorum L. 393.
— transcaucasica Grossh. 411; tab. 23.
Rubus L. 10, 5.
— abchaziensis Sudre 45.
— abnormis Sudre 42.
— arcticus L. 12; tab. 1.
— caesius L. 56; tab. 5.
— candicans Weihe 29; tab. 2.
— cartalinicus Juz. 31; tab. 3.
— caucasicus Focke 48; tab. 4.
— caucasigenus (Sudre) Juz. 37.
— chamaemorus L. 11; tab. 1.
— crataegifolius Bge. 15; tab. 1.
— Cyri Juz. 25; tab. 25.
— discernendus Sudre 26.
— dolichocarpus Juz. 46; tab. 4.
— georgicus Focke 33.
— hirtus Waldst. et Kit. 54; tab. 5.
— humulifolius C. A. Mey. 15; tab. 1
— hyrcanus Juz. 30.
— ibericus Juz. 26.
— idaeus L. 16; tab. 1.
— Komarovii Nakai 20.
— lanuginosus Stev. ex Ser. 46.
— lepidulus (Sudre) Juz. 37.
— leptostemon Juz. 41.
— Lloydianus Genev. 38.
— Miszczenkoi Juz. 34.
— moschus Juz. 49; tab. 5.
— nessensis W. Hall. 23; tab. 2.
— ochthodes Juz. 47; tab. 4.
— ossicus Juz. 32.
— persicus Boiss. 40.
— peruncinatus (Sudre) Juz. 33.
— piceetorum Juz. 41.
— platyphyllus C. Koch 50.
— ponticus Juz. 53, 607.
— Raddeanus Focke 39; tab. 3.
— sachalinensis Lévl. 19.
— sanguineus Friv. 24; tab. 2.

— saxatilis L. 14; tab. 1.
— serpens Weihe 55.
— stellatus Smith 13.
— tomentosus Borkh. 37; tab. 3.
— Woronowii Sudre 45.
Rudbeckia L. 25, 539.
— hirta L. 539.
— laciniata L. 540.
Rulac Adans. 14, 621.
Rumex L. 5, 444.
— acetosa L. 451; tab. 26.
— acetosella L. 449.
— acetoselloides Baill. 450.
— acutatus L. 456; tab. 26.
— alpinus L. 468; tab. 26.
— amurensis Fr. Schmidt ex Maxim. 479; tab. 27.
— angustifolius Campd. 469; tab. 26.
— aquaticus L. 467; tab. 26.
— arcticus Trautv. 469.
— arifolius All. 455.
— armenus C. Koch 467.
— aschabadensis Losinsk. 465; 715.
— aureostigmaticus Kom. 450.
— confertus Willd. 463; tab. 26.
— conglomeratus Murr. 470; tab. 26.
— crispus L. 461; tab. 26.
— domesticus Hartm. 458; tab. 26.
— Fauriei Rech. f. 461.
— Fischeri Reichb. 464.
— foveolatus Losinsk. 478, 717; tab. 27.
— Gmelinii Turcz. 457; tab. 26.
— graminifolius Lamb. 450.
— Halaczii Rech. 475; tab. 27.
— hydrolapathum Huds. 462; tab. 26.
— jakutensis Kom. 457.
— kamtschadalus Kom. 459.
— Komarovii Schischk. et Serg. 460.
— maritimus L. 481; tab. 27.
— Marschallianus Reichb. 479; tab. 27.
— obtusifolius L. 477; tab. 27.
— pamiricus Rech. f. 458; tab. 26.
— patientia L. 464; tab. 26.
— Paulsenianus Rech. f. 466; tab. 26.
— pseudonatronatus Borb. 462; tab. 26.
— pulcher L. 476; tab. 27.
— Rechingerianus Losinsk. 465, 715; tab. 26.
— Regelii Fr. Schmidt 474.
— reticulatus Bess. 475; tab. 27.
— rossicus Murb. 480; tab. 27.
— sanguineus L. 473; tab. 26.
— Schischkinii Losinsk. 468; 716.
— sibiricus Hult. 459.
— stenophyllus Ledeb. 473; tab. 27.
— syriacus Meisn. 477; tab. 27.
— thyrsiflorus Fingerh. 452; tab. 26.
— tianschanicus Losinsk. 466, 716.
— tuberosus L. 455; tab. 26.
— ucranicus Fisch. 480; tab. 27.
— ussuriensis Losinsk. 474, 717.
Rumia Hoffm. 16, 358.
— crithmifolia (Willd.) K.-Pol. 358; tab. 25, 30.
Rumia Ledeb. 16, 349.
Ruppia L. 1, 261.
— maritima L. 262.
— spiralis L. 262.
Ruscus Tourn. 4, 443.
— hypoglossum L. 444; tab. 26.
— hypophyllum L. 444; tab. 26.
— hyrcanus Woronow 447; tab. 26.
— ponticus Woronow 444; tab. 26.
Russowia C. Winkl. 28, 250.
— sogdiana (Bge.) B. Fedtsch. 252; tab. 12.
Ruta L. 14, 199.
— graveolens L. 199; tab. 11.
Ruyschiana Boehr. ex Mill. 20, 439, 471.

Sabulina Reichb. 6, 487.
Saccharum L. 2, 8.
— spontaneum L. 9; tab. 1.
Sageretia Brongn. 14, 638.
— laetevirens (Kom.) Gontsch. 641; tab. 34.
Sagina Druce 6, 482.
Sagina L. 6, 471.
— apetala Ard. 474; tab. 25.
— litoralis Hult. 473; tab. 25.
— maxima A. Gray 473; tab. 25.
— nivalis (Lindblad) Fries 475; tab. 25.
— nodosa (L.) Fenzl 472.
— oxysepala Boiss. 475; tab. 25.
— procumbens L. 474.
— saginoides (L.) Dalla Torre 472.
Sagittaria L. 1, 287.
— graminea Michx. 291.
— natans Pall. 291; tab. 14.
— sagittifolia L. 287; tab. 14.
— trifolia L. 288; tab. 14.
Salicornia L. 6, 172.
— herbacea L. 172; tab. 8.
Salix L. 5, 24.
— abscondita Laksch. 77.
— acmophylla Boiss. 194.
— acutifolia Willd. 188; tab. 9.
— aegyptiaca L. 94; tab. 5.

Salix alatavica Kar. et Kir. ex Stschegl. 60; tab. 4.
— alba L. 188; tab. 9.
— Alberti Rgl. 172.
— alifera Goerz 100.
— anadyrensis Floder. 69.
— anomala E. Wolf 209.
— apoda Trautv. 117.
— arbuscula L. 79; tab. 6.
— arctica Pall. 44; tab. 3.
— argyracea E. Wolf 143.
— aurita L. 101; tab. 6.
— australior Anderss. 191.
— babylonica L. 196; tab. 9.
— Barclayi Anderss. 121; tab. 4.
— Bebbiana Sarg. 108; tab. 5.
— berberifolia Pall. 55; tab. 3.
— Blakii Goerz 162.
— borealis Fries 87; tab. 6.
— bachypoda (Trautv. et Mey.) Kom. 122; 709.
— brevijulis Turcz. 80; tab. 6.
— burjatica Nas. 137.
— caesia Vill. 177.
— caprea L. 90; tab..6.
— cardiophylla Trautv. et Mey. 207.
— caspica Pall. 157; tab. 8.
— caucasica Anderss. 101.
— Chamissonis Anderss. 49; tab. 1.
— chilkoana Sukacz. 213.
— chlorostachya Turcz. 76; tab. 5.
— cinerea L. 99; tab. 5.
— coaetanea (Hartm.) Floder. 93.
— coerulea E. Wolf 159; tab. 8.
— cuneata Turcz. 50; tab. 3.
— daghestanica Goerz 103.
— dahurica Turcz. 155; tab. 8.
— daphnoides Vill. 180; tab. 9.
— dasyclados Wimm. 147; tab. 7.
— Daviesii Boiss. 195.
— divaricata Pall. 147; tab. 3.
— dolichostyla Seemen 198.
— erythrocarpa Kom. 55; tab. 2.
— euapiculata Nas. 192, 713.
— Fedtschenkoi Goerz 118.
— ferganensis Nas. 176, 713.
— fragilis L. 202.
— fulcrata Anderss. 75.
— fumosa Turcz. 51; tab. 3.
— fuscata Goerz 100.
— fuscescens Anderss. 113; tab. 5.
— glandulifera Floder. 62.
— glauca L. 58; tab. 5.
— gracilistyla Miq. 129; tab. 7.
— hastata L. 116; tab. 4.
— herbacea L. 36; tab. 2.
— holargyrea Bornm. et Goerz 174.
— Hookeriana Barr. 63.
— Hultenii Floder. 92.
— iliensis Rgl. 111.
— integra Thunb. 179; tab. 8.
— issykiensis Goerz 160, 712.
— jacutica Nas. 142, 711.
— Kirilowiana Stschegl. 191.
— Kochiana Trautv. 178.
— kolymensis Seemen 75; tab. 6.
— Komarovii E. Wolf 165; tab. 8.
— koreensis Anderss. 201.
— Korshinskyi Goerz 176.
— Krylovii E. Wolf 65; tab. 4.
— kurilensis Koidz. 34; tab. 1.
— Kusnetzowii Laksch. 98.
— lanata L. 61; tab. 5.
— lapponum L. 65.
— lepidostachys Seemen 175.
— leptoclados Anderss. 83.
— liliputa Nas. 37, 707; tab. 2.
— linearifolia E. Wolf 169.
— Lipskyi (Goerz) Nas. 170, 712.
— Litwinowii Goerz 120, 708.
— livida Wahl 105; tab. 6.
— macilenta Anderrs. 83.
— macropoda Stschegl. 209.
— macrostachya E. Wolf 172.
— margaritifera E. Wolf 173.
— Matsudana Koidz. 198.
— Maximoviczii Kom. 207; tab. 9.
— Medwedewii Dode 186.
— mezereoides E. Wolf 84.
— micans Anderss. 190.
— Michelsonii Goerz 158, 711.
— microstachya Turcz. 163.
— mongolica Siuzew 156.
— myrsinites L. 49; tab. 3.
— myrtilloides L. 112.
— Niedzwieckii Goerz 161.
— nigricans (Smith) Enand. 85; tab. 6.
— nipponica Franch. et Sav. 186.
— nummularia Anderss. 38; tab. 2.
— nyiwensis Kimura 211.
— oblongifolia Trautv. et Mey. 72.
— Oleninii Nas. 93, 708.
— Olgae Rgl. 170.
— opaca Anderss. ex Seemen 148.
— orbicularis Anderss. 33; tab. 1.
— orotchonorum Kimura 212.
— ovalifolia Trautv. 51; tab. 1
— oxica Dode 193.

— oxycarpa Anderss. 175.
— Palibinii Goerz 102.
— Pallasii Anderss. 45.
— pantosericea Goerz 104.
— paracaucasica Goerz 103.
— parallelinervis Floder. 73.
— pentandra L. 205; tab. 9.
— persica Boiss. 195.
— phlebophylla Anderss. 56; tab. 1.
— phylicifolia L. 71; tab. 5.
— Pierotii Miq. 128.
— podophylla Anderss. 77.
— polaris Wahl. 40; tab. 2.
— pororaica Kimura 211.
— Prszewalskii E. Wolf 169.
— pseudalba E. Wolf 171.
— pseudolinearis Nas. 137; tab. 7.
— psiloides (Floder.) Kom. 117.
— pulchra Cham. 46; tab. 3.
— purpurea L. 153.
— pycnostachya Anderss. 163.
— pyrolifolia Ledeb. 115; tab. 4.
— Raddeana Laksch. 92, 707.
— reptans Rupr. 60; tab. 5.
— reticulata L. 32; tab. 1.
— rhamnifolia Pall. 120; tab. 4.
— Richardsonii Hook. 63.
— rorida Laksch. 182; tab. 9.
— rosmarinifolia L. 123; tab. 7.
— rossica Nas. 135; tab. 7.
— rotundifolia Trautv. 39; tab. 2.
— rufescens Turcz. 137.
— sachalinensis Fr. Schmidt 145.
— sajanensis Nas. 141, 710; tab. 7.
— salviifoliae affinis Bge. 208.
— saravschanica Rgl. 160.
— saxatilis Turcz. ex Ledeb. 52; tab. 3.
— schugnanica Goerz 126.
— Seemannii Rydb. 59.
— semiviminalis E. Wolf 143.
— serrulatifolia E. Wolf 165.
— sibirica Pall. 125; tab. 7.
— sitchensis Sanson ex Bong. 127; tab. 7.
— Siuzewii Seemen 144; tab. 7.
— songarica Anderss. 204; tab. 9.
— speciosa Hook. et Arn. 66; tab. 4.
— spinidens E. Wolf 161; tab. 8.
— splendens Turcz. 136.
— Starkeana Willd. 106.
— stipulifera Floder. 59.
— stoloniferoides Kimura 213.
— strobilacea (E. Wolf) Nas. 141, 710.
— taraikensis Kimura 210.
— tenuifolia Turcz. ex Laksch. 154; tab. 8.
— tenuijulis Ledeb. 158.
— tianschanica Rgl. 84.
— tontomussirensis Koidz. 74.
— torulosa Trautv. 45; tab. 3.
— triandra L. 184; tab. 9.
— turanica Nas. 138, 709.
— Turczaninowii Laksch. 37; tab. 2.
— udensis Trautv. et Mey. 146.
— veriviminalis Nas. 134.
— vestita Pursch 34; tab. 1.
— viminalis L. 132.
— viridula Anderss. ex A. Gray 119; tab. 4.
— Wilhelmsiana M. B. 164; tab. 8.
— xerophila Floder. 107.
Salsola L. 6, 200.
— affinis C. A. Mey. 235.
— Androssovii Litv. 223.
— aperta Pauls. 222; tab. 12.
— aralensis Iljin 231.
— arbuscula Pall. 237; tab. 11.
— arbusculiformis Drob. 239; tab. 11.
— Aucheri Bge. 261; tab. 13.
— Baranovii Iljin 244, 876.
— brachiata Pall. 236.
— bucharica Iljin 225.
— cana C. Koch 260.
— carinata C. A. Mey. 257; tab. 11.
— chiwensis M. Pop. 241.
— collina Pall. 215; tab. 11.
— crassa M. B. 226; tab. 11.
— daghestanica (Turcz.) Lipsky 264.
— dendroides Pall. 255.
— deserticola Iljin 223.
— dschungarica Iljin 252.
— ericoides M. B. 256.
— foliosa (L.) Schrad. 265; tab. 13.
— forcipitata Iljin 263; tab. 13.
— gemmascens Pall. 259; tab. 13.
— glauca M. B. 246.
— gossypina Bge. 263; tab. 12.
— heptapotamica Iljin 233.
— hispidula Bge. 247.
— iliensis Lipsky, 248; tab. 12.
— incanescens C. A. Mey. 254.
— intricata Iljin 231.
— kali L. 214; tab. 1.
— kasakorum Iljin 225.
— Komarovii Iljin 221; tab. 12.
— Korshinskyi Drob. 230.
— lachnophylla Iljin 232, 876.
— lanata Pall. 224; tab. 11.
— laricifolia (Turcz.) Litv. 239.
— laricina Pall. 252.
— longistylosa Iljin 225; tab. 12.

Salsola maracandica Iljin 240.
— macera Litv. 254.
— Minkvitziae Korov. 232.
— monoptera Bge. 215; tab. 11.
— montana Litv. 240.
— mutica C. A. Mey. 220.
— nitraria Pall. 253.
— nodulosa (Moq.) Iljin 258; tab. 13.
— Olgae Iljin 235.
— Paletzkiana Litv. 243.
— Paulsenii Litv. 216.
— pellucida Litv. 217; tab. 11.
— praecox Litv. 216.
— Richteri Kar. 242; tab. 11.
— rigida Pall. 251.
— rosacea L. 219; tab. 12.
— Roshevitzii Iljin 244.
— ruthenica Iljin 212; tab. 11.
— sclerantha C. A. Mey. 257; tab. 11.
— soda L. 220.
— sogdiana Bge. 221; tab. 12.
— stellulata Korov. 262, 877; tab. 13.
— subaphylla C. A. Mey. 245; tab. 11.
— subcrassa M. Pop. 229, 875.
— Takhtadshjanii Iljin 261.
— Tamamschjanae Iljin 218.
— tamariscina Pall. 218.
— tragus L. 213; tab. 11.
— transhyrcanica Iljin 243; tab. 11.
— transoxana Iljin 230.
— turcomanica Litv. 234.
— turgaica Iljin 232.
— Vvedenskyi Iljin et M. Pop. 262, 876; tab. 13.
Salvia L. 21. 244.
— adenostachya Juz. 269.
— aequidens Botsch. 287.
— aethiopis L. 318.
— Alexandri Pobed. 263, 653.
— amasiaca Freyn et Bornm. 361; tab. 17.
— Andreji Pobed. 323, 659.
— armeniaca (Bordz.) Grossh. 355; tab. 16.
— austriaca Jacq. 353.
— baldshuanica Lipsky 362; tab. 18.
— Beckeri Trautv. 333.
— brachyantha (Bordz.) Pobed. 331, 660; tab. 15.
— campylodonta Botsch. 288.
— canescens C. A. Mey. 308.
— ceratophylla L. 330.
— chloroleuca Rech. f. et Aell. 327; tab. 15.
— compar Trautv. 304; tab. 14.
— daghestanica Sosn. 309.
— Demetrii Juz. 270.
— deserta Schang. 346.
— Drobovii Botsch. 286.
— dumetorum Andrz. 337
— Fominii Grossh. 329.
— Forskaehlei L. 282.
— fugax Pobed. 348, 663.
— Caredji Troitzky 264.
— glabricaulis Pobed. 278, 656.
— glutinosa L. 281.
— grandiflora Etl. 258.
— Grossheimii Sosn. 315.
— Gontscharovii Kudr. 307.
— hajastanica Pobed. 316, 658.
— horminum L. 291.
— insignis Kudr. 334.
— intercedens Pobed. 297, 657; tab. 17.
— karabachensis Pobed. 317, 659.
— Komarovii Pobed. 277, 655.
— kopetdaghensis Kudr. 313.
— Kuznetzovii Sosn. 339; tab. 16.
— lilacinocoerulea Nevski 279.
— limbata C. A. Mey. 325.
— Linczevskii Kudr. 314.
— Lipskyi Pobed. 273, 654.
— macrosiphon Boiss. 302.
— Margaritae Botsch. 285.
— nachiczevanica Pobed. 301, 657.
— nemorosa L. 344.
— nutans L. 349.
— officinalis L. 257.
— pachystachya Trautv. 261.
— plebeja R. Br. 290; tab. 17.
— pratensis L. 335.
— Prilipkoana Grossh. et Sosn. 328.
— ringens Sibth. et Smith 266.
— rosifolia Smith 265.
— scabiosifolia Lam. 268.
— Schmalhausenii Rgl. 284; tab. 14.
— sclarea L. 310.
— semilanata Czerniak. 331.
— seravschanica Rgl. et Schmalh. 306 tab. 15.
— Sibthorpii Smith ex Sibth. 342.
— spinosa L. 300.
— splendens Ker.-Gawl. 358.
— stepposa Schost. 338; tab. 16.
— submutica Botsch. et Vved. 278.
— syriaca L. 298.
— tesquicola Klok. et Pobed. 345, 662.
— Trautvetteri Rgl. 272; tab. 14.
— trigonocalyx Woronow 260.
— turcomanica Pobed. 341, 661.
— verbascifolia M. B. 322.

— verbenaca L. 356.
— verticillata L. 359.
— viridis L. 295.
— virgata Jacq. 340.
— xanthocheila Boiss. 324.
Salvinia Micheli 1, 89.
— natans (L.) All. 89; tab. 3.
Sambucus L. 23, 422.
— coreana (Nakai) Kom. et Aliss. 430.
— ebulus L. 427.
— kamtschatica E. Wolf 437.
— latipinna Nakai 429.
— manshurica Kitag. 436.
— nigra L. 424.
— racemosa L. 431.
— sachalinensis Pojark. 439, 726.
— sibirica Nakai 434.
— Sieboldiana (Miq.) Schwer. 441.
— Tigranii Troitzky 433.
Sameraria Desv. 8, 229.
— Aitchisonii (Korsh.) B. Fedtsch. 234; tab. 13.
— armena (L.) Desv. 232; tab. 13.
— bullata (Aitch. et Hemsl.) B. Fedtsch. 234; tab. 13.
— canaliculata Vass. 233, 642; tab. 13.
— cardiocarpa Trautv. 232; tab. 13.
— deserti N. Busch 232; 642.
— glastifolia (Fisch. et Mey.) Boiss. 231; tab. 13.
— Litvinovii N. Busch 233, 642.
— sclerocarpa Bordz. 231.
— turcomanica (Korsh.) B. Fedtsch. 233; tab. 13.
Samolus L. 18, 253.
— valerandii L. 254.
Sanguisorba L. 10, 421.
— alpina Bge. 428; tab. 26.
— glandulosa Kom. 424.
— magnifica Schischk. et Kom. 424.
— officinalis L. 422; tab. 26.
— parviflora (Maxim.) Takeda 427; tab. 26.
— polygama Nyl. 423.
— riparia Juz. 423, 637.
— sitchensis C. A. Mey. 428; tab. 26.
— tenuifolia Fisch. ex Link 427.
Sanicula L. 16, 61.
— chinensis Bge. 66.
— europaea L. 62.
— rubriflora Fr. Schmidt 62.
Santolina L. 26, 125.
— chamaecyparissus L. 126.
— viridis Willd. 127.
Sanvitalia Gualt. 25, 534.
— procumbens Lam. 534.
Saponaria L. 6, 861.
— cerastoides Fisch. 868; tab. 55.
— glutinosa M. B. 863.
— Griffithiana Boiss. 864; tab. 55.
— officinalis L. 863.
— orientalis L. 869.
— parvula Bge. 867.
— prostrata Willd. 867.
— Sewerzowii Rgl. et Schmalh. 867; tab. 55.
— viscosa C. A. Mey. 686; tab. 55.
Saposhnikovia Schischk. 17, 54, 359.
— divaricata (Turcz.) Schischk. 54; tab. 5.
Sarcozygium Bge. 14, 184.
Sarothamnus Wimm. 11, 93.
— scoparius (L.) Wimm. ex Koch 93; tab. 5.
Sasa Makino et Shibata 2, 729.
— japonica (Sieb. et Zucc.) Makino 731.
— kurilensis (Rupr.) Makino 730.
— paniculata (Fr. Schmidt) Makino et Schibata 731.
— spiculosa (Fr. Schmidt) Makino 730.
Satureia L. 21, 413.
— altaica Boriss. 419.
— bzybica Woronow 424.
— confinis Boriss. 422.
— hortensis L. 415.
— intermedia C. A. Mey. 423; tab. 22.
— laxiflora C. Koch 416.
— macrantha C. A. Mey. 425.
— montana L. 420.
— mutica Fisch. et Mey. 421.
— pachyphylla C. Koch 416.
— spicigera (C. Koch) Boiss. 425; tab. 22.
— subdentata Boriss. 422.
— taurica Velen. 420.
Saussurea DC. 27, 361.
— acuminata Turcz. ex Fisch. et Mey. 454; tab. 33.
— ajanensis (Rgl.) Lipsch. 441.
— alata DC. 526.
— Alberti Rgl. et C. Winkl. 426.
— alpina (L.) DC. 489.
— amara (L.) DC. 520.
— ambigua Kryl. ex Serg. 496.
— amurensis Turcz. 456.
— angustifolia (Willd.) DC. 498.
— Asbukinii Iljin 387.
— baicalensis (Adam) Robins. 393.
— blanda Schrenk 407.
— caespitosa Iljin 417.

Saussurea cana Ledeb. 413.
— canescens C. Winkl. 418.
— caprifolia Iljin et Zapr. 519.
— carduicephala (Iljin) Iljin 382.
— ceterahifolia Lipsch. 532.
— chondrilloides C. Winkl. 512.
— congesta Turcz. 482.
— controversa DC. 438.
— coronata Schrenk 406.
— crepidifolia Turcz. 527; tab. 39.
— daurica Adam 507; tab. 37.
— discolor (Willd.) DC. 439.
— Dorogostaiskii Palib. 394.
— dubia Freyn 460.
— duiensis Fr. Schmidt 458.
— elata Ledeb. 422.
— elegans Ledeb. 423.
— elongata DC. 444.
— esthonica Baer ex Rupr. 493.
— Faminziniana Krassn. 520.
— Fauriei Franch. 476.
— foliosa Ledeb. 479.
— Frolovii Ledeb. 384.
— glacialis Herd. 389.
— gnaphalodes (Royle) Sch. Bip. ex Klatt 390.
— grandifolia Maxim. 464.
— hypargyrea Lipsch. et Vved. 440.
— involucrata Kar. et Kir. ex Maxim. 383.
— ispajensis Iljin 409.
— Jadrinzevii Kryl. 429.
— karaartscha Saposhn. 511; tab. 38.
— kaschgarica Rupr. 509.
— Kitamurana Miyabe et Tatewaki 432; tab. 31.
— Krylovii Schischk. et Serg. 396.
— kurilensis Tatewaki 452; tab. 32.
— Kuschakeviczii C. Winkl. 502.
— laciniata Ledeb. 529.
— Larionovii C. Winkl. 425.
— latifolia Ledeb. 478.
— lenensis M. Pop. ex Lipsch. 441.
— leucophylla Schrenk 399.
— lomatolepis Lipsch. 412.
— manshurica Kom. 371.
— masarica Lipsky 513; tab. 30.
— Maximoviczii Herd. 448.
— Mikeschinii Iljin 408.
— neopulchella Lipsch. 525.
— neoserrata Nakai 475; tab. 34.
— Ninae Iljin 404.
— nuda Ledeb. 515.
— nupuripoensis Miyabe et Miyake 450.
— odontolepis Sch. Bip. ex Herd. 428.
— ovata Benth. 503.
— oxyodonta Hult. 488.
— × paradoxa Lipsch. 506.
— parviflora (Poir.) DC. 472; tab. 34.
— Poljakovii Glehn 483; tab. 35.
— porcellanea Lipsch. 431.
— Porcii Degen 474.
— Pricei Simps. 480.
— prostrata C. Winkl. 534.
— pseudoangustifolia Lipsch. 501.
— pseudosquarrosa M. Pop. et Lipsch. 437.
— pseudo-Tilesii Lipsch. 487; tab. 36.
— pulchella Fisch. 522.
— pulviniformis C. Winkl. 427.
— recurvata (Maxim.) Lipsch. 445.
— Riederi Herd. 453; tab. 32.
— rigida Ledeb. 403.
— robusta Ledeb. 533.
— runcinata DC. 528.
— sachalinensis Fr. Schmidt 455; tab. 33.
— sajanensis Gudoschn. 433; tab. 31.
— Salemannii C. Winkl. 421; tab. 30.
— salicifolia (L.) DC. 410.
— salsa (Pall.) Spreng. 504; tab. 37.
— saxosa Lipsch. 497.
— Schanginiana (Wydl.) Fisch. ex Herd. 397.
— serratuloides Turcz. 416.
— shiretokoensis Sugaw. 459.
— sinuata Kom. 468.
— Soczavae Lipsch. 486; tab. 36.
— sordida Kar. et Kir. 400.
— sovietica Kom. 434.
— splendens Kom. 443.
— squarrosa Turcz. 436.
— Stubendorfii Herd. 495.
— subtriangulata Kom. 465.
— Sukaczevii Lipsch. 492; tab. 35.
— sulcata Iljin 388.
— Sumneviczii Serg. 414.
— tadshikorum Iljin 514.
— Tilesii Ledeb. 184; tab. 36.
— tomentosa Kom. 430; tab. 31.
— triangulata Trautv. et Mey. 469.
— tschuktschorum Lipsch. 518.
— turgaiensis B. Fedtsch. 508; tab. 38.
— umbrosa Kom. 462.
— × uralensis Lipsch. 442.
— ussuriensis Maxim. 446.
— viscida Hult. 500.
— Vvedenskyi Lipsch. 420.
Saussuria Moench 20, 282.
Savastana Schrank 2, 59.
Saxifraga L. 9, 138.

— abchasica Oett. 191; tab. 11.
— adenophora C. Koch 179.
— adscendens L. 166; tab. 9.
— aizoides L. 186.
— Alberti Rgl. et Schmalh. 194; tab. 11.
— anadyrensis Losinsk. 180, 487; tab. 10.
— androsacea L. 174; tab. 10.
— astilbeoides Losinsk. 149, 486; tab. 8.
— bracteata D. Don 173.
— bronchialis L. 183; tab. 10.
— caespitosa L. 175; tab. 10.
— calycina Sternb. 151; tab. 8.
— carinata Oett. 193.
— cartilaginea Willd. ex Sternb. 186; tab. 11.
— caucasica Somm. et Lévier 191.
— cernua L. 168.
— cherlerioides D. Don 185; tab. 10.
— colchica Albov 189; tab. 11.
— columnaris Schmalh. 194; tab. 11.
— coriifolia (Somm. et Lévier) Grossh. 163.
— cortusifolia Sieb. et Zucc. 198.
— cymbalaria L. 165; tab. 9.
— dahurica Willd. 151; tab. 8.
— Desoulavyi Oett. 192.
— Dinnikii Schmalh. 197; tab. 11.
— Eschscholzii Sternb. 185.
— exarata Vill. 178; tab. 10.
— exilis Steph. 173.
— firma Litv. 184, 488; tab. 10.
— flagellaris Willd. ex Sternb. 161; tab. 9.
— foliolosa R. Br. 157; tab. 8.
— grandipetala (Engl. et Irmsch.) Losinsk. 151.
— granulata L. 167; tab. 9.
— hieracifolia Waldst. et Kit. 157.
— hirculus L. 160.
— Huetiana Boiss. 164; tab. 9.
— irrigua M. B. 167; tab. 9.
— juniperifolia Adam 188; tab. 11.
— Kolenatiana Rgl. 187; tab. 11.
— Komarovii Losinsk. 62, 486; tab. 9.
— Korshinskyi Kom. 146; tab. 8.
— Kotschyi Boiss. 193.
— Kusnezowiana Oett. 192.
— lactea Turcz. 175.
— laevis M. B. 189; tab. 11.
— manshuriensis (Engl.) Kom. 149; tab. 8
— melaleuca Fisch. 156.
— Merckii Fisch. 159.
— mollis Smith 172; tab. 9.
— moschata Wulf. 177; tab. 10.
— Nelsoniana D. Don 148.
— nivalis L. 152.
— nudicaulis D. Don 145; tab. 8.
— oblongifolia Nakai 198.
— oppositifolia L. 197.
— parnassioides Rgl. et Schmalh. 160.
— pontica Albov 178.
— pseudolaevis Oett. 190; tab. 11.
— punctata L. 147; tab. 8.
— purpurascens Kom. 149; tab. 8.
— Redowskiana Sternb. 150; tab. 8.
— Redowskii Adam 158; tab. 8.
— reniformis Ohwi 147; tab. 8.
— rivularis L. 172; tab. 9.
— sachalinensis Fr. Schmidt 156.
— scleropoda Somm. et Lév. 190.
— serpyllifolia Pursh 163.
— setigera Pursh 162; tab. 9.
— sibirica L. 171; tab. 9.
— Sieversiana Sternb. 146.
— sileniflora Sternb. 174.
— spinulosa Adam 183; tab. 10.
— stellaris L. 158.
— subverticillata Boiss. 188; tab. 11.
— tenuis (Wahl.) H. Smith 155.
— terektensis Bge. 176.
— Tilingiana Rgl. et Til. 155.
— tridactylites L. 165; tab. 9.
— verticillata Losinsk. 180; 487.
Scabiosa L. 24, 56.
— alpestris Kar. et Kir. 63; tab. 4.
— amoena Jacq. f. 80.
— argentea L. 66.
— austro-altaica Bobr. 62, 457; tab. 7.
— bipinnata C. Koch 77; tab. 6.
— caucasica M. B. 63; tab. 4.
— colchica Stev. 81; tab. 6.
— columbaria L. 84.
— comosa Fisch. ex Roem. et Schult. 89; tab. 6, 7.
— correvoniana Somm. et Lévier 79.
— georgica Sulak. 76.
— gumbetica Boiss. 65; tab. 5.
— hyrcanica Stev. 79.
— imeretica (Somm. et Lévier) Sulak. 78.
— isetensis L. 61.
— lachnophylla Kitag. ex Nakai 86.
— lucida Vill. 85.
— meskhetica Schchian 85.
— micrantha Desf. 70; tab. 6.
— ochroleuca L. 75.
— Olgae Albov 65; tab. 5.
— Olivieri Coult. 72.
— Owerinii Boiss. 83.
— persica Boiss. 70.
— rotata M. B. 71; tab. 6.

Scabiosa soongorica Schrenk 64; tab. 4.
— Sosnowskyi Sulak. 78.
— ucranica L. 69.
— Ulugbekii Zak. 475.
— Velenovskiana Bobr. 82; tab. 6.
Scaligeria DC. 16, 204.
— alaica (Lipsky) Korov. 210.
— allioides (Rgl. et Schmalh.) Boiss. 212.
— bucharica Korov. 16, 207.
— conica Korov. 209.
— glaucescens (DC.) Boiss. 213.
— ferganensis Lipsky 215.
— hirtula (Rgl. et Schmalh.) Lipsky 214.
— Knorringiana Korov. 208.
— kopetdaghensis (Korov.) Schischk. 207; tab. 34.
— Korovinii Bobr. 211, 593.
— Korshinskyi (Lipsky) Korov. 216.
— Lipskyi Korov. 208.
— platyphylla Korov. 20, 592.
— polycarpa Korov. 214.
— samarcandica Korov. 212.
— setacea (Schrenk) Korov. 217.
— tschimganica Korov. 217.
— transcaspica Korov. 216.
— ugamica Korov. 211.
Scandicium Thell. 16, 146.
Scandix L. 16, 139.
— Aucheri Boiss. 147; tab. 10.
— falcata Lond. 145; tab. 10.
— iberica M. B. 142; tab. 10.
— pecten-Veneris L. 141.
— persica Mart. 141.
— stellata Soland. 146; tab. 10.
Scaphospermum Korov. 17, 292, 358.
— asiaticum Korov. 292, 358.
Scaposi Hoffm. 26, 750.
Scariola F. W. Schmidt 29, 317.
— Albertoregelia (C. Winkl.) Kirp. 322.
— orientalis (Boiss.) Soják 320; tab. 23.
— viminea (L.) F. W. Schmidt 318; tab. 23.
Schanginia C. A. Mey. 6, 177, 182.
Schanginia Sievers. ex Pall. 15, 279.
Schellhammeria Moench 3, 162.
Scheuchzeria L. 1, 277.
— palustris L. 278; tab. 10.
Schischkinia Iljin 28, 350.
— albispina (Bge.) Iljin 352; tab. 20.
Schismus Beauv. 2, 365.
— arabicus Nees 365; tab. 27.
— calycinus (L.) Duv.-Jouve 365; tab. 27.
— minutus (Hoffm.) Roem. et Schult. 366.
Schiverekia Andrz. 8, 368.
— podolica Andrz. 371.
Schizachne Hack. 2, 543.
— callosa (Turcz.) Roshev. 543.
— Komarovii Roshev. 544.
Schizandra Rich. 7, 566.
— chinensis (Turcz.) Baill. 566; tab. 39.
Schizogyne Cass. 25, 433.
Schizonepeta Briq. 20, 282.
— annua (Pall.) Schischk. 284.
— multifida (L.) Briq. 283; tab. 14.
Schizopepon Maxim. 24, 96.
— bryoniifolius Maxim. 96.
Schlosseria Vucot. 17, 168.
Schmalhausenia C. Winkl. 27, 360.
— nidulans (Rgl.) Petrak 361.
Schoberia C. A. Mey. 6, 182, 191.
Schoenus L. 3, 98.
— ferrugineus L. 101; tab. 5.
— nigricans L. 101; tab. 5.
Schraderia Medic. 21, 363.
— acetabulosa (Vahl) Pobed. 369; tab. 18.
— bucharica (M. Pop.) Nevski 364; tab. 18.
— dracocephaloides (Boiss.) Pobed. 363; tab. 19.
— Korolkovii (Rgl. et Schmalh.) Pobed. 373; tab. 19.
Schrenkia Fisch. et Mey. 16, 192.
— Golickeana (Rgl. et Schmalh.) B. Fedtsch. 195.
— insignis Lipsky 193.
— involucrata Rgl. et Schmalh. 196.
— Kultiassovii Korov. 197.
— papillaris Rgl. et Schmalh. 194.
— pungens Rgl. et Schmalh. 197.
— vaginata (Ledeb.) Fisch. et Mey. 196.
Schtschurovskia Rgl. 16, 186.
— meifolia Rgl. et Schmalh. 187; tab. 14a.
— pentaceros (Korov.) Schischk. 188.
Schudia Mol. 16, 148.
Schultzia Spreng. 16, 540.
— albiflora (Kar. et Kir.) M. Pop. 541; tab. 34.
— crinita (Pall.) Spreng. 540; tab. 34.
Schumannia Kuntze 17, 147.
— Karelinii (Bge.) Korov. 148; tab. 6. tab. 23.
Schumeria Iljin 28, 305.
— latifolia (Boiss.) Iljin 307.
— Litwinowii (Iljin) Iljin 308.
Schweykerta C. C. Gmel. 18, 643.
Scilla L. 4, 369.
— armena Grossh. 377.
— atropatana Grossh. 372, 740; tab. 44.

— autumnalis L. 373; tab. 9.
— bifolia L. 373; tab. 9.
— bucharica Dessjatova 379; tab. 9.
— caucasica Miscz. 376.
— diziensis Grossh. 376.
— Hohenackeri Fisch. et Mey. 375; tab. 9.
— japonica (Thunb.) Baker 372; tab. 9.
— Misczenkoana Grossh. 378.
— monanthos C. Koch 377.
— nivalis Boiss. 374.
— puschkinioides Rgl. 378.
— Raevskiana Rgl. 379; tab. 9.
— Rosenii C. Koch 374; tab. 9.
— sibirica Andrevs 376; tab. 9.
— Winogradowii Sosn. 375; tab. 9.
Scirpus L. 3, 42.
— avatschensis Kom. 46; tab. 3.
— bucharicus Roshev. 55; tab. 4.
— depauperatus Kom. 52, 4.
— Ehrenbergii W. Beck. 51; tab. 4.
— erectus Poir. 54; tab. 4.
— Komarovii Roshev. 54; tab. 4.
— lacustris L. 47; tab. 4.
— litoralis Schrad. 52, 4.
— lupulinus (Nees) Roshev. 53; tab. 4.
— melanospermus C. A. Mey. 54; tab. 4.
— mucronatus L. 51; tab. 4.
— planiculmis Fr. Schmidt 48; tab. 4.
— radicans Schkuhr 45; tab. 3.
— setaceus L. 46; tab. 4.
— silvaticus L. 44; tab. 3.
— supinus L. 53; tab. 4.
— Tabernaemontani C. C. Gmel. 47; tab. 4.
— triqueter L. 48; tab. 4.
— Wichurai W. Beck. 45; tab. 3.
Sclarea Moench 21, 291.
Scleranthus L. 6, 546.
— annuus L. 547; tab. 31.
— perennis L. 547; tab. 31.
— polycarpus L. 548.
— uncinatus Schur 548; tab. 31.
Sclerocaryopsis Brand 19, 416.
Sclerochloa Beauv. 2, 364.
— dura (L.) Beauv. 364; tab. 27.
Sclerocyathium Prokh. 14, 336.
Sclerophyllum Gaudin 29, 595, 673.
Scleropoa Griseb. 2, 541.
— rigida (L.) Griseb. 542; tab. 41.
— Woronowii Hack. 541; tab. 41.
— rigescens (Trin.) Grossh. 542; tab. 41.
Scolochloa Link 2, 448.
— festucacea Link 448.
Scolymus L. 29, 11.
— hispanicus L. 12.
— maculatus L. 13.
Scopolia Jacq. 22, 99.
— carniolica Jacq. 100; tab. 4.
Scopolina Schult. 22, 99.
Scordium Gilib. 20, 39.
Scorodonia Adans. 20, 39.
Scorodosma Bge. 17, 62, 72.
Scorpioides Tourn. 13, 242.
Scorpiurus L. 13, 242.
— minima Losinsk. 242, 552.
— subvillosa L. 242.
Scorpius Loisel. 13, 242.
Scorzonera L. 29, 27.
— acanthoclada Franch. 89.
— alaica Lipsch. 64.
— Albertoregelia C. Winkl. 64.
— albicaulis Bge. 61; tab. 3.
— armeniaca (Boiss. et Huet) Boiss. 41; tab. 6.
— austriaca Willd. 70.
— baldshuanica Lipsch. 66.
— bicolor Freyn et Sint. 55.
— Biebersteinii Lipsch. 103; tab. 6.
— bracteosa C. Winkl. 62.
— Bungei Krasch. et Lipsch. 107.
— calcitrapifolia Vahl 39.
— cana (C. A. Mey.) O. Hoffm. 44.
— chantavica Pavl. 88.
— circumflexa Krasch. et Lipsch. 109.
— crassifolia Krasch. et Lipsch. 95.
— crispa M. B. 68; tab. 4.
— curvata (Popl.) Lipsch. 72.
— dianthoides (Lipsch. et Krasch.) Lipsch. 72.
— ensifolia M. B. 102.
— ferganica Krasch. 94.
— filifolia Boiss. 73.
— Franchetii Lipsch. 81.
— gageoides Boiss. 108.
— gracilis Lipsch. 85, 721.
— Grigoraschvilii (Sosn.) Lipsch. 48.
— Grossheimii Lipsch. et Vass. 44, 718.
— hispanica L. 87.
— hissarica C. Winkl. 97.
— humilis L. 78.
— Idae (Sosn.) Lipsch. 42.
— iliensis Krasch. 83.
— inconspicua Lipsch. ex Pavl. 87.
— Ketzkhovelii Sosn. 98.
— Kirpicznikovii Lipsch. 47, 719.
— Koslovskyi Sosn. 76.
— kuhistanica M. Pop. 90.
— lachnostegia (Woronow) Lipsch. 37.
— laciniata L. 38.

Scorzonera lanata (L.) Hoffm. 105.
— latifolia (Fisch. et Mey.) DC. 99.
— leptophylla (DC.) Krasch. et Lipsch. 50; tab. 3.
— Lipskyi Lipsch. 74.
— Litvinovii Krasch. et Lipsch. 110.
— Meyeri (C. Koch) Lipsch. 46; tab. 2.
— mollis M. B. 49.
— mongolica Maxim. 80.
— ovata Trautv. 95.
— oxiana M. Pop. 110.
— papposa DC. 96.
— parviflora Jacq. 79.
— Petrovii Lipsch. 66.
— pratorum (Krasch.) Stank. 69.
— pseudodivaricata Lipsch. 91.
— pseudolanata Grossh. 105.
— pubescens DC. 86.
— pulchra Lomak. 100.
— purpurea L. 56; tab. 4.
— pusilla Pall. 92.
— Raddeana C. Winkl. 54.
— radiata Fisch. 77; tab. 5.
— ramosissima DC. 104.
— rigida Auch. ex DC. 75.
— rosea Waldst. et Kit. 58.
— Ruprechtiana Lipsch. et Krasch. 69.
— Safievii Grossh. 101.
— Schischkinii Lipsch. et Vass. 42, 716.
— Seidlitzii Boiss. 75; tab. 5.
— sericeo-lanata (Bge.) Krasch. et Lipsch. 107.
— songorica (Kar. et Kir.) Lipsch. et Vass. 40; tab. 2.
— Sosnovskyi Lipsch. 98.
— stricta Hornem. 85.
— subacaulis Lipsch. 67.
— suberosa C. Koch 52.
— tadshikorum Krasch. et Lipsch. 93.
— tau-saghyz Lipsch. et Bosse 59.
— tragopogonoides Rgl. et Schmalh. 65.
— transiliensis M. Pop. 84.
— tuberosa Pall. 106.
— turcomanica Krasch. et Lipsch. 53.
— turkestanica Franch. 82.
— Turkeviczii Krasch. et Lipsch. 52.
Scorzoneroides Moench 29, 207.
Scrophularia L. 22, 229.
— alata Gilib. 270.
— altaica Murr. 261; tab. 11.
— amgunensis F. Schmidt 265.
— amplexicaulis Benth. 252.
— armeniaca Bordz. 286.
— atropatana Grossh. 279.
— canescens Bong. 298; tab. 13.
— canina L. 301.
— Charadzeae Kem-Nath. 278.
— chlorantha Kotschy et Boiss. 258.
— chrysantha Jaub. et Spach 249.
— cretacea Fisch. 297.
— czapandaghii B. Fedtsch. 304.
— Czernjakowskiana B. Fedtsch. 272.
— decipiens Boiss. et Kotschy 292.
— dissecta (B. Fedtsch.) Gorschk. 300; tab. 13.
— divaricata Ledeb. 256.
— donetzica Kotov 276.
— exilis Popl. 289.
— Fedtschenkoi Gorschk. 293.
— frigida Boiss. 281.
— Goldeana Juz. 278.
— Gontscharovii Gorschk. 295.
— Grayana Maxim. ex Kom. 271.
— Grossheimii Schischk. 290.
— haematantha Boiss. et Heldr. 296.
— heucheriiflora Schrenk 260; tab. 11.
— hyrcana Grossh. 250; tab. 10.
— ilvensis C. Koch 255; tab. 11.
— imerethica Kem.-Nath. 279.
— incisa Weinm. 307.
— integrifolia Pavl. 282.
— kabadianensis B. Fedtsch. 305.
— Kiriloviana Schischk. 306.
— Kotschyana Benth. 248.
— lateriflora Trautv. 246.
— leucoclada Bge. 297.
— Litwinowii B. Fedtsch. 281.
— lunariifolia Boiss. et Bal. 250.
— macrobotrys Ledeb. 266.
— mandshurica Maxim. 262.
— Maximowiczii Gorschk. 262.
— minima M. B. 275; tab. 12.
— mollis Somm. et Lévier 257.
— multicaulis Turcz. 295; tab. 12.
— nachitschevanica Grossh. 280.
— nervosa Benth. 274.
— Nikitinii Gorschk. 247.
— nodosa L. 269.
— Oldhami Oliv. 270; tab. 11.
— Olgae Grossh. 286.
— olympica Boiss. 288.
— orientalis L. 273.
— pamiro-alaica Gorschk. 294.
— peregrina L. 258.
— pruinosa Boiss. 300; tab. 13.
— rostrata Boiss. et Buhse 287.
— rupestris M. B. ex Willd. 277.
— Ruprechtii Boiss. 288; tab. 12.

— rutifolia Boiss. 285.
— sangtodensis B. Fedtsch. 306.
— sareptana Kleop. 276.
— Scopolii Hoppe ex Pers. 259.
— Sprengeriana Somm. et Lévier 257.
— striata Boiss. 291.
— tadshicorum Gontsch. 247; tab. 10.
— thesioides Boiss. et Buhse 303; tab. 13.
— turcomanica Bornm. et Sint. ex Rech. 304.
— variegata M. B. 302.
— vernalis L. 251; tab. 10.
— verticillata Gontsch. et Grig. 246.
— xanthoglossa Boiss. 291; tab. 12.
— zaravschanica Gorschk. et Zak. 293.
— zuvandica Grossh. 299.
Scutellaria L. 20, 72.
— adenostegia Briq. 168; tab. 10.
— adsurgens M. Pop. 176.
— Albertii Juz. 147.
— albida L. 111.
— Alexeenkoi Juz. 197, 513.
— altaica Fisch. ex Sweet 184.
— altissima L. 105; tab. 6.
— amphichlora Juz. 166.
— Androssovii Juz. 139.
— andrachnoides Vved. 219, 514.
— angrenica Juz. et Vved. 175, 511.
— Anitae Juz. 201; tab. 12.
— araratica Grossh. 167.
— araxensis Grossh. 128.
— artvinensis Grossh. 162.
— baicalensis Georgi 103; tab. 5.
— baldshuanica Nevski ex Juz. 220; tab. 13.
— bucharica Juz. 169.
— Catharinae Juz. 145.
— chenopodiifolia Juz. 119, 508.
— Chitrovoi Juz. 181.
— cisvolgensis Juz. 182.
— colpodea Nevski 115.
— comosa Juz. 137; tab. 8.
— cordifrons Juz. 175.
— creticola Juz. 180.
— cristata M. Pop. 116; tab. 7.
— daghestanica Grossh. 155.
— darriensis Grossh. 165.
— darvasica Juz. 207.
— dependens Maxim. 98; tab. 6.
— dubia Taliev et Schirj. 97.
— Fedtschenkoi Bornm. 209.
— filicaulis Rgl. 189; tab. 11.
— flagellulata Juz. 136, 508.
— galericulata L. 90.
— glabrata Vved. 124.
— Gontscharovii Juz. 169.
— grandiflora Sims 132; tab. 8.
— granulosa Juz. 156.
— Grossheimiana Juz. 162.
— guttata Nevski ex Juz. 223; tab. 13.
— haematochlora Juz. 170.
— hastifolia L. 97; tab. 5.
— Helenae Albov 114.
— heterochroa Juz. 152.
— heterotricha Juz. et Vved. 212.
— hirtella Juz. 152.
— hissarica B. Fedtsch. 210; tab. 13.
— holosericea Gontsch. ex Juz. 217.
— hypopolia Juz. 151.
— Ikonnikovii Juz. 96; tab. 5.
— immaculata Nevski ex Juz. 223.
— intermedia M. Pop. 123.
— irregularis Juz. 185.
— Iskanderi Juz. 172.
— jodudiana B. Fedtsch. 206; tab. 12.
— Juzepczukii Gontsch. 119.
— karatavica Juz. 198; tab. 11.
— Karjaginii Grossh. 163; tab. 9.
— karkaralensis Juz. 134.
— Knorringiae Juz. 186.
— Krasevii Kom. et I. Schischk. 92.
— Krylovii Juz. 149.
— kugarti Juz. 190.
— Kurssanovii Pavl. 141.
— lanipes Juz. 196, 513.
— leptosiphon Nevski 225.
— leptostegia Juz. 132.
— Linczevskii Juz. 194.
— Lipskyi Juz. 213.
— Litwinowii Bornm. et Sint. 120; tab. 7.
— luteo-coerulea Bornm. et Sint. 121.
— megalodonta Juz. 221, 515.
— mesostegia Juz. 141.
— microdasys Juz. 139.
— microphysa Juz. 205.
— Minkwitziae Juz. 510.
— mongolica Sobol. 185.
— moniliorrhiza Kom. 102.
— navicularis Juz. 118; tab. 7.
— nepetoides M. Pop. 135.
— Nevskii Juz. et Vved. 211.
— novorossica Juz. 153.
— ocellata Juz. 198; tab. 11.
— oligodonta Juz. 187.
— orbicularis Bge. 222.
— oroephila Grossh. 154.
— orientalis L. 157.
— oschtenica Juz. 155, 510.

Scutellaria oxyphylla Juz. 182; tab. 9.
— oxystegia Juz. 138.
— pacifica Juz. 107; tab. 6.
— pallida M. B. 111; tab. 6.
— pamirica Juz. 202; tab. 12.
— Paulsenii Briq. 188.
— phyllostachya Juz. 177.
— physocalyx Rgl. et Schmalh. 201.
— picta Juz. 136; tab. 8.
— platystegia Juz. 161.
— poecilantha Nevski ex Juz. 221.
— polyodon Juz. 153; tab. 9.
— polyphylla Juz. 102.
— polytricha Juz. 205.
— pontica C. Koch 113.
— Popovii Vved. 190, 511.
— Prilipkoana Grossh. 164.
— Przewalskii Juz. 148.
— pycnoclada Juz. 172; tab. 10.
— Raddeana Juz. 130.
— ramosissima M. Pop. 122; tab. 7.
— Regeliana Nakai 95.
— rhomboidalis Grossh. 164.
— rubromaculata Juz. et Vved. 224.
— schachristanica Juz. 207; tab. 12.
— schugnanica B. Fedtsch. 212.
— scordiifolia Fisch. ex Schrank 99; tab. 5.
— Sedelmeyeriae Juz. 158.
— sevanensis Sosn. 178.
— shikokiana Makino 108.
— Sieversii Bge. 150; tab. 10.
— soongorica Juz. 148, 509.
— Sosnovskyi Takht. 165; tab. 9.
— squarrosa Nevski 124.
— Stevenii Juz. 130.
— striatella Gontsch. 116; tab. 7.
— subcaespitosa Pavl. 195.
— subcordata Juz. 194, 512.
— supina L. 183.
— talassica Juz. 193.
— Taquetii Lévl. et Vaniot 100; tab. 5.
— Tatianae Juz. 130.
— taurica Juz. 129; tab. 8.
— Titovii Juz. 142.
— toguztoraviensis Juz. 196.
— Tournefortii Benth. 106; tab. 6.
— transiliensis Juz. 146; tab. 10.
— tschimganica Juz. 171, 510; tab. 10.
— tuminensis Nakai 101.
— turgaica Juz. 133.
— tuvensis Juz. 133.
— urticifolia Juz. et Vved. 171, 510.
— ussuriensis (Rgl.) Kudo 106.
— velutina Juz. et Vved. 211, 514.
— verna Bess. 179.
— villosissima Gontsch. ex Jus. 214.
— Woronowii Juz. 112.
— xanthosiphon Juz. 186.
— yezoënsis Kudo 95.
Sczukinia Turcz. 18, 625.
Secale L. 2, 666.
— anatolicum Boiss. 668.
— cereale L. 667.
— Kuprijanovii Grossh. 668; tab. 47a.
— silvestre Host 666, tab. 47a.
— Vavilovii Grossh. 668.
Securigera DC. 11, 281.
— securidaca (L.) Degen et Döerfl. 281; tab. 18.
Securinega Somm. 14, 282.
— suffruticosa (Pall.) Rend. 283.
Sedum L. 9, 45.
— acre L. 88.
— aetnense Tineo 98.
— aizoon L. 67.
— Alberti Rgl. 85.
— alboroseum Baker 63.
— album L. 82.
— anacampseros L. 66.
— annuum L. 95.
— atratum L. 93.
— bucharicum Boriss. 91, 481.
— caucasicum (Grossh.) Boriss. 57, 480; tab. 4.
— corymbosum Grossh. 93.
— cyaneum J. Rudolph 65.
— eupatorioides Kom. 63.
— Ewersii Ledeb. 64.
— fabaria Koch 66.
— gracile C. A. Mey. 83.
— hispanicum L. 89; tab. 5.
— hybridum L. 70, tab. 4.
— hyperaizoon Kom. 68.
— involucratum M. B. 78.
— kamtschaticum Fisch. 69.
— lenkoranicum Grossh. 84; tab. 5.
— Listoniae Vis. 79.
— litorale Kom. 71.
— lydium Boiss. 81.
— maximum (L.) Suter 55.
— Middendorfianum Maxim. 69.
— mugodsharicum Boriss. 54, 479.
— nanum Boiss. 95.
— obtusifolium C. A. Mey. 79.
— oppositifolium Sims 77.
— pallescens Freyn 62; tab. 4.
— pallidum M. B. 92; tab. 5.

— parvistamineum V. Petrov 55.
— pentapetalum Boriss. 90, 481; tab. 5.
— pluricaule Kudo 65; tab. 4.
— polytrichoides Hemsl. ex Forbes et Hemsl. 89.
— populifolium Pall. 67; tab. 4.
— purpureum (L.) Schult. 53.
— reflexum L. 86.
— rubens L. 96.
— rubrum (L.) Thell. 97; tab. 5.
— Selskianum Rgl. et Maak 71.
— sexangulare L. 87.
— spurium M. B. 74.
— Stevenianum Rouy et Camus 72.
— stoloniferum S. G. Gmel. 73; tab. 5.
— subulatum (C. A. Mey.) Boiss. 84; tab. 5.
— telephium L. 56.
— tenellum M. B. 80.
— tetramerum Trautv. 98; tab. 5.
— ussuriense Kom. 61.
— verticillatum L. 58.
— villosum L. 94.
— viviparum Maxim. 61.
Seidlitzia Bge. 6, 275.
— florida (M. B.) Boiss. 276.
— rosmarinus (Ehrh.) Bge. 275.
Selaginella Spring 1, 123.
— Aitchisonii Hieron. 125; tab. 6.
— borealis (Kaulf.) Rupr. 125; tab. 6.
— helvetica (L.) Link 126; tab. 6.
— involvensis Spring 126; tab. 6.
— Rossii (Baker) Warburg 126; tab. 6.
— sanguinolenta (L.) Spring 125; tab. 6.
— selaginoides (L.) Link 124; tab. 6.
— sibirica (Milde) Hieron. 124; tab. 6.
Selinopsis Coss. et Dur. ex Munby 16, 385.
Selinum L. 17, 168.
Selinum L. 16, 560.
— carvifolia L. 560; tab. 34.
— Kultiassovii Korov. 561, 604.
— Popovii (Korov.) Schischk. 563.
— tianschanicum Korov. 562.
Semenovia Rgl. et Herd. 17, 223.
Semiaquilegia Makino 7, 75.
— manshurica Kom. 75.
Sempervivum L. 9, 15.
— caucasicum Rupr. ex Boiss. 18; tab. 2.
— glabrifolium Boriss. 22, 471.
— globiferum L. 21.
— pumilum M. B. 16.
— ruthenicum (Koch) Schnittsp. et Lehm. 21.
— soboliferum Sims 23; tab. 2.
— tectorum L. 17.
Senecillis Gaertn. 26, 788, 836, 844.
Senecio L. 26, 699.
— ambraceus Turcz. ex DC. 715.
— amurensis Schischk. 757, 885.
— aquilonaris Schischk. 763, 884.
— arcticus Rupr. 752.
— argunensis Turcz. 716.
— asiaticus Schischk. et Serg. 762.
— atripurpureus (Ledeb.) B. Fedtsch. 771; tab. 35.
— Besserianus Mind. 768.
— borystenicus Andrz. ex DC. 718.
— Buschianus Sosn. 726.
— Calvertii Boiss. 780.
— campester (Retz.) DC. 753.
— cannabifolius Less. 747.
— capitatus (Wahl.) Steud. 768.
— carniolicus Willd. 748.
— carpathicus Herbich 750.
— caucasigenus Schischk. 769.
— cineraria DC. 720.
— cladobotrys Ledeb. 759.
— Czernjaevii Mind. 754.
— dubius Ledeb. 781; tab. 36.
— erraticus Bertol. 717.
— erucifolius L. 710.
— ferganensis Schischk. 720, 881.
— flammeus Turcz. ex DC. 767.
— fluviatilis Wallr. 745.
— Franchetii C. Winkl. 729.
— Fuchsii C. C. Gmel. 745.
— grandidentatus Ledeb. 711; tab. 33.
— hieraciiformis Kom. 765.
— Igoschinae Schischk. 763, 885.
— Iljinii Schischk. 743, 882.
— Jacobaea L. 715.
— jacuticus Schischk. 779.
— jailicola Juz. 757.
— karelinioides C. Winkl. 751.
— Karjaginii Sof. 770; tab. 35
— Kawakamii Makino 761.
— Kolenatianus C. A. Mey. 728.
— Korshinskyi Krasch. 712.
— Krascheninnikovii Schischk. 783; tab. 36.
— kubensis Grossh. 724.
— lapsanoides DC. 744.
— lenensis Schischk. 758, 885.
— Lipskyi Lomak. 722.
— Litvinovii Schischk. 747.
— Lorentii Hochst. 721.
— macrophyllum M. B. 730.
— Massagetovii Schischk. 722, 882.
— nemorensis L. 742.

Senecio Noëanus Rupr. 787.
— Olgae Rgl. et Schmalh. 728.
— Othonnae M. B. 741.
— paludosus L. 735.
— pandurifolius C. Koch 724.
— papposus (Reichb.) Less. 760.
— paucifolius J. G. Gmel. 733.
— paucilobus DC. 741.
— Paulsenii O. Hoffm. ex Pauls. 730.
— platyphylloides Somm. et Levier 740.
— Pojarkoviae Schischk. 739.
— polycephalus Ledeb. 785.
— porphyranthus Schischk. 766.
— Pricei Simps. 775.
— propinquus Schischk. 743.
— pseudoarnica Less. 723.
— pseudoaurantiacus Kom. 777.
— pseudoorientalis Schischk. 727.
— pyroglossus Kar. et Kir. 770.
— quinqueligulatus C. Winkl. 744.
— racemosus (M. B.) DC. 731; tab. 34.
— racemulifer Pavl. 735.
— Renardii C. Winkl. 749; tab. 33.
— resedifolius Less. 750.
— Reverdattoi K. Sobol. 778.
— rhombifolius (Willd.) Sch. Bip. 739.
— rivularis (Waldst. et Kit.) DC. 765.
— Saposhnikovii Krasch. 717.
— Schischkinianus Sof. 719.
— Schvetzovii Korsh. 733.
— sichotensis Kom. 764.
— Sosnovskyi Sof. 784; tab. 36.
— subalpinus W. D. J. Koch 726.
— subdentatus Ledeb. 786.
— subfloccosus Schischk. 760.
— subfrigidus Kom. 777.
— subscaposus Kom. 766.
— succisifolius Kom. 758.
— Sukaczevii Schischk. 762.
— Sumneviczii Schischk. et Serg. 777.
— sylvaticus L. 782.
— taraxacifolius (M. B.) DC. 725; tab. 34.
— tataricus Less. 736.
— thyrsophorus C. Koch 732.
— tianschanicus Rgl. et Schmalh. 749.
— Tichomirovii Schischk. 772, 886.
— tundricola Tolm. 776.
— Turczaninovii DC. 775.
— umbrosus Waldst. et Kit. 734.
— Veresczaginii Schischk. et Serg. 764.
— vernalis Waldst. et Kit. 783.
— viscosus L. 779.
— vulgaris L. 780.
Serapios L. 4, 721.
— vomeracea (Burb.) Briq. 721; tab. 43.
Sergia Fed. 24, 420, 474.
— Regelii (Trautv.) Fed. 421.
— Sewerzowii (Rgl.) Fed. 420; tab. 25.
Serianthes Benth. ex Hook. 11, 10.
Seridia Juss. 28, 564.
Seriphidium Bess. 26, 562.
Serratula L. 28. 52.
Serratula L. 28, 259.
— alatavica C. A. Mey. 290.
— algida Iljin 283; tab. 15.
— angulata Kar. et Kir. 272.
— aphyllopoda Iljin 289.
— Biebersteiniana (Iljin ex Grossh.) Takht. 275; tab. 14.
— cardunculus (Pall.) Schischk. 280; tab. 16.
— caucasica Boiss. 301.
— centauroides L. 273.
— chartacea C. Winkl. 286; tab. 15.
— coriacea Fisch. et Mey. 293; tab. 16.
— coronata L. 268.
— dissecta Ledeb. 271; tab. 15.
— dshungarica Iljin 284.
— erucifolia (L.) Boriss. 270.
— Gmelinii Tausch 276.
— hastifolia Kult. et Korov. 292.
— heterophylla (L.) Desf. 279.
— inermis Gilib. 265.
— kirghisorum Iljin 283.
— Komarovii Iljin 273.
— lancifolia Zak. 291.
— lyratifolia Schrenk 296; tab. 15.
— marginata Tausch 281.
— Modesti Boriss. 298, 606.
— procumbens Rgl. 287; tab. 14.
— quinquifolia M. B. 294; tab. 14.
— radiata (Waldst. et Kit.) M. B. 274; tab. 17.
— serratuloides (Fisch. et Mey.) Takht. 300.
— sogdiana Bge. 289; tab. 16.
— suffruticosa Schrenk 288.
— tanaitica P. Smirn. 278.
— tianschanica Saposhn. et Nikit. 297.
— transcaucasica (Bornm.) Sosn. ex Grossh. 285; tab. 14.
Sesamum L. 23, 12.
— orientale L. 12.
Seseli L. 16, 483.
— Abolinii (Korov.) Schischk. 505; tab. 34.
— aemulans M. Pop. 511.
— Alexeenkoi Lipsky 514.
— Andronakii Woronow 502, 603.
— annuum L. 490.

— arenarium M. B. 501.
— asperulum (Trautv.) Schischk. 520; tab. 34.
— campestre Bess. 499.
— coronatum Ledeb. 519.
— cuneifolium M. B. 516.
— dichotomum Pall. 514.
— elegans Schischk. 495.
— eriocarpum (Schrenk) B. Fedtsch. 503; tab. 33.
— eriocephalum (Pall.) Schischk. 518; tab. 34.
— fasciculatum Korov. 507.
— foliosum (Somm. et Lévier) Manden. 524; tab. 34.
— giganteum Lipsky 510.
— glabratum Willd. ex Schult. 496.
— grandivittatum (Somm. et Lévier) Schischk. 491.
— gummiferum Pall. 497.
— hippomarathrum Jacq. 513.
— iliense (Rgl. et Schmalh.) Lipsky 509; tab. 32.
— incanum (Steph.) B. Fedtsch. 504; tab. 32.
— jomuticum Schischk. 517; tab. 33.
— karatavicum Schischk. 515.
— Korovinii Schischk. 508.
— Ledebouri G. Don 513.
— Lehmannianum (Bge.) Boiss. 506.
— Lehmannii Degen 498.
— leptocladum Woronow 503.
— macrophyllum Rgl. et Schmalh. 523; tab. 34.
— Pallasii Bess. 492.
— pauciradiatum Schischk. 500, 602.
— petraeum M. B. 497.
— peucedanifolium (Spreng.) Bess. 502.
— peucedanoides (M. B.) K.-Pol. 495.
— platyphyllum (Schrenk) O. et B. Fedtsch. 517; tab. 34.
— ponticum Lipsky 499.
— rupicola Woronow 514.
— sessiliflorum Schrenk 520; tab. 32; 34.
— songoricum Schischk. 505, 602.
— squarrosum Schischk. 511.
— strictum Ledeb. 489.
— tenuisectum Rgl. et Schmalh. 509.
— turbinatum Korov. 507.
— Valentinae M. Pop. 512; tab. 33.
— varium Trev. 491.
Seselopsis Schischk. 16, 414.
— tianschanicum Schischk. 415; tab. 30.
Sesleria Scop. 2, 299.
— autumnalis (Scop.) F. Schultz. 301; tab. 23.
— coerulea (L.) Ard. 300; tab. 23.
— Hefleriana Schur 300.
— phleoides Stev. ex Roem. et Schult. 301; tab. 23.
Sestinia Boiss. 20, 488.
Setaria Beauv. 2, 38.
— glauca (L.) Beauv. 39; tab. 2.
— intermedia (Roth) Roem. et Schult. 40.
— italica (L.) Beauv. 43; tab. 2.
— verticillata (L.) Beauv. 39; tab. 2.
— viridis (L.) Beauv. 40; tab. 2.
Seutera Reichb. 18, 710.
— Wilfordii (Franch. et Savat.) Pobed. 713; tab. 39.
Sewerzowia Rgl. et Schmalh. 12, 298.
Sherardia L. 23, 176.
— arvensis L. 176.
Sibbaldia L. 10, 223.
— macrophylla Turcz. 225, 614.
— Olgae Juz. et Ovcz. 228, 614.
— parviflora Willd. 227.
— procumbens L. 224; tab. 17.
— semiglabra C. A. Mey. 226.
Sibbaldianthe Juz. 10, 229, 615.
— adpressa (Bge.) Juz. 230; tab. 17.
Sibiraea Maxim. 9, 306.
— altaensis (Laxm.) Schneid. 306; tab. 18.
— tianschanica (Krassn.) Pojark. 309, 491.
Sicyos L. 24, 123.
— angulata L. 124.
Sida L. 15, 149.
— spinosa L. 150; tab. 2.
Sideritis L. 20, 253.
— Balansae Boiss. 272; tab. 18.
— catillaris Juz. 261; tab. 17.
— chlorostegia Juz. 260.
— comosa (Rochel) Stank. 271; tab. 18.
— conferta Juz. 266.
— euxina Juz. 266.
— imbrex Juz. 264; tab. 17.
— Marschalliana Juz. 263, 507.
— montana L. 268; tab. 18.
— taurica Steph. ex Willd. 256; tab. 17.
Siebera J. Gay 27, 67.
— pungens (Lam.) DC. 68; tab. 5.
Siebera Hoppe 27, 67.
Sieberia Spreng. 27, 62.
Siegesbeckia L. 25, 536.
— orientalis L. 536; tab. 34.
Sieglingia Bernh. 2, 306.
— decumbens (L.) Bernh. 306; tab. 23.
Sieversia Willd. 10, 242.

Sieversia pentapetala (L.) Greene 243; tab. 18.
— pusilla (Gaertn.) Hult. 244; tab. 18.
Silaum Mill. 16, 545.
Silaus Bernh. 16, 545.
— Besseri DC. 546.
— Popovii Korov. 548; tab. 35.
— pratense (Grantz) Bess. 546.
— Rubtzovii Schischk. 547; tab. 35.
Silenanthe Griseb. et Schenk 6, 704.
Silenastrum Rouy 6, 602.
Silene L. 6, 517.
— acaulis L. 628.
— adenopetala Raik. 605.
— Alexandrae B. Keller 646.
— alpicola Schischk. 674.
— altaica Pers. 646.
— anglica L. 612; tab. 37.
— anisoloba Schrenk 642.
— apetala Willd. 614.
— araratica Schischk. 672.
— arenosa C. Koch 682; tab. 41.
— armena Boiss. 639.
— armeria L. 615.
— artwinensis Schischk. 606.
— Aucheriana Boiss. 657.
— balchaschensis Schischk. 618.
— baldshuanica B. Fedtsch. 643.
— baschkirorum Janisch. 687.
— Bobrovii Schischk. 679, 889.
— Bornmülleri Freyn 658.
— brachypetala Robill et Cast. ex DC. 613.
— brahuica Boiss. 653.
— Brotherana Somm. et Lévier 653.
— bucharica M. Pop. 643.
— bupleuroides L. 637.
— caespitosa Stev. 655.
— capitellata Boiss. 681.
— caucasica (Bge.) Boiss. 668.
— cephalantha Boiss. 654.
— chaetodonta Boiss. 681; tab. 41.
— chamarensis Turcz. 626; tab. 38.
— chlorantha (Willd.) Ehrh. 616.
— chlorifolia Smith 640.
— chloropetala Rupr. 623.
— claviformis Litv. 651.
— commelinifolia Boiss. 609.
— commutata Guss. 597.
— compacta Fisch. 615.
— conformifolia Preobr. 604; tab. 36.
— conica L. 690; tab. 35.
— coniflora Nees ex Otth. 691.
— conoidea L. 690; tab. 35.
— cretacea Fisch. 647.
— crispans Litv. 659.
— Cserei Baumg. 598; tab. 35.
— cyri Schischk. 688; tab. 41.
— dagestanica Rupr. 620; tab. 37.
— densiflora D'Urv. 686.
— depressa M. B. 667.
— dianthoides Pers. 629; tab. 38.
— dichotoma Ehrh. 610; tab. 37.
— eremitica Boiss. 623.
— euxina Rupr. 612.
— foliosa Maxim. 625.
— gasimailikensis B. Fedtsch. 636.
— Gawrilowii Krassn. 617.
— Gebleriana Schrenk 676; tab. 40.
— glaucescens Schischk. 618.
— graminifolia Otth. 625.
— Grossheimii Schischk. 667.
— guntensis B. Fedtsch. 672; tab. 39.
— Hellmannii Claus 688.
— heptapotamica Schischk. 651.
— holopetala Bge. 677.
— humilis C. A. Mey. 662.
— iberica M. B. 610; tab. 37.
— incurvifolia Kar. et Kir. 668; tab. 39.
— indeprensa Schischk. 659.
— ispirensis Boiss. 657.
— italica (L.) Pers. 675.
— jenisseensis Willd. 627; tab. 38.
— karaczukuri B. Fedtsch. 671; tab. 39.
— Komarovii Schischk. 678.
— koreana Kom. 624.
— Korshinskyi Schischk. 673; 889.
— kubanensis Somm. et Lévier 663.
— Kudrjaschevii Schischk. 644, 888.
— kungessana B. Fedtsch. 679.
— Kuschakewiczii Rgl. 670; tab. 39.
— lacera (Stev.) Sims. 602.
— lasiantha C. Koch 632; tab. 38.
— latifolia (Mill.) Rendle et Britt. 596; tab. 35.
— lazica Boiss. 609.
— leptocaulis Schischk. 645.
— linearifolia Otth. 663.
— lithophila Kar. et Kir. 618.
— Litwinowii Schischk. 637.
— longidens Schischk. 639.
— longiflora Ehrh. 636.
— lychnidea C. A. Mey. 662.
— macrostyla Maxim. 624.
— Marcowiczii Schischk. 630.
— Marschallii C. A. Mey. 635.
— media (Litv.) Kleop. 689.
— Meyeri Fenzl 671.
— Michelsonii Preobr. 605; tab. 36.

— multifida (Adam) Rohrb. 601.
— multiflora (Ehrh.) Pers. 617.
— nana Kar. et Kir. 682; tab. 41.
— Nevskii Schischk. 673.
— nutans L. 674.
— obovata Schischk. 680.
— obtusidentata B. Fedtsch. et M. Pop. 644.
— odoratissima Bge. 641.
— Olgiana B. Fedtsch. 641.
— oriena Schischk. 603; tab. 36.
— pachyneura Schischk. 660.
— pamirensis (H. Winkl.) Preobr. 670; tab. 39.
— parviflora (Ehrh.) Pers. 689.
— pauciflora Ledeb. 626.
— pendula L. 614.
— physocalyx Ledeb. 606.
— plurifolia Schischk. 642.
— polaris Kleop. 687, 890.
— Popovii Schischk. 659; tab. 39.
— praemixta M. Pop. 620.
— Prilipkoana Schischk. 658, 888.
— procumbens Murr. 601; tab. 37.
— propinqua Schischk. 635.
— pruinosa Boiss. 656.
— pseudotenuis Schischk. 678; tab. 40.
— pseudotites Bess. ex Reichb. 686.
— pygmaea Adam 661.
— Raddeana Trautv. 605; tab. 36.
— repens Patr. 654.
— rupestris L. 630.
— Ruprechtii Schischk. 631.
— samarkandensis Preobr. 604; tab. 36.
— sarawschanica Rgl. et Schmalh. 645.
— scabrifolia Kom. 677.
— Schafta S. G. Gmel. ex Hohen. 669.
— schugnanica B. Fedtsch. 680.
— Semenovii Rgl. 651; tab. 40.
— sibirica (L.) Pers. 676; tab. 40.
— sisianica Boiss. et Buhse 660.
— solenantha Trautv. 638.
— spergulifolia (Desf.) M. B. 652.
— stenophylla Ledeb. 627; tab. 38.
— suffrutescens M. B. 647.
— supina M. B. 656.
— swertiifolia Boiss. 641.
— tachtensis Franch. 638.
— talyschensis Schischk. 611.
— tatarica (L.) Pers. 619.
— Tatianae Schischk. 668.
— tenella C. A. Mey. 639.
— Thirkeana C. Koch 611.
— tomentella Schischk. 648.
— trajectorum Kom. 679; tab. 40.
— turcomanica Schischk. 603; tab. 36.
— turgida M. B. ex Bge. 664.
— uniflora Roth 597; tab. 35.
— viridiflora L. 674.
— wolgensis (Willd.) Bess. ex Spreng. 685.
Silybum Adans. 28, 227.
— marianum (L.) Gaertn. 227.
Simlera Bub. 25, 342.
Sinapis L. 8, 467.
— alba L. 468.
— arvensis L. 467.
Siphonostegia Benth. 22, 795.
— chinensis Benth. 796.
Sisarum Mill. 16, 458, 463.
Sison L. 16, 374.
— amomum L. 375.
Sisymbrium L. 8, 38.
— altissimum L. 49.
— bilobum (C. Koch) Grossh. 49; tab. 2.
— brachycarpum (N. Busch) Vass. 45. 634.
— brassiciforme C. A. Mey. 43.
— confertum Stev. 51.
— daghestanicum Vass. 50, 635.
— elatum C. Koch 42.
— erucastrifolium (Rupr.) Trautv. 45.
— heteromallum C. A. Mey. 43.
— hirsutum Lag. ex DC. 52.
— irio L. 44; tab. 2.
— isfarense Vass. 48, 635.
— Lipskyi N. Busch 46.
— Loeselii L. 44.
— luteum (Maxim.) O. E. Schulz 42.
— officinale (L.) Scop. 52.
— orientale L. 50; tab. 2.
— polymorphum (Murr.) Roth 46; tab. 3.
— runcinatum Lag. ex DC. 51; tab. 2.
— strictissimum L. 41.
— subspinescens (Fisch. et Mey.) Bge. 48; tab. 3.
— subtilissimum M. Pop. 55.
— turcomanicum Litv. 45.
— volgense M. B. ex Ledeb. 47.
Sisyrinchium L. 4, 577.
— angustifolium Mill. 578.
Sium L. 16, 459.
— latifolium L. 459.
— medium Fisch. et Mey. 465.
— sisaroides DC. 463; tab. 31.
— sisarum L. 464.
— suave Walt. 460; tab. 31.
— tenue Kom. 463; tab. 31.
Skimmia Thunb. 14, 233, 710.
— repens Nakai 710.
Smelovskia C. A. Mey. 8, 89.

Smelovskia alba (Pall.) Rgl. 89.
— aspleniifolia Turcz. 91.
— calycina (Steph.) C. A. Mey. 90.
— inopinata Kom. 91.
Smilacina Desf. 4, 451.
— dahurica Turcz. 451.
— hirta Maxim. 452.
— trifoliata Desf. 452; tab. 27.
Smilax L. 4, 474.
— excelsa L. 474; tab. 29.
— Oldhamii Miq. 475.
Smirnovia Bge. 11, 315.
— turkestana Bge. 315; tab. 21.
Smyrniopsis Boiss. 16, 222.
— armena Schischk. 222.
Smyrnium L. 16, 218.
— cordifolium Boiss. 221.
— perfoliatum L. 221; tab. 16.
Sobaria Haw. 9, 166.
Sobolevskia M. B. 9, 36.
— caucasica N. Busch 37.
— clavata (Boiss.) Fenzl 38.
— lithophila M. B. 37.
— truncata N. Busch 37.
Soja Moench 13, 527.
Solanopsis Boern. 22, 8, 42.
Solanum L. 22, 42.
Solanum L. 22, 3.
— alatum Moench 33.
— asiae-mediae Pojark. 21; tab. 1.
— decipiens Opiz 27.
— depilatum Kitag. 17.
— dulcamara L. 12.
— judaicum Bess. 28; tab. 3.
— Kieseritzkii C. A. Mey. 10; tab. 1.
— litorale Raab 16.
— luteum Mill. 36.
— marinum (Bav.) Pojark. 16.
— megacarpum Koidz. 20.
— melongena L. 39.
— nigrum L. 25.
— Olgae Pojark. 30; tab. 1.
— persicum Willd. ex Roem. et Schult. 19.
— pseudoflavum Pojark. 35.
— pseudopersicum Pojark. 18.
— rostratum Dun. 41.
— septemlobum Bge. 11.
— transcaucasicum Pojark. 29; tab. 2.
— tuberosum L. 8.
— Woronowii Pojark. 34; tab. 2.
— Zelenetzkii Pojark. 32; tab. 2.
Soldanella L. 18, 249.
— montana Mikan 250.
Solenanthus Ledeb. 19, 638.
— Biebersteinii DC. 655.
— brachystemon Fisch. et Mey. 656.
— circinnatus Ledeb. 642; tab. 37.
— hirsutus Rgl. 650; tab. 37.
— karateginus Lipsky 653.
— kokanicus Rgl. 651.
— plantaginifolius Lipsky 648.
— stamineus (Desf.) Wettst. 652.
— turkestanicus (Rgl. et Smirn.) Kusn. 644; tab. 36.
Solidago L. 25, 30.
— alpestris Waldst. et Kit. apud Willd. 39.
— armena Kem.-Nath. ex Grossh. 41.
— canadensis L. 48.
— caucasica Kem.-Nath. 40.
— compacta Turcz. 47; tab. 1.
— cuprea Juz. 46, 577.
— dahurica Kitag. 42.
— Gebleri Juz. 43, 576.
— jailarum Juz. 39, 575.
— kuhistanica M. Pop. 44, 576.
— kurilensis Juz. 45, 577.
— lapponica With. 41; tab. 1.
— pacifica Juz. 44, 576; tab. 1.
— spiraeifolia Fisch. ex Herd. 45.
— taurica Juz. 36, 575.
— turfosa Woronow ex Grossh. 36.
— virgaurea L. 34; tab. 1.
Solstitiaria Hill 28, 370, 570.
Sonchus L. 29, 244.
— arvensis L. 249; tab. 16.
— asper (L.) Hill 256; tab. 16.
— oleraceus L. 254; tab. 16.
— palustris L. 247; tab. 16.
— Sosnowskyi Schchian 248; tab. 16.
— transcaspicus Nevski 245; tab. 15.
Sophiopsis O. E. Schulz 8, 86.
— annua (Rupr.) O. E. Schulz 87.
— flavissima (Kar. et Kir.) O. E. Schulz 88.
— mongolica (Kom.) N. Busch 87.
— sisymbrioides (Rgl. et Herd.) O. E. Schulz 86.
Sophora L. 11, 23.
— alopecuroides L. 25.
— flavescens Ait. 28.
— Griffithii Stocks 25.
— japonica L. 24.
— pachycarpa C. A. Mey. 27; tab. 2.
— Prodanii Anderss. 26.
Soranthus Ledeb. 17, 142.
— Meyeri Ledeb. 142; tab. 23.
Sorbaria A. Br. 9, 312.
— Olgae Zinserl. 313.
— Pallasii (G. Don) Pojark. 315; tab. 19.

— rhoifolia Kom. 315.
— sorbifolia (L.) A. Br. 313.
Sorbus L. 9, 372.
— Albovii Zinserl. 390, 493.
— amurensis Koehne 381; tab. 23.
— anadyrensis Kom. 383.
— aria Crantz 397.
— armeniaca Hedl. 396; tab. 25.
— aucuparia L. 376.
— Baldaccii Degen et Fritsch 399; tab. 26.
— Boissieri Schneid. 382.
— Bushiana Zinserl. 392, 495; tab. 26.
— caucasica Zinserl. 395; tab. 25.
— colchica Zinserl. 391, 494; tab. 25.
— commixta Hedl. 382.
— domestica L. 374.
— dualis Zinserl. 402, 498.
— fennica (Kalm) Fries 405.
— glabrata Hedl. 377.
— graeca (Spach) Hedl. 397; tab. 26.
— kamtschatcensis Kom. 383.
— Kusnetzovii Zinserl. 397, 496; tab. 26.
— migarica Zinserl. 398, 496; tab. 26.
— obtusidentata Zinserl. 400, 498; tab. 26.
— persica Hedl. 401; tab. 25.
— polaris Koehne 378; tab. 24.
— sambucifolia Roem. 375.
— scandica Fries 396; tab. 25.
— schemachensis Zinserl. 392, 495; tab. 26.
— Schneideriana Koehne 376; tab. 23.
— sibirica Hedl. 378.
— subfusca (Ledeb.) Boiss. 390; tab. 25.
— subtomentosa (Albov) Zinserl. 391, 494; tab. 25.
— taurica Zinserl. 400, 497; tab. 26.
— tianschanica Rupr. 384; tab. 24.
— torminalis (L.) Crantz 405; tab. 26.
— turcica Zinserl. 399, 497; tab. 26.
— turkestanica (Franch.) Hedl. 401; tab. 25.
— velutina (Albov) Schneid. 391; tab. 25.
— Woronowii Zinserl. 395, 495; tab. 25.
Sorghum Pers. 2, 16.
— bicolor (L.) Moench 20.
— cernuum Host 20.
— halepense (L.) Pers. 21, tab. 1.
— japonicum (Hack.) Roshev. 20, 739.
— saccharatum (L.) Pers. 21.
— sudanense (Piper) Stapf 22.
— technicum (Koern.) Roshev. 20, 739.
— vulgare Pers. 19.
Sosnovskya Takht. 28, 370, 477.
Soyeria Monn. 29, 595, 606, 622, 627.
Spanioptilon Less. 28, 52.
Spanzium Griseb. 6, 861.
Sparganium L. 1, 217.
— affine Schnizl. 224; tab. 11.
— Friesii Beurl. 222; tab. 11.
— Glehnii Meinsh. 225.
— glomeratum Laest. 224; tab. 11.
— hyperboreum Laest. 226; tab. 11.
— japonicum Roth 222; tab. 11.
— microcarpum (Neum.) Čelak. 221; tab. 11.
— minimum Hill 225; tab. 11.
— neglectum Beeby 220; tab. 11.
— polyedrum Aschers. et Graebn. 219; tab. 11.
— simplex Huds. 223; tab. 11.
— stenophyllum Maxim. 221; tab. 11.
— stoloniferum Buch.-Ham. 219; tab. 11.
Spartium L. 11, 54.
— junceum L. 54.
Spathipappus Tzvel. 26, 261, 875.
— Griffithii (Clarke) Tzvel. 262; tab. 10.
Specularia A. DC. 24, 427.
Spergula L. 6, 550.
— linicola Boreau 552; tab. 32.
— maxima Weihe 552; tab. 32.
— pentandra L. 555.
— sativa Boenn. 551; tab. 32.
— vernalis Willd. 555; tab. 33.
— vulgaris Boenn. 551; tab. 32.
Spergularia (Pers.) J. et C. Presl 6, 556.
— campestris (L.) Aschers. 558; tab. 33.
— diandra (Guss.) Heldr. et Sart. 557; tab. 33.
— marginata (DC.) Kitt. 557; tab. 33.
— salina J. et C. Presl 651; tab. 33.
Spergulastrum Rich. 6, 394.
Spermatura Reichb. 16, 148.
Sphaerocephalus Kuntze 27, 2, 40.
Sphaerophysa DC. 11, 312.
— salsula (Pall.) DC. 312.
Sphallerocarpus Bess. ex DC. 16, 118.
— gracilis (Bess.) K.-Pol. 119; tab. 15.
Sphenocarpus Korov. 16, 525.
— eryngioides Korov. 525.
Sphenoclea Gaertn. 24, 449.
— zeylanica Gaertn. 449; tab. 27.
Sphenopus Trin. 2, 339.
— divaricatus (Gouan) Reichb. 339; tab. 27.
Sphondylium Mill. 17, 223.
Spielmannia Cuss. ex Juss. 16, 349.
Spilacron Cass. 28, 370.
Spinacia L. 6, 75.
— oleracea L. 75; tab. 2.
— tetrandra Stev. 77; tab. 2.
— turkestanica Iljin 76; tab. 2.
Spinovitis Romanet 14, 675.

Spiraea L. 9, 283.
— alpina Pall. 298; tab. 17.
— aquilegifolia Pall. 305; tab. 18.
— baldschuanica B. Fedtsch. 290.
— Beauverdiana Schneid. 289; tab. 17.
— betulifolia Pall. 288; tab. 17.
— chamaedryfolia L. 291; tab. 17.
— crenata L. 301; tab. 18.
— dahurica Maxim. 297; tab. 17.
— elegans Pojark. 293, 490; tab. 17.
— ferganensis Pojark. 303, 491.
— flexuosa Fisch. ex Cambess. 291; tab. 17.
— humilis Pojark. 287, 489.
— hypericifolia L. 303.
— lasiocarpa Kar. et Kir. 302; tab. 18.
— media F. Schmidt 294.
— pilosa Franch. 300; tab. 17.
— pubescens Turcz. 301.
— salicifolia L. 286.
— sericea Turcz. 297.
— tianschanica Pojark. 299, 490.
— trilobata L. 299; tab. 17.
— ussuriensis Pojark. 292, 489; tab. 17.
Spiraeanthus Maxim. 9, 316.
— Schrenkiana (Fisch. et Mey.) Maxim. 316; tab. 19.
Spiranthes Rich. 4, 637.
— amoena (M. B.) Spreng. 638; tab. 39.
— autumnalis (Balb.) L. 637; tab. 39.
— sinensis (Pers.) Ames 639; tab. 39.
Spirodela Schleid. 3, 492.
— polyrrhiza (L.) Schleid. 492; tab. 26.
Spirorhynchus Kar. et Kir. 8, 241.
— sabulosus Kar. et Kir. 241; tab. 20.
Spirostegia Ivanina 22, 512, 818.
— bucharica (B. Fedtsch.) Ivanina 513; tab. 25.
Spodiopogon Trin. 2, 11.
— sibiricus Trin. 11; tab. 1.
Sporobolus R. Br. 2, 163.
— indicus (L.) R. Br. 163; tab. 12.
Sprygina M. Pop. 8, 498.
— Winkleri M. Pop. 498.
Sredinskya (Stein) Fed. 18, 202.
— grandis (Trautv.) Fed. 202; tab. 10.
Stachyopsis M. Pop. et Vved. 21, 108.
— lamiiflora (Rupr.) M. Pop. et Vved. 109.
— marrubioides (Rgl.) Ik.-Gal. 110.
— oblongata (Schrenk) M. Pop. et Vved. 110.
Stachys L. 21, 194.
— alpina L. 208.
— angustifolia M. B. 225.
— annua L. 234.
— araxina Kapell. 227.
— arvensis L. 237.
— atherocalyx C. Koch 221; tab. 12.
— baicalensis Fisch. ex Benth. 219; tab. 12.
— Balansae Boiss. et Kotschy ex Boiss. 204; tab. 10.
— Boissieri Kapell. 229.
— chinensis Bge. 219.
— cretica L. 201; tab. 10.
— Czernjaevii Schost. 221.
— Fominii Sosn. 233.
— fruticulosa M. B. 225.
— germanica L. 201.
— Grossheimii Kapell. 227.
— heterodonta Zefir. 203.
— hissarica Rgl. 210; tab. 11.
— iberica M. B. 222.
— inflata Benth. 231.
— intermedia Ait. 204.
— iranica Rech. f. 212.
— japonica Miq. 218.
— Komarovii Knorr. 218; tab. 12.
— lanata Jacq. 200.
— lavandulifolia Vahl 228; tab. 13.
— macrophylla Albov 208.
— maeotica Postr. 217.
— maritima L. 230.
— odontophylla Freyn 228.
— palustris L. 216.
— Paulii Grossh. 229.
— persica J. F. Gmel. ex C. A. Mey. 207 tab. 10.
— pseudofloccosa Knorr. 210, 653.
— pubescens Ten. 231.
— recta L. 220.
— setifera C. A. Mey. 211; tab. 11.
— Sosnowskyi Kapell. 226.
— spectabiliformis Kapell. 207.
— spectabilis Choisy ex DC. 202.
— Stchtschegleewii Sosn. 232.
— sylvatica L. 215.
— talyschensis Kapell. 222.
— terekensis Knorr. 203.
— trapezuntea Boiss. 216.
— trinervis Aitch. et Hemsl. 233.
— tschatkalensis Knorr. 211.
— turcomanica Trautv. 232; tab. 13.
— turkestanica (Rgl.) M. Pop. 209; tab. 11.
— wolgensis Vilensky 217.
Staflinus Raf. 17, 288.
Staphylea L. 14, 578.
— colchica Stev. 578; tab. 31.
— pinnata L. 579; tab. 31.
Staphylodendron Scop. 14, 578.

Statice L. 18, 408, 411.
Staurophragma Fisch. et Mey. 22, 174.
— natolicum Fisch. et Mey. 174.
Stechmannia DC. 27, 647.
Steenhammera Reichb. 19, 238, 242.
Steirostemon Phil. 18, 253.
Stellaria L. 5, 389.
— Alexeenkoana Schischk. 422, 882.
— alsine Grimm 409.
— alsinoides Boiss. et Buhse 423.
— amblyosepala Schrenk 398.
— anagalloides C. A. Mey. ex Rupr. 410.
— arctica Schischk. 418, 881.
— brachypetala Bge. 405.
— Bungeana Fenzl 395; tab. 20.
— calycantha (Ledeb.) Bong. 404, tab. 21.
— cherleriae (Fisch.) Will. 420; tab. 21.
— ciliatosepala Trautv. 415.
— crassifolia Ehrh. 402.
— dahurica Willd. ex Schlecht. 415.
— dichotoma L. 397; tab. 20.
— dicranoides Fenzl 421; tab. 20.
— diffusa Willd. ex Schlecht. 410.
— discolor Turcz. 408.
— ebracteata Kom. 404.
— Edwardsii R. Br. 415; tab. 21.
— Eschscholtziana Fenzl 417.
— Fenzlii Rgl. 411.
— Fischeriana Ser. 417.
— fontana M. Pop. 407, 881.
— graminea L. 404.
— hebecalyx Fentzl 408.
— holostea L. 402; tab. 20.
— humifusa Roth 403; tab. 21.
— imbricata Bge. 406.
— irrigua Bge. 409.
— jaluana Nakai 405.
— karatavica (Lipsch.) Schischk. 398, 880.
— Kotschyana Fenzl 422.
— Laxmannii Fisch. ex Ser. 416.
— Martjanovii Kryl. 421.
— media (L.) Cyr. 395; tab. 20.
— neglecta Weihe 396.
— nemorum L. 394.
— pallida (Dum.) Pire 397.
— palustris Ehrh. 406.
— peduncularis Bge. 411; tab. 21.
— persica Boiss. 407.
— petraea Bge. 419.
— radians L. 418; tab. 20.
— ruscifolia Pall. ex Schlecht. 417; tab. 21.
— schugnanica Schischk. 422, 882.
— sibirica (Rgl. et Til.) Schischk. 420.
— soongorica Roshev. 412, 882; tab. 21.
— turkestanica Schischk. 401; tab. 21.
— umbellata Turcz. 409.
— Winkleri (Briq.) Schischk. 403.
Stellera L. 15, 502.
— chamaejasme L. 502.
Stellera Turcz. 18, 625.
Stelleropsis Pobed. 15, 503.
— altaica (Thieb.) Pobed. 504.
— Antoninae Pobed. 509.
— caucasica Pobed. 510.
— iranica Pobed. 509.
— issykkulensis Pobed. 505.
— Magakjanii (Sosn.) Pobed. 510.
— tarbagataica Pobed. 504; tab. 24.
— tianschanica Pobed. 506; tab. 24.
— turcomanica (Czerniak.) Pobed. 506 tab. 24.
Stenactis Cass. 25, 191.
Stenanthium Kunth 4, 6.
— sachalinense Fr. Schmidt 6; tab. 1.
Stenarrhena D. Don 21, 298.
Stenocephala Benth. 27, 647.
Stenocoelium Ledeb. 16, 469.
— athamantoides (M. B.) Ledeb. 470; tab. 30.
— trichocarpum Schrank 470.
Stenodiptera K.-Pol. 16, 123.
Stenolophus Cass. 28, 370, 440, 449.
Stenophragma Čelak. 8, 76.
Stenoselenium Turcz. 19, 182.
— saxatile (Pall.) Turcz. 183.
Stenotaenia Boiss. 17, 260.
— daralaghezica (Takht.) Schischk. 260; tab. 23.
Stephanocaryum M. Pop. 19, 521.
— Olgae (B. Fedtsch.) M. Pop. 522.
Steptorhamphus Bge. 29, 324.
— crambifolius Bge. 326; tab. 19.
— crassicaulis (Trautv.) Kirp. 327; tab. 19.
— Czerepanovii Kirp. 334; tab. 19.
— Linczevskii Kirp. 333; tab. 19.
— persicus (Boiss.) O. et B. Fedtsch. 329; tab. 19.
— petraeus (Fisch. et Mey.) Grossh. 331; tab. 19.
— tuberosus (Jacq.) Grossh. 330; tab. 19.
Sterigma DC. 8, 317.
Sterigmostemum M. B. 8, 317.
— acanthocarpum Fisch. et Mey. 319.
— tomentosum M. B. 318; tab. 17.
— torulosum (M. B.) Stapf 318.
Sternbergia Waldst. et Kit. 4, 485.
— colchiciflora Waldst. et Kit. 486.
— Fischeriana (Herb.) Roem. 486.

Sternbergia lutea (L.) Ker.-Gawl. 489; tab. 30.
Stevenia Adam et Fisch. 8, 197.
— alyssoides Adam et Fisch. 198.
— axillaris (Kom.) N. Busch 199.
— cheiranthoides DC. 198.
Steveniella Schlechter 4, 675.
— satyrioides (Stev.) Schlechter 675; tab. 43.
Stevia Cav. 25, 18.
— ovata Willd. 18.
Stipa L. 2, 79.
— aktauensis Roshev. 85.
— anomala P. Smirn. 96, 740.
— araxensis Grossh. 97.
— armeniaca P. Smirn. 92, 740.
— baicalensis Roshev. 110; tab. 7.
— breviflora Griseb. 91.
— Bungeana Trin. ex Bge. 106.
— canescens P. Smirn. 101, 741.
— capillata L. 109.
— caucasica Schmalh. 89; tab. 7.
— consanguinea Trin. et Rupr. 104.
— crassiculmis P. Smirn. 99.
— daghestanica (Westb.) Grossh. 106.
— dasyphylla Czern. 100.
— densiflora P. Smirn. 110.
— effusa (Maxim.) Nakai ex Honda 85.
— gigantea Lag. 105.
— glareosa P. Smirn. 89.
— gracilis Roshev. 92.
— grandis P. Smirn. 109; tab. 7.
— Hohenackeriana Trin. 92.
— holosericea Trin. 105, tab. 7.
— Iljinii Roshev. 101.
— Joannis Čelak. 96.
— karataviensis Roshev. 88.
— kirghisorum P. Smirn. 95.
— Korshinskyi Roshev. 104.
— Krascheninnikowii Roshev. 98.
— Krylovii Roshev. 112.
— Lessingiana Trin. 93.
— lingua Junge 86.
— Lipskyi Roshev. 87.
— lithophila P. Smirn. 98, 741.
— longiplumosa Roshev. 87.
— macroglossa P. Smirn. 94.
— magnifica Junge 86.
— margelanica P. Smirn. 111.
— orientalis Trin. 90; tab. 7.
— Ovczinnikovii Roshev. 87.
— paradoxa (Junge) P. Smirn. 99
— pontica P. Smirn. 102.
— pseudocapillata Roshev. 103.
— pulcherrima C. Koch 98.
— Regeliana Hack. 84; tab. 6.
— Richteriana Kar. et Kir. 90.
— sareptana A. Beck 111.
— sibirica (L.) Lam. 85; tab. 6.
— Spiridonovii Roshev. 104.
— stenophylla Czern. 95.
— Szowitsiana Trin. 91.
— tianschanica Roshev. 88.
— tortilis Desf. 103; tab. 6.
— trichoides P. Smirn. 94; tab. 7.
— turcomanica P. Smirn. 100.
— turkestanica Hack. 93.
— ucrainica P. Smirn. 97; tab. 7.
— Zalesskii Wilensky 102.
Stipularia Haw. 6, 556.
Stizolobium P. Browne 13, 530.
— utile (Wall.) Ditm. 530.
Stizolophus Cass. 28, 364.
— balsamita (Lam.) Cass. ex Takht. 366.
— coronopifolius (Lam.) Cass. 366; tab. 27.
Strabonia DC. 25, 487.
Stratiotes L. 1, 296.
— aloides L. 296; tab. 10.
Streptoloma Bge. 8, 305.
— desertorum Bge. 305.
Streptopus Rich. 4, 455.
— amplexifolius (L.) DC. 455; tab. 27.
— streptopoides (Ledeb.) Nels. et Macbr. 456; tab. 27.
Stroganovia Kar. et Kir. 8, 524.
— affghana (Boiss.) Pavl. 529.
— brachyata Kar. et Kir. 533.
— cardiophylla Pavl. 527.
— gracilis Pavl. 531.
— intermedia Kar. et Kir. 532.
— Litvinovii Lipsky 533.
— paniculata Rgl. et Schmalh. 527.
— persica N. Busch 528.
— robusta Pavl. 535.
— sagittata Kar. et Kir. 530.
— subalpina (Kom.) Thell. 530.
— Trautvetteri Botsch. 534, 652.
Strophiostoma Turcz. 19, 358.
Struthiopteris Haller 1, 29.
— filicastrum All. 29; tab. 1.
Stubendorffia Schrenk 8, 538.
— aptera Lipsky 540.
— Lipskyi N. Busch 541.
— orientalis Schrenk 539.
— subdidyma N. Busch 540.
Suaeda Forsk. 6, 174.
— acuminata (C. A. Mey.) Moq. 187; tab. 9.
— altissima (L.) Pall. 180; tab. 9.
— arcuata Bge. 182.

— baccifera Pall. 184; tab. 1.
— confusa Iljin 183.
— corniculata (C. A. Mey.) Bge. 195; tab. 9.
— dendroides (C. A. Mey.) Moq. 181.
— drepanophylla Litv. 196; tab. 9.
— eltonica Iljin 189.
— glauca Bge. 178; tab. 9.
— heterophylla (Kar. et Kir.) Bge. 197; tab. 9.
— Kossinskyi Iljin 197.
— linifolia Pall. 177; tab. 9.
— Lipskyi Litv. 183.
— maritima (L.) Dum. 193.
— microphylla Pall. 179.
— microsperma (C. A. Mey.) Fenzl 188.
— Olufsenii Pauls. 195; tab. 9.
— paradoxa Bge. 178.
— physophora Pall. 190.
— prostrata Pall. 194; tab. 9.
— pterantha (Kar. et Kir.) Bge. 187.
— pygmaea (Kar. et Kir.) Iljin 183.
— salsa (L.) Pall. 191; tab. 9.
— transoxana (Bge.) Boiss. 188.
— turkestanica Litv. 189.
— ussuriensis Iljin 192.
Suaeda Moq. 6, 177, 182, 191.
Subularia L. 8, 605.
— aquatica L. 606.
Succisa Neck. 24, 52.
— pratensis Moench 52; tab. 4.
Succisella G. Beck 24, 55.
— inflexa (Kluk) G. Beck 55; tab. 4.
Succisocrepis Fourr. 29, 595.
Suchtelenia Kar. 19, 680.
— calycina (C. A. Mey.) DC. 681; tab. 40.
Sumbulus H. Reinsch. 17, 62, 121.
Svyda Opiz 17, 331.
Swertia L. 18, 629.
— Aucheri Boiss. 636.
— connata Schrenk 632.
— erosula Gontsch. 638.
— graciliflora Gontsch. 636.
— iberica Fisch. et Mey. 630.
— lactea Bge. 637.
— marginata Schrenk 635.
— obtusa Ledeb. 631; tab. 34.
— perennis L. 18, 630; tab. 34.
— veratroides Maxim. ex Kom. 635.
Sykoraea Opiz 24, 133.
Sympegma Bge. 6, 353.
— Regelii Bge. 353; tab. 14.
Symphyandra A. DC. 24, 331.
— armena (Stev.) A. DC. 338.
— daralaghezica Grossh. 339.
— lazica Boiss. et Bal. 336.
lezgina Alexeenko ex Lipsky 341; tab. 20.
— pendula (M. B.) A. DC. 336; tab. 20.
— transcaucasica (Somm. et Levier) Grossh. 337.
— zangezura Lipsky 340; tab. 20.
Symphyllocarpus Maxim. 25, 298.
— exilis Maxim. 298.
Symphyoloma C. A. Mey. 17, 222.
— graveolens C. A. Mey. 223.
Symphytum L. 19, 279.
— asperum Lepech. 287.
— caucasicum M. B. 286.
— cordatum Waldst. et Kit. 281.
— grandiflorum DC. 283.
— officinale L. 289.
— orientale L. 285.
— peregrinum Ledeb. 289.
— tauricum Willd. 284.
— tuberosum L. 282.
Symplocarpus Salisb. 3, 481.
— foetidus (L.) Salisb. 481; tab. 26.
Sympodium C. Koch 16, 396.
Syncodon Fourr. 24, 133, 198.
Syneilesis Maxim. 26, 697.
— aconitifolia Maxim. 698.
Synosma Raf. 26, 683.
Synsiphon Rgl. 4, 23.
Synurus Iljin 28, 301.
— deltoides (Ait.) Nakai 302; tab. 17.
Syreitschikovia Pavl. 28, 303.
— spinulosa (Franch.) Pavl. 305.
— tenuifolia (Bong.) Pavl. 304.
Syrenia Andrz. 8, 128.
— angustifolia (Ehrh.) Reichb. 129.
— macrocarpa Vass. 130, 640.
— sessiliflora Ledeb. 129.
— siliculosa (M. B.) Andrz. 130.
— Talievii Klok. 128.
Syringa L. 18, 502.
— chinensis Willd. 509.
— Josikaea Jacq. f. 504; tab. 26.
— persica L. 509.
— robusta Nakai 505; tab. 26.
— villosa Vahl 504; tab. 26.
— vulgaris L. 506.
Szovitsia Fisch. et Mey. 16, 364.
— callicarpa Fisch. et Mey. 364; tab. 28, 30.

Tacamahaca Spach 5, 236.
Taeniopetalum Vis. 17, 168, 197.
Tagetes L. 25, 569.
— erecta L. 571.

Tagetes minuta L. 570.
— patula L. 571.
Tamarix L. 15, 290.
— affinis Bge. 317.
— altaica Niedz. 321.
— Androssowii Litv. 300.
— aralensis Bge. 318.
— araratica (Bge.) Gorschk. 299; tab. 15.
— arcenthoides Bge. 312.
— Bungei Boiss. 309.
— elongata Ledeb. 297; tab. 15.
— Ewersmannii C. Presl ex Bge. 316.
— florida Bge. 306; tab. 15.
— gracilis Willd. 307.
— hispida Willd. 308; tab. 16.
— Hohenackeri Bge. 305; tab. 15.
— karakalensis Freyn 313.
— Karelinii Bge. 315; tab. 16.
— kasakhorum Gorschk. 314.
— Komarovii Gorschk. 317.
— Korolkovii Rgl. et Schmalh. 316.
— Kotschyi Bge. 301.
— laxa Willd. 302; tab. 15.
— leptopetala Bge. 308.
— leptostachys Bge. 310; tab. 16.
— Litwinowii Gorschk. 299; tab. 15.
— Meyeri Boiss. 296; tab. 15.
— ostandra Bge. 295.
— passerinoides Delile 314.
— ramosissima Ledeb. 311; tab. 16.
— tetrandra Pall. 298.
Tamus L. 4, 497.
— communis L. 498.
Tanacetum L. 26, 317.
— abrotanifolium (L.) Druce 329.
— achilleifolium (M. B.) Sch. Bip. 351.
— Akinfievii (Alexeenko) Tzvel. 357; tab. 14.
— bipinnatum (L.) Sch. Bip. 324.
— boreale Fisch. ex DC. 327.
— canescens DC. 336.
— chiliophyllum (Fisch. et Mey.) Sch. Bip. 352.
— crassipes (Stschegl.) Tzvel. 338.
— duderanum (Boiss.) Tzvel. 332.
— heterophyllum Boiss. 333.
— Karelinii Tzvel. 344.
— Kittaryanum (C. A. Mey.) Tzvel. 345.
— longipedunculatum (Sosn.) Tzvel. 354.
— millefolium (L.) Tzvel. 348.
— myriophyllum Willd. 331.
— odessanum (Klok.) Tzvel. 348.
— oligocephalum (DC.) Sch. Bip. 354.
— Paczoskii (Zefir.) Tzvel. 349.
— pseudoachillea C. Winkl. 328.
— santolina C. Winkl. 343
— saxicolum (Krasch.) Tzvel. 342.
— sclerophyllum (Krasch.) Tzvel. 347.
— scopulorum (Krasch.) Tzvel. 342.
— tabrisianum (Boiss.) Sosn. et Takht. 335.
— tamrutense (Sosn.) Sosn. 334.
— tanacetoides (DC.) Tzvel. 337.
— tenuissimum (Trautv.) Grossh. 355; tab. 14.
— turcomanicum (Krasch.) Tzvel. 358.
— turlanicum (Pavl.) Tzvel. 341.
— ulutavicum Tzvel. 350, 878.
— uniflorum (Fisch. et Mey.) Sch. Bip. 356.
— uralense (Krasch.) Tzvel. 346.
— vulgare L. 326.
— Walteri (C. Winkl.) Tzvel. 359.
Tapeinanthus Boiss. 20, 232.
Taphrospermum C. A. Mey. 8, 32.
— altaicum C. A. Mey. 33.
— platypetalum Schrenk 33.
Taraxacum Wigg. 29, 405.
— acricorne Dahlst. 462.
— aksaicum Schischk. 551.
— alaicum Schischk. 551, 751.
— alatavicum Schischk. 477, 731.
— albescens Dahlst. 552.
— almaatense Schischk. 434, 728.
— alpinum Hegetschw. et Heer 441.
— altaicum Schischk. 513.
— anchorifolium Kom. 462.
— Androssovii Schischk. 502.
— arcticum (Trautv.) Dahlst. 526.
— armeniacum Schischk. 501.
— aschabadense Schischk. 487.
— atrans Schischk. 510, 739.
— atratum Schischk. 524, 743.
— badachschanicum Schischk. 508, 737.
— baicalense Schischk. 539, 747.
— balticum Dahlst. 519.
— bessarabicum (Hornem.) Hand.-Mazz. 533.
— bicorne Dahlst. 478.
— Botschantzevii Schischk. 480, 732.
— brachyceras Dahlst. 473.
— brevicorne Dahlst. 472.
— brevicorniculatum V. Korol. 492.
— brevirostre Hand.-Mazz. 547.
— Butkovii S. Kovalevsk. 490.
— calcareum V. Korol. 553.
— ceratophorum (Ledeb.) DC. 460.
— chamissonis Greene 465.
— chirieanum Kitam. 452.
— ciscaucasicum Schischk. 532, 745.

— collinum DC. 539.
— comitans S. Kovalevsk. 482.
— compactum Schischk. 512.
— complicatum S. Kovalevsk. 435.
— confusum Schischk. 473.
— contristans S. Kovalevsk. 477.
— coreanum Nakai 444.
— crepidiforme DC. 543; tab. 25.
— croceum Dahlst. 450.
— czuense Schischk. 541.
— daralagesicum Schischk. 534, 745.
— dealbatum Hand.-Mazz. 541.
— decolorans Dahlst. 517.
— desertorum Schischk. 497.
— dilutum Dahlst. 468.
— dissectum (Ledeb.) Ledeb. 538.
— dissimile Dahlst. 495.
— ecornutum S. Kovalevsk. 435.
— elongatum S. Kovalevsk. 479.
— eriobasis S. Kovalevsk. 553.
— eriopodum (D. Don) DC. 515.
— erostre Zak. 549.
— erythrospermum Andrz. 498.
— evittatum Dahlst. 463.
— Fedtschenkoi Hand.-Mazz. 536.
— fontanum Hand.-Mazz. 441.
— glabellum Schischk. 546, 750.
— glabrum DC. 523.
— glaucanthum (Ledeb.) DC. 475.
— glaucivirens Schischk. 476, 730.
— Gnezdilloi S. Kovalevsk. 554.
— Goloskokovii Schischk. 544, 748.
— grandisquamatum Koidz. 454.
— Grossheimii Schischk. 437.
— heptapotamicum Schischk. 544, 748.
— Hjeltii (Dahlst.) Dahlst. 469.
— holophyllum Schischk. 484, 733.
— Hultenii Dahlst. 460.
— hybernum Stev. 530.
— Ikonnikovii Schischk. 507, 736.
— intercedens Markl. 500.
— Juzepczukii Schischk. 479, 731.
— kamtschaticum Dahlst. 527; tab. 26.
— karatavicum Pavl. 488, 734.
— ketoiense Tatewaki et Kitam. 456.
— kirghizicum Schischk. 511, 740.
— kljutschevskoanum Kom. 463.
— Klokovii Litvinenko 476.
— Kojimae Kitam. 466.
— kok-saghyz Rodin 491.
— koraginense Kom. 464.
— koraginicola Kom. 467.
— lapponicum Kihlm. ex Hand.-Mazz. 450.
— lateritium Dahlst. 505.
— latisquameum Dahlst. 465.
— leptoceras Dahlst. 468.
— leucanthum (Ledeb.) Ledeb. 540.
— lilacinum Krassn. ex Schischk. 529.
— Linczevskyi Schischk. 554, 751.
— Lipskyi Schischk. 481.
— lissocarpum (Dahlst.) Dahlst. 518.
— Litwinowii Schischk. 555, 752.
— longicorne Dahlst. 461.
— longipes Kom. 458.
— longipyramidatum Schischk. 489, 735.
— longirostre Schischk. 493, 735.
— luridum Hagl. 555.
— Lyngeanum Hagl. 520.
— lyratum (Ledeb.) DC. 550.
— macilentum Dahlst. 467.
— macrochlamideum S. Kovalevsk. 438.
— macrolepium Schischk. 438, 729.
— magnum V. Korol. 515.
— majus Schischk. 437, 729.
— Malaisei Dahlst. 459.
— maracandicum S. Kovalevsk. 485.
— mexicanum DC. 444.
— microspermum Schischk. 487.
— minutilobum M. Pop. ex S. Kovalevsk. 548.
— Miyake Kitam. 454.
— modestum Schischk. 509, 738.
— mongolicum Hand.-Mazz. 506.
— monochlamydeum Hand.-Mazz. 483.
— montanum (C. A. Mey.) DC. 430; tab. 25.
— multiscaposum Schischk. 480.
— muricatum Schischk. 556.
— murmanicum Orlova 470, 730.
— nairoense Koidz. 453.
— natschikense Kom. 448.
— neo-lobulatum Van Soest 490.
— neo-sachalinense Koidz. 446.
— Nevskii Juz. 486.
— nigricans (Kit.) Reichb. 440.
— Nikitinii Schischk. 482.
— nivale Lange ex Kihlm. 522.
— norvegicum (Dahlst.) Dahlst. 470.
— novae-zemliae Holmboe 471.
— nuratavicum Schischk. 486, 733.
— obliquum (Fries) Dahlst. 494.
— officinale Wigg. 433.
— Ohwianum Kitam. 443.
— oliganthum Schott et Kotschy ex Hand.-Mazz. 533.
— oschense Schischk. 557, 752.
— pamiricum Schischk. 557, 753.
— perlatescens Dahlst. 459.

Taraxacum perpusillum Schischk. 516, 742.
— pineticola Klok. 499.
— pingue Schischk. 513.
— platyceras Dahlst. 463.
— platycranum Dahlst. 440.
— platylepium Dahlst. 520.
— platypecidum Diels 442.
— Pobedimoviae Schischk. 531, 744.
— Pojarkoviae Schischk. 546, 749; tab. 26.
— porphyranthum Boiss. 528.
— praecox Schischk. 483, 732.
— praestans Lindb. f. 439.
— praticola Schischk. 501.
— Printzii Dahlst. 496.
— pseudoalpinum Schischk. 449.
— pseudolasianthum Koidz. 453.
— pseudo-dissimile Van Soest 492.
— pseudoglabrum Dahlst. 448.
— pseudominutilobum S. Kovalevsk. 546.
— pseudonigricans Hand.-Mazz. 558.
— pseudoroseum Schischk. 528, 744.
— repandum Pavl. 502.
— rubidum Schischk. 549, 750; tab. 26.
— rubiginans Dahlst. 504.
— Rubtzovii Schischk. 515, 741.
— rufum Dahlst. 505.
— sachalinense Kitam. 455.
— sagittifolium Lindl. f. ex Dahlst. 451.
— Saposhnikovii Schischk. 509, 739.
— Schelkovnikovii Schischk. 474.
— Schischkinii V. Korol. 449.
— schugnanicum Schischk. 514, 741.
— seravschanicum Schischk. 558, 753.
— serotinum (Waldst. et Kit.) Poir. 431.
— shikotanense Kitam. 455.
— shimushirense Tatewaki et Kitam. 457.
— shumushuense Kitam. 457.
— sibiricum Dahlst. 526.
— simulum Brenn. 471.
— sinicum Kitag. 542.
— songoricum Schischk. 512, 740.
— Stanjukoviczii Schischk. 508, 737.
— stenolepium Hand.-Mazz. 535.
— stenolobum Stschegl. 496.
— strobilocephalum S. Kovalevsk. 485.
— subglaciale Schischk. 524, 743.
— suecicum Hagl. 517; tab. 25.
— Sumneviczii Schischk. 503.
— syriacum Boiss. 430.
— tadshikorum Ovcz. 488, 734.
— Tatewakii Kitam. 446.
— tauricum Kotov 499, 736; tab. 26.
— tenuisectum Somm. et Lévier 559.
— tianschanicum Pavl. 503.
— tundricola Hand.-Mazz. 522.
— turgaicum Schischk. 436, 728.
— turiense Orlova 521, 742.
— Tzvelevii Schischk. 545, 749.
— ussuriense Kom. 445.
— varsobicum Schischk. 537, 747.
— Vassilczenkoi Schischk. 511, 739.
— Voronovii Schischk. 536, 746.
— vulcanorum Koidz. 469.
— xerophilum Markl. 494.
— Yamamotoi Koidz. 464.
— yetrofuense Kitam. 456.
Tauscheria Fisch. 8, 228.
— lasiocarpa Fisch. 228; tab. 13.
— oblonga Vass. 229, 643; tab. 13.
Taxus L. 1, 131.
— baccata L. 131.
— cuspidata Sieb. et Zucc. 132; tab. 8.
Teesdalia R. Br. 8, 593.
— nudicaulis (L.) R. Br. 593.
Telekia Baumg. 25, 510.
— speciosa (Schreb.) Baumg. 510.
Telephium L. 6, 562.
— oligosperma Steud. ex Boiss. 563; tab. 32.
— orientale Boiss. 563; tab. 32.
Teline Medic. 1, 70.
— monspessulana (L.) Koch 70; tab. 5.
Teloxis Moq. 6, 47.
Tenageia Dum. 3, 515.
Tetracme Bge. 8, 299.
— pamirica Vass. 300, 648; tab. 19.
— quadricornis (Steph.) Bge. 300; tab. 19.
— recurvata Bge. 301.
Tetracmidion Korsh. 8, 301.
— bucharicum Korsh. 302; tab.19, tab. 20.
Tetradiclis Stev. 14, 151.
— tenella (Ehrenb.) Litv. 151.
Tetragonolobus Scop. 11, 297.
— purpureus Moench 298.
— siliquosus (L.) Roth 297; tab. 19.
Tetrahit Gilib. 21, 111.
Tetramorphaea DC. 28, 370, 576.
Teucrium L. 20, 39.
— × Alexeenkoanum Juz. 64, 506.
— botrys L. 50.
— canum Fisch. et Mey. 63.
— chamaedrys L. 55; tab. 4.
— excelsum Juz. 62, 505.
— Fischeri Juz. 57.
— hircanicum L. 45; tab. 3.
— jailae Juz. 65; tab. 4.
— krymense Juz. 61; tab. 4.
— multinodum (Bordz.) Juz. 53.

— nuchense C. Koch 52; tab. 4.
— orientale L. 42; tab. 3.
— pannonicum A. Kerner 67.
— parviflorum Schreb. 44.
— polium L. 68.
— praemontanum Klok. 66, 506.
— pulchrius Juz. 57.
— scordioides Schreb. 50; tab. 3.
— scordium L. 46; tab. 3.
— syspirense C. Koch 58.
— Taylori Boiss. 43.
— trapezunticum (Reichb. f.) Juz. 54.
— ussuriense Kom. 528.
Thaeniatherum Nevski 2, 719.
— asperum (Simk.) Nevski 719; tab. 50.
— crinitum (Schreb.) Nevski 719; tab. 50.
Thalictrum L. 7, 510.
— alpinum L. 519; tab. 35.
— angustifolium L. 527.
— aquilegifolium L. 512; tab. 35.
— baicalense Turcz. 515; tab. 35.
— contortum L. 513; tab. 35.
— filamentosum Maxim. 516.
— flavum L. 527.
— foetidum L. 520; tab. 35.
— isopyroides C. A. Mey. 523; tab. 35.
— minus L. 524.
— petaloideum L. 518; tab. 35.
— podolicum Lecoyer 525.
— sachalinense Lecoyer 514.
— simplex L. 526.
— sparsiflorum Turcz. ex Fisch. et Mey. 513.
— squarrosum Steph. ex Willd. 523.
— sultanabadense Stapf 518; tab. 35.
— triternatum Rupr. 517.
— tuberiferum Maxim. 515.
Thea L. 15, 198.
— sinensis L. 198.
Thellungiella O. E. Schulz 8, 75.
— halophila (C. A. Mey.) O. E. Schulz 76.
— salsuginea (Pall.) O. E. Schulz 75.
Thelycrania (Dum.) Fourr. 17, 331.
— alba (L.) Pojark. 344; tab. 24.
— australis (C. A. Mey.) Sanadze 337.
— brachypoda (C. A. Mey.) Pojark. 333.
— darvasica Pojark. 343.
— iberica (Woronow) Pojark. 338.
— Koenigii (Schneid.) Sanadze 334; tab. 24.
— Meyeri Pojark. 341.
— sanguinea (L.) Fourr. 336.
— stolonifera (Michx.) Pojark. 346; tab. 24.
Thelygonum L. 6, 372.
— cynocrambe L. 372.
Theodorea Cass. 27, 361, 520.
Thermopsis R. Br. 11, 37.
— alpina (Pall.) Ledeb. 38; tab. 3.
— alterniflora Rgl. et Schmalh. 41.
— dolichocarpa V. Nikit. 42.
— fabacea (Pall.) DC. 43.
— lanceolata R. Br. 39.
— turkestanica Gand. 40.
Therorhodion Small 18, 58.
Thesium L. 5, 412.
— alatavicum Kar. et Kir. 421; tab. 24.
— alpinum L. 430; tab. 24.
— brachyphyllum Boiss. 430; tab. 24.
— chinense Turcz. 418; tab. 24.
— ebracteatum Hayne 419
— ferganense Bobr. 428.
— Gontscharovii Bobr. 428; tab. 24.
— Kotschyanum Boiss. 416; tab. 24.
— linifolium Schrank 419; tab. 24.
— longifolium Turcz. 423; tab. 24.
— maritimum C. A. Mey. 415.
— Minkwitzianum B. Fedtsch. 415; tab. 24.
— multicaule Ledeb. 424.
— procumbens C. A. Mey. 418; tab. 24.
— ramosissimum Bobr. 429; tab. 24.
— ramosum Hayne 427; tab. 24.
— refractum C. A. Mey. 421.
— repens Ledeb. 420.
— rupestre Ledeb. 422.
— saxatile Turcz. ex DC. 422; tab. 24.
— Szovitsii DC. 417; tab. 24.
Thladiantha Bge. 24, 94.
— dubia Bge. 95.
Thlaspi L. 8, 576.
— annuum C. Koch 596.
— armenum N. Busch 582.
— arvense L. 581; tab. 33.
— cochleariforme DC. 587.
— exauriculatum Kom. 587.
— ferganense N. Busch 589.
— Freynii N. Busch 588; tab. 33.
— Huetii Boiss. 581; tab. 33.
— Kotschyanum Boiss. et Hohen. 585.
— macranthum N. Busch 588.
— orbiculatum Stev. 581; tab. 33.
— perfoliatum L .585; tab. 33.
— platycarpum Fisch. et Mey. 582; tab. 33.
— praecox Wulf. 587.
— pumilum Ledeb. 589; tab. 33.
— roseolum N. Busch 586.
— rostratum N. Busch 589.
— Szovitsianum Boiss. 585.
— umbellatum Stev. 585.
Thuja Tourn. 1, 192

Thuja occidentalis L. 192.
Thuspeinantha Durand 20, 232.
— persica (Boiss.) Brig. 232.
Thylacosperma Lindl. 6, 545.
Thylacospermum Fenzl 6, 545.
— caespitosum (Chamb.) Schischk. 546; tab. 31.
Thymelaea Endl. 15, 482.
— passerina (L.) Coss. et Germ. 483.
Thymus L. 21, 470.
— alatauensis (Klok. et Schost.) Klok. 535.
— altaicus Klok. et Schost. 540.
— alternans Klok. 481.
— amictus Klok. 495.
— amurensis Klok. 519; tab. 29.
— ararati-minoris Klok. et Schost. 557.
— armeniacus Klok. et Schost. 556.
— Arsenijevii Klok. 531.
— Aschurbajevii Klok. 544.
— asiaticus Serg. 535.
— baschkiriensis Klok. et Schost. 562.
— binervulatus Klok. et Schost. 565; tab. 30.
— bituminosus Klok. 524.
— borysthenicus Klok. et Schost. 581; tab. 31.
— bucharicus Klok. 543.
— bulgaricus (Domin et Podp.) Ronn. 497.
— Buschianus Klok. et Schost. 478.
— calcareus Klok. et Schost. 575.
— Callieri Borb. 493; tab. 28.
— caucasicus Willd. 477.
— chankoanus Klok. 521.
— circumcinctus Klok. 481; tab. 27.
— collinus M. B. 552.
— coriifolius Ronn. 505.
— crebrifolius Klok. 528.
— crenulatus Klok. 527.
— cretaceus Klok. et Schost. 576; tab. 31.
— cuneatus Klok. 588.
— curtus Klok. 522.
— Czernjaevii Klok. 506; tab. 28.
— dagestanicus Klok. et Schost. 571.
— Desjatoviae Ronn. 560.
— diminutus Klok. 545.
— dimorphus Klok. et Schost. 500; tab. 28.
— disjunctus Klok. 483.
— diversifolius Klok. 532.
— Dubjanskii Klok. et Schost. 577; tab. 31.
— Dzevanovskyi Klok. et Schost. 510.
— Elisabethae Klok. et Schost. 498.
— eltonicus Klok. et Schost. 587.
— enervius Klok. 488.
— eravinensis Serg. 525.
— eremita Klok. 586.
— eriophorus Ronn. 550.
— eubajcalensis Klok. 526.
— eupatoriensis Klok. et Schost. 570.
— extremus Klok. 533.
— Fedtschenkoi Ronn. 551.
— flexilis Klok. 520.
— Fominii Klok. et Schost. 558.
— glacialis Klok. 507.
— graniticus Klok. et Schost. 574.
— Grossheimii Ronn. 560.
— guberlinensis Iljin 563; tab. 30.
— Hadzhievii Grossh. 578.
— helendzhicus Klok. et Schost. 580; tab. 31.
— hirsutus M. B. 492; tab. 28.
— hirticaulis Klok. 536.
— Iljinii Klok. et Schost. 541.
— inaequalis Klok. 523.
— incertus Klok. 588.
— irtyschensis Klok. 589.
— jajlae (Klok. et Schost.) Stank. 499.
— jenisseensis Iljin 546.
— kalmiussicus Klok. et Schost. 575.
— karamarianicus Klok. et Schost. 508.
— Karjaginii Grossh. 573.
— kasakstanicus Klok. et Schost. 586.
— kirgisorum Dub. 585.
— kjapazi Grossh. 559.
— Klokovii (Ronn.) Schost. 501.
— Komarovii Serg. 484.
— Kotschyanus Boiss. et Hohen. 550; tab. 30.
— ladjanuricus Kem.-Nath. 570.
— lanulosus Klok. 585.
— latifolius (Bess.) Andrz. 509.
— Lavrenkoanus Klok. 512.
— Lipskyi Klok. et Schost. 572.
— litoralis Klok. et Schost. 498.
— Loevyanus Opiz 507.
— majkopensis Klok. et Schost. 580.
— Marschallianus Willd. 511.
— markhotensis Maleev 496.
— migricus Klok. et Schost. 551.
— minussinensis Serg. 524.
— moldavicus Klok. et Schost. 573.
— mongolicus Klok. 540.
— mugodzharicus Klok. et Schost. 561; tab. 30.
— narymensis Serg. 528.
— nerczensis Klok. 590.
— nervulosus Klok. 522.
— nummularius M. B. 476; tab. 27.
— ochotensis Klok. 520.

— oxyodontus Klok. 541.
— Pallasianus H. Br. 582; tab. 31.
— pannonicus All. 509.
— pastoralis Iljin 510.
— paucifolius Klok. 539.
— petraeus Serg. 590.
— phyllopodus Klok. 526.
— podolicus Klok. et Schost. 495.
— proximus Serg. 545.
— Przewalskii Kom. 502.
— pseudograniticus Klok. et Schost. 574.
— pseudohumillimus Klok. et Schost. 578.
— pseudonummularius Klok. et Schost. 476.
— pseudopulegioides Klok. et Schost. 487.
— pulchellus C. A. Mey. 489; tab. 27.
— pulegioides L. 487.
— rariflorus C. Koch 555; tab. 30.
— rasitatus Klok. 562.
— Reverdattoanus Serg. 533.
— roseus Schipcz. 564.
— Schischkinii Serg. 528.
— semiglaber Klok. 482.
— serpyllum L. 537; tab. 29.
— sibiricus Klok. et Schost. 539.
— Sokolovii Klok. 532.
— Sosnowskyi Grossh. 572.
— stepposus Klok. et Schost. 512; tab. 28.
— subalpestris Klok. 485.
— subarcticus Klok. et Schost. 538; tab. 29.
— Talievii Klok. et Schost. 534.
— tauricus Klok. et Schost. 579.
— tiflisiensis Klok. et Schost. 497.
— tonsilis Klok. 537.
— transcaspicus Klok. et Schost. 555.
— transcaucasicus Ronn. 558.
— Trautvetteri Klok. et Schost. 559.
— Turczaninovii Serg. 501.
— ucrainicus (Klok. et Schost.) Klok. 486.
— ussuriensis Klok. 519.
— Zelenetzkyi Klok. et Schost. 494.
— zeravshanicus Klok. 542; tab. 29.
— zheguliensis Klok. et Schost. 563.
— ziaratinus Klok. et Serg. 557.
Thyrsanthus Schrank 18, 268.
Thysanocarex Boern. 3, 160.
Thysselinum Adans. 16, 560.
Thysselinum Hoffm. 17, 168, 201.
Tianschaniella B. Fedtsch. 19, 523.
— umbellifera B. Fedtsch. ex M. Pop. 523.
Tilia L. 15, 4.
— amurensis Rupr. 16.
— begoniifolia Stev. 13.
— caucasica Rupr. 11.
— cordata Mill. 20.
— cordifolia Bess. 14.
— dasystyla Stev. 10.
— mandshurica Rupr. 8.
— multiflora Ledeb. 13.
— sibirica Fisch. ex Bayer 18.
— Taquetii Schneid. 17.
— tomentosa Moench 7.
Tilingia Rgl. 16, 549.
Tillaea L. 9, 11.
— alata Viv. 12; tab. 2.
— aquatica L. 12; tab. 2.
— Vaillantii Willd. 11; tab. 2.
Timouria Roshev. 2, 118.
— Saposhnikowii Roshev. 118; tab. 5.
Tissa Adans. 6, 556.
Tithymalus Scop. 14, 308.
Tofieldia Huds. 4, 3.
— calyculata Wahl. 5.
— cernua Smith 4; tab. 1.
— nutans Willd. ex Schult. 4.
— palustris Huds. 3.
Toisusu Kimura 5, 24.
Tomanthea DC. 28, 359.
— Aucheri DC. 362.
— daralaghezica (Fomin) Takht. 364.
— phaeopappa (DC.) Takht. et Czer. 363; tab. 21.
— spectabilis (Fisch. et Mey.) Takht. 361.
Tomiophyllum Fourr. 22, 229.
Tommasinia Boiss. 17, 38.
Tordylium L. 17, 276.
— Komarovii Manden. 278.
— maximum L. 277; tab. 22.
Torilis Adans. 16, 153.
— arvensis (Huds.) Link 152.
— heterophylla Guss. 160; tab. 11.
— japonica (Houtt.) DC. 154.
— leptophylla (L.) Reichb. 163; tab. 11.
— nodosa (L.) Gaertn. 162.
— radiata Moench 160.
— tenella (Del.) Reichb. 161.
— ucrainica Spreng. 156; tab. 11.
— xanthotricha (Stev.) Schischk. 164.
Torresia Ruiz. et Pav. 2, 59.
Torularia O. E. Schulz 8, 59.
— adpressa (Trautv.) O. E. Schulz 64; tab. 4.
— brachycarpa Vass. 66, 635.
— brevipes (Kar. et Kir.) O. E. Schulz 65.
— contortuplicata (Steph.) O. E. Schulz 62.
— glandulosa (Kar. et Kir.) Vass. 69.
— humilis (C. A. Mey.) O. E. Schulz 65.
— Korolkovii (Rgl. et Schmalh.) O. E. Schulz 62.

Torularia mollipila (Maxim.) O. E. Schulz 66; tab. 4.
— sulphurea (Korsh.) O. E. Schulz 63; tab. 4.
— sumbarensis (Lipsky) O. E. Schulz 64; tab. 4.
— torulosa (Desf.) O. E. Schulz 61.
Torulinium Desv. 3, 24.
— caucasicum Palla 24; tab. 1.
Tournefortia L. 19, 116.
— sibirica L. 119.
— sogdiana (Bge.) M. Pop. 120; tab. 6.
Tournesol Adans. 14, 288.
Tournesolia Scop. 14, 288.
Toxicodendron Mill. 14, 531.
— orientale Greene 535.
— radicans (L.) Kuntze 535; tab. 29.
— trichocarpum (Miq.) Kuntze 538.
— vernicifluum (Stokes) Lincz. 537; tab. 29.
— vernix (L.) Kuntze 538; tab. 29.
Tozzia L. 22, 554.
— carpathica Woloszcz. 554.
Trachelanthus Kuntze 19, 620.
— hissaricus Lipsky 621; tab. 34.
— Korolkovii Lipsky 622; tab. 34.
Trachomitum Woodson 18, 652.
— armenum Pobed. 657; tab. 35.
— lancifolium (Russan.) Pobed. 658.
— Russanovii Pobed. 654; tab. 35.
— sarmatiense Woodson 653.
— scabrum (Russan.) Pobed. 659.
— tauricum Pobed. 657; tab. 35.
Trachycarpus H. Wendl. 3, 476.
— excelsa (Thunb.) H. Wendl. 476; tab. 25.
Trachydium Lindl. 16, 246.
— dichotomum Korov. 249.
— kopetdaghense Korov. 246; tab. 15.
Trachynia Link 2, 596.
— distachya (L.) Link 596; tab. 44.
Trachyspermum Link 16, 378.
— ammi (L.) Sprague 379.
Trachystemon D. Don 19, 291.
— orientale (L.) D. Don 292.
Tragacantha Mill. 12, 318.
Tragiopsis Pom. 16, 378.
Tragium Spreng. 16, 422, 435.
Tragopogon L. 29, 115.
— acanthocarpus Boiss. 158.
— alaicus S. Nikit. 191.
— altaicus S. Nikit. 147.
— armeniacus Kuthath. 160.
— badachschanicus Boriss. 147, 723; tab. 9.
— bjelorussicus Artemcz. 143; tab. 11.
— borysthenicus Artemcz. 174.
— brevirostris DC. 169.
— buphtalmoides (DC.) Boiss. 156.
— capitatus S. Nikit. 132; tab. 7.
— Charadzeae Kuthath. 177.
— colchicus Albov 171.
— collinus DC. 186.
— coloratus C. A. Mey. 185.
— conduplicatus S. Nikit. 189.
— cretaceus S. Nikit. 152.
— daghestanicus (Artemcz.) Kuthath. 174.
— dasyrhynchus Artemcz. 165.
— donetzicus Artemcz. 184.
— Dubianskyi Krasch. et S. Nikit. 180.
— dubius Scop. 131.
— elatior Stev. 168.
— elongatus S. Nikit. 192.
— filifolius Rehm. ex Boiss. 171.
— gaudanicus Boriss. 191.
— Gorskianus Reichb. 163.
— gracilis D. Don 150.
— graminifolius DC. 169.
— heterospermus Schweigg. 164; tab. 7.
— Idae Kuthath. 159.
— Karelinii S. Nikit. 148.
— Karjaginii Kuthath. 157.
— kasahstanicus S. Nikit. 195; tab. 8.
— Kemulariae Kuthath. 136.
— Ketzkhovelii Kuthath. 155.
— kopetdaghensis Boriss. 179; tab. 10.
— Krascheninnikovii S. Nikit. 140.
— Kultiassovii M. Pop. ex S. Nikit. 195.
— latifolius Boiss. 158.
— lithuanicus (DC.) Boriss. 162.
— macropogon C. A. Mey. 139; tab. 8.
— Makaschwilii Kuthath. 159.
— malikus S. Nikit. 194.
— marginatus Boiss. 175.
— marginifolius Pavl. 188.
— maturatus Boriss. 177.
— meskheticus Kuthath. 154.
— montanus S. Nikit. 190.
— nachitschevanicus Kuthath. 185.
— orientalis L. 142.
— paradoxus S. Nikit. 134.
— plantagineus Boiss. et Huet 155.
— podolicus (Bess. ex DC.) Artemcz. 167.
— porrifolium L. 196.
— pratensis L. 193.
— pteudomajor S. Nikit. 135.
— pusillus M. B. 153.
— reticulatus Boiss. et Huet 176.
— ruber S. G. Gmel. 187; tab. 11.
— ruthenicus Bess. ex Krasch. et S. Nikit. 183.

— sabulosus Krasch. et S. Nikit. 178; tab. 9.
— scoparius S. Nikit. 182.
— segetus Kuthath. 138.
— serawschanicus S. Nikit. 135.
— serotinus Sosn. 170.
— sibiricus Ganesch. 151; tab. 7.
— songoricus S. Nikit. 145.
— Sosnovskyi Kuthath. 137.
— stepposus (S. Nikit.) Stank. 172.
— subalpinus S. Nikit. 151.
— tanaiticus Artemcz. 163.
— tasch-kala Kuthath. 157.
— tomentosulus Boriss. 148.
— trachycarpus S. Nikit. 144.
— tuberosus C. Koch 152.
— turkestanicus S. Nikit. 146.
— ucrainicus Artemcz. 173.
— volgensis S. Nikit. 166; tab. 10.
— Vvedenskyi M. Pop. ex Pavl. 138.
Tragopyrum M. B. 5, 501.
Tragoselinum Mill. 16, 422, 427.
Tragus Hall. 2, 23.
— racemosus (L.) Desf. 23; tab. 1.
Trapa L. 15, 638.
— amurensis Fler. 659.
— astrachanica (Fler.) N. Wint. 653; tab. 33.
— carinthiaca (G. Beck) V. Vassil. 661; tab. 33.
— colchica Albov 655.
— conocarpa (Aresch.) Fler. 649; tab. 32.
— cruciata (Glück) V. Vassil. 650; tab. 32.
— europaea Fler. 662; tab. 33.
— hyrcana Woronow 654.
— Komarovii V. Vassil 647, 693; tab. 32.
— Korshinskyi V. Vassil. 648, 694; tab. 32.
— Litwinowii V. Vassil. 649, 694; tab. 32.
— longicornis V. Vassil. 661, 698; tab. 33.
— maeotica Woronow 654.
— Maleevii V. Vassil. 655, 696; tab. 33.
— manshurica Fler. 660.
— Maximowiczii Korsh. 643; tab. 32.
— natans L. 652.
— pectinata V. Vassil. 659, 697; tab. 33.
— Potaninii V. Vassil. 647, 693; tab. 32.
— rossica V. Vassil. 651, 695; tab. 33.
— septentrionalis V. Vassil. 651, 695; tab. 32.
— sibirica Fler. 656.
— Sprygínii V. Vassil. 660, 697; tab. 33.
— Tranzschelii V. Vassil. 644, 692; tab. 32.
— tuberculifera V. Vassil. 644, 692; tab. 32.
Trapella Oliv. 23, 14.
— sinensis Oliv. 14; tab. 2.
Traunsteinera Reichb. f. 4, 671.
— globosa (L.) Reichb. f. 671; tab. 41.
— sphaerica (M. B.) Schlecht. 672; tab. 41.
Trautvetteria Fisch. et Mey. 7, 509.
— japonica Sieb. et Zucc. 509.
Triadenum Raf. 15, 201.
— japonicum (Blume) Makino 202; tab. 9.
Tribulus L. 14, 194.
— macropterus Boiss. 194.
— terrestris L. 195.
Trichanthemis Rgl. et Schmalh. 26, 273.
— aulieatensis (B. Fedtsch.) Krasch. 277; tab. 27.
— aurea Krasch. 281.
— Butkovii S. Kovalevsk. 282.
— karataviensis Rgl. et Schmalh. 278; tab. 12.
— Litwinowii (Krasch.) Tzvel. 283.
— paradoxos (C. Winkl.) Tzvel. 281.
— radiata Krasch. et Vved. 280; tab. 11
Trichasma Walp. 11, 51.
Trichaurus Arn. 15, 291, 314.
Trichelostylis Lest. 3, 95.
Trichera Schrad. 24, 12, 13.
Trichocarpus Neck. 10, 600.
Trichochiton Kom. 8, 322.
— inconspicuum Kom. 323; tab. 21.
Trichocrepis Vis. 29, 699, 708.
Trichodesma R. Br. 19, 690.
— incanum (Bge.) DC. 690.
Trichodium Michx. 2, 174.
Tricholepis DC. 28, 258.
— trichocephala Lincz. 258.
Trichomanes L. 1, 17.
— parvulum Poir. 18; tab. 1.
Trichophorum Pers. 3, 37.
— alpinum (L.) Pers. 38; tab. 3.
— caespitosum (L.) Hartm. 37; tab. 3.
— pumilum (Vahl) Schinz et Thell. 38; tab. 3.
Tridactylina (DC.) Sch. Bip. 26, 389.
— Kirilowii (Turcz.) Sch. Bip. 390; tab. 17.
Trientalis L. 18, 269.
— arctica Fisch. ex Hook. 271.
— europaea L. 270.
Trifolium L. 1, 189.
— alexandrinum L. 255.
— alpestre L. 247.
— ambiguum M. B. 207.
— angulatum Waldst. et Kit. 214.
— angustifolium L. 240.

Trifolium apertum Bobr. 254, 391; tab. 16.
— arvense L. 257.
— Bordzilowskyi Grossh. 206.
— borysthenicum Grun. 251.
— campestre Schreb. 223.
— canescens Willd. 232.
— caucasicum Tausch 234; tab. 15.
— cyathiferum Lindl. 199.
— diffusum Ehrh. 253.
— dubium Sibth. 224.
— echinatum M. B. 240.
— Elisabethae Grossh. 208.
— eximium Steph. ex DC. 202; tab. 14.
— expansum Waldst. et Kit. 250.
— fontanum Bobr. 251, 391.
— fragiferum L. 226.
— glomeratum L. 215.
— hirtum All. 256.
— hybridum L. 212.
— incarnatum L. 243.
— lappaceum M. B. 257.
— leucanthum M. B. 238; tab. 16.
— lupinaster L. 200; tab. 13.
— maritimum Huds. 234; tab. 16.
— medium L. 246.
— micranthum Viv. 225.
— Molineri Balb. 242.
— montanum L. 206.
— neglectum C. A. Mey. 226.
— ochroleucum Huds. 237.
— pacificum Bobr. 202; tab. 13.
— pannonicum Jacq. 233.
— parviflorum Ehrh. 214.
— phleoides Pourr. 243.
— physodes Stev. 228.
— polyphyllum C. A. Mey. 205; tab. 14.
— pratense L. 248.
— Raddeanum Trautv. 228.
— repens L. 211.
— resupinatum L. 230.
— rubens L. 247.
— rytidosemium Boiss. et Hohen. 219.
— sativum Crome 252.
— scabrum L. 244.
— Sebastianii Savi 224.
— seravschanicum Ovcz. 252, 391.
— spadiceum L. 220.
— speciosum Willd. 221.
— spumosum L. 217.
— squarrosum L. 239.
— stellatum L. 241.
— stipitatum Boiss. et Bal. 218.
— strepens Crantz 222.
— striatum L. 245.
— subterraneum L. 261.
— suffocatum L. 216.
— tomentosum L. 231.
— trichocephalum M. B. 233; tab. 15.
— tumens Stev. 229.
— vesiculosum Savi 217.
Triganthemum Gib. et Belli 11, 216.
Triglochin L. 1, 276.
— maritima L. 276; tab. 10.
— palustris L. 277; tab. 10.
Trigonella L. 11, 102.
— adscendens Afan. et Gontsch. 126; tab. 9
— arcuata C. A. Mey. 113; tab. 8.
— astroides Fisch. et Mey. 114; tab. 8.
— badachschanica Afan. 125; tab. 9.
— biflora Griseb. 124; tab. 9.
— brachycarpa (Fisch.) G. Moris 123; tab. 9.
— calliceras Fisch. 108; tab. 8.
— capitata Boiss. 118; tab. 9.
— coerulea (Desr.) Ser. 116; tab. 9.
— coerulescens (M. B.) Halacsy 118; tab. 9.
— cretacea (M. B.) Grossh. 120; tab. 9.
— Fischeriana Ser. 113; tab. 8.
— foenum-graecum L. 119; tab. 9.
— geminiflora Bge. 115; tab. 8.
— gladiata Stev. 119; tab. 9.
— Gordijevii (Kom.) Grossh. 129; tab. 9.
— grandiflora Bge. 106; tab. 8.
— Korschinskyi Grossh. 127; tab. 9.
— laxiflora Aitch. et Baker 126; tab. 9.
— Lipskyi Širj. 125; tab. 9.
— monspeliaca L. 116; tab. 8.
— montana C. A. Mey. 114; tab. 8.
— Noëana Boiss. 115; tab. 8.
— orthoceras Kar. et Kir. 115.
— pamirica Boriss. 124; tab. 9.
— platycarpos L. 128; tab. 9.
— Popovii Korov. 128; tab. 9.
— procumbens (Bess.) Reichb. 117; tab. 9.
— radiata (L.) Boiss. 123; tab. 9.
— ruthenica L. 127; tab. 9.
— spicata Sibth. et Sm. 111; tab. 8.
— strangulata Boiss. 110.
— tenuis Fisch. 112; tab. 8.
— torulosa Griseb. 107; tab. 8.
— turkmena M. Pop. 106; tab. 8.
— verae Širj. 106; tab. 8.
Trigonocaryum Trautv. 19, 297.
— involucratum (Stev.) Kusn. 297.
Trigonotis Stev. 19, 260.
— koreana Nakai 261; tab. 14.
— myosotidea Maxim. 263; tab. 14.

— peduncularis (Trev.) Benth. ex S. Moore et Baker 264; tab. 14.
— radicans (Turcz.) Stev. 262.
Trillium L. 4, 473.
— kamtschaticum Pall. 473; tab. 29.
— rhombifolium Kom. 474, 748.
Trimorpha Cass. 25, 191.
Trimorphaea Cass. 25, 191.
Trinia Hoffm. 16, 349.
— hispida Hoffm. 355.
— Kitaibelii M. B. 357; tab. 24.
— leiogona (C. A. Mey.) B. Fedtsch. 357; tab. 25.
— multicaulis (Poir.) Schischk. 352; tab. 24.
— muricata Godet 356.
— polyclada Schischk. 351.
— Stankovii Schischk. 352.
— ucrainica Schischk. 351.
Triosteon Adans. 23, 458.
Triosteospermum Post et Kuntze 23, 458.
Triosteum L. 23, 458.
— sinuatum Maxim. 458.
Tripleurospermum Sch. Bip. 26, 157.
— ambiguum (Ledeb.) Franch. et Savat. 168; tab. 7.
— breviradiatum (Ledeb.) Pobed. 178.
— caucasicum (Willd.) Hayek 161.
— colchicum (Manden.) Pobed. 174.
— decipens (Fisch. et Mey.) Bornm. 182.
— disciforme (C. A. Mey.) Sch. Bip. 181; tab. 6.
— elongatum (Fisch. et Mey.) Bornm. 176; tab. 5.
— Grossheimii (Fed.) Pobed. 165.
— inodorum (L.) Sch. Bip. 175.
— Karjaginii (Manden. et Sof.) Pobed. 167.
— limosum (Maxim.) Pobed. 177; tab. 6.
— maritimum (L.) Koch 177.
— parviflorum (Willd.) Pobed. 183.
— phaeocephalum (Rupr.) Pobed. 171.
— rupestre (Somm. et Lévier) Pobed. 173.
— sevanense (Manden.) Pobed. 182; tab. 7.
— subnivale Pobed. 162; tab. 5.
— subpolare Pobed. 172.
— Szowitzii (DC.) Pobed. 166; tab. 5.
— Tchichatchevii (Boiss.) Bornm. 174.
— tetragonospermum (Fr. Schmidt) Pobed. 170.
— transcaucasicum (Manden.) Pobed. 167.
Triplocentron Cass. 28, 370, 570.
Tripogon Roth ex Roem. et Schult. 2, 287.
— chinensis (Franch.) Hack. 287; tab. 22.
Tripolium Nees 25, 183.
— vulgare Nees 184; tab. 4.
Tripolium (Astropolium) Nutt. 25, 185.
Tripolium (Oxytripolium) DC. 25, 185.
Trisetum Pers. 2, 252.
— altaicum (Steph.) Roshev. 254; tab. 18.
— Cavanillesii Trin. 257; tab. 18.
— distichophyllum (Vill.) Beauv. 256; tab. 18.
— lineare (Forsk.) Boiss. 257; tab. 18.
— pratense Pers. 253; tab. 18.
— rigidum (M. B.) Roem. et Schult. 256; tab. 18.
— seravschanicum Roshev. 255; tab. 18.
— sibiricum Rupr. 253; tab. 18.
— spicatum (L.) Richt. 255; tab. 18.
— subalpestre (Hartm.) Neum. 254.
— virescens (Rgl.) B. Fedtsch. 255; tab. 18.
Triticum L. 2, 675.
— aestivum L. 687; tab. 48.
— armeniacum (Stolet.) Nevski 683.
— carthlicum Nevski 685.
— compactum Host 687; tab. 48.
— dicoccoides (Koern.) Aaronsohn 681.
— durum Desf. 685; tab. 48.
— macha Dekapr. et Menabde 686.
— monococcum L. 681; tab. 48.
— polonicum L. 688; tab. 48.
— spelta L. 687; tab. 48.
— Thaoudar Reut. 677; tab. 48.
— Timofeevii Zhuk. 682.
— turgidum L. 684; tab. 48.
— volgense (Flaksb.) Nevski 683; tab. 48.
Trixago Stev. 22, 658.
Trochostigma Sieb. et Zucc. 15, 186.
Trollius L. 7, 42.
— altaicus C. A. Mey. 45.
— asiaticus L. 49.
— chartosepalus Schipcz. 53.
— chinensis Bge. 47.
— dschungaricus Rgl. 44; tab. 5.
— europaeus L. 45.
— japonicus Miq. 50.
— Ledebouri Reichb. 47.
— patulus Salisb. 46.
— Riederianus Fisch. et Mey. 50.
— sibiricus (Rgl. et Til.) Schipcz. 48.
Tropaeolum L. 14, 84.
— major L. 84.
— minus L. 84.
Tulipa L. 4, 320.
— Alberti Rgl. 334.
— altaica Pall. ex Spreng. 346.
— anisophylla Vved. 348.
— Behmiana Rgl. 341.
— Biebersteiniana Schult. 354.

Tulipa biflora Pall. 358.
— bifloriformis Vved. 358.
— Borszowii Rgl. 342.
— brachystemon Rgl. 346.
— Buhseana Boiss. 356.
— Callieri Halacsy et Lévier 360.
— carinata Vved. 332.
— caucasica Lipsky 353.
— dasystemon Rgl. 361.
— dasystemonoides Vved. 361.
— dubia Vved. 347.
— Eichleri Rgl. 336.
— ferganica Vved. 348.
— Florenskyi Woronov 331.
— Fosteriana Irving 332.
— Greigii Rgl. 335.
— heteropetala Ledeb. 363.
— heterophylla (Rgl.) Bakerm 364.
— hissarica M. Pop. et Vved. 350.
— Hoogiana B. Fedtsch. 330.
— iliensis Rgl. 347.
— ingens Hoog 332.
— Julia C. Koch 330.
— Kaufmanniana Rgl. 351.
— koktebelica Junge 359.
— Kolpakovskiana Rgl. 345.
— Korolkovii Rgl. 342.
— Korshinskyi Vved. 349.
— Krauseana Rgl. 351.
— kuschkensis B. Fedtsch. 329.
— lanata Rgl. 331.
— Lehmanniana Merckl. 341.
— linifolia Rgl. 340.
— Maximowiczii Rgl. 339.
— Micheliana Hoog 336.
— mogoltavica M. Pop. et Vved. 335; tab. 20.
— nitida Hoog 343.
— orithyioides Vved. 360.
— Ostrovskiana Rgl. 344.
— patens J. Agardh ex Schult. 355.
— platystemon Vved. 346.
— polychroma Stapf 359.
— praestans Hoog 333.
— Regelii Krasn. 352.
— rosea Vved. 343.
— Schmidtii Fomin 339.
— Schrenkii Rgl. 343.
— silvestris L. 353.
— sogdiana Bge. 355; tab. 20.
— subpraestans Vved. 334.
— tarda Stapf 362; tab. 21.
— tetraphylla Rgl. 350.
— tianschanica Rgl. 349.
— Tubergeniana Hoog 329.
— turcomanica B. Fedtsch. 357.
— turkestanica Rgl. 357.
— uniflora (L.) Bess. ex Baker 363.
— violacea Boiss. et Buhse 352.
— Wilsoniana Hoog 340.
— Zenaidae Vved. 345.
Tunica (Hallier) Scop. 6, 777.
— pachygona Fisch. et Mey. 778.
— saxifraga (L.) Scop. 777; tab. 55.
— stricta (Bge.) Fisch. et Mey. 778; tab. 55.
Turaniphytum Poljak. 26, 632, 880.
— eranthemum (Bge.) Poljak. 635; tab. 30.
— kopetdaghense Poljak. 636, 881.
Turczaninowia DC. 25, 136.
— fastigiata (Fisch.) DC. 137.
Turgenia Hoffm. 16, 174.
— latifolia (L.) Hoffm. 174.
Turritis (Tourn.) L. 8, 171.
— glabra L. 171.
Tussilago L. 26, 641, 645.
— farfara L. 641.
Typha L. 1, 210.
— angustata Bory et Chaub. 215; tab. 10.
— angustifolia L. 215; tab. 10.
— elephantina Roxb. 212.
— latifolia L. 210; tab. 10.
— Laxmannii Lepech. 212.
— minima Funck 216; tab. 10.
— orientalis C. Presl 211.
— Veresczaginii Kryl. et Schischk. 211.
Typhoides Moench 2, 54.

Uechtritzia Freyn 28, 595.
— armena Freyn 596.
— kokanica (Rgl. et Schmalh.) Pobed. 597; tab. 28.
Ugamia Pavl. 26, 271.
— angrenica (Krasch.) Tzvel. 272.
Ulex L. 11, 72.
— europaeus L. 72.
Ulmaria Moench 10, 284.
Ulmus L. 5, 360.
— Androssowii Litv. 370.
— celtidea (Rogow.) Litv. 364; tab. 20.
— densa Litv. 369.
— elliptica C. Koch 372; tab. 20.
— foliacea Gilib. 364; tab. 20.
— laciniata (Trautv.) Mayr 372; tab. 20.
— laevis Pall. 363; tab. 20.
— macrocarpa Hance 373; tab. 20.
— pinnato-ramosa Dieck ex Koehne 370.
— propinqua Koidz. 366; tab. 20.

— pumila L. 369; tab. 20.
— scabra Mill. 371; tab. 20.
— suberosa Moench 365.
Umbilicus DC. 9, 13.
— oppositifolius Ledeb. 14; tab. 2.
Ungernia Bge. 4, 481.
— ferganica Vved. 484.
— minor Vved. 484.
— Severtzovii (Rgl.) B. Fedtsch. 485.
— tadshikorum Vved. 483.
— trisphaera Bge. 482.
— Victoris Vved. 483.
Uraspermum Nutt. 16, 148.
Urospermum Scop. 29, 228
— picroides (L.) Desf. 229; tab. 12.
Urosteima Bge. 18, 672.
Urtica L. 5, 386.
— angustifolia Fisch. ex Hornem. 393; tab. 21.
— cannabina L. 388; tab. 21.
— cyanescens Kom. 392, 714.
— dioica L. 392.
— kioviensis Rogov. 391.
— laetevirens Maxim. 388.
— pilulifera L. 387; tab. 21.
— platyphylla Wedd. 391; tab. 21.
— pubescens Ledeb. 394.
— urens L. 387.
Utricularia L. 23, 123.
— Bremii Heer 125.
— intermedia Hayne 124.
— japonica Makino 126.
— major Schmidel 127.
— minor L. 125
— vulgaris L. 127.

Vaccaria Medic. 6, 802.
— segetalis (Neck.) Garcke 802; tab. 55.
Vaccinium L. 18, 94.
— arctostaphylos L. 102.
— hirtum Thunb. 95.
— myrtillus L. 97.
— ovalifolium Smith 99.
— praestans Lamb. 100.
— uliginosum L. 96.
— vitis-idaea L. 100.
Vahlenbergella Blytt 6, 714.
Vahlodea Fries 2, 241.
— atropurpurea (Wahl.) Fries 242; tab. 18.
— paramushirensis (Kudo) Roshev. 242, 750.
Valantia L. 23, 381.
— muralis L. 381.
Valeriana L. 23, 594, 587.
— ajanensis (Rgl. et Til.) Kom. 637; tab. 32.
— alliariifolia Adam 602.
— alpestris Stev. 607.
— amurensis P. Smirn. ex Kom. 635; tab. 32.
— capitata Pall. 616; tab. 30.
— cardamines M. B. 620; tab. 31.
— chionophila M. Pop. et Kult. 601; tab. 29.
— colchica Utkin 632; tab. 32.
— daghestanica Rupr. ex Boiss. 624; tab. 31.
— dioica L. 611.
— dubia Bge. 630.
— eriophylla (Ledeb.) Utkin 619; tab. 29.
— Fedtschenkoi Coincy 639; tab. 30.
— ficariifolia Boiss. 618; tab. 31.
— Jelenevskyi P. Smirn. 609; tab. 29.
— leucophaea DC. 601.
— Martjanovii Kryl. 638; tab. 30.
— officinalis L. 625.
— petrophila Bge. 637; tab. 30.
— saxicola C. A. Mey. 608; tab. 29.
— simplicifolia Kabath 612.
— sisymbriifolia Vahl 617, tab. 31.
— tiliifolia Troitzky 605.
— tripteris L. 609.
— tuberosa L. 600.
— Turczaninovii Grub. 615; tab. 30.
Valerianella Mill. 23, 642.
— amblyotis Fisch. et Mey. 673; tab. 34.
— anodon Lincz. 656; tab. 33.
— brachystephana (Ten.) Bertol. 675.
— carinata Loisel. 670; tab. 34.
— corniculata C. A. Mey. 657; tab. 33.
— coronata (L.) DC. 677; tab. 34.
— costata (Stev.) Betcke 664; tab. 34.
— cymbocarpa C. A. Mey. 658; tab. 33.
— dactylophylla Boiss. et Hohen. 651; tab. 33.
— dentata (L.) Poll. 667; tab. 34.
— diodon Boiss. 653; tab. 33.
— Dufresnia Bge. 679; tab. 34.
— echinata (L.) DC. 662; tab. 33.
— Kotschyi Boiss. 678.
— kulabensis Lipsky ex Lincz. 654; tab. 33.
— lasiocarpa (Stev.) Betcke 676; tab. 34.
— leiocarpa (C. Koch) Kuntze 680.
— Lipskyi Lincz. 673, 739; tab. 34.
— locusta (L.) Betcke 663; tab. 34.
— muricata (Stev.) W. Baxt. 668; tab. 34.
— oxyrrhyncha Fisch. et Mey. 655; tab. 33.
— plagiostephana Fisch. et Mey. 659; tab. 33.

Valerianella platycarpa Trautv. 674; tab. 34.
— pontica Lipsky 676; tab. 34.
— pumila (L.) DC. 674; tab. 34.
— rimosa Bast. 672; tab. 34.
— sclerocarpa Fisch. et Mey. 656; tab. 33.
— Szovitsiana Fisch. et Mey. 660; tab. 33.
— triplaris Boiss. et Buhse 652; tab. 34.
— tuberculata Boiss. 647; tab. 33.
— turgida (Stev.) Betcke 671; tab. 34.
— turkestanica Rgl. et Schmalh. 652; tab. 33.
— uncinata (M. B.) Dufr. 648; tab. 33.
— vesicaria (L.) Moench 680; tab. 34.
— Vvedenskyi Lincz. 661, 739; tab. 34.
Valerianella Moench 23, 587.
Vallisneria L. 1, 295.
— spiralis L. 295; tab. 10.
Vandellia L. 22, 326.
— diffusa L. 326.
Varthemia DC. 25, 486.
— persica DC. 486.
Vavilovia Fed. 13. 521.
Velezia L. 6, 869.
— rigida L. 869; tab. 45.
Venilia Fourr. 22, 229.
Ventenata Koel. 2, 257.
— dubia (Leers) F. Schultz 258; tab. 18.
Veratrum L. 4, 10.
— calyciflorum Kom. 13, 733.
— dahuricum (Turcz.) Loes. 12; tab. 1.
— Lobelianum Bernh. 13.
— Maackii Rgl. 11; tab. 1.
— Misae (Širj.) Loes. 14.
— nigrum L. 11.
— oxysepalum Turcz. 14.
Verbascum L. 22, 122.
— alpigenum C. Koch 154.
— artvinense Wulff 138.
— bactrianum Bge. 132.
— banaticum Rochel 134.
— blattaria L. 167.
— cedreti Boiss. 152.
— cheiranthifolium Boiss. 136.
— eriorhabdon Boiss. 141.
— erivanicum Wulff 152.
— flavidum (Boiss.) Freyn et Bornm. 170.
— flexuosum Wulff 147.
— formosum Fisch. 157.
— georgicum Benth. 124.
— glomeratum Boiss. 131.
— gnaphaloides M. B. 140.
— gossypinum M. B. 145.
— hajastanicum Bordz. 145.
— laxum Filar. et Jav. 148.
— lychnitis L. 142.
— macrocarpum Boiss. 164.
— megaphlomos (Boiss. et Heldr.) Halacsy 135.
— nigrum L. 149.
— oreophilum C. Koch 163.
— orientale M. B. 147.
— ovalifolium Donn 156.
— paniculatum Wulff 153.
— phlomoides L. 123.
— phoeniceum L. 168.
— pinnatifidum Vahl 137.
— punalense Boiss. et Buhse 161.
— pyramidatum M. B. 162; tab. 6.
— saccatum C. Koch 158; tab. 5.
— sessiliflorum Murb. 125.
— sinuatum L. 144.
— songoricum Schrenk ex Fisch. et Mey. 133.
— speciosum Schrad. 135.
— spectabile M. B. 162; tab. 6.
— stachydiforme Boiss. et Buhse 139.
— Szovitsianum Boiss. 151.
— thapsiforme Schrad. 126.
— thapsus L. 128.
— transcaucasicum Wulff 154.
— turcomanicum Murb. 143.
— turkestanicum Franch. 140.
— varians Freyn et Sint. 146.
— Wilhelmsianum C. Koch 150.
Verbascum Kuntze 22, 170.
Verbena L. 19, 693.
— bracteosa Michx. 694.
— hastata L. 696.
— officinalis L. 694.
— supina L. 695.
Veronica L. 22, 329.
— acinifolia L. 429; tab. 18.
— agrestis L. 408.
— alatavica M. Pop. 385.
— albanica C. Koch 397.
— alpina L. 481.
— americana (Raf.) Schweinitz ex Benth. 477.
— amoena Stev. 407; tab. 17.
— anagallidiformis Boreau 473; tab. 21.
— anagallis-aquatica L. 469.
— anagalloides Guss. 470; tab. 21.
— aphylla L. 451.
— arceutobia Woronow 439.
— arenosa (Serg.) Boriss. 390.
— argute-serrata Rgl. et Schmalh. 394.
— armena Boiss. et Huet 467; tab. 20.
— arvensis L. 418; tab. 18.
— austriaca L. 438.

— Bachofenii Heuff. 369.
— Baranetzkii Bordz. 465.
— Barrelieri Schult. 383.
— Baumgartenii Roem. et Schult. 451.
— beccabunga L. 475; tab. 21.
— beccabungoides Bornm. 478.
— bellidifolia Juz. 379.
— bellidioides L. 482.
— biloba L. 392; tab. 16.
— Bobrovii Nevski 479; tab. 21.
— Bornmülleri Hausskn. 395.
— bucharica B. Fedtsch. 399; tab. 16.
— callitrichoides Kom. 454.
— campylopoda Boiss. 397.
— capillipes Nevski 400.
— cardiocarpa (Kar. et Kir.) Walp. 404; tab. 16.
— caucasica M. B. 439; tab. 20.
— cerasifolia Monjuschko 499.
— ceratocarpa C. A. Mey. 424; tab. 17.
— chamaedrys L. 430.
— chantavica Pavl. 393.
— Charadzeae Kem.-Nath. 361.
— ciliata Fisch. 490; tab. 22.
— crista-galli Stev. 413; tab. 18.
— cymbalaria Bod. 417.
— Czerniakowskiana Monjuschko 446.
— daghestanica Trautv. 493.
— dahurica Stev. 372.
— densiflora Ledeb. 486; tab. 22.
— dentata Fr. Schmidt 435.
— denudata Albov 444.
— didyma Ten. 409; tab. 17.
— Dillenii Crantz 420; tab. 18.
— Fedtschenkoi Boriss. 491.
— filifolia Lipsky 445; tab. 20.
— filiformis Smith 423.
— fruticans Jacq. 484.
— fruticulosa L. 483.
— galathica Boiss. 450.
— gentianoides Vahl 356; tab. 14.
— glabrifolia Boriss. 461; tab. 20.
— grandiflora Gaertn. 452.
— Gorbunovii Gontsch. 489; tab. 22.
— hederifolia L. 414; tab. 18.
— hololeuca Juz. 380.
— humifusa Dickson 366.
— imeretica Kem.-Nath. 358.
— incana L. 377.
— intercedens Bornm. 404.
— karatavica Pavl. ex Nevski 395; tab. 16.
— Kemulariae Kuthath. 358.
— khorossanica Czernjak. 448.
— Komarovii Monjuschko 377.
— kopetdaghensis B. Fedtsch. 460.
— Krylovii Schischk. 436; tab. 19.
— kurdica Benth. 443.
— laeta Kar. et Kir. 389; tab. 15.
— linariifolia Pall. ex Link 386; tab. 15.
— longifolia L. 367.
— Lütkeana Rupr. 485; tab. 22.
— lysimachioides Boiss. 480.
— macrostemon Bge. 485.
— macrostemonoides Zak. 487.
— maxima Mill. 456.
— melissifolia Desf. ex Poir. 423; tab. 19.
— Michauxii Lam. 479.
— microcarpa Boiss. 467.
— minima C. Koch 426.
— minuta C. A. Mey. 459.
— montana L. 455.
— monticola Trautv. 362.
— montioides Boiss. 478.
— multifida L. 444.
— Nevskii Boriss. 396.
— nigricans C. Koch 433.
— officinalis L. 449.
— olgensis Kom. 370.
— oltensis Woronow 466.
— opaca Fries 410.
— orchidea Crantz 384.
— orientalis Mill. 440.
— oxycarpa Boiss. 480.
— peduncularis M. B. 462.
— peregrina L. 420.
— perpusilla Boiss. 425.
— persica Poir. 411; tab. 17.
— petraea (M. B.) Stev. 464.
— pinnata L. 391.
— poljensis Murb. 475.
— Porphyriana Pavl. 382.
— praecox All. 406.
— propinqua Boriss. 464.
— prostrata L. 437.
— ramosissima Boriss. 398.
— Riederiana Gand. ex Herd. 367.
— rubrifolia Boiss. 396; tab. 18.
— Ruprechtii Lipsky 492.
— sachalinensis Boriss. 496; tab. 23.
— sajanensis Printz 375.
— scardica Griseb. 474.
— schistosa E. Busch 362; tab. 14.
— Schmidtiana Rgl. 364; tab. 14.
— scutellata L. 453.
— septentrionalis Boriss. 369; tab. 15.
— serpyllifolia L. 365.
— serpylloides Rgl. 488.
— sibirica L. 495; tab. 23.

Veronica spicata L. 381.
— spuria L. 376.
— Stelleri Pall. ex Link 363.
— stylophora M. Pop. 400; tab. 16.
— subsessilis (Miq.) Carr. 371.
— taurica Willd. 443; tab. 19.
— teberdensis (Kem.-Nath.) Boriss. 493.
— telephiifolia Vahl 461.
— tenuissima Boriss. 403; tab. 16.
— teucrium L. 434.
— tianschanica Lincz. 488.
— tripartita Boriss. 447.
— triphyllos L. 405; tab. 17.
— tubiflora Fisch. et Mey. 494; tab. 23.
— turkmenorum B. Fedtsch. 422.
— umbrosa M. B. 432.
— verna L. 421; tab. 18.
Veronicastrum Heist. ex Fabr. 22, 329, 356, 494.
Veronicella Fourr. 22, 329, 355.
Verutina Cass. 28, 370, 570.
Vesicaria Adans. 8, 361.
— graeca Reut. 362; tab. 22.
— leiocarpa (Trautv.) N. Buch 362.
Vesicaria L. 11, 216, 226.
Viburnum L. 23, 443.
— burejaeticus Rgl. et Herd. 448.
— furcatum Blume ex Maxim. 445.
— lantana L. 447.
— mongolicum (Pall.) Rehd. 450.
— opulus L. 456.
— orientale Pall. 452.
— Sargentii Koehne 454.
— Wrightii Miq. 451.
Vicia L. 13, 406.
— abbreviata Fisch. ex Spreng. 428; tab. 25.
— alpestris Stev. 435.
— amoena Fisch. 444; tab. 26.
— amphicarpa Dorthes 465.
— amurensis Oett. 445.
— angustifolia L. 464.
— antiqua Grossh. 441.
— baicalensis (Turcz.) B. Fedtsch. 424; tab. 24.
— Balansae Boiss. 457.
— biennis L. 437.
— bithynica L. 472; tab. 25.
— Boissieri Freyn 440.
— calcarata Desf. 418.
— cassubica L. 427.
— ciceroidea Boiss. 449.
— ciliatula Lipsky 471.
— cordata Wulf. 463.
— costata Ledeb. 449; tab. 25.
— cracca L. 436.
— crocea (Desf.) B. Fedtsch. 425.
— dadianorum Somm. et Lévier 429.
— dalmatica A. Kerner 439.
— dasycarpa Ten. 451.
— dumetorum L. 422.
— ecirrhosa Rupr. ex Boiss. 430.
— elegans Guss. 439.
— ervilia (L.) Willd. 407.
— faba L. 474.
— Fedtschenkoana V. Nikit. 441, 555.
— geminiflora Trautv. 443; tab. 25.
— gracilior M. Pop. 467.
— gracilis Loisel. 411.
— grandiflora Scop. 459.
— Grossheimii Ekvtim. 438.
— hajastana Grossh. 470.
— hirsuta (L.) S. F. Gray 408.
— hololasia Woronow 452; tab. 26.
— hybrida L. 469; tab. 23.
— hyrcanica Fisch. et Mey. 465.
— iberica Grossh. 442.
— incisa M. B. 463.
— iranica Boiss. 427.
— japonica A. Gray 445; tab. 24.
— kokanica Rgl. et Schmalh. 446; tab. 23.
— lathyroides L. 457.
— lilacina Ledeb. 436.
— lutea L. 468.
— macrophylla (Maxim.) B. Fedtsch. 437.
— megalosperma M. B. 467.
— megalotropis Ledeb. 442; tab. 24.
— Meyeri Boiss. 403; tab. 25.
— Michauxii Spreng. 468.
— monanthos (L.) Desf. 409.
— multicaulis Ledeb. 443.
— narbonensis L. 473.
— olbiensis Reut. 458.
— pannonica Crantz 470.
— paucijuga (Trautv.) B. Fedtsch. 411; tab. 23.
— peregrina L. 466.
— picta Fisch. et Mey. 451; tab. 26.
— pilosa M. B. 464.
— pisiformis L. 422.
— pseudorobus Fisch. et Mey. 421; tab. 26.
— pubescens (DC.) Boiss. 412.
— ramuliflora (Maxim.) Ohwi 424.
— sativa L. 460.
— saxatilis (Vent.) Tropea 458.
— sepium L. 455.
— Semenovii (Rgl. et Herd.) B. Fedtsch. 430; tab. 23.
— semiglabra Rupr. ex Boiss. 431.

— serratifolia Jacq. 474.
— silvatica L. 429.
— Sosnovskyi Ekvtim. 432.
— subvillosa (Ledeb.) Trautv. 426.
— tenuifolia Roth 440.
— tetrasperma (L.) Moench 410.
— truncatula Fisch. 456.
— unijuga A. Br. 423; tab. 24.
— variabilis Freyn et Sint. 438.
— variegata Willd. 431; tab. 26.
— venosa (Willd.) Maxim. 425; tab. 24.
— venulosa Boiss. et Hohen. 421; tab. 23.
— villosa Roth 450.
Vicoa Cass. 25, 477.
Vigna Savi 13, 538.
— sinensis Endl. 538.
Vignantha Schur 3, 201.
Vignea Raf. 3, 125.
Vilfa Adans. 2, 177.
Vinca L. 18, 646.
— erecta Rgl. et Schmalh. 650; tab. 35.
— herbacea Waldst. et Kit. 649.
— major L. 649.
— minor L. 647.
— pubescens Urv. 648.
Vincetoxicum Moench 18, 674.
Viola L. 15, 350.
— acuminata Ledeb. 394.
— acutifolia (Kar. et Kir.) W. Beck. 450.
— alba Bess. 18, 366; tab. 18.
— alaica Vved. 429.
— Alexandrowiana (W. Beck.) Juz. 412.
— Alisoviana Kiss. 419.
— allochroa Botsch. 444.
— altaica Ker.-Gawl. 476; tab. 22.
— ambigua Waldst. et Kit. 370; tab. 18.
— amurica W. Beck. 440.
— arvensis Murr. 468.
— atroviolacea W. Beck. 459; tab. 22.
— austro-ussuriensis Kom. 395; tab. 19.
— avatschensis W. Beck. et Hult. 445.
— biflora L. 442; tab. 18.
— brachyceras Turcz. 405.
— brachysepala Maxim. 377; tab. 19.
— canina L. 398; tab. 18.
— carnosula W. Beck. 409.
— caucasica Kolenati ex Rupr. 444; tab. 21.
— chaerophylloides (Rgl.) W. Beck. 433.
— collina Bess. 373; tab. 18.
— crassa Makino 445.
— crassicornis W. Beck. et Hult. 410.
— cretacea Klok. 471, 686.
— dactyloides Roem. et Schult. 434; tab. 20.
— declinata Waldst. et Kit. 457.
— disjuncta W. Beck. 458; tab. 22.
— dissecta Ledeb. 432; tab. 20.
— dolichocentra Botsch. 429.
— elatior Fries 389.
— Elisabethae Klok. 465, 684.
— epipsila Ledeb. 404.
— Fedtschenkoana W. Beck. 386; tab. 19.
— Fischeri W. Beck. 451.
— glabella Nutt. 449; tab. 18.
— Gmeliniana Roem. et Schult. 426.
— hirta L. 369; tab. 18.
— hirtipes S. Moore 413.
— hissarica Juz. 422, 682.
— Hultenii W. Beck. 406.
— ignobilis Rupr. 363.
— incisa Turcz. 431.
— ircutiana Turcz. 415.
— isopetala Juz. 384, 681.
— Jordanii Hanry 392.
— kamtschadalorum W. Beck. et Hult. 438; tab. 21.
— karakalensis Klok. 473, 687; tab. 22.
— Kitaibeliana Roem. et Schult. 469.
— Kupfferi Klok. 460, 683.
— Kusnezowiana W. Beck. 407; tab. 18, 20.
— Langsdorfii Fisch. 437.
— Lavrenkoana Klok. 472; tab. 22.
— littoralis Spreng. 460.
— macroceras Bge. 417; tab. 20.
— mandshurica W. Beck. 425; tab. 20.
— matutina Klok. 467, 685.
— Mauritii Tepl. 379; tab. 19.
— Meyeriana (Rupr.) Klok. 479; tab. 22.
— minuta M. B. 478.
— mirabilis L. 375.
— modestula Klok. 474, 687; tab. 22.
— monochroa Klok. 477, 688; tab. 22.
— montana L. 396.
— Muehldorfii Kiss. 450; tab. 21.
— nemausensis Jord. 470.
— occulta Lehm. 475; tab. 22.
— odorata L. 363; tab. 18.
— oreades M. B. 478.
— orientalis (Maxim.) W. Beck. 448; tab. 21.
— orthoceras Ledeb. 456.
— oxycentra Juz. 430, 682.
— pacifica Juz. 411.
— palustris L. 403; tab. 18.
— parvula Tineo 473.
— Patrinii Ging. 421.
— persicifolia Roth 390.
— phalacrocarpa Maxim. 411; tab. 20.

Viola prionantha Bge. 419.
— pumila Chaix 392.
— pyrenaica Ram. 374.
— Raddeana Rgl. 441; tab. 21.
— repens Turcz. 405.
— Riviniana Reichb. 381.
— rupestris F. M. Schmidt 388.
— sacchalinensis De Boiss. 387; tab. 19.
— Selkirkii Pursh 408.
— Sieheana W. Beck. 383.
— silvestriformis W. Beck. 380; tab. 19.
— Sintenisii W. Beck. 386.
— somchetica C. Koch 418.
— suavis M. B. 365; tab. 18.
— sylvestris Lam. 378.
— tanaitica Grosset 382.
— tarbagataica Klok. 458, 683.
— tianschanica Maxim. 427.
— trichosepala (W. Beck.) Juz. 416.
— tricolor L. 462; tab. 18.
— Turczaninowii Juz. 395.
— turkestanica Rgl. et Schmalh. 428.
— uliginosa Bess. 401; tab. 18.
— uniflora L. 447.
— ursina Kom. 437.
— variegata Fisch. ex Link 414; tab. 18, 20.
— verecunda A. Gray 439; tab. 18.
— vespertina Klok. 466, 684.
Viscaria Roehl. 6, 576.
— alpina (L.) G. Don 577; tab. 45.
— viscosa (Scop.) Aschers. 576; tab. 45.
Viscum L. 5, 408.
— album L. 411; tab. 23.
— coloratum (Kom.) Nakai 411.
Visiania DC. 18, 518.
Visnaga Gaertn. 16, 381.
Vitex L. 19, 698.
— agnus-castus L. 698.
Vitis L. 14, 675.
— amurensis Rupr. 700; tab. 38.
— Berlandieri Planch. 683; tab. 38.
— labrusca L. 678; tab. 38.
— rupestris Scheele 684; tab. 38.
— silvestris J. F. Gmel. 697; tab. 39.
— Thunbergii Sieb. et Zucc. 680; tab. 38.
— viniferi L. 686; tab. 38.
— vulpina L. 685; tab. 38.
Vitmannia Endl. 6, 370.
Vitmannia Turr. 6, 370.
Vleckia Raf. 20, 273, 275.
Volvulus Medic. 19, 33.
Vulpia J. F. Gmel. 2, 535.
— ciliata (Danth.) Link 536.
— dertonensis (All.) Volkart ex Schinz et Keller 539; tab. 41.
— myuros (L.) J. F. Gmel. 539; tab. 41.
— persica (Boiss. et Buhse) V. Krecz. et Bobr. 535; tab. 41.
Vvedenskia Korov. 17, 1.

Wahlbergella Fries 6, 714.
Waldheimia Kar. et Kir. 26, 263.
— glabra (Decne.) Rgl. 270.
— Stoliczkae (Clarke) Ostenf. 268.
— tomentosa (Decne.) Rgl. 267.
— transalaica Tzvel. 269, 879; tab. 11.
— tridactylites Kar. et Kir. 269.
Waldschmidtia Bluff 18, 643.
Waldsteinia Willd. 10, 239.
— geoides Willd. 240.
— ternata (Steph.) Fritsch 240; tab. 18.
Washingtonia Raf. 16, 148.
Washingtonia H. Wendl. 3, 473.
— filamentosa (H. Wendl.) Kuntze 473; tab. 25.
— robusta H. Wendl. 474.
Weigela Thunb. 23, 574.
— Middendorffiana (Carr.) C. Koch 582; tab. 27.
— praecox (Lemoine) Bailey 575; tab. 27.
— suavis (Kom.) Bailey 581; tab. 27.
Weingaertneria Bernh. 2, 240.
Weitenwebera Opiz 24, 133, 198.
Wendia Hoffm. 17, 251.
Whitleya Sweet 22, 99.
Wibelia Gaertn. 29, 594, 687.
Wiborgia Moench 11, 75.
Wiedemannia Fisch. et Mey. 21, 142.
— multifida (L.) Benth. 142.
Willemetia Neck. 29, 586.
— tuberosa Fisch. et Mey. ex DC. 587; tab. 29.
Winklera Rgl. 8, 545.
— patrinioides Rgl. 546.
— silaifolia Korsh. 546.
Wistaria Nutt. 11, 304.
— sinensis (Sims) DC. 11, 305.
Wolffia Horkel et Schleid. 3, 494.
— arrhiza (L.) Wimm. 494.
Woodsia R. Br. 1, 19.
— alpina (Bolt.) S. F. Gray 23.
— fragilis (Trev.) Moore 20.
— glabella R. Br. 22.
— gracillima C. Chr. 23.
— ilvensis R. Br. 23.
— macrochlaena Mett. ex Kuhn 21; tab. 1.
— manchuriensis Hook. 21.

— polystichoides Eaton 21.
— sinuata (Hook.) Christ 22.
— subcordata Turcz. 22; tab. 1.
Woronowia Juz. 10, 250, 615.
— speciosa (Albov) Juz. 250; tab. 19.
Wydleria DC. 16, 373.
Wylia Hoffm. 16, 139, 145.

Xanthium L. 25, 521.
— californicus Greene 526.
— ociidentale Bertero 525.
— riparium Itz. et Hertsch. 529.
— sibiricum Patr. 524.
— spinosum L. 523; tab. 32.
— strumarium L. 524; tab. 32.
Xanthogalum Avé-Lall. 17, 38.
— purpurascens Avé-Lall. 38; tab. 3, 23.
— Sachokianum Karjag. 39.
— Tatianae (Bordz.) Schischk. 40.
Xanthophtalmum Sch. Bip. 26, 131, 136.
Xanthopsis (DC.) C. Koch 28, 370, 474.
Xanthoselinum Schur 17, 168, 197.
Xeranthemum L. 27, 59.
— annuum L. 62.
— cylindraceum Smith 66.
— inapertum (L.) Mill. 64; tab. 4.
— longipapposum Fisch. et Mey. 65.
— squarrosum Boiss. 63.
Xeroloma Cass. 27, 59, 66.
Xiphion Medic. 4, 554.
Xiphocoma Stev. 7, 351, 483.
Xylanthemum Tzvel. 26, 284, 877.
— Fischerae (Aitch. et Hemsl.) Tzvel. 287.
— pamiricum (Hoffm.) Tzvel. 286.
— rupestre (M. Pop. et Nevski) Tzvel. 287.

Youngia Cass. 29, 375, 595, 650.
— altaica (Babc. et Stebbins) Czer. 383.
— diversifolia (Ledeb.) Ledeb. 380.
— seravschanica (B. Fedtsch.) Babc. et Stebbins 386; tab. 30.
— stemona (Turcz.) Ledeb. 378; tab. 30.
— tenuicaulis (Babc. et Stebbins) Czer. 385.
— tenuifolia (Willd.) Babc. et Stebbins 381.

Zacintha Mill. 29, 715.
— verrucosa Gaertn. 716.
Zannichellia L. 1, 263.
— major Boenn. 265.
— palustris L. 264.
— pedunculata Reichb. 264.
Zapania Lam. 19, 697.
Zea L. 2, 4.
— mays L. 4.
Zelkova Spach 5, 375.
— carpinifolia (Pall.) Dipp. 376; tab. 20.
Zeravschania Korov. 16, 412.
— Regeliana Korov. 412.
Zerna Panz. 2, 555.
Zeuxine Lindl. 4, 640.
— strateumatica (L.) Schlechter 640; tab. 39.
Zinnia L. 25, 532.
— elegans Jacq. 532.
Zizania L. 2, 45.
— aquatica L. 46.
— latifolia Turcz. 46; tab. 3.
Ziziphora L. 21, 381.
— Biebersteiniana Grossh. 390.
— borzhomica Juz. 401.
— Brantii C. Koch 393.
— brevicalyx Juz. 394, 664.
— Bungeana Juz. 386, 664.
— capitata L. 405; tab. 20.
— capitellata Juz. 406, 671.
— clinopodioides Lam. 398; tab. 20.
— denticulata Juz. 402, 670.
— Galinae Juz. 395, 666.
— interrupta Juz. 396, 667.
— pamiroalaica Juz. 399.
— persica Bge. 410.
— Puschkinii Adam 404; tab. 20.
— Raddei Juz. 400, 669.
— rigida (Boiss.) H. Br. 389.
— serpyllacea M. B. 387.
— taurica M. B. 407; tab. 20.
— tenuior L. 408; tab. 20.
— tomentosa Juz. 399, 668.
— turcomanica Juz. 395, 665.
— Woronowii Maleev 403.
Zizyphus Mill. 14, 636.
— jujuba Mill. 637; tab. 34.
Zoegea L. 28, 246.
— baldshuanica C. Winkl. 248; tab. 12.
— purpurea Fresen. 249.
Zollikoferia DC. 29, 239.
Zygophyllum L. 14, 152.
— atriplicoides Fisch. 191.
— balchaschense Boriss. 171, 725.
— brachypterum Kar. et Kir. 165.
— bucharicum B. Fedtsch. 188; tab. 10.
— cuspidatum Boriss. 174, 726.
— darvasicum Boriss. 190, 729.
— Dielsianum M. Pop. 169.
— Eichwaldii C. A. Mey. 179; tab. 9.
— eurypterum Boiss. 189; tab. 10.
— fabago L. 159; tab. 9.
— fabagoides M. Pop. 178.

Zygophyllum ferganense (Drob.) Boriss. 184; tab. 10.
— furcatum C. A. Mey. 172.
— Gontscharovii Boriss. 190, 729; tab. 10.
— iliense M. Pop. 176.
— karatavicum Boriss. 175, 726; tab. 10.
— kaschgaricum Boriss. 187, 728.
— kegense Boriss. 181, 727.
— kopalense Boriss. 173, 726.
— latifolium Schrenk 167.
— Lehmannianum Bge. 175.
— macropodum Boriss. 161, 724.
— macropterum C. A. Mey. 182; tab. 10.
— megacarpum Boriss. 192, **730**; tab. 10.
— melongena Bge. 193.
— microcarpum Boriss. 168, **724**.
— miniatum Cham.. et Schlecht. 166; tab. 9.
— obliquum M. Pop. 162.
— ovigerum Fisch. et Mey. 180; tab. 9.
— oxianum Boriss. 160, 723; tab. 9.
— oxycarpum M. Pop. 172; tab. 10.
— portulacoides Cham. 158.
— Potaninii Maxim. 183.
— pterocarpum Bge. 177.
— Rosovii Bge. 167.
— stenopterum Schrenk 181; tab. 9.
— subtrijugum C. A. Mey. 170.
— taldy-kurganicum Boriss. 178, 727.
— turcomanicum Fisch. et Mey. 168.
Zornia Moench 20, 439.
Zosimia Hoffm. 17, 266.
— absinthifolia (Vent.) Link 267.
— tordyloides Korov. 267.
Zostera L. 1, 266.
— japonica Aschers. et Graebn. 268.
— marina L. 266; tab. 12.
— minor (Cavol.) Nolte 267; tab. 12.
— pacifica S. Wats. 267.
Zygadenus Rich. 4, 9.
— sibiricus (L.) A. Gray 9; tab. 1.

ALPHABETICAL LIST OF AUTHORS OF THE TAXA[1]

INDEX ALPHABETICUS AUCTORUM AD TAXA ADDUCTORUM

The present list includes all surnames cited after taxa in the "Index generum specierumque generalis."

The abbreviations or (very rarely !) the full surnames are arranged strictly in alphabetical order. Periods, hyphens, and other diacritical marks are not taken into consideration, consequently, each abbreviation is taken as one word, regardless of whether it actually consists of one, two, or many words. For example, Buch.-Ham., for the purpose of alphabetizing, is taken as one word (Buchham).

If the abbreviated or full surname of an author is conventionally used with his initials, then in this list it should be searched for in the reverse order–the surname, followed by the initials. Thus, for example, abbreviations such as R. Br., M. Pop., and O.F. Muell. should be searched for respectively under Br. R., Pop. M., and Muell. O.F.

For each author abbreviation, the full surname and also the initials are given, except in a very few cases where it was not possible to confirm the initials.

The present list is considerably more complete than the one published earlier. Moreover, compared to this latest list, it contained fewer addenda and corrigenda. In accordance with long tradition in our own literature on systematics, certain surnames are given not strictly in the spirit of the recommendation on this subject in the International Code of Botanical Nomenclature. As a case in point, the names Bunge and Regel are not given in full but in abbreviated form–Bge. and Rgl.; likewise, the abbreviation Bieb. is used in the traditional form—M.B.

Aar. — Aaronsohn A.
Abbe — Abbe E. C.
Abol. — Abolin — Аболин Р. И.
Abramov — Абрамов И. И.
Abrams — Abrams Le Roy.
Abrom. — Abromeit J.
Acht. — Achtarov (Achtaroff) B. — Ахтаров Б. (Болгария).
Achund. — Achundov — Ахундов Г. Ф.
Achverd. — Achverdov — Ахвердов А. А.
Adam (Adams) — Адам (Адамс) М. И.
Adamov. — Adamović L.
Adams — см. Adam.
Adams J. — Adams J. E.
Adans. — Adanson M.
Adolf — Адольф Н. А.
Aell. — Aellen P.
Afan. — Afanassiev — Афанасьев К. С.
Afz. — Afzelius A.
Afz. P. — Afzelius P.
Agadzh. — Agadzhanov — Агаджанов С. Д.
Agardh — Agardh K. A.
Agardh J. — Agardh J. G.
Agg. — Aggeenko — Аггеенко В. Н.
Airy-Shaw — Airy-Shaw H. K.
Ait. — Aiton W.
Aitch. — Aitchison J. E. T.
Aitch. et Baker — Aitchison J. E. T. et Baker J. G.
Aitch. et Hemsl. — Aitchison J E. T. et Hemsley W. B.

[1]Compiled by M.E. Kirpicznikov.

[2]Cf. "Index of the authors of taxa" in: *Botanicheskie Materialy Gerbariya Botanicheskogo Instituta im. V.A. Komarova Akademii Nauk SSSR. Ukazateli i bibliografiya k tomam 1-20* [Botanical materials of the Herbarium of the V.L. Komarov Botanical Institute of the Academy of Sciences of the USSR. Indexes and bibliography to Vols. 1-20]. Izd. Akad. Nauk SSSR, Moscow-Leningrad, 1962.

Ait. f. — Aiton W. T.
Akinf. — Akinfiew — Акинфиев И. Я.
Albert — Albert A.
Albert et Jahand. — Albert A. et Jahandiez E.
Albov (Alboff, Albow) — Альбов Н. М.
Alech. — Alechin — Алёхин В. В.
Alef. — Alefeld F. G. C.
Alexand. — Alexander J. E.
Alexandrov — Александров А. П.
Alexe(j)enko — Алексеенко Ф. Н.
Alexe(j)enko M. — Алексеенко (Олексиенко) М. И.
Aliev — Алиев Д. А.
Aliss. — Alissova — Алисова-Клобукова Е. Н.
All. — Allioni C.
Almq. — Almquist S. O. I.
Almq. E. — Almquist E.
Alst. — Alston A. H. G.
Ambrosi — Ambrosi F.
Ames — Ames O.
Ames L. — Ames L. M.
Amman — Амман И.
Anders. E. — Anderson E. S.
Andersen — Andersen J. C.
Anderss. — Andersson N. J.
Anders. T. — Anderson T.
Andr. — Andrews H. C.
André — André E.
Andrejev (w) — Андреев В. Н.
Andres — Andres H.
Andron. — Andronov — Андронов Н. М.
Andross. — Androssow — Андросов Н. В.
Andrz. — Andrzejowski — Анджейовский (Анджиевский) А. Л.
Angelis — Angelis P.
Ångstr. — Ångström J.
Ant. — Antoine F.
Antrop. V. et V. — Antropov V. et V. — Антропов В. И. и Антропова В. Ф.
Arcang. — Arcangeli G.
Ard. — Arduino P.
Arèn. — Arènes J.
Aresch. — Areschoug F. W. C.
Arn. — Arnott G. A. W.
Arnold — Arnold J. F. X.
Arrab. — см. Vell. P.
Arrhen. — Arrhenius J. P. (1811—1889).
Arrhen. A. — Arrhenius J. I. A. (1858—1950).
Arrhen. et Fries — Arrhenius J. P. et Fries E. M.
Artemcz. — Artemczuk — Артемчук И. В.
Arv.-Touv. — Arvet-Touvet J. M. C.
Ashe — Ashe W. W.
Aschers. — Ascherson P. F. A.
Aschers. et Graebn. — Ascherson P. F. A. et Graebner P.
Aschers. et Magnus — Ascherson P. F. A. et Magnus P. W.
Aschers. et Sint. — Ascherson P. F. A. et Sintenis P. E. E.
Asker. — Askerova — Аскерова Р. К.
Asso — Asso y del Rio I. J. de.
Atkins. — Atkinson G. F.
Aubl. — Aublet J. B. C. F.
Auch. — Aucher-Eloy P. M. R.
Audub. — Audubon J. J.
Aust. — Austin C. F.
Avdul. — Avdulov — Авдулов Н. П.
Avé-Lall. — Avé-Lallemant J. L. E. — Аяллеман Э. Х.
Avet. — Avetissjan — Аветисян Е. М.
Avror. — Avrorin — Аврорин Н. А.
Aznav. — Aznavour G. V.
Aznav. et Bornm. — Aznavour G. V. et Bornmüller J. F. N.

Baagoe — Baagoe J. S.
Bab. — Babington C. C.
Babc. — Babcock E. B.
Babc. et Stebb. — Babcock E. B. et Stebbins G. L., Ir.
Bachteev — Бахтеев Ф. Х.
Baenitz — Baenitz K. G.
Baer — Бэр К. М.
Baer et Helmers. — Baer et Helmersen — Бэр К. М. и Гельмерсен Г. П.
Bailey — Bailey L. H.
Baill. — Baillon H. E.
Bajt. — Bajtenov — Байтенов М.
Baker — Baker J. G.
Baker f. — Baker E. G.
Baker H. — Baker H. G.
Baker M. — Baker M. S.
Bal. — Balansa B.
Balb. — Balbis G. B.
Bald. — Baldinger E. G. (1738—1804)
Baldacci — Baldacci A. (1867—1950)
Balf. — Balfour J. H.
Balf. f. — Balfour J. B.
Balk. — Balkovsky — Балковский Б. Е
Ball — Ball J.
Ball C. — Ball C. R.
Ball P. — Ball P. W.
Banks — Banks J.

Banks et Soland. — Banks J. et Solander D. C.
Baran. A. — Baranov A. — Баранов А. И.
Baran. P. — Baranov P. — Баранов П. А.
Baran. V. — Baranov V. — Баранов В. И.
Barbar. — Barbaricz — Барбарич А. И.
Baron — Baron P. A.
Baroni — Baroni E.
Barrate — Barrate G.
Barratt — Barratt J.
Barton — Barton B. S.
Barton W. — Barton W. P. C.
Bartl. — Bartling F. G.
Bartlett — Bartlett H. H.
Bartr. — Bartram W.
Basil. — Bas(z)ilevskaja — Базилевская Н. А.
Basin. — Basiner — Базинер Ф. И.
Bassi — Bassi F.
Bast. — Bastard T.
Batal. — Batalin — Баталин А. Ф.
Batsch — Batsch A. J. G. K.
Batt. — Battandier J. A.
Bauh. — Bauhin, Caspar.
Baumg. — Baumgarten J. C. G.
Baxt. — Baxter W.
Bayer — Bayer J. N.
Beal — Beal W. J.
Bean — Bean W. J.
Beauv. — Beauvois de Palisot A. M. F. J.
Beauverd — Beauverd G.
Bebb — Bebb M. S.
Becc. — Beccari O.
Bechst. — Bechstein J. M.
Beck — Beck L. C.
Beck. A. — Becker A. — Беккер А. К.
Beck G. — Beck v. Mannagetta G.
Beck. J. — Becker J.
Beck L. — Beck L. W.
Beckm. — Beckmann J.
Beckm. K. — Beckmann K. L.
Beck. W. — Becker, Wilhelm.
Bedd. — Beddome R. H.
Beeby — Beeby W. H.
Beetle — Beetle A. A.
Bég. — Béguinot A.
Beger — Beger H. K. E.
Bég. et Vacc. — Béguinot A. et Vaccari A.
Behr. — Behrendsen W.
Beissn. — Beissner L.
Beket. — Beketow — Бекетов А. Н.
Bell. — Bellardi C. A. L.
Belli — Belli (C.) S.
Benn. — Bennett J. J. (1801—1876)
Benn. A. — Bennett A. (1843—1929)
Bensin — Bensin B. M.
Benson — Benson G. T.
Benson B. — Benson B. W.
Benson L. — Benson L. D.
Benth. — Bentham G.
Benth. et Hook. f. — Bentham G. et Hooker J. D.
Bercht. — Berchtold F. v.
Bercht. et Presl — Berchtold F. et Presl J. S.
Berg. — Bergius P. J.
Berger — Berger A.
Bergg. — Berggren S.
Bergg. et Wahl. — Berggren S. et Wahlenberg G.
Berg O. — Berg O. K.
Berl. — Berlandier J. L.
Berlin — Berlin J. A.
Bernh. — Bernhardi J. J.
Bertero — Bertero C. G.
Berth. — Berthelot S.
Bertol. — Bertoloni A.
Bess. — Besser W. S. J. G. v. — Бессер В. Г.
Bessey — Bessey C. E.
Betcke — Betcke E. F.
Beurl. — Beurling P. J.
Beyr. — Beyrich H. K.
Bge. — Bunge A. A. v. — Бунге А. А.
Biatz. — Biatzovsky.
Bickis — Бицкис Я.
Bickn. — Bicknell E. P.
Biehl. — Biehler J. F. T.
Bien. — Bienert — Бинерт Т.
Bigel. — Bigelow J.
Billot — Billot P. C.
Bilyk — Билик Г. И.
Bisch. — Bischoff G. W.
Bitt. — Bitter F. A. G.
Biv.-Bern. — Bivona-Bernardi A.
Bjornstr. — Björström F. J.
Blackw. — Blackwell E.
Blake — Blake S. F.
Blakel. — Blakelock R. A.
Blanco — Blanco M.
Blin. — Blinovskij — Блиновский К. В.
Blin. et Shal. — Blinovskij et Shalyt — Блиновский К. В. и Шалыт М. С.
Błocki — Błocki B.
Błoński — Błoński F.
Bluff — Bluff M. J.
Bluff et Fingerh. — Bluff M. J. et Fingerhuth K. A.
Bluk. — Bluket — Блукет Н. А.
Blum — Блум (Блюм) Ф. К.

Blume — Blume K. L. v.
Blytt — Blytt M. N.
Bobr. — Bobrov — Бобров Е. Г.
Bobr. A. — Bobrov A. — Бобров А. Е.
Bod. — Bodard P. H. H.
Boeck. — Boeckeler (Böckeler) J. O.
Boehm. — Boehmer (Böhmer) G. R.
Boenn. — Boenninghausen C. M. F. v.
Boern. — Boerner (Börner) C.
Bogdan A. — Богдан А. В.
Bogusl. — Boguslaw — Богуслав И. А.
Bois — Bois D. G. J. M.
Boiss. — Boissier, Pierre-Edmond.
Boiss. et Bal. — Boissier E. et Balansa B.
Boiss. et Buhse — Boissier E. et Buhse F. A.
Boiss. et Hausskn. — Boissier E. et Haussknecht H. C.
Boiss. et Heldr. — Boissier E. et Heldreich T.
Boiss. et Hohen. — Boissier E. et Hohenacker R. F.
Boiss. et Huet — Boissier E. et Huet du Pavillon A.
Boiss. et Kotschy — Boissier E. et Kotschy T.
Boiss. et Noë — Boissier E. et Noë F. W.
Boiss. et Reut. — Boissier E. et Reuter G. F.
Boissieu — Boissieu H. de.
Bojer — Bojer W. B.
Bolle — Bolle K. A.
Bolle F. — Bolle F.
Bolt. — Bolton J.
Bolus — Bolus H.
Bolus L. — Bolus H. M. L.
Bonati — Bonati G.
Bondar. — Bondarenko — Бондаренко О. Н.
Bong. — Bongard H. G. v. — Бонгард Г. Г.
Bong. et Mey. — Bongard et Meyer C. A. — Бонгард Г. Г. и Мейер К. А.
Bonnet — Bonnet E.
Bonnier — Bonnier G.
Bonpl. — Bonpland A. J. A.
Booth — Booth W. B.
Booth W. — Booth W. E.
Boott — Boott F.
Bor — Bor N. L.
Borb. — Borbás V. v.
Borb. et Degen — Borbás V. et Degen A.
Bordèr. — Bordères-Rey O.
Bordèr. et Gauss. — Bordères-Rey O. et Gaussen H.
Bordz. — Bordzilowski — Бордзиловский Е. И.
Boreau — Boreau A.
Boriss. — Borissova — Борисова (Бекряшева) А. Г.
Boriss. et Korol. A. — Borissova et Koroleva A. — Борисова А. Г. и Королева А. С.
Boriss. et Schischk. — Borissova et Schischkin — Борисова А. Г. и Шишкин Б. К.
Borkh. — Borkhausen M. B.
Börn. — см. Boern.
Bornm. — Bornmüller J. F. N.
Bornm. et Gauba — Bornmüller J. F. N. et Gauba E.
Bornm. et Sint. — Bornmüller J. F. N. et Sintenis P. E. E.
Bornm. et Vierh. — Bornmüller J. F. N. et Vierhapper F.
Bornm. et Woronow — Bornmüller J. F. N. et Woronow G.
Boros — Boros A.
Borszcz. — Borszczow — Борщов И. Г.
Bory — Bory de St. Vincent J. B. G. M.
Bory et Chaub. — Bory de St. Vincent J. B. G. M. et Chaubard L. A.
Bosc — Bosc L. A. G.
Bosch — Bosch R. B. van den.
Botsch. — Botschantzev — Бочанцев В. П.
Botsch. et Karav. — Botschantzev et Karavaev — Бочанцев В. П. и Караваев М. Н.
Botsch. et Vved. — Botschantzev et Vvedensky — Бочанцев В. П. и Введенский А. И.
Bouché — Bouché C. D.
Boud. — Boudier E.
Bouleng. — Boulenger G. A.
Boulos — Boulos L.
Bourgeau — Bourgeau E.
Br. A. — Braun A. C. H.
Br. A. et Milde — Braun A. C. H. et Milde C. A. J.
Brand — Brand A.
Brandza — Brandza D.
Br.-Bl. — Braun-Blanquet J.
Bréb. — Brébisson L. A. de.
Bref. — Brefeld O.
Brenn. — Brenner M.
Br. F. — Braun F. A. v.
Br. H. — Braun H.
Brign. — Brignoli di Brunnhoff G.
Briq. — Briquet J. I.
Britt. — Britton N. L.
Britt. D. — Britton D.
Britt. E. — Britton E. G.
Britten — Britten J.
Britten et Rendle — Britten J. et Rendle A. B.
Britt. et Br. — Britton N. L. et Brown A. C. H.
Britt. et Rose — Britton N. L. et Rose J. N.

Britt. et Rusby — Britton N. et Rusby H. H.
Britt., Sterns et Pogg. — Britton N. L., Sterns E. E. et Poggenburg J. F.
Br. N. E. — Brown N. E.
Brockmann — Brockmann-Jerosch H.
Bromf. — Bromfield W. A.
Brongn. — Brongniart A. T.
Brongn. et Griseb. — Brongniart A. T. et Grisebach A. H. R.
Brot. — Brotero F. de A(d)vellar.
Br. P. — Browne P.
Br. R. — Brown R.
Brzez. — Brzezicki (Brshezicki, Brzhezitzky) — Бржезицкий М. В.
Bub. — Bubani P.
Buchenau — Buchenau F. G. P.
Buch.-Ham. — Buchanan F. (позднее Lord Hamilton).
Buek — Buek H. W.
Buerg. — Buerger F.
Buhse — Бузе Ф. А.
Buia — Buia A.
Bull. — Bulliard P.
Bull W. — Bull W.
Burch. — Burchell W. J.
Bureau — Bureau E.
Bureau et Franch. — Bureau E. et Franchet A.
Burg. — Burgess E. S.
Burgsd. — Burgsdorff F. A. L. v.
Burk. — Burkill I. H.
Burm. — Burman J.
Burm. f. — Burman N. L.
Burn. — Burnat E.
Burn. et Gremli — Burnat E. et Gremli A.
Bus. — Buser R.
Busch E. — Буш Е. А.
Busch N. — Буш Н. А.
Bush — Bush B. F.
Butk. — Butkov — Бутков А. Я.
Buxb. — Buxbaum J. C. — Буксбаум И. Х.
Bykov — Быков Б. А.

Cabrera — Cabrera A. L.
Cajand. — Cajander A. K.
Cald. — Caldesi L.
Calest. — Calestani V.
Call. — Callier A. C.
Cambess. — Cambessèdes J.
C. A. Mey. — см. Mey. C. A.
Camp — Camp W. H.
Campd. — Campderá F.
Camus A. — Camus, Aimée.
Camus E. G. — Camus, Edmond Gustave.
Card. — Cardot I.
Carey — Carey W.
Carey J. — Carey J.
Carr. — Carrière E. A.
Carruth. — Carruthers W.
Caruel — Caruel T.
Casar. — Casaretto G.
Casp. — Caspary J. X. R.
Cass. — Cassini A. H. G. de.
Cast. — Castiglioni L.
Castagne — Castagne L.
Cav. — Cavanilles A. J.
Cavara — Cavara F.
Cavill. — Cavillier F.
Cavol. — Cavolini F.
Čelak. — Čelakovsky L. J.
Čelak. f. — Čelakovsky L. F.
Cerv. — Cervantes V.
Cesati — Cesati V. de.
Chabert — Chabert A.
Chaix — Chaix D.
Chal. — Chalilov — Халилов Э. Х.
Cham. — Chamisso A. v. (Chamisseau de Boncourt L. C. A.).
Cham. et Schlecht. — Chamisso A. et Schlechtendal D. F. L.
Champ. — Champion J. G.
Chang — Chang Chiao-chien.
Char. — Charadze — Харадзе А. Л.
Charkev. — Charkevicz — Харкевич С. С.
Chase — Chase E.
Châtel. — Châtelain J. J.
Chaub. — Chaubard L. A.
Chav. — Chavannes E.
Chenev. — Chenevard P.
Cheng W.-C. — Cheng Wan-chun.
Chevall. — Chevallier F. F.
Chevall. A. — Chevallier A. J. B.
Ching R.-C. — Ching Ren-chang.
Chinth. — Chinthibidze — Хинтибидзе Л.
Chiov. — Chiovenda E.
Chitrowo — Хитрово В. Н.
Chitt. F. — Chittenden F. J.
Chodat — Chodat R. H.
Choisy — Choisy J. D.
Chor. — Choroschkov — Хорошков А. А.
Chr. C. — Christensen, Carl.
Christ — Christ H.
Christ D. — Christ D. H.
Chrshan. — Chrshanovski — Хржановский В. Г.
Chrtek — Chrtek J.
Chun W.-Y. — Chun Woon-young.
Church — Church G. L.
Cirillo — Cirillo D. M. L.

Clairv. — Clairville J. P. de.
Clarion — Clarion J.
Clark — Clark R. B.
Clarke — Clarke, Charles B.
Claus — Клаус К. К. (К. Ф.)
Clausen — Clausen J. C.
Clausen R. — Clausen R. T.
Clayt. — Clayton J.
Clementi — Clementi G.
Clements — Clements F. E.
Clerc — Клер О. Е.
Clokey — Clokey J. W.
Clos — Clos D.
Clus. — Clusius, Carolius (Charles de l'Ecluse).
Clute — Clute W. N.
Cogn. — Cogniaux C. A.
Coincy — Coincy A. de.
Comber — Comber H. F.
Comm. — Commerson P.
Conr. — Conrad S. W. (1779—1831)
Conr. J. — Conrad J.
Conrath — Conrath P. (1861—1931)
Conrath et Freyn — Conrath P. et Freyn J.
Constance — Constance L.
Conti — Conti P.
Copel. — Copeland E. B.
Copel. f. — Copeland H. F.
Coq. — см. Montbr.
Core — Core E. L.
Cornaz — Cornaz E.
Corner — Corner E. J. H.
Correa — Correa (Correia) J. F.
Coss. — Cosson E. St.-Ch.
Coss. et Bal. — Cosson E. St.-Ch. et Balansa B.
Coss. et Durieu — Cosson E. St.-Ch. et Durieu de Maisonneuve M. C.
Coss. et Germ. — Cosson E. St.-Ch. et Germain de Saint-Pierre J. N. E.
Coss. et Kral. — Cosson E. St.-Ch. et Kralik L.
Coste — Coste J. F.
Coste H. — Coste H. J.
Coult. — Coulter J. M.
Coult. et Evans — Coulter J. M. et Evans W. H.
Coult. et Rose — Coulter J. M. et Rose J. N.
Court. — Courtois R. J.
Cov. — Coville F. V.
Covas — Covas G.
Cowan — Cowan J. M.
Craib — Craib W. G.
Crantz — Crantz H. J. N. v.
Crép. — Crépin F.
Cretz. — Cretzoiu P.
Croizat — Croizat L. C. M.
Crome — Crome G. E. W.
Cronq. — Cronquist A. J.
Cruchet — Cruchet D.
Cuf. — Cufodontis G.
Cuming — Cuming H.
Cunn. A. — Cunningham A.
Cunn. A. M. — Cunningham A. M.
Cunn. R. — Cunningham R.
Curt. — Curtis W.
Curt. M. — Curtis M. A.
Cuss. — Cusson P.
Cust. — Custer J. L.
Cyr. — Cyrillo D.
Czako — Czako K.
Czecz. — Czeczott H.
Czefr. — Czefranova — **Цефранова З. В.**
Czer. — Czerepanov — **Черепанов С. К.**
Czerep. — см. Czer.
Czern. — Czernajew (Czernjaew, Tschernajeff) — **Черняев В. М.**
Czerniak. — Czerniakowska — **Черняковская Е. Г.**
Czernova — **Чернова Н. М.**
Czetz — Czetz A.
Czevr. — Czevrenidi — **Чевренидн С. Х.**
Czuk. — Czukavina — **Чукавина А. П.**

Dahl — Dahl O. C.
Dahlst. — Dahlstedt G. A. H.
Dalla Torre — De Dalla Torre K. W. (C. G.) v.
Dalla Torre et Harms — De Dalla Torre K. W. et Harms H.
Dalla Torre et Sarnth. — De Dalla Torre K. W. et Sarnthein L.
Damm. — Dammer U.
Dandy — Dandy J. E.
Dandy et Wilm. — Dandy J. E. et Wilmott A. J.
Danguy — Danguy P.
Daniel — Daniel L.
Daniels — Daniels F. P.
Danil. — Danilov — **Данилов А. Д.**
Danser — Danser B. H.
Danth. — Danthoine F.
Daveau — Daveau J.
Davenp. — Davenport G. E.
Davids. J. — Davidson J. F.
Davis P. — Davis P. H.
DC. — De Candolle, Augustin P.
DC. A. — De Candolle, Alphonse.

DC. C. — De Candolle, Casimir.
Deane — Deane W.
Deane H. et Maid. — Deane H. et Maiden J. H.
Deb. — Debeaux J. O.
Déchy — Déchy M. v.
Decne. — Decaisne J.
Degen — Degen A. v.
Degen et Doerfl. — Degen A. et Dörfler I.
Degen et Zahn — Degen A. et Zahn C. H.
Degl. — Degland J. V. Y.
Dehnh. — Dehnhardt F.
Dekapr. — Dekaprelevicz — Декапрелевич Л. Л.
Dekapr. et Menabde — Dekaprelevicz et Menabde — Декапрелевич Л. Л. и Менабде В. Л.
De Lann. — De Lannoy.
Delarb. — Delarbre A.
Delavay — Delavay J. M.
Deless. — Delessert B.
Delile — Delile A. R.
Deloney — Делоне Л. Н.
Delp. — Delpino F.
Demid. — Demidow — Демидов П. А.
Demjan. — Demjanenko — Демьяненко О. Н.
Deniss. — Denissova — Денисова Г. А.
Derg. — Derganc L.
Déségl. — Déséglise P. A.
Desf. — Desfontaines R. L.
Desjat., Des.-Shost. — см. Schost.
DesMoul. — DesMoulins (Desmoulins) C.
Desr. — Desrousseaux L. A. J.
Desv. — Desvaux A. N.
Desv. E. — Desveaux E. E.
Dew. — Dewey C.
De Wild. — De Wildeman E.
Dicks. — Dickson J.
Didrichs. — Didrichsen P.
Dieck G. — Dieck G. (по другим данным Dieck E.).
Diels — Diels F. L. E.
Diels et Gilg — Diels F. L. E. et Gilg E. F.
Dierb. — Dierbach J. H.
Dietr. — Dietrich F. G. (1768—1850).
Dietr. A. — Dietrich A. (1795—1856).
Dietr. D. — Dietrich D. N. F. (1799—1888).
Dill. — Dillenius J. J.
Dipp. — Dippel L.
Ditmer — Дитмер Э. Э.
Dmitr. — Dmitriev(w) — Дмитриев В. Д.
Dmitr. A. — Dmitrieva — Дмитриева А. А.
Dobroch. — Dobrochotova — Доброхотова К. В.
Dobrocz. — Dobroczajeva — Доброчаева-Ковальчук Д. Н.
Doctur. — Docturowsky — Доктуровский В.
Doctur. et Fler. — Docturowsky et Flerow — Доктуровский В. С. и Флеров А. Ф.
Dode — Dode L. A.
Doell — Doell (Döll) J. C.
Doerfl. — Doerfler (Dörfler) I.
Doluch. — Doluchanov — Долуханов А. Г.
Domin — Domin K.
Domke — Domke W.
Don D. — Don, David.
Don G. — Don, George.
Donn — Donn J.
Dörfl. — см. Doerfl.
Dorth. — Dorthes J. A.
Dostál — Dostál J.
Dougl. — Douglas D.
Downar — Довнар Н. В.
Drake — Drake del Castillo E.
Drej. — Drejer S. T. N.
Drev. — Dreves J. F. P.
Drob. — Drobow — Дробов В. П.
Druce — Druce G. C.
Drude — Drude C. G. O.
Drumm. — Drummond J. L.
Drumm. J. — Drummond J. R.
Drumm. J. et Hutch. — Drummond J. R. et Hutchinson J.
Drumm. R. — Drummond R. B.
Drury — Drury W. H.
Drury et Rollins — Drury et Rollins R. C.
Dryand. — Dryander J. C.
Düben — см. Dueben.
Dubjan. — Dubjansky — Дубянский В. А.
Dubois — Dubois F. N. A.
Duby — Duby J. E.
Duch. — Duchesne A. N. (1747—1827).
Duchartre — Duchartre P. E. S. (1811—1894).
Ducomm. — Ducommun J. C.
Dueben — Dueben (Düben) M. W.
Duf. — Dufour J.-M. L.
Dufr. — Dufresne P.
Duham. — Duhamel du Monceau H. L.
Dulac — Dulac J.
Dum. — Dumortier B. C. J. du.
Dumb. — Dumbadze — Думбадзе Т. А.
Dum.-Cours. — Dumont de Courset G. L. M.
Dümm. — Dümmer R. A.
Dun. — Dunal M. F.
Dunn — Dunn S. T.
Dunn D. — Dunn D. B.
Durand — Durand E. M.

Durande — Durande J. F.
Durand E. J. — Durand E. J.
Durand E. J. et Barrate — Durand E. J. et Barrate G.
Durand H. — Durand H.
Durand L. — Durand L.
Durand T. — Durand T.
Durand T. et Schinz — Durand T. et Schinz H.
Durazz. — Durazzini A.
Durieu — Durieu de Maissonneuve M. C.
DuRoi — DuRoi J. P.
D'Urv. — Dumont D'Urville J. S. C.
Dusen — Dusen P. K. H.
Duthie — Duthie J. F.
Duthie et Atkins. — Duthie J. F. et Atkinson G. F.
Duval — Duval C. J.
Duv.-Jouve — Duval-Jouve J.
Dwig. — Dwigubsky — Двигубский И. А.
Dykes — Dykes W. R.
Dyl. — Dylis — Дылис Н. В.
Dzhang. — Dzhangurasov — Джангуразов Ф. Х.

Eastw. — Eastwood A.
Eat. — Eaton A. (1776—1842).
Eat. A. A. — Eaton A. A. (1865—1908)
Eat. D. — Eaton D. C. (1834—1895).
Eat. et Wright — Eaton A. et Wright J.
Eat. H. — Eaton H. H. (1809—1832).
Eat. R. — Eaton R. J.
Eberm. — Ebermaier K. H.
Eckl. — Ecklon C. F.
Eckl. et Zeyh. — Ecklon C. F. et Zeyher K.
Edelb. — Edelberg L.
Edgew. — Edgeworth M. P.
Edw. E. — Edwards E.
Egor. — Egorova — Егорова Т. В.
Ehrenb. — Ehrenberg C. G.
Ehrend. — Ehrendorfer F.
Ehrh. — Ehrhart F.
Eichl. — Eichler A. W.
Eichw. — Eichwald — Эйхвальд Э.
Eichw. K. — Eichwald K. — Эйхвальд К. Ю.
Eig — Eig A.
Ekim. — Ekimov — Екимов В. П.
Ekman — Ekman E. L.
Ekstam — Ekstam O.
Ekvtim. — Ekvtimischvili — Еквтимишвили М.
Elfstr. — Elfstrand M.
Elk. — Elkan L.
Ell. — Elliott S.
Elmer — Elmer A. D. E.
Elwes — Elwes H. J.
Emberg. — Emberger L.
Emberg. et Maire — Emberger L. et Maire R.
Enand. — Enander S. J.
Endl. — Endlicher S. L.
Engelm. — Engelmann G.
Engl. — Engler H. G. Adolf.
Engl. et Irmsch. — Engler H. G. Adolf et Irmscher E.
Engl. et Prantl — Engler H. G. Adolf et Prantl K. A. E.
Erdn. — Erdner E.
Erman — Erman A.
Eschsch. — Eschscholtz — Эшшольц И. Ф.
Etl. — Etlinger A. E.
Ettingsh. — Ettingshausen K. v.
Evans — Evans W. H. (1868—1959)
Evans A. H. — Evans A. H. (1855—1943).
Eversm. — Eversmann — Эверсманн Э.

Fabr. — Fabricius P. K.
Falc. — Falconer W.
Falk — Фальк И. П.
Farr — Farr E. M.
Farrer — Farrer R. J.
Farw. — Farwell O. A.
Fawc. — Fawcett W.
Faye — Faye L.
Fed. — Fedorov An. — Федоров Ан. А.
Fed. Al. — см. Theod. Al.
Fedde — Fedde F. K. G.
Fedde et Schust. — Fedde F. K. G. et Schuster J.
Fed. et Losinsk. — Fedorov An. et Losinskaja — Федоров Ан. А. и Лозинская А.
Fed. et Rzazade — Fedorov An. et Rzazade — Федоров Ан. А. и Рзазаде Р. Я.
Fedoss. M. et S. — Fedossejew M. et S. — Федосеевы М. и С.
Fedtsch. B. — Fedtschenko B. — Федченко Б. А.
Fedtsch. B. et Basil. — Fedtschenko B. et Basilevskaja — Федченко Б. А. и Базилевская Н. А.
Fedtsch. B. et Bobr. — Fedtschenko B. et Bobrov — Федченко Б. А. и Бобров Е. Г.
Fedtsch. B. et Fler. — Fedtschenko B. et Flerow — Федченко и Флеров А. Ф.
Fedtsch. O. — Fedtschenko O. — Федченко О. А.
Fedtsch. O. et B. — Fedtschenko O. et Fedtschenko B. — Федченко О. А. и Б. А.

Fée — Fée A. L. A.
Feer — Feer H.
Felix — Felix J.
Fellm. — Fellmann — Фельман Я.
Fenzl — Fenzl E.
Fern. — Fernald M. L.
Fern. et Weath. — Fernald M. L. et Weatherby C. A.
Fern. et Wieg. — Fernald M. L. et Wiegand K. McK.
Fieb. — Fieber F. X.
Field. — Fielding H. B.
Field. et Gardn. — Fielding H. B. et Gardner G.
Filar. F. — Filarszky F.
Filat. — Filatova — Филатова Н. С.
Filov — Филов А. И.
Fin. — Finet A.
Fin. et Gagnep. — Finet A. et Gagnepain F.
Fingerh. — Fingerhuth K. A.
Fiori — Fiori A.
Fiori et Paol. — Fiori A. et Paoletti G.
Fisch. — Fischer F. E. L. v. — Фишер Ф. Б.
Fisch. et Avé-Lall. — Fischer et Avé-Lallemant — Фишер Ф. Б. и Ляллеман Э. Х.
Fisch. et Mey. — Fischer et Meyer C. A. — Фишер Ф. Б. и Мейер К. А.
Fisch. et Trautv. — Fischer et Trautvetter - Фишер Ф. Б. и Траутфеттер Э. Р.
Fisjun — Фисюн В. В.
Fitzp. H. — Fitzpatrick H. M.
Flaksb. — Flaksberger — Фляксбергер К. А.
Fleisch. — Fleischer J. G.
Fleisch. B. — Fleischer B.
Fleisch. F. — Fleischer F.
Fleischm. H. — Fleischmann H.
Fler. — Flerow — Флеров А. Ф.
Fletch. — Fletcher H. R.
Floder. B. — Floderus B. G. O.
Florin — Florin C. R.
Fluegge — Fluegge (Flügge) J.
Focke — Focke W. O.
Foëx — Foëx G.
Fomin — Фомин А. В.
Fomin et Woronow — Фомин А. В. и Воронов Ю. Н.
Fomin et Zahn — Фомин А. В. и Цан К. Г.
Font Quer — Font Quer P. (Font y Quer P.).
Forbes — Forbes J. (1773—1861).
Forbes et Hemsl. — Forbes F. B. et Hemsley W. B.
Forbes F. — Forbes F. B. (1839—1908).
Form. — Formánek E.
Forrest — Forrest G.
Forsell. — Forselles J. H.
Forsk. — Forskål P.
Forst. — Forster J. R. (1729—1798).
Forst. E. — Forster E. (1765—1849).
Forst. et Forst. f. — Forster J. R. et Forster G.
Forst. f. — Forster G. (J. G. A.) (1754—1794).
Forst. T. — Forster T. F. (1761—1825).
Fosb. — Fosberg F. R.
Fost. — Foster R. C.
Fouc. — Foucaud J.
Foug. — Fougeroux de Bondaroy A. D.
Fourn. — Fournier E.
Fourn. P. — Fournier P.
Fourr. — Fourreau J. P.
Franch. — Franchet A. R.
Franch. et Savat. — Franchet A. R. et Savatier L.
Frankl. — Franklin J.
Fresen. — Fresenius J. B. G. W.
Freyer — Freyer J. G.
Freyn — Freyn J.
Freyn et Sint. — Freyn J. et Sintenis P. E. E.
Fries — Fries E. M.
Fries Th. — Fries, Th. Magnus.
Fritsch — Fritsch K.
Friv. — Frivaldzky v. Frivald I. (E.)
Froehl. — Froehlich (Fröhlich) J. A. v.
Froehl. Ant. — Froehlich (Fröhlich), Antonîn.
Fröl. — см. Froehl.
Fuck. — Fuckel L.
Funck — Funck H. C.
Fryer — Fryer A.
Furumi — Furumi M.
Fuss — Fuss M.
Futó — Futó M.

Gabr. — Gabrieljan — Габриэлян Э. Ц.
Gaertn. — Gaertner J. — Гертнер И.
Gaertn. f. — Gaertner K. F. v.
Gaertn., Mey. et Scherb. — Gaertner P. G., Meyer B. et Scherbius J.
Gaertn. P. — Gaertner P. G.
Gagnep. — Gagnepain F.
Galeotti — Galeotti H. G.
Galushko — Галушко А. И.
Gamajun. — Gamajunova — Гамаюнова А. П.
Gamble — Gamble J. S.
Gams — Gams H.
Gand. — Gandoger M.

Ganesch. — Ganeschin — Ганешин С. С.
Ganesch. et Zahn — Ganeschin et Zahn — Ганешин С. С. и Цан К. Г.
Garcke — Garcke F. A.
Gardner — Gardner G.
Gars. — Garsault F. A. de.
Gasp. — Gasparrini G.
Gatsch. — Gatschetschiladze — Гачечиладзе К. А.
Gauba — Gauba E.
Gaudich. — Gaudichaud-Beaupré C.
Gaudin — Gaudin J. F. G. P.
Gauss. — Gaussen H.
Gavr. — Gavrilenko — Гавриленко Б. Д.
Gay C. — Gay C.
Gay F. — Gay F.
Gay J. — Gay J. E.
Gáyer — Gáyer G.
Gelert — Gelert O.
Genev. — Genevier L. Gaston.
Georgi — Georgi J. G. — Георги И. И.
Georgiev — Georgiev S. — Георгиев Ст. (Болгария).
Germ. — Germain de Saint-Pierre J. N. E.
Gesn. — Gesner K.
Gibelli — Gibelli G.
Gibelli et Belli — Gibelli G. et Belli (C.) S.
Gibs. — Gibson G. S.
Giesenh. — Giesenhagen K.
Gilg — Gilg E. F.
Gilib. — Gilibert J. E. — Жилибер(т) И. Э.
Gill. — Gillies J.
Gillet — Gillet C. C.
Gill. et Hook. — Gillies J. et Hooker W. J.
Gillet J. — Gillet J.
Gilli — Gilli A.
Gilm. — Gilmour J. S. L.
Ging. — Gingins de Lassaraz F.
Girard — Girard F. de.
Giroux — Giroux M.
Glas. — Glasau F.
Gleas. — Gleason H. A.
Gled. — Gleditsch J. G.
Glehn — Глен П. П.
Glück — Glück (C. M.) H.
Gmel. C. C. — Gmelin C. C. (K. C.).
Gmel. J. F. — Gmelin J. F.
Gmel. J. G. — Gmelin J. G. — Гмелин И. Г. (Гмелин-старший)
Gmel. P. F. — Gmelin P. F.
Gmel. S. G. — Gmelin S. G. — Гмелин С. Г. (Гмелин-младший)
Gochn. — Gochnat F. C.
Godet — Godet C. H.
Godr. — Godron D. A.
Godwinski — Годвинский М. И.
Goerz — Goerz R.
Goir. — Goiran A.
Goldb. — Goldbach — Гольдбах Л. Ф.
Goldie — Goldie J.
Golenk. — Golenkin — Голенкин М. И.
Golits. — Golitsin — Голицын С. В.
Golosk. — Goloskokov — Голоскоков В. П.
Golosk. et Vass. — Goloskokov et Vassilczenko — Голоскоков В. П. и Васильченко И. Т.
Golubk. — Golubkova — Голубкова В. Ф.
Gombocz — Gombocz E.
Gontsch. — Gontscharow — Гончаров Н. Ф.
Gontsch. et Novopokr. — Gontscharow et Novopokrovsky — Гончаров Н. Ф. и Новопокровский И. В.
Good. — Goodenough S.
Gord. — Gordon G.
Gorodk. — Gorodkov(w) — Городков Б. Н.
Gorschk. — Gorschkova — Горшкова С. Г.
Gorschk. et Pjat. — Gorschkova et Pjataeva — Горшкова С. Г. и Пятаева А. Д.
Gorschk. et Zak. — Gorschkova et Zakirov Горшкова С. Г. и Закиров К. З.
Gorski — Gorski S.
Gort. — Gorter D. de — Гортер Д.
Görz — см. Goerz.
Gouan — Gouan A.
Govor. — Gov(w)oruchin — Говорухин В. С.
Grab. — Grabowski H. E.
Graebn. — Graebner K. O. R. P. P.
Graells — Graells Mariano de la Paz.
Graf — Graf S.
Graf R. — Graf R.
Grah. — Graham R.
Grah. J. — Graham J.
Grande — Grande L.
Grant — Grant A. L.
Gray A. — Gray, Asa.
Gray F. W. — Gray F. W.
Gray S. F. — Gray, Samuel F.
Grec. — Grecescu D.
Green — Green M. L.
Greene — Greene E. L.
Gremli — Gremli A.
Gren. — Grenier J. C. M.
Gren. et Godr. — Grenier J. C. M. et Godron D. A.
Grev. — Greville R. K.
Griff. — Griffith W.
Grig. — Grigorjev — Григорьев Ю. С.

Grimm — Grimm J. F. K.
Grinev. — см. Hryn.
Grint. — Grinţescu G. P.
Griseb. — Grisebach A. H. R.
Griseb. et Schenk — Grisebach A. H. R. et Schenk A.
Gronov. — Gronovius J. F.
Grosser — Grosser W.
Grosset — Гроссет Г. Э.
Gross H. — Gross, Hugo.
Grossh. — Grossheim — Гроссгейм А. А.
Grossh. et Gontsch. — Grossheim et Gontscharow — Гроссгейм А. А. и Гончаров Н. Ф.
Grossh. et Schischk. — Grossheim et Schischkin — Гроссгейм А. А. и Шишкин Б. К.
Grub. — Grubov — Грубов В. И.
Grudz. — Grudzinskaja — Грудзинская И. А.
Grufb. — Grufberg J. O.
Grumm-Grzhim. et Semen. — Grumm-Gr(s)zhimailo et Semenov(w)-Tjan-Schansky — Грумм-Гржимайло Г. Е. и Семенов-Тянь-Шанский В. П.
Grun. — Gruner — Грунер Л. Ф.
Gudoschn. — Gudoschnikov — Гудошников С. В.
Gueldenst. — Gueldenstaedt (Güldenstaedt) — И. А. Ги(ю)льденштедт.
Guenth. — Guenther K. C.
Guenth., Grab. et Wimm. — Guenther K. C., Grabowski H. E. et Wimmer C. F. H.
Guerke — Gürke R. L. A. M.
Guett. — Guettard J. E.
Gugl. — Gugler W.
Guill. — Guillemin (J. P.) A.
Guillaum. — Guillaumin A.
Guillem. — Guillemeau J. L. M.
Guill., Perr. et Rich. — Guillemin A., Perrottet G. S. et Richard A.
Guin. — Guinochet M.
Güldenst. — см. Gueldenst.
Gunn. — Gunnerus J. E.
Gunnars. — Gunnarson J. G.
Günth. — см. Guenth.
Gürke — см. Guerke.
Guss. — Gussone G.
Gusul. — Guşuleac M.

Haage — Haage F. A.
Habl. — Hablizl C. L. — Габлиц К. И.
Hack. — Hackel E.
Hacq. — Hacquet B.
Hadač — Hadač E.
Haenke — Haenke T.
Hagerup — Hagerup O.
Hagl. — Haglund G. E.
Hagstr. — Hagström J. O.
Halacsy — Halácsy E. v.
Halacsy et Doerfl. — Halácsy E. et Doerfler I.
Hall. — Haller, Albert v. (1708—1777).
Hall C. — Hall C. C.
Hall. f. — Haller, Albert v. (1758—1823).
Hall H. M. — Hall H. M.
Hall H. M. et Clements — Hall H. M. et Clements F. E.
Hallier — Hallier E.
Hallier f. — Hallier J.
Hall W. — Hall, William.
Hamilt. A. — Hamilton A.
Hampe — Hampe E.
Hance — Hance H. F.
Hand.-Mazz. — Handel-Mazzetti H. v.
Hanelt — Hanelt P.
Hanst. — Hanstein J.
Hara — Hara H.
Harms — Harms H. A. T.
Harrison H. — Harrison H. Heslop.
Harrison J. W. — Harrison J. W. Heslop.
Hartm. — Hartmann F. X. v. (1737—1791).
Hartm. C. — Hartman C. (K.) J. (1790—1849).
Hartm. f. — Hartman C. (1824—1884).
Hartw. — Hartweg K. T.
Hartwig — Hartwig A. K. J.
Hartwiss — Гартвис Н. А.
Harv. — Harvey W. H.
Harz — Harz, Karl O.
Harz K. — Harz, Kurt.
Hasselq. — Hasselquist F.
Hassk. — Hasskarl J. K.
Hausm. — Hausmann F. v.
Hausskn. — Haussknecht H. C. (H. K.).
Hausskn. et Bornm. — Haussknecht H. C. et Bornmüller J. F. N.
Hausskn. et Schulz — Haussknecht H. C. et Schulz O. E.
Haw. — Haworth A. H.
Hayata — Hayata B.
Hayek — Hayek A. v.
Hayek et Bornm. — Hayek A. et Bornmüller J. F. N.
Haynald — Haynald L.
Hayne — Hayne F. G.
H. B. K. — Humboldt F. A., Bonpland A. J. A. et Kunth C. S.
Hedl. — Hedlung J. T.
Hedr. — Hedrick U. P.

Heer — Heer O. v.
Hegelm. — Hegelmaier C. F.
Hegetschw. — Hegetschweiler J.
Hegetschw. et Heer — Hegetschweiler J. et Heer O.
Hegi — Hegi G.
Heimerl — Heimerl A.
Heist. — Heister L.
Heldr. — Heldreich T. v.
Hellen. — Hellenius J.
Heller — Heller A. A.
Helm — Helm F. G. — Гельм Г. Ф.
Helm J. — Helm J.
Helmers. — Helmersen — Гельмерсен Г. П.
Hemsl. — Hemsley W. B.
Henckel — Henckel v. Donnersmarck L. V. F.
Henders. — Henderson G.
Henders. et Hume — Henderson G. et Hume A. O.
Henders. L. — Henderson L. F.
Henn. — Hennings P.
Henrard — Henrard J. T.
Henry A. — Henry A.
Herb. — Herbert W.
Herbich — Herbich F.
Herd. — Herder F. G. T. M. v. — Гердер Ф. Е.
Hérincq — Hérincq F.
Hérit. — см. L'Hér.
Herm. F. — Hermann F. J.
Herrm. — Herrmann J.
Herter — Herter W. G.
Hertsch — Hertsch H.
Heuff. — Heuffel J.
Heurck — Heurck H. van.
Heynh. — Heynhold G.
Heyw. — Heywood V. H.
Hiern — Hiern W. P.
Hieron. — Hieronymus G. H. E. W.
Hiit. — Hiitonen I. (H. I. A.).
Hildebr. — Hildebrand F. H. G.
Hill — Hill J.
Hill E. — Hill E. J.
Hillebr. — Hillebrand W.
Hiltebr. — Hiltebrandt F.
Himmelb. — Himmelbaur W.
Hiroe — Hiroe M.
Hitchc. — Hitchcock A. S.
Hitchc. C. — Hitchcock C. L.
Hjelt — Hjelt H.
Hochr. — Hochreutiner B. P. G.
Hochst. — Hochstetter C. F.
Hochst. et Steud. — Hochstetter C. F. et Steudel E. G.
Hocking — Hocking G. M.
Hoeck — Hoeck F.
Hoefft — Гефт Ф. А.
Hoffm. — Hoffmann — Гофман Г. Ф.
Hoffmgg. — Hoffmannsegg J. C. v.
Hoffmgg. et Link — Hoffmannsegg J. C. et Link J. H. F.
Hoffm. K. — Hoffmann K.
Hoffm. O. — Hoffmann K. A. O.
Hoffm. R. — Hoffmann R.
Hogg — Hogg (20-е годы XIX в.).
Hohen. — Hohenacker R. F. — Гогенакер Ф. Р.
Holandre — Holandre J. J. J.
Holl — Holl F.
Hollick — Hollick C. A.
Holm — Holm, Theodor **(1732—1794)** — см. Holmsk.
Holmb. — Holmberg O. R.
Holmboe — Holmboe J.
Holm H. T. — Holm, Herman Theodor (1854—1932, США).
Holmsk. — Holmskiold T.
Holub — Holub J.
Holuby — Holuby J. F.
Honck. — Honckeny G. A.
Honda — Honda M.
Hoog — Hoog J.
Hook. — Hooker W. J.
Hook. et Arn. — Hooker W. J. et Arnott G. A. W.
Hook. f. — Hooker J. D.
Hook. f. et Benth. — Hooker J. D. et Bentham G.
Hook. f. et Thoms. — Hooker J. D. et Thomson T.
Hoppe — Hoppe D. H.
Hoppe et Hornsch. — Hoppe D. H. et Hornschuch C. F.
Horan. — Horaninow — Горянинов П. Ф.
Horkel — Horkel J.
Horm. — Hormuzaki C.
Hornem. — Hornemann J. W.
Hornsch. — Hornschuch C. F.
Horsf. — Horsfield T.
Hort. — hortorum (садов), hortulanorum (садовников).
Horvát — Horvát A. O.
Host — Host N. T.
House — House H. D.
Houtt. — Houttuyn M.
Houtte — Houtte van L. B.
Houz. — Houzeau de Lehaie J.
Howe — Howe E. C.

Howell — Howell T. J.
Howell J. — Howell J. T.
Howell S. — Howell S. R.
Hruby — Hruby J.
Hrubý K. — Hrubý K.
Hryn. — Hryniewiecki — Гриневецкий Б. Б.
Hu — Hu Hsen-hsu.
Hubb. C. E. — Hubbard C. E.
Hubb. F. T. — Hubbard F. T.
Huds. — Hudson W.
Huelph. — Huelphers (Hülphers) K. A.
Huels. — Hülsen R.
Huet A. — Huet du Pavillon, Alfred.
Huet E. — Huet du Pavillon, Edouard.
Hu et Chun — Hu Hsen-hsu et Chun Woon-young.
Hughes — Hughes D. K.
Hull — Hull J.
Hüls. — см. Huels.
Hult. — Hultén E.
Humb. — Humboldt F. A. (F. W. H. A.) v.
Humb. et Bonpl. — Humboldt F. A. et Bonpland A. J. A.
Hume — Hume A. O.
Hu S.-Y. — Hu Shiu-ying.
Husn. — Husnot T.
Hutch. — Hutchinson, John.
Huter — Huter R.
Huter, Porta et Rigo — Huter R., Porta P. et Rigo G.
Huth — Huth E.
Hyl. — Hylander N.

Ik.-Gal. — Ikonnikov-Galitzky — Иконников-Галицкий Н. П.
Ik.-Gal. et Nevski — Ikonnikov-Galitzky et Nevski — Иконников-Галицкий Н. П. и Невский С. А.
Iljin — Ильин М. М.
Iljin et Korov. — Iljin et Korovin — Ильин М. М. и Коровин Е. П.
Iljin et Krasch. — Iljin et Krascheninnikov — Ильин М. М. и Крашенинников И. М.
Iljinsk. — Iljinski — Ильинский А. П.
Iljinsk. I. — Iljinskaja I. — Ильинская И. А.
Illar. — Illarionova — Илларионова Н. Б.
Illitsch. — Illitschevsky — Илличевский С.
Irmisch — Irmisch T.
Irmsch. — Irmscher E.
Irving — Irving W.
Issaiev — Исаев Я.
Itz. — Itzigsohn H.
Itz. et Hertsch — Itzigsohn H. et Hertsch H.
Ivanina — Иванина Л. И.
Ivanitzk. — Ivanitzkaja — Иваницкая А.
Ivanova — Иванова Н. А.
Ives — Ives E.

Jabr.-Kolak. — Jabrova-Kolakovskaja — Яброва-Колаковская В. С.
Jacks. — Jackson B. D.
Jacq. — Jacquin N. J. v.
Jacq. f. — Jacquin J. F. v.
Jacquem. — Jacquemont V.
Jacques — Jacques A.
Jahand. — Jahandiez E.
Jahand. et Maire — Jahandiez E. et Maire R.
Jakubz. — Jakubziner — Якубцинер М. М.
Jalas — Jalas J.
Jan — Jan G.
Janch. — Janchen E.
Jancz. — Janczewski v. Glinka E.
Janisch. — Janischewsky — Янишевский Д. Е.
Janka — Janka v. Bules V.
Jans. — Jansen P.
Jarm. — Jarmolenko — Ярмоленко А. В.
Jas. — Jasiewicz A.
Jaub. — Jaubert H. F.
Jaub. et Spach — Jaubert H. F. et Spach E.
Jaume — Jaume Saint-Hilaire J. H.
Jáv. — Jávorka S.
Jav. et Csap. — Jávorka et Csapody V.
Jeps. — Jepson W. L.
Jess. — Jessen C. (K.) F. W.
Jir. — Jirásek V.
Joh. K. — Johansson K.
Johnst. — Johnston I. M.
Jones — Jones M. E.
Jones F. — Jones F. F.
Jones Fr. — Jones Fr. B.
Jones G. — Jones G. N.
Jones K. — Jones K.
Jord. — Jordan A.
Jordanov — Jordanov D. — Йорданов, Даки (Болгария).
Jord. D. — Jordan D. S.
Jord. et Fourr. — Jordan A. et Fourreau J. P.
Jouin — Jouin V.
Jundz. — Jundzill (Jundsiłł) S. B.
Junge — Junge P.
Junger — Junger E.
Jung W. — Jung W.
Juratzka — Juratzka J.
Jurtz. — Jurtzev — Юрцев Б. А.
Jusl. — Juslenius A. D.
Juss. — Jussieu Antoine Laurent de.

Juss. Adr. — Jussieu Adrien de.
Juss. B. — Jussieu Bernard de.
Juxip — Юксип А. Я.
Juz. — Juzepczuk — Юзепчук С. В.
Juz. et Pozd. — Juzepczuk et Pozdeeva — Юзепчук С. В. и Поздеева Н. Г.
Juz. et Serg. E. — Juzepczuck et Sergievskaja E. — Юзепчук С. В. и Сергиевская Е. В.
Juz. et Vved. — Juzepczuk et Vvedensky — Юзепчук С. В. и Введенский А. И.
Juz. et Vysok. — Juzepczuk et Vysokoostrovskaja — Юзепчук С. В. и Высокоостровская И. Б.

Kaban. — Kabanov — Кабанов Н. Е.
Kabath — Kabath H.
Kabulov — Кабулов Д. Т.
Kalen. — Kaleniczenko — Калениченко И. О.
Kalm — Kalm P.
Kamien. — Kamienski — Каменский Ф. М.
Kanitz — Kanitz A.
Kantsch. Z. — Kantschaweli — Канчавели З. А.
Kapell. — Kapeller — Капеллер О. А.
Kar. — Karelin — Карелин Г. С.
Karav. — Karavaev — Караваев М. Н.
Kar. et Kir. — Karelin et Kirilow — Карелин Г. С. и Кирилов И. П.
Karjag. — Karjagin — Карягин И. И.
Karst. — Karsten G. K. W. H.
Kaschm. — Kaschmensky — Кашменский Б. Ф.
Katina — Катина З. Ф.
Kauffm. — Kauffmann — Кауфман Н. Н.
Kaulf. — Kaulfuss G. F.
Kazak. — Kazakevicz — Казакевич Л. И.
Kazar. — Kazarjan — Казарян Е. С. (Казарджян Э. С.).
Kearn. — Kearney T. H.
Keck — Keck D. D.
Keissl. — Keissler K.
Keller B. — Келлер Б. А.
Keller B. et Kom. N. — Keller et Komarov N. — Келлер Б. А. и Комаров Н. Ф.
Keller G. — Keller G.
Keller L. — Keller L.
Keller R. — Keller R.
Kellogg — Kellogg A.
Kem.-Nath. — Kemularia-Nathadze — Кемулярия-Натадзе Л. М.
Keng — Keng, Yi-li.
Ker-Gawl. — Bellenden Ker, John или (более верно!) John Ker Bellenden; до 1804 г. — John Gawler.
Kerner A. — Kerner v. Marilaun A. J.
Kerner J. — Kerner J.
Keys. — Keyserling A. F. M. L. v. — Кейзерлинг А. А.
Kickx — Kickx J.
Kier. — Kierulff (сотрудник А. Бунге).
Kihlm. — Kihlman A. O.
Kimura — Kimura Y.
King — King G.
King et Gamble — King G. et Gamble J. S.
Kir. — Kirilow — Кирилов И. П.
Kirchn. — Kirchner G.
Kirchn. E. — Kirchner E. O. O. v.
Kir. P. — Kirillow — Кирилов П. Е.
Kirp. — Kirpicznikov — Кирпичников М. Э.
Kirschl. — Kirschleger F.
Kiss — Kiss A.
Kit. — Kitaibel P.
Kitag. — Kitagawa M.
Kitam. — Kitamura S.
Kitt. — Kittel M. B.
Kjellm. — Kjellman F. R.
Kjerulff — см. Kier.
Klatt — Klatt F. W.
Kleop. — Kleopow — Клеопов Ю. Д.
Klinge — Klinge J. Ch. — Клинге И. Г.
Klinggr. — Klingraeff K. J. v.
Klinggr. H. — Klingräff (Klingraeff) H. v.
Klok. — Klokov — Клоков М. В.
Klok. et Sakalo — Klokov et Sakalo — Клоков М. В. и Сакало Д. И.
Klok. et Schost. — Klokov et Schostenko — Клоков М. В. и Шостенко Н. А.
Klotzsch — Krotzsch J. F.
Kluk — Kluk C.
Knab. — Knaben G.
Knaf — Knaf J.
Knapp — Knapp J. L.
Knapp F. — Knapp F. H.
Knapp J. A. — Knapp J. A.
Knapp J. H. — Knapp J. H.
Kneuck. — Kneucker A.
Knoop — Knoop J. H.
Knorr. — Knorring — Кнорринг-Неуструева О. Э.
Knorr. et Tamamsch. — Knorring et Tamamschjan — Кнорринг-Неуструева О. Э. и Тамамшян С. Г.
Knuth — Knuth R. G. P.
Knuth F. — Knuth F. M.
Koben. — Kobendza R.
Koch — Koch, Wilhelm Daniel Joseph.
Koch C. — Koch, Karl Heinrich Emil.

Koch et Sond. — Koch W. D. J. et Sonder O. W.
Kodama — Kodama S.
Koehne — Koehne B. A. E.
Koel. — Koeler G. L.
Koelle — Koelle J. L. C.
Koern. — Koernicke (Körnicke) F.
Koerte — Koerte (Körte) F.
Koidz. — Koidzumi G.
Koidz. H. — Koidzumi H.
Köie — Köie (Køie) M.
Kolak. — Kolakov(w)sky — Колаковский А. А.
Kolen. — Kolenati — Коленати Ф. А.
Kolesn. — Kolesnikov — Колесников Б. П.
Kolok. — Kolokolnikov — Колокольников Л. Б.
Kom. — Komarov — Комаров В. Л.
Kom. et Aliss. — Komarov et Alissova — Комаров В. Л. и Алисова-Клобукова Е. Н.
Kom. et Schischk. I. — Komarov et Schischkin I. — Комаров В. Л. и Шишкин И. К.
Kom. N. — Komarov N. — Комаров Н. Ф.
Kondr. — Kondratjuk — Кондратюк Е. Н.
Korcz. — Korczagin — Корчагин А. А.
Korkon. — Korkonossova — Корконосова Л. И.
Kornaś — Kornaś J.
Körnicke — см. Koern.
Kornilova — Корнилова В. С.
Korn.-Tr. — Kornuch-Trotzki — Корнух-Троцкий П. Я.
Korol. A. — Koroleva A. — Королева А. С.
Korol. K. — Koroleva K. — Королева К. М.
Korol. V. — Koroleva V. — Королева В. А.
Korotk. — Korotkova — Короткова Е. Е.
Korotky — Korotky — Короткий М. Ф.
Korov. — Korovin — Коровин Е. П.
Korov. et Miron. — Korovin et Mironov — Коровин Е. П. и Миронов Б. А.
Korov. et Pop. M. — Korovin et Popov M. — Коровин Е. П. и Попов М. Г.
Korsh. — Korshinsky — Коржинский С. И.
Körte — см. Koerte.
Koshewn. — Koshewnikow — Кожевников Д. А.
Koshewn. et Zing. — Koshewnikow et Zinger — Кожевников Д. А. и Цингер В. Я.
Koss. C. — Kossinsky C. — Косинский К. К.
Koss G. — Косс Ю. И.
Kossenko — Косенко И. С.
Kossych — Косых В. М.
Kost. — Kosteletzky V. F.
Kostina — Костина К. Ф.
Kostina et Rjab. — Kostina et Rjabov — Костина К. Ф. и Рябов И. Н.
Kotov — Котов М. И.
Kotschy — Kotschy T.
Kotschy et Boiss. — Kotschy T. et Boissier E.
Kotula — Kotula B.
Koval. — Kovalev — Ковалев Н. В.
Koval. et Ekim. — Kovalev et Ekimov — Ковалев Н. В. и Екимов В. П.
Koval. et Kostina — Kovalev et Kostina — Ковалев Н. В. и Костина К. Ф.
Kovalevsk. — Kovalevskaja — Ковалевская С. С.
Kozl. — Kozlowsky — Козловский В. Л.
Kozl. et Zahn — Kozlowsky et Zahn — Козловский В. Л. и Цан К. Г.
Kozl. N. — Kozlowskaja N. — Козловская Н. В.
K.-Pol. — Koso-Poljansky — Козо-Полянский Б. М.
K.-Pol. et Preobr. — Koso-Poljansky et Preobraschensky — Козо-Полянский Б. М. и Преображенский Г. А.
Kraenzl. — Kraenzlin (Kränzlin) F.
Krajina — Krajina V. J.
Kral. — Kralik L.
Kram. W. — Kramer W. H.
Kränzl. — см. Kraenzl.
Krasch. — Krascheninnikov — Крашенинников И. М.
Krasch. et Abol. — Krascheninnikov et Abolin — Крашенинников И. М. и Аболин Р. И.
Krasch. et Butk. — Krascheninnikov et Butkov — Крашенинников И. М. и Бутков А. Я.
Krasch. et Gontsch. — Krascheninnikov et Gontscharow — Крашенинников И. М. и Гончаров Н. Ф.
Krasch. et Iljin — Krascheninnikov et Iljin — Крашенинников И. М. и Ильин М. М.
Krasch. et Lincz. — Krascheninnikov et Linczevski — Крашенинников И. М. и Линчевский И. А.
Krasch. et Poljak. — Krascheninnikov et Poljakov — Крашенинников И. М. и П[illegible]ляков П. П.
Krasch. et Pop. M. — Kraschenin[illegible] Popov M. — Крашеннинико[illegible] Попов М. Г.
Krasch. et Poretzky — К[illegible]

Poretzky A. — Крашенинников И. М. и Порецкий А. С.
Krasch. et Rubtz. N. — Krascheninnikov et Rubtzov N. — Крашенинников И. М. и Рубцов Н. И.
Krasch. et Vved. — Krascheninnikov et Vvedensky — Крашенинников И. М. и Введенский А. И.
Krasch. et Zapr. — Krascheninnikov et Zaprjagaev — Крашенинников И. М. и Запрягаев Ф. Л.
Krasch., Pop. M. et Lincz. — Krascheninnikov, Popov M. et Linczevski — Крашенинников И. М., Попов М. Г. и Линчевский И. А.
Krasch., Pop. M. et Vved. — Krascheninnikov, Popov M. et Vvedenski — Крашенинников И. М., Попов М. Г. и Введенский А. И.
Krasn. — Kras(s)nov — Краснов А. Н.
Krause E. — Krause E. H. L.
Krause J. — Krause J. W.
Krause K. — Krause K.
Krause M. - Krause M.
Krecz. — Kreczetowicz L. — Кречетович Л. М.
Krecz. V. — Kreczetowicz V. — Кречетович В. И.
Krecz. V. et Bobr. — Kreczetowicz V. et Bobrov — Кречетович В. И. и Бобров Е. Г
Krecz. V. et Gontsch. — Kreczetowicz V. et Gontscharow — Кречетович В. И. и Гончаров Н. Ф.
Krecz. V. et Kuprian. — Kreczetowicz V. et Kuprianova — Кречетович В. И. и Куприянова Л. А.
Kreyer — Крейер Г. К.
Krivot. — Krivotulenko — Кривотуленко У. Ф
Krock. — Krocker A. J.
Krok — Krok T. O. B. N.
Kronf. — Kronfeld M.
Krug — Krug L.
Krug et Urb. — Krug L. et Urban I.
Kryl. — Krylov — Крылов П. Н.
Kryl. et Ivanitzk. — Krylov et Ivanitzkaja — Крылов П. Н. и Иваницкая А.
Kryl. et Schischk. — Krylov et Schischkin — Крылов П. Н. и Шишкин Б. К.
Kryl. et Steinb. — Krylov et Steinberg — Крылов П. Н. и Штейнберг Е. И.
ryl. et Sumn. — Krylov et Sumnevicz — Крылов П. Н. и Сумневич Г. П.

Krysht. — Kryshtofovicz — Криштофович А. Н.
Kubansk. — Kubanskaja — Кубанская З. В.
Kudo — Kudo Y.
Kudr. — Kudrjaschev — Кудряшев С. Н.
Kühl. — Kühlewein P. E.
Kuhn — Kuhn F. A. M. (M. F. A.).
Kük. — Kükenthal G.
Kulcz. — Kulczyński S.
Kult. — Kultias(s)ow — Культиасов М. В.
Kumin. — Kuminova — Куминова А. В.
Kümmerle — Kümmerle J. Béla.
Kümmerle et Degen — Kümmerle J. Béla et Degen A.
Kunth — Kunth C. (K.) S.
Kuntze — Kuntze C. (K.) E. O.
Kunze G. — Kunze G.
Kupff. — Kupffer — Купффер К. Р.
Kuprian. — Kupri(j)anova — Куприянова Л. А.
Kurtz — Kurtz F.
Kurz — Kurz (W.) S.
Kusn. — Kusne(t)zow — Кузнецов Н. И.
Kusn. I. — Kusne(t)zow I. — Кузнецов И. В.
Kusn. V. — Kusnezow V. (W.) — Кузнецов В. А.
Kuth. — Kuthatheladze — Кутателадзе Ш. И.
Kuth. A. — Kuthatheladze A. — Кутателадзе А. И.
Kuth. A. S. — Kuthatheladze A. S. — Кутателадзе А. С.
Kütz. — Kützing F. T.
Kuzen. — Kuzeneva — Кузенева О. И.

L. — Linné v. Carl (он же Linnaeus, Carolus).
Labill. — La Billardiere J. J. H. de.
Lacass. — Lacassagne M.
Ladau — Ladau J. F.
Laest. — Laestadius L. L.
Laff. — Laffon J. C.
Laff. et Miégev. — Laffon J. C. et Miégeville.
Lag. — Lagasca y Segura M.
Lagr.-Foss. — Lagrèze-Fossat A. R. A.
Laharpe — Laharpe (La Harpe) J. de.
Laksch. — Lakschev(w)itz — Лакшевиц П. А.
La Llave — La Llave P.
La Llave et Lex. — La Llave P. et Lexarza J.
Lam — Lam H. J.
Lam. — Lamarck, J. B. A. P. Monnet de.

Lamb. — Lambert A. B.
Lam. et DC. — Lamarck, J. B. A. P. Monnet de et DeCandolle A. P.
Lamotte — Lamotte M.
Lams.-Scribn. — см. Scribn.
Láng — Lang A. F.
Lange — Lange J. M. C.
Lang K. — Lang K. H.
Lang O. — Lang O. F.
Lang T. — Lang T.
Langsd. — Langsdorff G. H v. — Лангсдорф Г. И.
Lann. — см. De Lann.
Lapeyr. — Lapeyrouse P. P. de.
La Pyl. — La Pylaie A. J. M. B. de.
Larsen — Larsen K.
Larss. — Larsson L. M.
Lastey. — Lasteyrie C. P.
Latourr. — Latourrette M. A. L. Claret de.
Latsch. — Latschaschvili — Лачашвили И. Я.
Lauche — Lauche W.
Lauterb. — Lauterbach C. A. G.
Lav. — Lavallée A.
Lava — см. Ljava.
Lavr. — Lavrenko — Лавренко Е. М.
Lawr. — Lawrence J.
Lawr. G. — Lawrence G. H. M.
Lawr. W. — Lawrence W. J. C.
Laws. — Lawson P.
Laws. C. — Lawson C.
Laws. G. — Lawson G.
Laxm. — Laxmann E. — Лаксман Э. Г. (К А.).
Lecomte — Lecomte (M.) H.
Lecoq — Lecoq H.
Lecoq et Lamotte — Lecoq H. et Lamotte M.
Lecoy. — Lecoyer I.-C.
Ledeb. — Ledebour C. F. v. — Ледебур К. Ф.
Lee — Lee H. A.
Leers — Leers J. D.
Lee Shun — Lee Shun-ch'ing (Li Shun-ching).
Lehm. — Lehmann J. G. C.
Leichtl. — Leichtlin M.
Lej. — Lejeune A. L. S.
Lej. et Court. — Lejeune A. L. S. et Courtois R. J.
Lem. — Lemaire C. A.
Léman — Léman D. S.
Lemm. — Lemmon J. G.
Lemoine — Lemoine V.
Lemoine E. — Lemoine E.
Leon. — Leonard E. C.
Leonova — Леонова Т. Г
Lepech. — Lepechin — Лепехин И. И.
Lepeschk. — Лепешкин С. Н.
Lerch. — Lerchenfeld J.
Leroy — Leroy A.
Leroy J. — Leroy J. F.
Leroy L. — Leroy L.
Leskov — Лесков А. И.
Lesp. — Lespinasse (J. M.) G.
Lesq. — Lesquereux L.
Less. — Lessing C. F. — Лессинг Х. Ф.
Lest. — Lestiboudois F. J.
Lest. f. — Lestiboudois T.
Levier — Levier E.
Levin — Левин Э. Г.
Levina — Левина Р. Е.
Lévl. — Léveillé A. A. H.
Lévl. et Vaniot — Leveillé A. A. H. et. Vaniot E.
Lewton — Lewton F. L.
Lex. — Lexarza J.
Leyss. — Leysser F. W. v.
L. f. — Linné v. Carl (Линней-сын, сын Карла Линнея).
L.'Hér. — L'Héritier de Brutelle C. L.
Li H. L. — Li Hui-lin.
Libosch. — Liboschitz J. — Либошиц Иосиф (Осип) Я.
Lid — Lid J.
Liebl. — Lieblein F. K.
Liebm. — Liebmann F. M.
Lightf. — Lightfoot J.
Liljebl. — Liljeblad S.
Limpr. W. — Limpricht W.
Lincz. — Linczevski — Линчевский И. А.
Lincz. O. — Linczevski O. — Линчевский О. А.
Lindb. — Lindberg J. B. W.
Lindb. f. — Lindberg H.
Lindbl. — Lindblom A. E.
Lindblad — Lindblad M. A.
Lindeb. — Lindeberg C. J.
Lindem. — Lindemann — Линдеман(н) Э. Э.
Linden — Linden J.
Lindl. — Lindley J.
Lindm. — Lindman C. A. M.
Ling — Ling Yo(u)ng.
Lingelsh. — Lingelsheim A. v.
Link — Link J. H. F.
Link et Hoffmgg. — Link J. H. F. et Hoffmannsegg J. C. v.
Link et Otto — Link J. H. F. et Otto C. F.
Lipat. — Lipatova — Липатова В. В.
Lipsch. — Lipschitz — Липшиц С. Ю.
Lipsky — Липский В. И.

Litv. — Litvinov (Litwinow) — Литвинов Д. И.
Litv. et Zahn — Litvinov et Zahn — Литвинов Д. И. и Цан К. Г.
Litw. — см. Litv.
Ljava — Лява Я. И.
Lloyd — Lloyd, James.
Lloyd J. — Lloyd, John.
Lodd. — Loddiges C.
Loefl. — Loefling P.
Loennr. — Loennrot (Lönnrot) K. J.
Loes. — Loesener L. E. T.
Loes. f. — Loesener O.
Loisel. — Loiseleur-Deslongchamps J. L. A.
Lojac. — Lojacono, Pojero M.
Lomak. — Lomakin — Ломакин А. А.
Lonacz. — Lonaczewski — Лоначевский А.
Londes — Londes F. W. — Лондес Ф. В.
Lönnr. — см. Loennr.
Losinsk. — Losinskaja — Лозинская (Лозина—Лозинская) А. С.
Loud. — Loudon J. C.
Lour. — Loureiro J. de.
Löve Á. — Löve, Áskell.
Löve Á. et D. — Löve, Áskell et Löve, Doris.
Löve D. — Löve, Doris.
Lowe — Lowe R. T.
Lüb. — Lüben A.
Lucé — Lucé J. W. L. v. (Люце Иоанн-Вильгельм-Людвиг).
Lüdi W. — Lüdi W.
Ludw. — Ludwig C. G. (1709—1773).
Ludw. A. — Ludwig A.
Ludw. C. — Ludwig C. F.
Luerss. — Luerssen C.
Lukitsch. — Lukitscheva — Лукичева А. Н.
Lumn. — Lumnitzer S.
Lund — Lund N.
Lundstr. — Lundström A. N.
Lunell — Lunell J.
Lyka K. — Lyka, Károly.
Lynch — Lynch R. I.
Lynge — Lynge B.

Ma — Ma, Yu-chuan (Ma, Yü-ch'uan).
Maack — Маак Р. К.
Macbr. — Macbride J. F.
Macf. — Macfadyen J.
Mackenz. — Mackenzie K. K.
Mackenz. et Bush — Mackenzie K. K. et Bush B. F.
MacMill. — MacMillan C.
Macoun — Macoun, John.
Macoun J. — Macoun, James M.
Maek. — Maekawa T.
Maek. F. — Maekawa F.
Magne — Magne J. H.
Magnus — Magnus P. W.
Maguire — Maguire B.
Maid. — Maiden J. H.
Maire — Maire R.
Majlun — Майлун З. А
Majorov — Майоров А. А.
Makino — Makino T.
Makino et Nemoto — Makino T. et Nemoto K.
Makino et Shibata — Makino T. et Shibata K.
Makino et Yabe — Makino T. et Yabe Y.
Maleev — Малеев В. П.
Malme — Malme G. O. A.
Malmgr. — Malmgren A. J.
Malte — Malte M. O.
Maly — Maly J. K.
Malý K. — Malý K.
Malysch. — Malyschev — Малышев Л. И.
Malz. — Malzev(w) — Мальцев А. И.
Manden. — Mandenova — Манденова И. П.
Manden. et Scheng. — Mandenova et Schengelia — Манденова И. П. и Шенгелия Е. М.
Mandl — Mandl K.
Mandl et Muehld. — Mandl K. et Muehldorf A.
Mansf. — Mansfeld R.
Marc. — Marcowicz — Маркович В. В.
Marg. — Margittai A.
Markl. — Marklund G. G.
Marsh. — Marshall H.
Marss. — Marsson T.
Mart. C. — Martius (C.) K. F. P. v.
Martens — Martens M.
Martens et Galeotti — Martens M. et Galeotti H. G.
Mart. H. — Martius H. v. — Мартиус (Марциус) Г.
Masl. — Maslennikova — Масленникова.
Mason — Mason S. C.
Mason C. — Mason C. T.
Mason H. — Mason H. L.
Massalski — Массальский В. И.
Massart — Massart J.
Masson — Masson F.
Mast. — Masters M. T.
Math. — Mathias M. E.
Math. et Const. — Mathias M. E. et Constance L.

Matsum. — Matsumura J.
Mattei — Mattei G. E.
Mattf. — Mattfeld J.
Matuszk. — Matuszkiewicz W.
Matz. — Matzenko — Маценко А. Е.
Mauri — Mauri E.
Maw — Maw G.
Maxim. — Maximowicz — Максимович К. И.
Maxon — Maxon W. R.
Mayer E. — Mayer E.
Mayer J. — Mayer J. C. A.
Mayr — Mayr H.
M. B. — Bieberstein, Friedrich August Marschall v. — Биберштейн Ф. К.
Medic. — Medicus F. C. (Medikus F. K.).
Medw. — Medwedew — Медведев Я. С.
Meff. — Meffert — Мефферт-Абрамович В. В.
Meinsh. — Meinshausen — Мейнсгаузен К.
Meissn. — Meis(s)ner C. F.
Mela — Mela A. J.
Menabde — Менабде В. Л.
Menz. — Menzies A.
Mérat — Mérat F. V.
Merckl. — Mercklin — Мерклин К. Е.
Merr. — Merrill E. D.
Merr. et Chun — Merrill E. D. et Chun Woon-young.
Mert. — Mertens F. K.
Mert. et Koch — Mertens F. K. et Koch W. D. J.
Metc. — Metcalf F. P.
Mett. — Mettenius G. H.
Metzg. — Metzger J.
Mey. B. — Meyer B.
Mey. C. A. — Meyer C. A. — Мейер Карл Андреевич.
Mey. E. — Meyer, Ernst H. F.
Meyen — Meyen F. J. F.
Mey. F. G. — Meyer, Frederick G.
Mey. G. — Meyer, Georg F. W.
Mez — Mez K. C.
Michajl. — Michajlov(w)ski — Михайловский С. И.
Michal. — Michalet E.
Micheli — Micheli P. A.
Micheli M. — Micheli M.
Michels. — Michelson — Михельсон А. И.
Michx. — Michaux A.
Michx. f. — Michaux F. A.
Midd. — Middendorf A. T. v. — Миддендорф А. Ф.
Mieg — Mieg A.
Miégev. — Miégeville.
Miers — Miers J.
Mikan — Mikan J. C.
Mikesch. — Mikeschin — Микешин Г. В.
Milde — Milde C. A. J.
Mill. — Miller P.
Mill. G. — Miller G. S.
Millsp. — Millspaugh C. F.
Min. — Miniaev — Миняев Н. А.
Minder. — Minderova — Миндерова Е. В.
Minkw. — Minkwitz — Минквиц З. А.
Miq. — Miquel F. A. W.
Mirb. — Mirbel C. F. B. de.
Miron. — Mironov — Миронов Б. А.
Mirzoeva — Мирзоева Н. В.
Mischk. — Mischkin — Мишкин Б. А.
Miscz. — Misczenko — Мищенко П. И.
Mitch. — Mitchell J.
Mitf. — Mitford A. B. Freeman.
Mitt. — Mitterpacher L. v.
Miyabe — Miyabe K.
Miyabe et Kudo — Miyabe K. et Kudo Y.
Miyabe et Miyake — Miyabe K. et Miyake T.
Miyabe et Tatew. — Miyabe K. et Tatewaki M.
Miyake — Miyake T.
Miyoshi — Miyoshi M.
Mizg. — Mizgireva — Мизгирева О. Ф.
Mlokoss. — Mlokossewicz — Млокосевич Ю. Л.
Moehr. — Moehring P. H. G.
Moench — Moench K. (C.).
Moeschl — Möschl W.
Moessl. — Moessler J. C.
Moesz — Moesz G.
Moggr. — Moggridge J. T.
Mohr — Mohr D. M. H.
Mohr C. — Mohr C. T.
Mol. — Molina J. I. (G. I.).
Mold. — Moldenke H. N.
Monjuschko — Монюшко В. А.
Monn. — Monnier A.
Montbr. — Coquebert de Montbret G.
Montbr. et Auch. — Coquebert de Montbret G. et Aucher-Eloy P. M. R.
Monti — Monti G.
Moore — Moore T.
Moore A. — Moore A. H.
Moore H. — Moore H. E.
Moore S. — Le Marchant Moore, Spencer.
Moq. — Moquin-Tandon C. H. B. A.
Mor. — Moretti G.
Moric. — Moricand M. E.
Mori K. — Mori K.
Moris G. — Moris G. G.

Moritzi — Moritzi A.
Morong — Morong T.
Morr. — Morren C. F. A.
Morr. E. — Morren E. (C. J. E.).
Morr. et Decne. — Morren C. F. A. et Decaisne J.
Morris — Morris E. L.
Möschl. — см. **Moeschl.**
Mott. — Mottet S.
Muehl. — Muehlenberg (Mühlenberg) H. L.
Muehld. — Muehldorf (Mühldorf) A.
Muell. Arg. — Mueller (Müller Argoviensis), Jean.
Muell. Corn. — Mueller (Müller), Cornelius H.
Muell. F. — Mueller (Müller) F. v.
Muell. O. F. — Mueller (Müller), Otto F.
Muell. P. J. — Muller (Müller), Philipp J.
Muenchh. — Muenchhausen (Münchhausen) O. v.
Mühl. — см. **Muehl.**
Mühld. — см. Muehld.
Mulk. — Mulkidjanian — Мулкиджанян Я. И.
Müll. — см. Muell.
Munby — Munby G.
Münchh. — см. **Muenchh.**
Munro — Munro W.
Murav. — Muravjeva — Муравьева О. А.
Murb. — Murbeck S. S.
Murr. — Murray J. A.
Murr. A. — Murray A.
Muscheg. — Muschegjan — Мушегян А. М.
Muschl. — Muschler R.
Muschl. et Thell. — Muschler R. et Thellung A.
Muss.-Puschk. — Mussin-Puschkin — Мусин-Пушкин А. А.
Mutel — Mutel A.
Mutis — Mutis J. C.

Nábĕl. — Nábĕlek F.
Nabiev — Набиев М. М.
Naeg. — Naegeli C (Nägeli K.)
Naeg. et Peter — Naegeli C. et Peter G. A.
Naeg. O. — Naegeli (Nägeli) O.
Nakai — Nakai T.
Nakai et Koidz. — Nakai T. et Koidzumi G.
Nannf. — Nannfeldt J. A.
Nas. — Nasarow — Назаров М. И.
Nash — Nash G. V.
Nath. — Nathorst A. G.
Nathh. — Nathhorst T. E
Naud. — Naudin C.
Neck. — Necker N. J. de.
Nees — Nees v. Esenbeck C. G. D.
Neger F. — Neger F. W.
Neilr. — Neilreich A.
Nekr. — Nekrassov(w)a — Некрасова В. Л.
Nelmes — Nelmes E.
Nels. — Nelson A.
Nels. E. — Nelson E. E.
Nels. et Macbr. — Nelson A. et Macbride J.
Nemoto — Nemoto K.
Nendtv. — Nedtvich C. M.
Nenuk. — Nenukow — Ненюков Ф. С.
Nepli — Непли Г. Н.
Nestl. — Nestler C. G.
Neum. — Neumann K. G.
Neum. A. — Neumann A.
Neum. L. — Neuman L. M.
Neumay. — Neumayer J.
Neumay. H. — Neumayer H.
Nevski — Невский С. А.
Nevski et Roshev. — Nevski et Roshevitz — Невский С. А. и Рожевиц Р. Ю.
Newb. — Newberry J. S.
Newm. — Newmann E.
Neyg. — Neygenfind F. W.
Nichols. — Nicholson G.
Niedz. — Niedenzu F. J.
Nikif. — Nikiforova — Никифорова Н. Б.
Nikit. E. — Nikitina E. — Никитина Е. В.
Nikit. E. et Plotn. — Nikitina E. et Plotnikov — Никитина и Плотников И. А.
Nikit. S. — Nikitin S. — Никитин С. А.
Nikit. S. et Schischk. — Nikitin S. et Schischkin — Никитин и Шишкин Б. К.
Nikit. V. — Nikitin V. — Никитин В. А.
Nikit. V. V. — Nikitin V. V. — Никитин В. В.
Nocca — Nocca D.
Noë — Noë F. W.
Nolte — Nolte E. F.
Nordh. — Nordhagen R.
Nordm. — Nordmann A. v. — Нордман А. Д.
Norm. — Norman J. M.
Norm. C. — Norman C.
Norrl. — Norrlin J. P.
Noto — Notø A.
Noulet — Noulet J. B.
Novák — Novák F. A.
Novelli — Novelli N.
Novik. — Novikova — Новикова Н. Г.
Novopokr. — Novopokrovsky — Новопокровский И. В.
Novopokr. et Abramov — Novopokrovsky et

Abramov — Новопокровский И. В. и Абрамов И. И.
Novopokr. et Bogdan A. — Novopokrovsky et Bogdan A. — Новопокровский И. В. и Богдан А. В.
Novopokr. et Nepli — Novopokrovsky et Nepli — Новопокровский И. В. и Непли Г. Н.
Novopokr. et Nikit. V. V. — Novopokrovsky et Nikitin V. V. — Новопокровский И. В. и Никитин В. В.
Novopokr. et Pissjauk. — Novopokrovsky et Pissjaukova — Новопокровский И. В. и Письяукова В. В.
Novopokr. et Sidor. — Novopokrovsky et Sidorenko — Новопокровский И. В. и Сидоренко Г. Т.
Nutt. — Nuttall T.
Nyl. — Nylander F.
Nym. — Nyman C. F.

Oakes — Oakes W.
Oborny — Oborny A.
Oed. — Oeder G. C.
Oerst. — Oersted (Ørsted, Örsted) A. S.
Oett. — Oettel K. C.
Oetting. — Oettingen H. v. — Эттинген Г.
Ohlsen — Ohlsen A. R.
Ohwi — Ohwi J.
Olin — Olin J. H.
Oliv. — Oliver D.
Olivier — Olivier G. A.
Olney — Olney S. T.
Omang — Omang S. O. F.
Omelcz. — Omelczuk — Омельчук Т. Я.
O'Neill — O'Neill H. T.
Onno — Onno M.
Opiz — Opiz P. M.
Opperm. — Opperman — Опперман П. А.
Orcutt — Orcutt C. R.
Orlova — Орлова Н. И.
Orph. — Orphanides T. G.
Orr — Orr M. Y.
Orst. — см. Oerst.
Ort. — Ortega, C. Gómez.
Ortm. — Ortmann A.
Osbeck — Osbeck P.
Osborn — Osborn A.
Ostenf. — Ostenfeld C. E. H.
Ostenf. et Dahl — Ostenfeld C. E. H. et Dahl O. C.
Ostenf. et Resv. — Ostenfeld C. H. et Resvoll T. R.
Osten-Sacken — Остен-Сакен Ф. Р.
Osten-Sacken et Rupr. — Osten-Sacken et Ruprecht — Остен-Сакен Ф. Р. и Рупрехт Ф. И.
Ostrejko — Острейко С. А.
Otth — Otth C. (K.) A.
Otto F. — Otto (C.) F.
Ovcz. — Ovczinnikov — Овчинников П. Н.
Ovcz. et Czuk. — Ovczinnikov et Czukavina — Овчинников П. Н. и Чукавина А. П.
Ovcz. et Karav. — Ovczinnikov et Karavaev — Овчинников П. Н. и Караваев М. Н.
Ovcz. et Sidor. — Ovczinnikov et Sidorenko — Овчинников П. Н. и Сидоренко Г. Т.
Over — Over W. H.
Ower. — Owerin — Оверин А. П.
Ownbey — Ownbey F. M.
Ownbey G. — Ownbey G. B.

Pachom. — Pachomova — Пахомова М. Г.
Packer — Packer J. G.
Pacz. — Paczoski — Пачоский И. К.
Paill. — Paillot J.
Palib. — Palibin — Палибин И. В.
Pall. — Pallas — Паллас П. С.
Palla — Palla E.
Pallis — Pallis M.
Palmer E. J. — Palmer E. J.
Palmgr. — Palmgren A.
Pamp. — Pampanini R.
Panč. — Pančič (Pančić) J.
Pang. — Pangalo — Пангало К. И.
Panjut. — Panjutin — Панютин П. С.
Pant. — Pantocsek J.
Panz. — Panzer G. W. F.
Paol. — Paoletti G.
Parl. — Parlatore F.
Parodi — Parodi L. R.
Parrot — Паррот Ф. Е.
Parry C. — Parry C. C.
Parsa — Parsa A.
Pasch. — Pascher A.
Patrin — Patrin E. L. M. — Патрен (Патрэн) Е.
Patze — Patze C.
Patze, Mey. et Elk. — Patze C., Meyer E. et Elkan L.
Pau — Pau C.
Pauls. — Paulsen O. W.
Pav. — Pavon (Pavón) J. A.
Pavl. — Pavlov — Павлов Н. В.
Pavl. et Lipsch. — Pavlov et Lipschitz — Павлов Н. В. и Липшиц С. Ю.

Pavl. N. M. — Pavlova N. M. — Павлова Н. М.
Pavl. V. — Pavlov V. — Павлов В. Н.
Pawł. — Pawłowski B.
Pax — Pax F.
Paxt. — Paxton J.
Pays. — Payson E. B.
Pazij — Пазий В. К.
P. B., P. Beauv. — см. Beauv.
Pears. — Pearson G.
Peebl. — Peebles R. H.
Penn. — Pennell F. W.
Pénzes — Pénzes A.
Pérard — Pérard A.
Perciv. — Percival J.
Perf. — Perfiljew — Перфильев И. А.
Perkins — Perkins J. R.
Perr. — Perrottet G. S. (1793—1870).
Perrier A. — Perrier A.
Perrier E. — Perrier E.
Perrier et Song. — Perrier E. et Songeon A.
Pers. — Persoon C. H.
Peter — Peter G. A.
Peterm. — Petermann W. L.
Peter-Stib. — Peter-Stibal E.
Petitm. — Petitmengin M.
Petrak — Petrak F.
Petri — Petri F.
Petrovič — Petrovič S.
Petrov V. — Петров В. А.
Petunn. — Petunnikow — Петунников А. Н.
Pfeiff. — Pfeiffer L.
Pfeiff. N. — Pfeiffer N. E.
Pfitz. — Pfitzer E. H. H.
Phil. — Philippi R. A.
Philippow — Филиппов Ю.
Philips. — Philipson W. R.
Phillips — Phillips H.
Phillips E. — Phillips E. P.
Phillips L. — Phillips L. L.
Pichon — Pichon M.
Pierre — Pierre J. B. L.
Pilg. — Pilger R. K. F.
Pill. — Piller M.
Pill. et Mill. — Piller M. et Miller P.
Pill. et Mitt. — Piller M. et Mitterpacher L.
Pimen. — Pimenova N. — Пименова Н. В.
Pimen. L. — Pimenova L. — Пименова Л. В.
Pimen. M. — Pimenov M. — Пименов М. Т.
Pinzg. — Pinzger P.
Piper — Piper C. V.
Piré — Piré L. (L. A. H. J.).
Pissjauk. — Pissjaukova — Письяукова В. В.
Pjat. — Pjataeva — Пятаева А. Д.
Pjat. et Lincz. — Pjataeva et Linczevski — Пятаева А. Д. и Линчевский И. А.
Planch. — Planchon J. E.
Platsch. A. — Platschek A. — Плачек А.
Platsch. E. — Platschek E. — Плачек Е. М.
Pleijel — Pleijel C. G. V.
Plotn. — Plotnikov — Плотников И. А.
Plum. — Plumier C.
P. M. E. — см. Patze, Mey. et Elk.
Pobed. — Pobedimova — Победимова Е. Г.
Podp. — Podpěra J.
Poepp. — Poeppig E. F.
Poepp. et Endl. — Poeppig E. F. et Endlicher S. L.
Pogg. — Poggenburg J. F.
Pohl — Pohl J. B. E.
Pohle — Поле Р. Р.
Pohle et Zahn — Поле Р. Р. и Цан К. Г.
Poir. — Poiret J. L. M.
Poit. — Poiteau A.
Pojark. — Pojarkova — Пояркова А. И.
Pojark. et Zak. — Pojarkova et Zakirov — Пояркова А. И. и Закиров К. З.
Pojark. H. — Pojarkova H. — Пояркова Е. Н.
Poll. — Pollich J. A.
Polozh. — Polozhij — Положий А. В.
Poljak. — Poljakov P. — Поляков П. П.
Polun. — Polunin N. V.
Pom. — Pomel A.
Pont. — Pontedera G.
Pop. M. — Popov M. — Попов М. Г.
Popl. — Poplavskaja (Poplawska) — Поплавская Г. И.
Pop. M. — Popov M. — Попов М. Г.
Pop. M. et Baran. — Popov M. et Baranov P. — Попов М. Г. и Баранов П. А.
Pop. M. et Chrshan. — Popov M. et Chrshanovski — Попов М. Г. и Хржановский В. Г.
Pop. M. et Fedtsch. B. — Popov M. et Fedtschenko B. — Попов М. Г. и Федченко Б. А.
Pop. M. et Klok. — Popov M. et Klokov — Попов М. Г. и Клоков М. В.
Pop. M. et Kult. — Popov M. et Kultias(s)ov — Попов и Культиасов М. В.
Pop. M. et Vved. — Popov M. et Vvedensky — Попов М. Г. и Введенский А. И.
Pop. N. — Popov N. — Попов Н. П.
Pop. T. — Popova T. — Попова Т.
Porc. — Porcius F.
Porc. et Czetz — Porcius F. et Czetz A.

Poretzky A. — Порецкий А. С.
Pors. — Porsild M. P.
Pors. A. — Porsild A. E.
Porta — Porta P.
Porter — Porter T. C.
Post — Post G. E.
Postr. — Postriganj — Постригань С. А.
Pott — Pott J. F.
Pourr. — Pourret (de Figeac) P. A.
Pozd. — Pozdeeva — Поздеева Н. Г.
Prain — Prain D.
Prantl — Prantl K. A. E.
Preobr. — Preobraschensky — Преображенский Г. А.
Presc. — Prescott J. O.
Presl C. — Presl C. (K.) B.
Presl J. — Presl J. S.
Presl J. et C. — Presl J. S. et Presl C. (K.) B.
Priem. — Priemer F.
Pril. — Prilipko — Прилипко Л. И.
Printz — Printz H.
Pritz. E. — Pritzel E.
Pritz. G. — Pritzel G. A.
Prod. — Prodan J.
Prokh. — Prokhanov — Проханов Я. И.
Prokud. — Prokudin — Прокудин Ю. Н.
Prov. — Provancer L.
Pugsl. — Pugsley H. W.
Puring — Пуринг Н. И.
Pursch — Pursch F. T.
Pyl. — см. La Pyl.

Quick — Quick C. R.

Raab — Raab (первая четверть XIX в.).
Rabenh. — Rabenhorst G. L.
Racib. — Raciborski M.
Radde — Радде Г. И.
Radde-Fomin(a) — Радде-Фомина О.
Raddi — Raddi G.
Radf. — Radford A. E.
Radius — Radius J.
Raeusch. — Raeuschel E. A.
Raf. — Rafinesque-Schmaltz C. S.
Raik. — Raikova — Райкова И. А.
Ramat. — Ramatuelle (T. ?) de.
Ramond — Ramond de Carbonnières L. F. E.
Rand — Rand E. L.
Rand et Redf. — Rand E. L. et Redfield J. H.
Rapaics — Rapaics R.
Rasinš — Расиньш А. П.
Rasmuss. — Rasmussen R.
Rasp. — Raspail F. V.
Rau — Rau A.
Raunk. — Raunkiaer C.
Raup — Raup H. M.
Rayss — Rayss T.
Rebent. — Rebentisch J. F.
Rech. — Rechinger, Karl.
Rech. f. — Rechinger, Karl Heinz.
Rech. f. et Edelb. — Rechinger K. H. et Edelberg L.
Rech. f. et Köie — Rechinger K. H. et Køie (Köie) M.
Redf. — Redfield J. H.
Redouté — Redouté P. J.
Redow. — Redowsky (Redoffsky) I. — Редовский И.
Reed — Reed C. F.
Rees — Rees A.
Rehd. — Rehder A.
Rehd. et Wils. — Rehder A. et Wilson E. H.
Rehm. — Rehman(n) A.
Reichard — Reichard J. J.
Reichardt H. — Reichardt H. W.
Reichb. — Reichenbach, H. Gottlieb L.
Reichb. f. — Reichenbach, H. Gustav
Reinsch — Reinsch P.
Reinw. — Reinwardt K. G. K.
Remy — Remy J.
Remy E. — Remy E. A.
Renault — Renault P. A.
Renault B. — Renault B.
Rendle — Rendle A. B.
Renealme — Renea(u)lme P. de.
Resv. — Resvoll T. R.
Retz. — Retzius A. J.
Reuss G. — Reuss G.
Reut. — Reuter G. F.
Reverd. — Reverdatto — Ревердатто В. В.
Reyn. — Reynier A.
Rgl. — Regel E. A. v. — Регель Э. Л.
Rgl. C. — Regel C. — Регель К. В.
Rgl. et Herd. — Regel E. et Herder — Регель Э. Л. и Гердер Ф. Е.
Rgl. et Maack — Regel E. et Maack — Регель Э. Л. и Маак Р. К.
Rgl. et Maw — Regel E. et Maw G.
Rgl. et Radde — Regel E. et Radde — Регель Э. Л. и Радде Г. И.
Rgl. et Schmalh. — Regel E. et Schmalhausen — Регель Э. Л. и Шмальгаузен И. Ф.
Rgl. et Til. — Regel E. et Tiling — Регель Э. Л. и Тилинг Г.
Rgl. R. — Regel R. — Регель Р. Э.
Rgl. R. et Mlokoss. — Regel R. et Mlokos-

sewicz — Регель Р. Э. и Млокосевич Ю. Л.
Rich. — Richard L. C. M.
Rich. A. — Richard A.
Richards. — Richardson J.
Richt. A — Richter A.
Richt. Al. — Richter, Aladár.
Richt. K. — Richter K.
Richt. L. — Richter L.
Ricker — Ricker P. L.
Rickett — Rickett H. W.
Ridl. — Ridley H. N.
Rigo — Rigo G.
Rikli — Rikli M.
Risso — Risso J. A.
Riv. — Rivinus A. Q. (Bachmann).
Rivière A. et C. — Rivière Auguste et Charles.
Rjab. — Rjabov — Рябов И. Н.
Rjabk. — Rjabkova — Рябкова Л. С.
Robill. — Robillard D'Argentelle L. M. A.
Robill. et Cast. — Robillard D'Argentelle et ? Castiglioni L.
Robins. — Robinson B. L.
Rochel — Rochel A.
Rodin — Родин Л. Е.
Roehl. — Roehling J. C.
Roem. — Roemer J. J.
Roem. et Schult. — Roemer J. J. et Schultes J. A.
Roem. M. — Roemer M. J.
Rogow. — Rogowicz (Rogowitsch) — Рогович А. С.
Rohde — Rohde (начало XIX в.).
Rohl., Röhl. — см. Roehl.
Rohrb. — Rohrbach P.
Roldug. — Roldugin — Ролдугин И. И.
Rollins — Rollins R. C.
Romanet — Romanet du Gaillaud F.
Ronn. — Ronniger K.
Ronn. et Saars. — Ronniger K. et Saarsoo B.
Rose — Rose J. N.
Roshev. — Roshevitz — Рожевиц Р. Ю.
Roshk. — Roshkova — Рожкова О. И.
Rostk. — Rostkovius F. W. G. (Th.).
Rostk. et Schmidt — Rostkovius F. W. G. et Schmidt W. L. E.
Roth — Roth A. W.
Rothert — Rothert W. A.
Rothm. — Rothmaler W.
Rothr. — Rothrock J. T.
Rotsch. E. — Rotschild E. — Ротшильд Е. В.
Rottb. — Rottboel (Rottbøl) C. F.
Rouy — Rouy G. C. C.
Rouy et Fouc. — Rouy G. C. C. et Foucaud J.
Roxb. — Roxburg W.
Royle — Royle J. F.
Roz. — Rozier F.
Rozan. — Rozanova — Розанова М. А.
Rubn. — Rubner K.
Rubtz. — Rubtzov — Рубцов Г. А.
Rubtz. N. — Rubtzov N. — Рубцов Н. И.
Rudge — Rudge E.
Rudolph J. — Rudolph J. H. — Рудольф И. Г. (И. Я.).
Ruempl. — Ruempler (Rümpler) T.
Ruhl. — Ruhland W.
Ruhm. — Ruhmer G.
Ruiz — Ruiz López H.
Ruiz et Pav. — Ruiz López H. et Pavon J. A.
Rumpf (Rumph.) — Rumpf (Rumphius) G. E.
Rümpl. — см. **Ruempl.**
Rune — Rune O.
Rupp. — Ruppius (Rupp) H. B.
Rupr. — Ruprecht — Рупрехт Ф. И.
Rupr. et Maxim. — Ruprecht et Maximowicz — Рупрехт Ф. И. и Максимович К. И.
Rusby — Rusby H. H.
Russan. — Rus(s)anov — Русанов Ф. Н.
Russell N. — Russel N. H.
Rydb. — Rydberg P. A.
Ryl. — Rylands T. G.
Rzazade — Рзазаде Р. Я.

Saars. — Saarsoo B.
Sabine — Sabine J.
Sabir. — Sabirov — Сабиров Б. З.
Sacalo — см Sakalo.
Sachok. — Sachokia — Сахокия М. Ф.
Sadl. — Sadler J.
Sadl. M. — Sadler M.
Sael. — Saelan A. T.
Sael. E. — Saelan E. L.
Sael. T. — Saelan T.
Sael. T., Kihlm. et Hjelt — Saelan T., Kihlman A. et Hjelt H.
Sager. — Sageret A.
Sagor. — Sagorski E.
Sagor. et Schneid. G. — Sagorski E. et Schneider G.
Sakalo — Сакало Д. И.
Salisb. — Salisbury R. A.
Salm-Dyck — Salm-Reifferscheid-Dyck J.
Salmon — Salmon C. E.

Salzm. — Salzmann P.
Sam. — Samuelsson G.
Sambuk — Самбук Ф. В.
Sanadze — Санадзе К.
Sander — Sander H. F. C.
Sándor — Sândor I.
Sanio — Sanio K.
Sanson — Sanson C. A. (по другим данным Sanson M.) — Сансон (петербургский ботаник 30-х годов XIX в.)
Santi — Santi G.
Sap. — Saporta G.
Sapeg. — Sapeghin (Sapjegin) — Сапегин А. А.
Saposhn. — Saposhnikow — Сапожников В. В.
Saposhn. et Nikit. E. — Saposhnikow et Nikitina E. — Сапожников В. В. и Никитина Е. В.
Sarg. — Sargent C. S.
Sarnth. — Sarnthein L. v.
Sart. — Sartorelli G. B.
Satake — Satake Y.
Saund. — Saunders W. W.
Saut. — Sauter A. E.
Sauv. — Sauvage C.
Savage — Savage S.
Savat. — Savatier L.
Savi — Savi G.
Savi P. — Savi, Pietro.
Sav.-Rycz. — Savicz-Ryczegorsky — Савич (Савич-Рычегорский) В. М.
Savul. — Săvulescu T.
Savul. et Rayss — Săvulescu T. et Rayss T.
Scepin — Щепин К.
Schachn. — Schachnazarov — Шахназаров А. И.
Schack — Schack H.
Schaeff. — Schaeffer J. C.
Schag. — Schagerström J. A.
Schang. — Schangin — Шангин П. И.
Schau — Schau J. K.
Schauer — Schauer J. K.
Sch. Bip. — Schultz Bipontinus K. (C.) H.
Schchian — Шхиян А. С.
Scheele — Scheele G. H. A.
Schelk. — Schelkownikow — Шелковников А. Б.
Schelk. et Zahn — Schelkownikow et Zahn — Шелковников А. Б. и Цан К. Г
Scheng. — Schengelia — Шенгелия Е. М.
Schenk — Schenk A.
Schenk E. — Schenk E.
Schenk H. — Schenk H.
Scherb. — Scherbius J.
Scheuchz. — Scheuchzer J.
Scheutz — Scheutz N. I.
Schiede — Schiede C. J. W.
Schimp. — Schimper K. (C.) F.
Schimp. W. — Schimper W. P.
Schindl. — Schindler A. K.
Schinz — Schinz H.
Schinz et Keller — Schinz H. et Keller R.
Schinz et Thell. — Schinz H. et Thellung A.
Schipcz. — Schipczinski — Шипчинский Н. В.
Schir. — см. Širj.
Schischk. — Schischkin — Шишкин Б. К.
Schischk. et Serg. — Schischkin et Sergievskaja — Шишкин Б. К. и Сергиевская Л. П.
Schischk. I. — Schischkin I. — Шишкин И. К.
Schischk. I. et Kom. — Schischkin I. et Komarov — Шишкин И. К. и Комаров В. Л.
Schk. — Schkuhr C.
Schlecht. — Schlechtendal D. F. L. v.
Schlechter — Schlechter F. R. R.
Schlecht. et Cham. — Schlechtendal D. F. L. et Chamisso A.
Schleich. — Schleicher J. C.
Schleid. — Schleiden M. J.
Schljak. — Schljakov — Шляков Р. Н.
Schmalh. — Schmalhausen — Шмальгаузен И. Ф.
Schmidel — Schmidel C. C.
Schmidt — Schmidt, Franz (1751—1834).
Schmidt Fr. — Schmidt, Fridericus — Шмидт, Федор Богданович (1832—1908).
Schmidt Fr. et Trautv. — Schmidt Fr. et Trautvetter — Шмидт Ф. Б. и Траутфеттер Е. Р.
Schmidt F. W. — Schmidt, Franz Wilibald (1763—1796).
Schmidt J. Ant. — Schmidt, Johann Anton (1823—1905).
Schmidt J. K. — Schmidt, Johann Karl (1793—1850).
Schmidt W. — Schmidt, Wilhelm Ludwig Ewald (1804—1843).
Schneid. — Schneider C. K.
Schneid. G. — Schneider G.
Schneid. U. — Schneider U.
Schnittsp. — Schnittspahn G. F.
Schnizl. — Schnizlein A.
Schoenh. — Schoenheit (Schönheit) F. C. H.
Scholl. — Scholler F. A.
Scholz — Scholz J. B.

Schost. — Schostenko — Шостенко (Десятова-Шостенко) Н. А.
Schott — Schott H. W.
Schott et Kotschy — Schott H. W. et Kotschy T.
Schott, Nym. et Kotschy — Schott H. W., Nyman C. F. et Kotschy T.
Schousb. — Schousboe P. K. A.
Schrad. — Schrader H. A.
Schrank — Paula v. Schrank F.
Schreb. — Schreber J. C. D. v.
Schrenk — Schrenk A. G. v. — Шренк А. Г.
Schröding. — см. Schroeding.
Schroed. — Schroeder — Шредер Р. И.
Schroeding. — Schroedinger (Schrödinger) R.
Schroet. — Schroeter (Schröter) C.
Schtschegl. — см. Stschegl.
Schtscheglow — Щеглов Н. П.
Schtscherb. — Schtscherbina — Щербина А.
Schuebl. — Schuebler (Schübler) G.
Schuette — Schuette J. H.
Schult. — Schultes J. A.
Schult. f. — Schultes J. H.
Schultz C. H. — см. Sch. Bip.
Schultz F. — Schultz F. W.
Schulze M. — Schulze M.
Schulz O. E. — Schulz, Otto E.
Schum. — Schumacher H. C. F. (1757—1830).
Schumann — Schumann K. M. (1851—1904).
Schur — Schur P. J. F.
Schust. — Schuster J.
Schwarz O. — Schwarz, Otto.
Schweigg. — Schweigger A. F.
Schweigg. et Koerte — Schweigger A. F. et Körte F.
Schwein. — Schweinitz L. D. v.
Schweinf. — Schweinfurth G. A.
Schwer. — Schwerin F. v.
Scop. — Scopoli J. A. (G. A.).
Scribn. — Scribner F. Lamson.
Scribn. et Ball C. — Scribner F. Lamson et Ball C. R.
Scribn. et Merr. — Scribner F. Lamson et Merrill E. D.
Scribn. et Smith J. — Scribner F. Lamson et Smith J. G.
Seb. — Sebastiani A.
Sebeòk — Sebeòk de Szent-Miklòs A.
Seb. et Mauri — Sebastiani A. et Mauri E.
Seem. — Seemann B. C.
Seemen — Seemen K. O. v.
Seen. — Seenus J.
Seidl. — Seidlitz — Зейдлиц Н. К.
Selin — Selin G.
Sell. — Sello(w) F.
Semen. — Semenov(w)-Tjan-Schansky — Семенов-Тян-Шанский В. П.
Semen. N. — Semenova-Tjan-Schanskaja N. — Семенова-Тян-Шанская Н. З.
Semenova — Семенова М. Н.
Semiotr. — Semiotroczeva — Семиотрочева Н. Л.
Senjan.-Korcz. — Senjaninova-Korczagina — Сенянинова-Корчагина М. В.
Sennen — Sennen F.
Ser. — Seringe N. C.
Serebr. — Serebrjakova T. J. — Серебрякова Т. Я.
Serg. — Sergievskaja — Сергиевская Л. П.
Serg. E. — Sergievskaja — Сергиевская Е. В.
Servett. — Servettaz C.
S. et L. — см. Somm. et Levier.
S. et Z. — см. Siebold et Zucc.
Sevast. — Sevastianov (Sevastianof) — Севастьянов А. Ф.
Seym. E. — Seymour E. L. D.
Shaf. — Shafer J. A.
Shal. — Shalyt — Шалыт М. С.
Shap. — Shaparenko — Шапаренко К. К.
Shaw — Shaw G. R.
Shear — Shear C. L.
Sheld. E. — Sheldon E. P.
Sheld. J. — Sheldon J. L.
Sherff — Sherff E. E.
Shibata — Shibata K.
Sibth. — Sibthorp J.
Sibth. et Smith — Sibthorp J. et Smith J. E.
Sidor. — Sidorenko — Сидоренко Г. Т.
Sieb. — Sieber F. W. (1785—1844).
Siebold — Siebold P. F. v. (1796—1866).
Siebold et Zucc. — Siebold P. F. et Zuccarini J. G.
Sieg. — Siegesbeck J. G. — Сигезбек И. Г.
Siegfr. — Siegfried H.
Siehe — Siehe W.
Siemss. — Siemssen C. H.
Siev. — Sievers J. — Сиверс И.
Sijaz. — Si(j)azov — Сиязов М. М.
Sim T. — Sim T. R.
Simk. — Simkovicz L. (впоследствии Simonkai).
Simm. — Simmons H. G.
Simon. — Simonet M.
Simon. et Guin. — Simonet M. et Guinochet M.
Simonk. — Simonkai, Lajos tól (он же Simkovicz L.).

Simon-Louis — Simon-Louis L. L.
Simonova — Симонова О. А.
Simps. — Simpson N. D.
Sims — Sims J.
Sinsk. — Sinskaya (Sinskaja) — Синская Е. Н.
Sint. — Sintenis P. E. E.
Sirj. — Širjaev (Schirjaev) — Ширяев Г. И.
Siuzew — Сюзев П. В.
Skvorts. — Skvortsov (Skvortzow) — Скворцов Б. В.
Skvorts. A. — Skvortsov — Скворцов А. К.
Slob. — Slobodov — Слободов А. А.
Sloss. — Slosson M.
Small — Small J. K.
Smirn. — Smirnow — Смирнов С. М.
Smirn. A. — Smirnova A. — Смирнова А. Д.
Smirn. P. — Smirnov P. — Смирнов П. А.
Smith — Smith J. E.
Smith et Sowerby — Smith J. E. et Sowerby J.
Smith G. — Smith G. E.
Smith H. — Smith, Harry.
Smith J. — Smith, John.
Smith J. G. — Smith, Jared Gage.
Smith J. J. — Smith, Johannes Jacobus.
Smith W. W. — Smith, William Wright.
Smith W. W. et Farrer — Smith William W. et Farrer R. J.
Smith W. W. et Fletch. — Smith William W. et Fletcher H. R.
Smith W. W. et Forrest — Smith William W. et Forrest G.
Smoljan. — Smoljaninova — Смольянинова Л. А.
Snowd. — Snowden J. D.
Sobol. — Sobolewski G. — Соболевский Г. Ф.
Sobol. K.—Sobolevskaja K.—Соболевская К.
Socz. — Soczava — Сочава В. Б.
Socz. et Lipat. — Soczava et Lipatova — Сочава В. Б. и Липатова В. В.
Soest — Soest van J. L.
Sof. — Sofieva — Софиева Р. М.
Soják — Soják J.
Soland. — Solander D. C.
Solered. — Solereder H.
Solms — Solms-Laubach H. de.
Soltok. — Soltoković M.
Somm. — Sommier S.
Sommerf. — Sommerfelt S. C.
Somm. et Levier — Sommier S. et Levier E.
Sond. — Sonder O. W.
Song. — Songeon A.
Soó — Soó R.
Soó et Jáv. — Soó R. et Jávorka S.
Sosk. — Soskov — Сосков Ю. Д.
Sosn. — Sosnowsky — Сосновский Д. И.
Sosn. et Takht. — Sosnowsky et Takhtadjan — Сосновский и Тахтаджян А. Л.
Sowerby — Sowerby, James DeCarl.
Soy.-Willem. — Soyer-Willemet H. F.
Soy.-Willem. et Godr. — Soyer-Willemet H. F. et Godron D. A.
Spach — Spach E.
Speg. — Spegazzini C.
Spenn. — Spenner F. K. (C.) L.
Sprague — Sprague T. A.
Spreng. — Sprengel C. (K.) P. J.
Spring — Spring A. F.
Sprun. — Spruner W. v.
Spryg. — Sprygin — Спрыгин И. И.
Spryg. et Pop. M. — Sprygin et Popov M. — Спрыгин И. И. и Попов М. Г.
Sred. — Sredinsky — Срединский Н. К.
Stacey — Stacey J. W.
Stackh. — Stackhouse J.
Stadlm. — Stadlmann J.
Standl. — Standley P. C.
Stank. — Stankov — Станков С. С.
Stapf — Stapf O.
Stapf et Wettst. — Stapf O. et Wettstein R.
Stearn — Stearn W. T.
Stebb. — Stebbins G. L., Ir.
Stechm. — Stechmann J. P.
Stef. — Stefanoff (Stefanov) B. — Стефанов Б. (Болгария).
Stefánss. — Stefánsson S.
Stef. et Georgiev — Stefanoff B. et Georgiev S. — Стефанов Б. и Георгиев Ст. (Болгария).
Steffen — Steffen H.
Stein — Stein B.
Steinb. — Steinberg — Штейнберг Е. И.
Stell. — Steller G. W. — Стеллер Г.
Stenstr. — Stenström K. O. E.
Steph. — Stephan C. F. — Стефан Ф. Х.
Sternb. — Sternberg (C.) K. M. v.
Sternb. et Hoppe — Sternberg K. M. et Hoppe D. H.
Sterneck — Sterneck J.
Sterns E. — Sterns E. E.
Steud. — Steudel E. G.
Stev. — Steven Chr. v. — Стевен Х. Х.
Stev. et Fisch. — Steven et Fischer — Стевен Х. Х. и Фишер Ф. Б.
Steward — Steward A. N.
Stewart A. — Stewart A. N.

Stewart J. — Stewart J. L.
Steyerm. — Steyermark J. A.
St.-Hil. — Saint-Hilaire A. (A. F. C. P.) de.
Stiefelh. — Stiefelhagen H.
St.-Lag. — Saint-Lager J. B.
Stoerck — Stoerck (Störk) A.
Stojan. — Stojanoff (Stojanov) N. — Стоянов Н. (Болгария).
Stojan. et Stef. — Stojanoff N. et Stefanoff B. — Стоянов Н. и Стефанов Б. (Болгария).
Stokes — Stokes J.
Stolet. — Stoletova — Столетова Е. А.
Stone — Stone W.
Störk — см. Stoerck.
Strand — Strand B. J.
Strangw. — Strangways (40-е годы XIX в.).
Stremp. — Strempel K. (C.) F.
Stschegl. — Stscheglejew [Schtschegleev(w), Tchegleeff] — Щеглеев С. С.
Stschep. — St(s)chepotjev — Щепотьев Ф. Л.
Sturm — Sturm J. W.
St.-Yves — Saint-Yves A.
Sudre — Sudre H.
Sudw. — Sudworth G. B.
Sugaw. — Sugawara S.
Sukacz. — Sukaczev(w), Sukatschev(w) — Сукачев В. Н.
Sulak. — Sulakadze — Сулакадзе Т.
Sumn. — Sumnevicz — Сумневич Г. П.
Suter — Suter J. R.
Sutt. — Sutton C.
Sutul. — Sutulov — Сутулов А.
Svens. — Svenson H. K.
Sw. — Swartz O. P.
Swall. — Swallen J. R.
Sweet — Sweet R.
Swingle — Swingle W. T.
Syme — Syme Boswell J. T. J.
Syr. — Syreisczikov (Ssyreitschikov) — Сырейщиков Д. П.
Szabo — Szabó Z.
Szaf. — Szafer W.
Szaf., Kulcz. et Pawł. — Szafer W., Kulczyński S. et Pawłowski B.
Szov. — Szovits (Szowics) A. J. — Сович (Шович) О. И.
Szov. et Lang — Szovits et Lang — Сович О. И. и Ланг А. Ф.

Takeda — Takeda H.
Takeda et Miyake — Takeda H. et Miyake T.
Takht. — Takhtadjan (Takhtajan, Takhtadzhian, Tachtadshjan) — Тахтаджян А. Л.
Taliev(w) — Талиев В. И.
Tamamsch. — Tamamschjan — Тамамшян С. Г.
Tanaka — Tanaka, Yoshio.
Tanaka C. (T.) — Tanaka, Chozaburo (Tyozaburo).
Tanaka Y. — Tanaka, Yuichiro.
Tarass. — Tarassov — Тарасов Р. П.
Targ.-Tozz. — Targioni-Tozzetti O.
Tatarinow — Татаринов А. А.
Tatew. — Tatewaki M.
Tatew. et Ohwi — Tatewaki M. et Ohwi J.
Taub. — Taubert P. H.
Tausch — Tausch I. F.
Tchich. — Tchichatscheff (Tschihatchef) — Чихачев П. А.
Teijsm. — Teijsmann J. E.
Tek. — Tekutjev — Текутьев Г.
Ten. — Tenore M.
Tepl. — Teplouchow — Теплоухов Ф. А.
Teras. — Terasaki T.
Ter-Chatsch. — Ter-Chatschaturova — Тер-Хачатурова С. Я.
Terech. — Terechov — Терехов А. Ф.
Terr. — Terracciano A.
Terr. N. — Terracciano N.
Teysm. — см. Teijsm.
Thell. — Thellung A.
Theod. Al. — Theodorov (Fedorov) Al. — Федоров Ал. А.
Theod. Al., Fed. et Rzazade — Theodorov (Fedorov) Al., Fedorov An. et Rzazade — Федоров Ал. А., Федоров Ан. А. и Рзазаде Р. Я.
Thiéb. — Thiébaut J.
Thoms. — Thomson T.
Thoms. et Hook. f. — Thomson T. et Hooker J. D.
Thuill. — Thuillier J. L.
Thunb. — Thunberg C. P.
Thwaites — Thwaites G. H. K.
Tiegh. — Tieghem van P. E. L.
Til. — Tiling S. H. — Тилинг Г.
Timb. — Timbal-Lagrave E.
Timm — Timm J. C.
Tineo — Tineo V.
Titov — Титов В. С.
Tobl. — Tobler F.
Tod. — Todaro A.
Toepff. — Toepffer A.
Tolm. — Tolmat(s)chew — Толмачев А. И.
Torg. — Torges E.

Torn. — Torner E.
Torr. — Torrey J.
Torre — см. Dalla Torre
Torr. et Gray — Torrey J. et Gray, Asa.
Tourn. — Tournefort J. P. de.
Towns. — Townsend F.
Trab. — Trabut L.
Tratt. — Trattinick L.
Trautv. — Trautvetter E. R. v. — Траутфеттер Э. Р.
Trautv. et Mey. — Trautvetter et Meyer C. A. — Траутфеттер Э. Р. и Мейėр К. А.
Trel. — Trelease W.
Trev. — Treviranus L. C. (1799—1864).
Trevis. — Trevisan di San Leon V. B. A. (1818—1897).
Trimen — Trimen H.
Trin. — Trinius C. (K.) B. v. — Триниус К. А.
Trin. et Rupr. — Trinius et Ruprecht — Триниус К. А. и Рупрехт Ф. И.
Troitzk. — Troitzky N. — Троицкий Н. А.
Troitzk. Z. — Troitzkaja Z. — Троицкая З. Ф.
Tron — Трóн Е. Ж.
Tropea — Tropea C.
Tryon — Tryon R. M.
Tscherneva — Чернева О. В.
Tscherning — Tscherning J. W.
Tsyr. — Tsyrina — Цырина Т. С.
Tuckerm. — Tuckerman E.
Tul. — Tulasne L. R.
Turcz. — Turczaninow — Турчанинов Н. С.
Turkev. — Turkev(w)icz — Туркевич С. Ю.
Turp. — Turpin P. J. F.
Turra — Turra A.
Turrill — Turrill W. B.
Tutin — Tutin T. G.
Tuzs. — Tuzson J.
Tzvel. — Tzvelev — Цвелев Н. Н.

Ucria — Ucria B. (он же Michelangelo Aurifici).
Uechtr. — Uechtritz R. K. F. v.
Uechtr. et Koern. — Uechtritz R. K. F. et Koernicke (Körnicke) F.
Ueki — см. Uyeki.
Uglitzk. — Uglitzkich — Углицких А. Н.
Ugr. — Ugrinsky — Угринский К. А.
Ulbr. — Ulbrich E.
Uline — Uline E. B.
Underw. — Underwood L. M.
Underw. J. — Underwood J. K.
Ung.-Sternb. — Ungern-Sternberg F.
Urb. — Urban I.
Urv. — см. D'Urv.
Utkin — Уткин Л. А.
Uyeki — Uyeki H.

Vacc. — Vaccari A.
Vahl — Vahl, Martin.
Vahl J. — Vahl, Jens.
Vaill. — Vaillant S.
Vakul. — Vakulin — Вакулин Д. Я.
Vand. — Vandelli D.
Vandas — Vandas K.
Van (den) Bosch — см. Bosch.
Van Heurck — см. Heurck.
Van Houtte — см. Houtte.
Vaniot — Vaniot E.
Van Soest — см. Soest.
Van-Tiegh. — см. Tiegh.
Vasey — Vasey G.
Vasing. — Vasinger — Вазингер-Алекторова А. В.
Vass. — Vassilczenko — Васильченко И. Т.
Vass. et Dzhang. — Vassilczenko et Dzhangurazov — Васильченко И. Т. и Джангуразов Ф. Х.
Vass. et Vassil. L. — Vassilczenko et Vassiljeva L. — Васильченко И. Т. и Васильева Л. И.
Vassil. A. — Vassiljeva A. — Васильева А. Н.
Vassil. Ig. — Vassiljev Ig. — Васильев И. В.
Vassil. Ja. — Vassiljev Ja. — Васильев Я. Я.
Vassilkovsk. — Vassilkovskaja — Васильковская А. П.
Vassil. L. — Vassiljeva L. — Васильева Л. И.
Vassil. V. — Vassiljev V. — Васильев В. Н.
Vatke — Vatke G. K. (C.) W.
Vav. — Vavilov — Вавилов Н. И.
Veitch — Veitch J. G.
Velen. — Velenovský J.
Vell. — Velloso de Conceição J. M.
Vell. P. — Velloso P. (Antonius da Arrabida).
Vent. — Ventenat É. P.
Verlot — Verlot J. B.
Verm. — Vermeulen P.
Vest — Vest L. C. v.
Vidal — Vidal, Sebastian y Soler.
Vidal L. — Vidal L.
Vieill. — Vieillard E.
Vierh. — Vierhapper F.
Vikul. — Vikulova — Викулова Н. В.
Vilensky — см. Wilensky.

Vill. — Villar(s) D.
Vilm. — Vilmorin P. L. F. L. de.
Vis. — Visiani R. de.
Vitm. — Vitman F.
Viv. — Viviani D.
Volgun. — Volgunov — Волгунов Д. К.
Volk. — Volkens G. (1855—1917).
Volkart — Volkart A. (1873—1951).
Volk. et Schweinf. — Volkens G. et Schweinfurth G. A.
Volkova — Волкова Е. В.
Vollm. — Vollmann F.
Vorosch. — Voroschilov — Ворошилов В. Н.
Voss — Voss A.
Voss J. — Voss J. W.
Vuk. — Vukotinović L. F.
Vved. — Vvedensky — Введенский А. И.
Vved. et Ovcz. — Vvedensky et Ovczinnikov — Введенский А. И. и Овчинников П. Н.
Vysok. — Vysokoostrovskaja — Высокоостровская И. Б.

Wacht. — Wachter W. H.
Waga — Вага А. Я.
Wagner J. — Wagner, János.
Wahl — Wahl H. A.
Wahl. — Wahlenberg G.
Wahlb. — Wahlberg P. F.
Wahlb. J. — Wahlberg J.
Wahlenb. — см. Wahl.
Waitz — Waitz K. F.
Waldst. — Waldstein F. A. v.
Waldst. et Kit. — Waldstein F. A. et Kitaibel P.
Wall. — Wallich N.
Wallm. — Wallman J. H.
Wallr. — Wallroth K. (C.) F. W.
Walp. — Walpers W. G.
Walsh — Walsh R.
Walt. — Walter T.
Wanger. — Wangerin W.
Wangh. — Wangenheim F. A. J. v.
Warb. — Warburg, Otto.
Warb. E. — Warburg E. F.
Warb. O. — Warburg, Oscar E.
Ward G. — Ward G. H.
Warder — Warder J. A.
Wassil. — Wassili(j)ew — Васильев В. Ф.
Wats. — Watson P. W.
Wats. S. — Watson S.
Watt — Watt D. A. P.
Wawra — Wawra H.
Weath. — Weatherby C. A.
Web. — Weber G. H.
Web. A. — Weber A.
Webb — Webb P. B.
Webb et Berth. — Webb P. W. et Berthelot S.
Web. C. — Weber C. O.
Weberb. — Weberbauer A.
Web. et Mohr — Weber F. et Mohr D. M. H.
Web. F. — Weber F.
Web. F. E. — Weber F. E. — Вебер Ф. Э.
Web. J. — Weber J.
Wedd. — Weddell H. A.
Wehrh. — Wehrhahn H. R.
Weig. — Weigel C. E. v.
Weihe — Weihe C. (K.) E. A.
Weihe et Nees — Weihe C. E. A. et Nees v. Esenbeck C. G. D.
Weinm. — Weinmann — Вейнман И. А.
Weis(s) — Weis(s) F. W.
Welw. — Welwitsch F.
Wend. — Wenderoth G. W. F.
Wendelbo — Wendelbo P.
Wendl. — Wendland J. C. (1755—1828).
Wendl. H. — Wendland H. A. (1823—1903, внук J. C. Wendland).
Wendl. H. L. — Wendland H. L. (1792—1869, сын J. C. Wendland).
Wenz. — Wenzig T.
Wenzlaw. — Wenzlawowicz (Ventslavovich) — Венцлавович Ф. С.
Wesm. — Wesmael A.
Westb. — Westberg — Вестберг Г. Ф.
Westerl. — Westerlund C. A.
Wettst. — Wettstein R. v.
Weyrich — Вейрих Г.
White E. A. — White, Edward A.
Wib. — Wibel A. W. E. C.
Widd. — Widder F. J.
Widm. — Widmer E.
Wieg. — Wiegand K. McKay.
Wierzb. — Wierzbicki P.
Wigg. — Wiggers F. H.
Wight — Wight R.
Wight et Arn. — Wight R. et Arnott G. A. W.
Wight W. — Wight W. F.
Wight W. et Hedr. — Wight W. F. et Hedrick U. P.
Wiinst. — Wiinstedt K.
Wikstr. — Wikstroem (Wikström) J. E.
Wilcz. — Wilczyński T.
Wilensky — Виленский Д. Г.
Will. — Williams F. N.
Willd. — Willdenow C. L.

Willem. — Willemet P. R.
Williams L. — Williams L. O.
Willis — Willis O. R.
Willk. — Willkomm H. M.
Willk. et Lange — Willkomm H. M. et Lange J. M. C.
Wilm. A. — Wilmott A. J.
Willm. — Willmott E.
Wils. — Wilson E. H.
Wimm. — Wimmer C. F. H.
Wimm. et Grab. — Wimmer C. F. H. et Grabowski H. E.
Winkl. C. — Winkler C. — Винклер К. Ю.
Winkl. C. et Fedtsch. B. — Winkler C. et Fedtschenko B. — Винклер К. Ю. и Федченко Б. А.
Winkl. H. — Winkler H.
Wint. L. — Winter L.
Wint. N. — Winter N. — Винтер Н. А.
Wirs. — Wirsing A. L.
Wirtg. — Wirtgen P.
Wissjul. — Wissjulina — Висюлина Е. Д.
Witas. — Witasek J.
With. — Withering W.
Wittm. — Wittmack M. C. L.
Wittr. — Wittrock V. B.
Wittst. — Wittstein G. C.
Wlangali — Wlangali A. — Влангали А.
Wohlf. — Wohlfahrt R.
Wolf E. — Вольф Э. Л.
Wolff H. — Wolff H.
Wolfg. — Wolfgang J. F.
Wolf Th. — Wolf Th.
Wołoszcz. — Wołoszczak E.
Wołoszcz. et Zahn — Wołoszczak E. et Zahn C. H.
Wood A. — Wood A.
Wood J. B. — Wood J. B.
Woodson — Woodson R. E.
Woot. — Wooton E. O.
Woot. et Standl. — Wooton E. O. et Standley P. C.
Wormsk. — Wormskiold (Wormskjöld) M.
Woronow — Woronow G. — Воронов Ю. Н.
Woronow et Tron — Воронов Ю. Н. и Трон Е. Ж.
Woronow et Zahn — Woronow G. et Zahn C. H. — Воронов Ю. Н. и Цан К. Г.
Worosch. — см. Vorosch.
Woynar — Woynar H.
Wright C. H. — Wright, Charles H.
Wright J. — Wright, John.
Wróbl. — Wróblewski A.
Wu C. Y. — Wu C. Y.
Wulf. — Wulfen F. X. v.
Wulf E. — Wulf(f) E. — Вульф Е. В.
Wulff — Wulff J. C.
Wu W. — Wu, Wen-chen.
Wu Y. C. — Wu Y. C.
Wydl. — Wydler H.
Wyss. — Wyssotzky — Высоцкий Г. Н.

Yabe — Yabe Y.
Yunck. — Yuncker T. G.

Zab. — Zabel H.
Zagar. — Zagareli — Цагарели П. П.
Zagor. — Zagorodskich — Загородских П.
Zahar. — Zahariadi C.
Zahn — Zahn C. (K.) H.
Zak. — Zakirov — Закиров К. З.
Zak. et Novopokr. — Zakirov et Novopokrovsky — Закиров К. З. и Новопокровский И. В.
Zalessky — Залесский К. М.
Zalk. — Zalkind — Залкинд Ф. Л.
Zalzm. — Zalzmann P.
Zam. — Zāmelis (Zāmlis) — Замелис А.
Zapał. — Zapałowicz H.
Zapr. — Zaprjagaev — Запрягаев Ф. Л.
Zefir. — Zefirov — Зефиров Б. М.
Zelen. — Zelenetzky — Зеленецкий Н. М.
Zeyh. — Zeyher J. M.
Zeyh. K. — Zeyher K. L. P.
Zhuk. — Zhukovsky — Жуковский П. М.
Zimm. F. — Zimmermann, Ferdinand J. v.
Zimm. Fr. — Zimmermann, Friedrich.
Zimm. W. — Zimmerman W.
Zing. — Zinger W. — Цингер В. Я.
Zing. N. — Zinger N. — Цингер Н. В.
Zinn — Zinn J. G.
Zinserl. — Zinserling — Цинзерлинг Ю. Д.
Ziz — Ziz J. B.
Zoll. — Zollinger H.
Zoz — Зоз И. Г.
Zucc. — Zuccarini J. G.
Zum. — Zumaglini A. M.

ABBREVIATIONS

ABBREVIATIONS OF FLORISTIC REGIONS OF USSR*

Abbreviated name	Full name
	I. Arctic
1. Arc. Eur.	Arctic (European part)
2. Nov. Z.	Novaya Zemlya
3. Arc. Sib.	Arctic (Siberia)
4. Chuk.	Chukotka
5. An.	Anadyr
	II. European Part
6. Kar.-Lap.	Karelia-Lapland
7. Dv.-Pech.	Dvina-Pechora
8. Balt.	Baltic States Area
9. Lad.-Ilm.	Ladoga-Ilmen
10. U.V.	Upper Volga
11. V.-Kama	Volga-Kama
12. U. Dnp.	Upper Dnieper
13. M. Dep.	Middle Dnieper
14. V.-Don	Volga-Don
15. Transv.	Transvolga Area
16. U. Dns.	Upper Dniester
	III. Caucasus
17. Bes.	Bessarabia
18. Bl.	Black Sea Area
19. Crim.	Crimea
20. L. Don	Lower Don
21. L.V.	Lower Volga
22. Cisc.	Ciscaucasia
23. Dag.	Dagestan

*The list of regions reproduced here differs significantly from the one published in earlier volumes (cf. Volume VII).

Regions 17-21 of this list were earlier included in the European Part and regions 28 and 29 in Western Siberia—Gen. Ed.

24. W. Transc.		Western Transcaucasia
25. E. Transc.		Eastern Transcaucasia
26. S. Transc.		Southern Transcaucasia
27. Tal.		Talysh
28. Ob.		Ob region (from eastern slopes of Urals to Yenisei River)
29. U. Tob.		Upper Tobol
	IV. Western Siberia	
30. Irt.		Irtysh
31. Alt.		Altai
	V. Eastern Siberia	
32. Yenis.		Yenisei
33. Lena-Kol.		Lena-Kolyma
34. Ang. Say.		Angara River-Sayans
35. Dau.		Dauria
	VI. Far East	
36. Kamch.		Kamchatka
37. Okh.		Okhotsk
38. Ze.-Bu.		Zeya-Bureya
39. Uda		Uda River area
40. Uss.		Ussuri
41. Sakh.		Sakhalin
	VII. Soviet Central Asia	
42. Ar.-Casp.		Aral-Caspian
43. Balkh.		Lake Balkhash area
44. Dzu.-Tarb.		Dzungaria-Tarbatagai
45. Kyz. K.		Kazyl-Kum
46. Kara K.		Kara-Kum
47. Mtn. Turkm.		Montane Turkmenistan
48. Amu D.		Amu Darya foothills
49. Syr. D.		Syr Darya foothills
50. Pam.-Al.		Pamir-Alai
51. T. Sh.		Tien Shan

ABBREVIATIONS OF REGIONS USED TO INDICATE GENERAL DISTRIBUTION OF SPECIES IN "*FLORA OF USSR*"

I. Arc.	Arctic (Spitsbergen, Greenland and others)

II. Scand.	Scandinavia (Norway, Denmark, Sweden, Finland)
III. Centr. Eur.	Central Europe (Germany, Poland, Czechoslovakia, Hungary, Austria, Switzerland)
IV. Atl.-Eur.	Atlantic Europe (Netherlands, Belgium, England, France, Portugal)
V. Med.	Mediterranean (including North Africa)
(V'W., V"E.)	(V' Western, V" Eastern)
VI. Bal.-As. Min.	Balkan Peninsula and Asia Minor
VII. Arm.-Kurd.	Lesser Armenia and Kurdistan
VIII. Iran	Iran and Afghanistan
IX. Ind.-Him.	India, Pakistan and Nepal
X. Dzu.-Kash.	[Dzungaria-Kashgar area] Northwestern China (Sinkiang)
XI. Mong.	Mongolia
XII. Jap.-Ch.	Japan, China, Korean Peninsula and Northeastern China
XIII. Ber.	North American coast of the Bering Sea
XIV. N. Am.	North America (Canada and U.S.A.)
XV. Tib.	Tibet

OTHER GEOGRAPHICAL ABBREVIATIONS*

Afr.	Africa
Aust.	Australia
Centr.	Central
E	East(ern)
Gr.	Great, Greater
I	Island
Is.	Islands
Mt.	Mount
Mts.	Mountains
N.	North(ern)
R	River
S	South(ern)
W	West(ern)

*These abbreviations were published with many of the translated volumes but do not appear in this volume in the Russian original.

GENERAL ABBREVIATIONS

vn.	venchik	corolla
zv.	zavyaz	ovary
kz.	korzinka	capitulum
klk.	kolosok	spikelet
kr.	koren	root
krshch.	kornevishche	rhizome
l.	list, listya	leaf, leaves
lp.	lepestki	petals
lch.	listochek	leaflet
okolotsv.	okolotsvetnik	perianth
pl.	plod	fruit
pln.	pylnik	anther
pls.	plastinka	blade
pst.	pestik	pistil
pritsv.	pritsvetnik	bract
pritsvch.	pritsvetnichek	bractiole
prlst.	prilistnik	stipule
r.	rastenie	plant
rlts.	ryltse	stigma
s.	semya	seed
smk.	semyanka	achene
smpch.	semyapochka	ovule
st.	stebel	stem
stlb.	stolbik	style
stsv.	sotsvetie	inflorescence
tych.	tychinka	stamen
tsv.	tsvetok	flower
tsvn.	tsvetonozhka	pedicel
tsvtl.	tsvelolozhe	receptacle
chrsh.	chereshok	petiole
chrshch.	chereshochek	petiolule
chshl.	chashelistiki	sepals
chshch.	chashechka	calyx
chsh.	cheshuya	scale
f.	figura	figure, i.e., individual illustration on a plate

var.	varietas, raznovidnost	variety
b.ch.	bolsheyu chastyu	mostly
b.m.	bolee ili menee	more or less
vys.	vysota	height
dl.	dlina	length
shir.	shirina	width
grech.	grecheskii	Greek
lat.	latinskii	Latin
m.	metr	meter
sm.	santimetr	centimeter
mm.	millimetr	millimeter
⊙	odnoletnik	annual
⊙	dvuletnik	biennial
♃	mnogoletnik	perennial
ħ, ħ	kustarniki i derevya	shrubs and trees
g.	god	year
str.	stranitsa	page
Ic	icones, risunki	figures, illustrations
tab.	tabula, tablista	plate
Exs.	exsiccata	dried plants, i.e., herbaria, issued in sets of specimens representing standards of the given species
GRF	Gerbarii Russkoi Flory	Herbarium of Russian Flora
rn.	raion	region